The Subnuclear Series • Volume 46

Proceedings of the International School of Subnuclear Physics

PREDICTED AND TOTALLY UNEXPECTED IN THE ENERGY FRONTIER OPENED BY LHC

THE SUBNUCLEAR SERIES*

Series Editor: **ANTONINO ZICHICHI**, *European Physical Society, Geneva, Switzerland*

16. 1978 THE NEW ASPECTS OF SUBNUCLEAR PHYSICS
17. 1979 POINTLIKE STRUCTURES INSIDE AND OUTSIDE HADRONS
18. 1980 THE HIGH-ENERGY LIMIT
19. 1981 THE UNITY OF THE FUNDAMENTAL INTERACTIONS
20. 1982 GAUGE INTERACTIONS: Theory and Experiment
21. 1983 HOW FAR ARE WE FROM THE GAUGE FORCES?
22. 1984 QUARKS, LEPTONS, AND THEIR CONSTITUENTS
23. 1985 OLD AND NEW FORCES OF NATURE
24. 1986 THE SUPERWORLD I
25. 1987 THE SUPERWORLD II
26. 1988 THE SUPERWORLD III
27. 1989 THE CHALLENGING QUESTIONS
28. 1990 PHYSICS UP TO 200 TeV
29. 1991 PHYSICS AT THE HIGHEST ENERGY AND LUMINOSITY: To Understand the Origin of Mass
30. 1992 FROM SUPERSTRINGS TO THE REAL SUPERWORLD
31. 1993 FROM SUPERSYMMETRY TO THE ORIGIN OF SPACE–TIME
32. 1994 FROM SUPERSTRING TO PRESENT-DAY PHYSICS
33. 1995 VACUUM AND VACUA: The Physics of Nothing
34. 1996 EFFECTIVE THEORIES AND FUNDAMENTAL INTERACTIONS
35. 1997 HIGHLIGHTS OF SUBNUCLEAR PHYSICS: 50 Years Later
36. 1998 FROM THE PLANCK LENGTH TO THE HUBBLE RADIUS
37. 1999 BASICS AND HIGHLIGHTS IN FUNDAMENTAL PHYSICS
38. 2000 THEORY AND EXPERIMENT HEADING FOR NEW PHYSICS
39. 2001 NEW FIELDS AND STRINGS IN SUBNUCLEAR PHYSICS
40. 2002 FROM QUARKS AND GLUONS TO QUANTUM GRAVITY
41. 2003 FROM QUARKS TO BLACK HOLES: PROGRESS IN UNDERSTANDING THE LOGIC OF NATURE
42. 2004 HOW AND WHERE TO GO BEYOND THE STANDARD MODEL
43. 2005 TOWARDS NEW MILESTONES IN OUR QUEST TO GO BEYOND THE STANDARD MODEL
44. 2006 THE LOGIC OF NATURE, COMPLEXITY AND NEW PHYSICS: FROM QUARK-GLUON PLASMA TO SUPERSTRINGS, QUANTUM GRAVITY AND BEYOND
45. 2007 SEARCH FOR THE "TOTALLY UNEXPECTED" IN THE LHC ERA
46. 2009 PREDICTED AND TOTALLY UNEXPECTED IN THE ENERGY FRONTIER OPENED BY LHC

Volume 1 was published by W. A. Benjamin, Inc., New York; 2–8 and 11–12 by Academic Press, New York and London; 9–10 by Editrice Compositori, Bologna; 13–29 by Plenum Press, New York and London; 30–46 by World Scientific, Singapore.

*For the complete list of titles in this series, please go to http://www.worldscibooks.com/series/ss series.shtml

The Subnuclear Series • Volume 46

Proceedings of the International School of Subnuclear Physics

PREDICTED AND TOTALLY UNEXPECTED IN THE ENERGY FRONTIER OPENED BY LHC

Edited by

Antonino Zichichi

European Physical Society
Geneva, Switzerland

World Scientific

NEW JERSEY • LONDON • SINGAPORE • BEIJING • SHANGHAI • HONG KONG • TAIPEI • CHENNAI

Published by

World Scientific Publishing Co. Pte. Ltd.
5 Toh Tuck Link, Singapore 596224
USA office: 27 Warren Street, Suite 401-402, Hackensack, NJ 07601
UK office: 57 Shelton Street, Covent Garden, London WC2H 9HE

British Library Cataloguing-in-Publication Data
A catalogue record for this book is available from the British Library.

The Subnuclear Series — Vol. 46
PREDICTED AND TOTALLY UNEXPECTED IN THE ENERGY FRONTIER OPENED BY LHC
Proceedings of the International School of Subnuclear Physics

ISBN-13 978-981-4340-20-5
ISBN-10 981-4340-20-0

Printed in Singapore by World Scientific Printers.

PREFACE

In August/September 2008, a group of 104 physicists from 19 countries, working in 43 international laboratories located in 16 countries, have met in Erice to participate in the 46th Course of the International School of Subnuclear Physics that, this year, was dedicated to the memory of Sidney Coleman. The countries represented by the participants were: Armenia, China, Croazia, France, Germany, Greece, Hungary, India, Italy, Japan, the Netherlands, Pakistan, Poland, Romania, Russia, South Korea, Sweden, Ukraine, United Kingdom and the United States of America.

The School was sponsored by the Academies of Sciences of Estonia, Georgia, Lithuania, Russia and Ukraine; the Chinese Academy of Sciences; the Commission of the European Communities; the European Physical Society; the Italian Ministry of Education, University and Scientific Research; the Sicilian Regional Government; the Weizmann Institute of Science; the World Federation of Scientists and the World Laboratory.

The purpose of the School was to focus attention on the "Predicted and Totally Unexpected in The Energy Frontier opened by LHC". The most recent developments in Supersymmetry, Quantum Gravity and Extra-dimensions theories have been presented, together with selected QCD problems and specialized topics in other sectors of Subnuclear Physics. Experimental highlights from the most relevant sources of new data were also presented and discussed.

An original feature of the School, introduced in 1996, is a series of special sessions devoted to "New Talents", with the aim of encouraging and promoting young physicists to achieve recognition at an international level. This is a serious problem in Experimental Physics, where collaborations count several hundreds of participants and it is almost impossible for young fellows to be known. Even if with much less emphasis, the problem exists also in Theoretical Physics. So we decided to offer 27 young fellows the possibility to give an open presentation of the results of their studies, followed by a discussion. Among them four best "New Talents" have been selected for publication in this volume, and were given an award.

During the organization and the running of the Course, I enjoyed the collaboration of my colleague and friend, Gerardus 't Hooft, who shared with me the Directorship of the School. I would like to thank him, together with the group of invited scientists and all the people who have contributed to the success of the Course.

I hope the reader will enjoy the book as much as the students attending the lectures and the discussion sessions, which, as every year, have been the focal point of this School activity. Thanks to the work of the Scientific Secretaries, the discussions have been reproduced as faithfully as possible. At various stages of my work I have enjoyed the collaboration of many friends whose contributions have been extremely important for the School and are highly appreciated. I thank them most warmly. A final acknowledgement to all those in Erice, Bologna and Geneva, who have helped me in so many occasions and to whom I feel very indebted.

Antonino Zichichi

Geneva, October 2008

CONTENTS

QCD PROBLEMS

SEMINARS ON SPECIALIZED TOPICS

HIGHLIGHTS FROM LABORATORIES

«ETTORE MAJORANA» FOUNDATION AND CENTRE FOR SCIENTIFIC CULTURE

TO PAY A PERMANENT TRIBUTE TO GALILEO GALILEI, FOUNDER OF MODERN SCIENCE AND TO ENRICO FERMI, THE "ITALIAN NAVIGATOR", FATHER OF THE WEAK FORCES

INTERNATIONAL SCHOOL OF SUBNUCLEAR PHYSICS

PREDICTED AND TOTALLY UNEXPECTED IN THE ENERGY FRONTIER OPENED BY LHC

46th Course – **ERICE-SICILY: 29 AUGUST - 7 SEPTEMBER 2008**

Sponsored by the: • Italian Ministry of Education, University and Scientific Research • Sicilian Regional Government • • Academies of Sciences of Estonia, Georgia, Lithuania, Russia and Ukraine • Chinese Academy of Sciences • • Commission of the European Communities • European Physical Society • Italian National Institute for Nuclear Physics • • Weizmann Institute of Science • World Federation of Scientists • World Laboratory

PROGRAMME AND LECTURERS

HOMAGE TO SIDNEY COLEMAN

SID COLEMAN AND ERICE

Antonino Zichichi

CERN, Geneva, Switzerland
Enrico Fermi Centre, Rome, Italy
INFN and University of Bologna, Italy
Ettore Majorana Foundation and Centre for Scientific Culture

INTRODUCTION

This lecture is an attempt to recall the extraordinary, original and fascinating contributions by Sidney Coleman to the Erice Subnuclear Physics School. The lecture consists of three parts.

The **first part** contains the "starting point": when I "discovered" Sid Coleman. The year was 1965. This was a period when problems that are taken for granted today, were far from being understood. For example the apparent triumph of the S-matrix (the negation of quantum field theory, QFT) accompanied by the breaking of all symmetry operators (P, C, CP, T) cast serious doubts on QFT as the basic theoretical structure for all fundamental forces of Nature, whose fundamental example was QED. In this great theoretical confusion experimental physics was dominated by the Bubble-Chamber Technology which gave an enormous number of discoveries all in the field of new mesons and baryons. The last one, with strangeness −3, was named Ω^- because it was believed to be the last particle ever to be discovered, and therefore given the name of the last letter in the Greek alphabet. Why so many hadrons and only one heavy lepton, the muon? Would another heavy lepton have been seen if its mass was in the GeV range? The incredible quasi degeneracy in the ($\mu-\pi$) masses was the reason why the muon was all over the place.

The year 1965 brought an enormous number of mesons and baryons but not a single piece of nuclear antimatter, not even the simplest one, the antideuteron, $\bar{d}$. Are the nuclear forces described by a QFT? Or is the absence of $\bar{d}$ a sign to understand the reason why S-matrix is dominating and the symmetry operators are all broken?

It was during this exciting period that Sid Coleman became a star of the Subnuclear Physics School. The first part ends with a synthesis of the Standard Model whose roots are in the second part.

In **the second part** I have selected some crucial problems discussed in his superb series of lectures; a few pages per problem will – I hope – stimulate the young generation to reflect on many achievements taken for granted today despite their complex origin. The pencil marks on some pages have their origin in my reading these fascinating lectures. The selection of pages is intended to be a stimulus in order to read all lectures, which are the proof that each crucial step in building the Standard Model was identified and first discussed at the Erice Subnuclear Physics School by Sid Coleman. And this is a world record.

The third part is devoted to some pictures of Sid in Erice, including those taken when he was awarded the Gold Medal as "Best Lecturer" on the occasion of the 15th anniversary of the International School of Subnuclear Physics in 1979.

FIRST PART

ANTIPARTICLES AND ANTIMATTER: THE BASIC DIFFERENCE

Antonino Zichichi
CERN, Geneva, Switzerland
Enrico Fermi Centre, Rome, Italy
INFN and University of Bologna, Italy

«Those who say that antihydrogen is antimatter should realize that we are not made of hydrogen and we drink water, not liquid hydrogen». These are Dirac's own words to a group of physicists (Fig. 1) gathered around him, who, with a single equation [1], opened new horizons to human knowledge. To obtain water, hydrogen is not sufficient by itself. You also need oxygen whose nucleus is made of 8 protons and 8 neutrons. Hydrogen is the only element in Mendeleev's Table to be constituted of two charged particles, the electron and the proton, without any role being played by the Nuclear Forces. The first element on which Nuclear Forces come into play is the heavy hydrogen, whose nucleus, called deuteron, is made with one proton and one neutron. For these two particles to remain together the "nuclear glue" is needed. Starting from the heavy hydrogen, all the elements of the Table, to exist, must have their nuclei made with protons, neutrons and the nuclear glue. If these last two ingredients, the neutron and the nuclear glue, were not available, nothing but the "light" hydrogen could exist. Farewell water and farewell all material which we are familiar with.

In Dirac's famous statement, 70 years of theoretical and experimental discoveries are taken into consideration, with the conclusion that the existence of antimatter is supported exclusively on an experimental basis. In fact – as evidenced by T.D. Lee [2], – the CPT theorem is invalidated at the Planck Scale ($\cong 10^{19}$ GeV) where all Nature's Fundamental Forces converge. Since the Grand Unification is the source of everything, if CPT collapses at the energy level of the Grand Unification we can then bid farewell to all that derives from CPT.

A few words on the three Invariance operators of CPT. The first to be discovered was C: this tells us that the physical reality must remain invariable if we replace the charges additively conserved by their

THE FIRST PART

The formidable developments in our field played a crucial role in the establishment of the "Ettore Majorana Centre for Scientific Culture". The lectures by Sid span fourteen years, from 1966 to 1979. Let me quote Sid's words [1]: «This was a great time to be a high-energy theorist, the period of the famous triumph of quantum field theory. And what a triumph it was, in the old sense of the world: a glorious victory parade, full of wonderful things brought back from far places to make the spectator gasp with awe and laugh with joy».

This victory gave us the Standard Model. Since this is a School I will close this first part with a few words on the Standard Model.

Let me go back to how I "discovered" Sidney Coleman in 1965. As mentioned in the Introduction, it happened in the midst of the great confusion generated by the apparent dominance of the S-matrix and the breaking of all Symmetry operators (P, C, PC, T). The School started in 1963 and during the years 1963–1966 the crises of QED and of QFT were, from the experimental point of view, corroborated for instance by the absence of the first example of antinuclear matter, $\bar{d}$, the antideuteron. The existence of the antideuteron was expected to be the proof that the nuclear forces acting between two antinucleons were producing the same binding energy as two nucleons. This could be granted if the nuclear forces were described by a QFT. But this was not the case and the S matrix was the negation of QFT. The existence of the antiparticles could not guarantee the existence of the antinuclei, having the same masses as the nuclei. This confusion is still going on at present, as testified by the famous Dirac statement [2] recently reported in my lecture celebrating the 50th anniversary of Karlsruhe (the first page of this lecture is reproduced on page 6). Sid was very interested to know if the $\bar{d}$ was there. It was searched for at the 10^{-7} level for the $\left(\bar{d}/\pi^-\right)$ ratio and not found to be there. The reason being that an order of magnitude of accuracy in fighting against background was needed. In order to discover the antideuteron it was necessary to have a new high precision Time-of-Flight electronic circuit capable of reaching the $\left(\bar{d}/\pi^-\right)$ ratio of 10^{-8}. Let me quote Luciano Maiani: «The Discovery of the antideuteron gave more confidence to the search for a field theoretical basis for the strong interactions, that today seems so obvious to us» [3].

I was very interested in the theoretical problems concerning the existence of a relativistic quantum field theory (RQFT) whose guide was QED since one of my first experiments was to check QED in the field outside the electron and the photon, i.e. in muon physics, via the high precision determination of the muon anomalous magnetic moment: $(g-2)_\mu$. All these problems are now taken for granted but were open to front line research 50 years ago.

Some of you know that the problem of why we should believe in RQFT has been discussed at this School on many occasions by the world leader Arthur S. Wightman.

As all of you know I am not a theorist. How is it that our School is heavily oriented towards theoretical fellows, one of which is Sidney Coleman? The answer is in my origin. I had the privilege of being a pupil of Lord Patrick Maynard Stuart Blackett to whom the building where we are is dedicated.

Blackett used to tell us young fellows, «*We experimentalists are not like theorists: the originality of an idea is not for being printed in a paper, but for being shown in the implementation of an original experiment*». This explains the difference between theory and experiment.

And now the reason why we have to invite our theoretical friends to speak. Here comes the second part of Blackett's teaching: «*We must know what boils in the brain of our theoretical friends. Not because they are able to predict our physics discoveries, but just for intellectual pleasure*». Sid Coleman was an incredible source of "intellectual pleasure".

Let me recall when at CERN he visited my experiment called PAPLEP (Proton antiproton annihilation into Lepton pairs) whose purpose was to see if a lepton heavier than the muon was produced. This new lepton was called HL [4] and was being searched for by looking for acoplanar $(e\mu)$ pairs in the final state of $(\bar{p}p)$ annihilation. The point being that if its mass was for example at 1 GeV it would not had been seen, the incredible miracle of the muon-pion small mass difference being the only reason why there are so many muons around. Having found no HL at CERN, I moved to Frascati where the new (e^+e^-) machine (ADONE) was being implemented. Sid came there once and was encouraging me to go on proposing to increase the ADONE energy to higher levels. The HL was found at SLAC following exactly the same strategy [($e\mu$) acoplanar pairs] but giving to HL the name τ. Sid phoned to congratulate me on the final discovery many years after PAPLEP at CERN and reminded me of what Rabi said in Erice, on the occasion of the tenth anniversary of the Subnuclear Physics School, concerning those who open a new field in physics. Sid asked me to have Rabi's statement reproduced in bronze in the Rabi Institute Hall (see photo on page 12).

From 1965 on, during long conversations in Erice or via transatlantic phone calls, I discussed with Sid what was *Past*, *Passing* and *to Come* in Subnuclear Physics. Needless to say I am very grateful to Sid for his words in the Preface of his book where his Erice lectures have been collected [1].

And now let me close this first part with a few words on the Standard Model.

A FEW WORDS ON THE STANDARD MODEL

At present, the Standard Model is the most formidable synthesis of all phenomena known to exist, from the inner structure of a proton to the extreme border of the universe. The Standard Model is based on:

i) matter particles (quarks or leptons), fermions, and therefore described by the Dirac spinor function ψ ; these particles have spin ½ ;

ii) gauge particles [the photon (g), the weak bosons ($W^{\pm}$, Z^0) and the gluons (g)], all being of vector nature; these particles have spin 1;

iii) scalar particles, with imaginary masses; these particles have zero spin. They cause Spontaneous Symmetry Breaking (SSB), thus providing real masses to matter particles (spin ½) and gauge particles (spin 1) in addition to themselves (spin zero particles), called Higgs particles. We denote these particles by the letter H.

All fundamental interactions in the Standard Model are based on simple Feynman diagrams of which the one below is, at the tree level, the prototype:

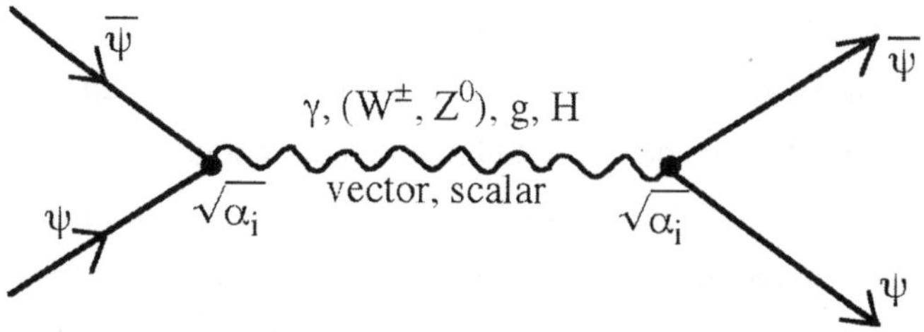

(1)

The basic ingredient of the Standard Model is the covariant derivative, which must have a term for each local gauge invariance; in fact, the way a gauge force is generated is through the substitution of the standard derivative ∂_μ with the covariant derivative D_μ.

Since the local gauge invariances are U(1), SU(2) and SU(3), there will be three gauge couplings g_1, g_2, g_3, respectively, with

$$\alpha_i = \frac{g_i^2}{4\pi} \quad (i = 1, 2, 3).$$

Each symmetry group has its generators

$$\begin{aligned} U(1) &\rightarrow Y \\ SU(2) &\rightarrow \tau_i = \tau_1, \tau_2, \tau_3 \\ SU(3) &\rightarrow \lambda_a = \lambda_1, \lambda_2 \dots \lambda_8 . \end{aligned}$$

Each gauge invariance generates a gauge boson

$$\begin{aligned} U(1) &\rightarrow B_\mu \\ SU(2) &\rightarrow W_\mu^i \\ SU(3) &\rightarrow G_\mu^a \end{aligned}$$

The covariant derivative of the Standard Model is therefore:

$$D_\mu = \partial_\mu - ig_1 \frac{Y}{2} B_\mu - ig_2 \frac{\tau^i}{2} W_\mu^i - ig_3 \frac{\lambda^a}{2} G_\mu^a .$$

The photon, described by the field A_μ, is the result of a mixing between the field B_μ and the neutral component of the weak boson field, W_μ^3 ; the values of the gauge couplings are not given by theory but by experiment, as well as all masses. More precisely, the Standard Model needs the following inputs:

3 Gauge couplings: g_1 , g_2 , g_3 .

1 Non-perturbative vacuum-angle parameters: θ_3 for the non-Abelian gauge forces SU(3).

6 Quark masses (which range from few MeV up to 180 GeV for the top quark).

3 Charged lepton masses (which range from 0.5 MeV up to 1.77 GeV for the third lepton).

3 Neutral lepton masses (which are very small, not zero, and with very small mass differences $\Delta m^2 \simeq 10^{-3}$ eV^2).

4 Mixing angle parameters in the quark sector (which come from the experimentally measured transitions across quark flavour states, including a phase angle to describe CP violation in the flavour-changing charged currents).

4 (Presumably) mixing angle parameters in the lepton sector.

2 Weak boson mass parameters (which can be taken as $M_{W^\pm}$ and M_{Higgs}).

These 26 parameters are the reason why we need to search for new physics beyond the Standard Model. But how could we start this search if we were unable to find a firm basis for the structure of the Standard Model? This firm basis could indeed be achieved and the key-source was the effective theoretical understanding of renormalization. Note that, if gravitational forces are included, there are two more parameters in a theory beyond the Standard Model: i.e. the Newton constant and the cosmological constant.

More details in Reference [5].

I.I. Rabi - 1972

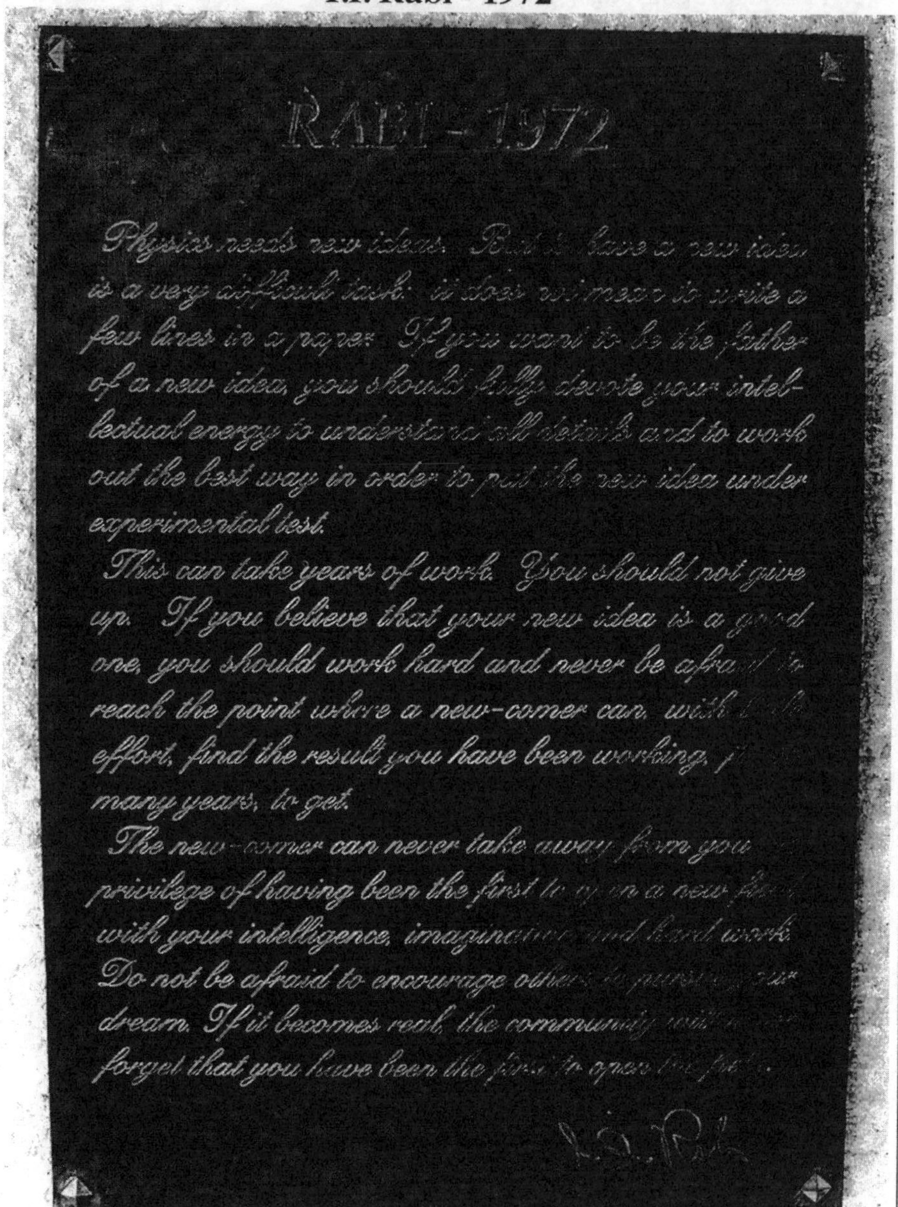

Bronze plate in the I.I. Rabi Institute Hall.

SECOND PART

I have selected some crucial problems discussed in the superb series of Coleman lectures [1]; a few pages per problem will – I hope – stimulate the young generations to reflect on many achievements taken for granted today despite their complex origin. The pencil marks on some pages have their origin in my reading these fascinating lectures. The selection of pages is intended to be a stimulus in order to read all lectures, which are the proof that each crucial step in building the Standard Model was identified and first discussed at the Erice Subnuclear Physics School by Sid Coleman. And this is a world record.

ASPECTS OF
SYMMETRY

CAMBRIDGE
UNIVERSITY PRESS

ISBN 0-521-31827-0
90000
9 780521 318273

Preface

I first came to Erice in 1966, to lecture at the fourth of the annual schools on subnuclear physics organized by Nino Zichichi. I was charmed by the beauty of Erice, fascinated by the thick layers of Sicilian culture and history, and terrified by the iron rule with which Nino kept students and faculty in line. In a word, I was won over, and I returned to Erice every year or two thereafter, to talk of what was past, or passing, or to come, at least insofar as it touched on subnuclear theory. Eight of these lectures, or more properly lecture series, are collected here.

No attempt has been made to bring the lectures up to date. Typographical errors, when spotted, have been corrected, and references to works to be published have been changed to references to published works. (I thank Hugh Osborne for taking on this dull task.) Otherwise, these are unaltered reprints of the original publications.

Numerous debts are acknowledged in the individual lectures, but there is one overriding debt that must be acknowledged here. None of this would have existed were it not for Nino Zichichi. Of course, he is the creator and director of the subnuclear school, and of the International Center for Scientific Culture 'Ettore Majorana' which encompasses it, but, more than that, he is personally responsible for each and every one of these lectures. The lecture notes would never have been written were it not for his blandishments and threats, transmitted in a fusillade of urgent cablegrams and transatlantic phone calls at odd hours of the morning. This book may be the least of his many accomplishments, but one of his accomplishments it is, and it should be counted as such.

Finally: These lectures span fourteen years, from 1966 to 1979. This was a great time to be a high-energy theorist, the period of the famous triumph of quantum field theory. And what a triumph it was, in the old sense of the word: a glorious victory parade, full of wonderful things brought back from far places to make the spectator gasp with awe and laugh with joy. I hope some of that awe and joy has been captured here.

Harvard 1984 Sidney Coleman

Acknowledgements

An Introduction to Unitary Symmetry (1966) was originally published in *Strong and Weak Interactions – Present Problems* (Academic Press, New York and London, 1966) © Academic Press, and is reproduced with their permission.

Soft Pions (1967) was originally published in *Hadrons and their Interactions* (Academic Press, New York and London, 1968) © Academic Press 1968 and is reproduced with their permission.

Dilatations (1971) and *Renormalization and Symmetry* (1971) were originally published in *Properties of the Fundamental Interactions* (Editrice Compositori, Bologna, 1973) © Editrice Compositori 1973, and are reproduced with their permission. The work was supported in part by Air Force Office of Scientific Research under contract no. F44620-70-C-0030 and the National Science Foundation under grant no. GP-30819X.

Secret Symmetry (1973) was originally published in *Laws of Hadronic Matter* (Academic Press, London and New York, 1975) © Ettore Majorana Centre and is reproduced with permission of the copyright holder and the publisher. The work was supported in part by the National Science Foundation under grant no. GP-30819X.

Classical Lumps and their Quantum Descendants (1975) was originally published in *New Phenomena in Subnuclear Physics* (Plenum Publishing Co., New York, 1976) © S. Coleman 1976 and is reproduced with permission of the copyright holder and the publisher. The work was supported in part by the National Science Foundation under grant no. MPS75-20427.

The Uses of Instantons (1977) was originally published in *The Whys of Subnuclear Physics* (Plenum Publishing Co., New York, 1979) © Plenum Press 1979 and is reproduced with their permission.

1/N (1979) was originally published in *Pointlike Structures Inside and Outside Hadrons* (Plenum Publishing Co., New York, 1982) © Plenum Press 1982 and is reproduced with their permission. The work was supported in part by the Department of Energy under contract DE-AC03-76SF00515 and by the National Science Foundation under grant no. PHY77-22864.

brought back from far places to make the spectator gasp [illegible] laugh with joy. I hope some of that awe and joy has been [illegible]

Harvard 1984 Sidney [illegible]

Acknowledgements

[illegible]

1

An introduction to unitary symmetry

(1966)

1 The search for higher symmetries

1.1 *The eight-baryon puzzle*

Let us begin with a very simple observation: there are eight baryons. By this I mean there are eight positive-parity particles with spin one-half and nucleon number one (the nucleons, the Λ, the Σs, and the Ξs), and that the masses of these particles are close together, all lying within 20% of their common mean mass. There are no other particles with the same parity, spin, and nucleon number which lie at all close in mass to these eight. Now, eight is a disquietingly large number of particles; although there is no fundamental reason why we should not have eight independent particles in a field theory, with eight arbitrary masses and eight sets of arbitrary coupling constants, still, it would be very pleasant if there were some way to reduce the number of entities. Two methods immediately suggest themselves.

(i) Perhaps some of the baryons are fundamental and the others are composite meson–baryon states. This was the central idea of the original Sakata model, where p, n, and Λ were taken as fundamental. Then the approximate mass degeneracy would arise because the masses of the mesons are small compared to the masses of the fundamental baryons. This idea has not met with much success.

(ii) Perhaps the strong interactions are approximately symmetric under a group larger than the ordinary isospin–hypercharge group. The eight baryons form a basis for an irreducible representation of this 'higher symmetry group'. If the symmetry were exact, the masses would be degenerate; because the symmetry is only approximate, the masses are split. (The situation envisioned here is something like that which prevails for isospin symmetry; if we turned off the weak interactions and electromagnetism, the world would be isospin symmetric, and all the particles in an isotopic multiplet would be degenerate; in fact, the world is not isospin symmetric, and the masses are slightly different.) This is sometimes phrased by saying that the strong-interaction Hamiltonian is a sum of two terms: one due to 'very strong interactions', which is symmetric under the higher symmetry group, and one due to 'medium-strong interactions', which is not.

The second viewpoint is one which has met with considerable success, and we intend to explore it in more detail here.

Now, what do we know about the structure of such a 'higher symmetry group'? Well, in the first place, there are several assumptions we can make, not because they are forced upon us by the problem, but because they are straightforward generalizations of what we know about the isospin group, and because they simplify the problem considerably. We assume, in the limit of exact symmetry, that:

(i) The group commutes with the Poincaré group.

(ii) The group acts on the Hilbert space of the states of the world as a group of unitary operators. (This is connected with conservation of probability.)

(iii) The one-particle states form an invariant subspace under the action of the group. The many-particle asymptotic states transform like tensor products of one-particle states.

(iv) The group commutes with the S-matrix.

A group which satisfies these conditions is usually called 'an internal symmetry group'. It is worth noting that in a field theory any group whose generators are obtained by integrating, over all space, the time components of a set of conserved Hermitian currents is an internal symmetry group.

In addition to these assumptions, there are several conditions forced on the group by the nature of the problem.

(i) The group turns baryon states at rest, with spin up, into baryon states at rest, with spin up. Therefore, it must have an eight-dimensional representation.

(ii) Since the whole point of the hypothesis is to explain the approximate equality of the baryon masses, this representation must be irreducible.

(iii) (This is a rather fine technical point, but it is necessary in order and

$$\Delta m^{(2)} = (\Sigma^+ + \Sigma^- - 2\Sigma^0)/\sqrt{6} \approx 0.7 \text{ MeV}.$$

Precisely the reverse of our naive prediction!

Clearly, then, there is something in the structure of the strong interactions that tends to enhance the isospin one part of the mass splitting (and perhaps, also, to suppress the isospin two part). The strong interactions are the sum of the very strong interactions and the medium-strong interactions. In the last section we derived sum rules for electromagnetic mass splittings by neglecting the effects of the medium-strong interactions, and obtained good agreement with experiment. So, we would be surprised if the medium-strong interactions had an important effect on the mass splittings. Thus, whatever it is that enhances the isospin-one mass splitting would be present even if we turned off the medium-strong interactions.

However, the very strong interactions do not know isotopic spin; they only know SU(3). They do not have the option of enhancing $I=1$ perturbations. They have only the option of enhancing the octet or the 27-plet part of the perturbation. If they enhanced the 27-plet, we would see enhancement of both isospins, for the 27-plet contains both neutral isotriplets and neutral isoquintuplets. The octet, however, contains only isotriplets.

Thus, we are forced to conclude that there is something in the structure of the very strong interactions that causes them to enhance the octet part of the electromagnetic mass splittings. This is called octet enhancement.

Notice that octet enhancement is a property of the very strong interactions, not of electromagnetism. The observed octet pattern of electromagnetic mass splittings has nothing at all to do with the transformation properties of the electromagnetic interaction. Once we have come to this discovery, it is very tempting to speculate that the same thing is true for the medium-strong interactions. Perhaps the observed octet pattern of medium-strong mass splittings is due, not to the transformation properties of these interactions, but to an octet enhancement mechanism. In fact, perhaps it is the same mechanism for both interactions – perhaps there is a mechanism for *universal octet enhancement.*

Is this a plausible idea or just a wild speculation? Well, one way of checking its plausibility is to construct dynamical models that embody universal octet enhancement. I will discuss three such models here.

(i) First, there is the tadpole model, proposed by Glashow and myself. This model assumes the existence of an octet of scalar mesons. (These need not be stable particles, of course; they would be resonances or anti-bound states. The last is a more likely situation for *s*-wave channels.) These are labeled with the names of the corresponding pseudoscalar mesons, with a prime. Thus, K′, $\bar{K}'$, π', η'. In a field theory involving scalar mesons there arises the possibility of a class of Feynman diagrams that do not otherwise occur. These are diagrams with only one external line. Fig. 3 shows such a diagram. For obvious reasons, these are called tadpole diagrams. In the absence of the symmetry-breaking interactions these diagrams vanish, for they represent a transition between a unitary octet (the scalar mesons) and a unitary singlet (the vacuum). However, if we turn on the medium-strong interactions we can make an η' tadpole, and if we turn on electromagnetism we can make a $\pi^{0\prime}$ tadpole. These diagrams, by themselves, are of very little interest, but they are important as internal parts of other diagrams. Fig. 4 shows two tadpoles occurring as internal parts of baryon self-energy diagrams. The tadpole on the left makes a contribution to the medium-strong self-mass of the Λ; the one on the right makes a contribution to the electromagnetic self-energy of the proton. Our fundamental (and completely unjustified) assumption is that these diagrams dominate symmetry breaking. Because the scalar mesons form an octet, we get universal octet dominance. If they formed a 27-plet, we would get 27-plet dominance.

(ii) Next, there is the mixing model, proposed by Sakurai, Radicati, Zanello and Picasso. These authors posit that the medium-strong mass splittings are dominated by diagrams that involve the mixing of a singlet vector meson and an octet vector meson. The electromagnetic splittings are, correspondingly, dominated by ρ^0–ω_1 mixing. These mixing matrix elements transform like the components of an octet; thus we obtain universal octet dominance.

Fig. 3

Fig. 4

(iii) Finally, there is the bootstrap model, introduced by Cutkosky and Tarjanne, and investigated in more detail by Dashen and Frautschi. Let us suppose we have a system of particles, interacting through some interaction with a high degree of symmetry, under which the particles form degenerate bound states. Let us introduce a new interaction which breaks the symmetry. Then the energies of the bound states will be changed by two mechanisms: the masses of the components will be changed, and the forces will also be changed. In lowest-order perturbation theory this can be written as

$$\delta m_B = A\,\delta m_c + f,$$

where m_B is a vector composed of the masses of the bound states, m_c is a vector composed of the masses of the components, f is a vector composed of the contribution of the change in the forces, and A is a matrix. Now let us suppose the problem is, in fact, a bootstrap. (The example studied by Dashen and Frautschi is the static baryon-octet resonance-decuplet bootstrap.) Then m_B and m_c are identical, and the equation becomes

$$(1-A)\,\delta m = f.$$

If A has an eigenvalue close to one, then there will be enhancement, independent of f. If this eigenvalue corresponds to a set of eigenvectors that transform like an octet, then it will be octet enhancement. Dashen and Frautschi have found, in the static bootstrap, that there *is* an octet of eigenvectors with eigenvalue close to one. This is a strong point in favour of the bootstrap model. In the other theories, octet enhancement is fed in at the beginning; in the bootstrap model it is calculated. The calculation could lead to 27-plet enhancement; it does not.

If universal octet enhancement was the only consequence of these models, they would not really be worth much; there would be hardly more conclusions than assumptions. However, all the models I have discussed possess a second feature which is most easily demonstrated for the tadpole model. Let us return to Fig. 4. The diagram on the left gives the principal part of the medium-strong mass splittings of the hadrons in terms of the strength of their coupling to η' and the magnitude of the η' tadpole. Conversely, we can deduce from the observed splittings the η' hadron–hadron coupling constants, to within an unknown multiplicative factor. Since η' and $\pi^{0\prime}$ are part of the same octet, we can then calculate the $\pi^{0\prime}$ hadron–hadron coupling constants, just by doing a unitary rotation. But if we know these, we know the values of the diagrams on the right, to within an unknown multiplicative constant.

Thus, we can calculate the principal part of the electromagnetic mass

2

Soft pions

(1967)

The purpose of these lectures is to explain certain techniques, developed in the last few years by Adler, Weisberger, Weinberg, and others, for the analysis of processes involving low-energy pions. I have tried, as far as possible, to make the lectures self-contained; the only background required of the reader is an understanding of field theory on the Feynman diagram level. In particular, no previous knowledge of current commutators or low-energy theorems is assumed.

In Sec. 1 the reduction formula is developed and some of its consequences discussed. Sec. 2 is a brief summary of the relevant parts of weak-interaction theory. Soft pions first appear in Sec. 3, a discussion of the Goldberger–Treiman relation. Sec. 4 is an analysis of the various definitions of PCAC. Sec. 5 is a discussion of Lagrangian models in general, and the gradient-coupling model in particular. In Sec. 6, Adler's rule for the emission of one soft pion is derived. The current commutation relations are introduced in Sec. 7. In Sec. 8 the formula for the *s*-wave pion–hadron scattering length is derived. In Sec. 9, the special case of pion–pion scattering is treated. In Sec. 10, a few remarks are made about leptonic decays of kaons.

I have done no original work in this field; most of what I know I have learned in conversations with S. Adler, H. Schnitzer, and S. Weinberg. Those who were lucky enough to attend Weinberg's Loeb lectures at Harvard will know how much these lectures owe to him. I am also indebted to J. Bernstein, who allowed me to read the manuscript of his forthcoming book on currents.

1 The reduction formula

Let us suppose we have a Lagrangian field theory, in which the Lagrange density, $\mathscr{L}$, depends only on a single scalar field, ϕ, and its where k is the four-momentum transfer, the us are the appropriate Dirac bispinors, and the three dots represents other terms that are of order k^2, and hence do not make a significant contribution to this process. (Although they are important in muon capture by nuclei, and also in high-energy neutrino reactions.) The A and V terms clearly come from the axial-vector and vector currents, respectively. Experimentally,

$$g_V(0) \approx 1,\ g_A(0)/g_V(0) \approx 1.25.$$

We shall often denote these quantities simply by g_V and g_A.

The second example is pion decay. Here, by parity, only the axial current can contribute. The relevant matrix element is

$$\langle 0|A_\mu(x)|\pi^-\rangle = (\text{K.F.})e^{-ip\cdot x}(ip_\mu F_\pi/\sqrt{2}). \tag{2.7}$$

where p is the pion four-momentum. Unlike the previous example, here the quantity F_π has no p^2 dependence, since p^2 is fixed at m_π^2. From the observed pion lifetime F_π is readily calculated to be 0.19 M_p.

If we take the divergence of Eq. (2.7), we find that

$$\langle 0|\partial^\mu A_\mu(x)|\pi^-\rangle = (\text{K.F.})m_\pi^2 F_\pi e^{-ip\cdot x}/\sqrt{2}. \tag{2.8}$$

The most important property of this equation is that the right-hand side is not zero! For, in view of the arguments of the proceeding section, this means that

$$\phi_{\pi^-} = \sqrt{2}\,\partial_\mu A^\mu/F_\pi m_\pi^2, \tag{2.9}$$

is a perfectly good pion field, suitable for use in the reduction formula. We shall begin to exploit this fact in the next section.

3 The Goldberger–Treiman relation and a first glance at PCAC

Let us define $g(k^2)$ by

$$\langle \text{p}|\phi_{\pi^-}(x)|\text{n}\rangle = (\text{K.F.})e^{-ik\cdot x}(k^2 - m_\pi^2)^{-1}\bar{u}_p\gamma_5 u_n g(k^2)\sqrt{2}. \tag{3.1}$$

The only thing we know about $g(k^2)$ experimentally is its value at m_π^2. For, by the arguments of Sec. 1, this must be the renormalized pion–nucleon coupling constant, the quantity we measure by extrapolating to the pion pole in nucleon–nucleon scattering, or by extrapolating to the nucleon pole in pion photoproduction. (The $\sqrt{2}$ in the formula is just an isospin factor.) Thus,

$$g(m_\pi^2) = g = 13.5. \tag{3.2}$$

Now let us return to the axial-vector contribution to neutron beta-decay. This time I will write out all the invariants.

$$\langle \text{p}|A_\mu(x)|\text{n}\rangle = i(\text{K.F.})e^{-ik\cdot x}\bar{u}_p \times [\gamma_\mu\gamma_5 g_A(k^2) + k_\mu\gamma_5 g_p(k^2) + \sigma_{\mu\nu}\gamma_5 k^\nu g_M(k^2)]u_n. \tag{3.3}$$

Taking the divergence, we find

$$\langle p|\partial^\mu A_\mu(x)|n\rangle = (\text{K.F.})e^{-ik\cdot x}\bar{u}_p\gamma_5 u_n[-2Mg_A(k^2)+k^2 g_p(k^2)], \quad (3.4)$$

where M is the nucleon mass. Comparing this with Eqs. (2.9) and (3.1), we obtain

$$\sqrt{2}\,g(k^2)/(k^2-m_\pi^2) = [-2Mg_A(k^2)+k^2 g_p(k^2)]\sqrt{2}/F_\pi m_\pi^2. \quad (3.5)$$

This equation is simply a consequence of our definition of ϕ_π, and is without predictive power unless supplemented by further assumptions. We now make such an assumption. We assume that $g(k^2)$ is 'slowly varying' over a distance of the order of m_π^2, that is to say, that

$$g(m_\pi^2) \approx g(0). \quad (3.6)$$

This is a special case of the PCAC hypothesis which we will discuss in more detail shortly. (PCAC is an acronym for 'partially conserved axial current'. We will explain this peculiar phrase in the next section.)

I stress that Eq. (3.6) is pure assumption. It would certainly be false, for example, if ϕ_π were like the peculiar fields involving Klein–Gordon operators which we constructed at the end of Sec. 1.

However, if we accept Eq. (3.6), then, evaluating Eq. (3.5) at $k^2=0$, we instantly obtain

$$F_\pi g \approx 2Mg_A. \quad (3.7)$$

This is the famous Goldberger–Treiman relation. It is in excellent agreement with experiment:

$$2.56M \approx 2.50M.$$

4 A hard look at PCAC

Many statements about the meaning of PCAC and the explanation of the Goldberger–Treiman relation exist in the literature and in the folklore of physics. Here are some of them:

1. 'The Goldberger–Treiman relation is just pololology. It is simply the statement that neutron beta-decay is dominated by the one-pion pole.' This is *wrong*. If the one-pion pole diagram (Fig. 4) made the only contribu-

Fig. 4

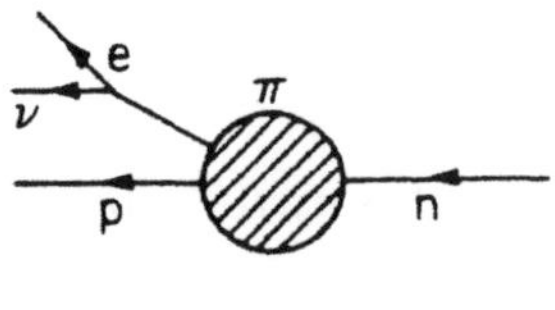

tion to neutron beta-decay, we would predict

$$g_{\rm p} = gF_\pi(k^2 - m_\pi^2)^{-1}, \tag{4.1}$$

and

$$g_{\rm A} = 0. \tag{4.2}$$

This is (1) not the Goldberger–Treiman relation and (2) in flat contradiction with experiment.

2. 'PCAC is the statement that the matrix elements of $\partial_\mu A^\mu$ are slowly varying.' This is almost right. (That is to say, it is very close to the viewpoint I will espouse in these lectures.) However, I would prefer to say rather that the matrix elements are *normally varying*: that, just as for S-matrix elements and electromagnetic form factors, about which we do have empirical information, the rate of variation is determined by the distance to the nearest singularity. Thus, in the particular case of $g(k^2)$, the nearest singularity is the beginning of the three-pion cut. This should induce a derivative on the order of $(9m_\pi^2)^{-1}$ at the origin, just as the ρ pole induces a derivative in the electromagnetic form factor of the order of m_ρ^{-2}. Thus the error made in extrapolating from zero to m_π^2 should be of the order of $(m_\pi^2/9m_\pi^2)$ or approximately 10%. The advantage of stating things this way is twofold. (1) It gives a clear idea of how to use PCAC in processes where the kinematics are more complicated than in neutron decay, and where the invariant functions may depend on several variables. (2) It emphasizes that $\partial^\mu A_\mu$ is in no way especially 'smooth'; the derivatives of its matrix elements are no smaller than those of many familiar operators. What is special is that the pion mass is small, compared to the characteristic masses of strong interaction physics; thus extrapolation over a distance of m_π^2 introduces only small errors.

3. 'PCAC is the statement that if the pion mass were zero the axial vector current would be exactly conserved.' This does not look at all like the preceding version of PCAC; nevertheless, it also leads to the Goldberger–Treiman relation. Let us see how this works.

Let us imagine ourselves in a world in which the pion mass is zero and the axial current is conserved. We can still define F_π, $G_{\rm A}$, etc. by Eqs. (2.8) and (3.3). (Although their physical meanings are, of course, quite different – F_π is now not the pion decay constant, but the electron decay constant.) Because the axial current is conserved, Eq. (3.4) becomes

$$-2Mg_{\rm A}(k^2) + k^2 g_{\rm p}(k^2) = 0. \tag{4.3}$$

However, one-pion exchange (Fig. 4 again!) now produces a pole in $g_{\rm p}$:

$$g_{\rm p}(k^2) = gF_\pi/k^2 + \text{non-singular terms}. \tag{4.4}$$

Thus, when we evaluate (4.3) at $k^2=0$, we obtain

$$-2Mg_A+gF_\pi=0, \tag{4.5}$$

which is the Goldberger–Treiman relation!

Of course, in the real world the pion mass is not zero and the axial current is not conserved, so Eq. (4.5) is not exact. However, the pion mass is small; thus if it is the only thing that keeps the axial current from being conserved, we might expect Eq. (4.5) to be accurate up to terms of the order of m_π^2. One might express this viewpoint by saying that the axial current is 'almost conserved'. Actually, the phrase used is 'partially conserved' – hence the term PCAC.

Note that although this formulation of PCAC is quite different from the preceding one, it leads to the same result and the same rough estimate of the error. Although in these lectures I will try and stick with the preceding formulation (principally because I believe it offers a clearer idea of the sources of possible errors in the more complicated applications which we shall do shortly), this way of looking at things is also a good one. Sometimes we will refer back to it, and it will give us new insights.

4. 'PCAC is the statement that the pion field is the divergence of the axial-vector current.' This statement occurs frequently in the literature. Unfortunately, it is completely free of content, since, as we have seen, it is true by definition. Of course, if we are working within the framework of a Lagrangian field theory with a specific strong-interaction Lagrangian, it is a well-defined statement to say that $\partial^\mu A_\mu$ is proportional to the canonical pion field that occurs in the Lagrangian. However, it is a statement without predictive power, since we know no more about the variation of the matrix elements of canonical fields than we do about those of any other local operator.

Nevertheless, we will sometimes have occasion to write down Lagrangian field theories which possess PCAC in this sense. Our motivation is this: when we make assumptions about the properties of the weak interaction currents, we do not have absolute freedom; we must always be sure that our assumptions do not contradict each other or the general principles of relativistic local field theory. The easiest way to check consistency is to construct a Lagrangian field theory that embodies all of our assumptions. However, since we are *only* using this theory to check consistency, there is no reason why the strong interaction coupling constants must be large. Indeed, we will usually take them to be small. In this case, we can calculate the matrix elements of the canonical pion field in lowest-order perturbation theory, and verify that the definition of PCAC given here implies the correct version (statement 2 above).

In the next section we will construct such a Lagrangian model to verify the consistency of the assumptions that lie behind the Goldberger–Treiman relation.

5 The gradient-coupling model

In the early days of pion physics, a popular model of pion decay was the one expressed in Fig. 5. In this model, F_π is clearly proportional to

Fig. 5

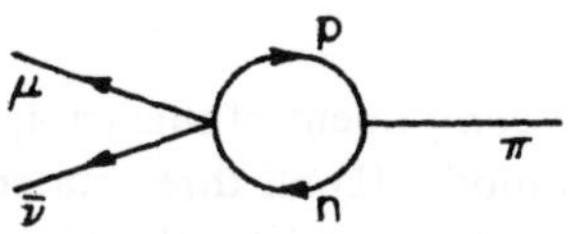

g. The Goldberger–Treiman relation, however, makes precisely the opposite assertion:

$$F_\pi = 2Mg_A/g. \tag{5.1}$$

This strongly suggests that the Goldberger–Treiman relation is a characteristically strong-interaction result – that it would not be true if the strong interactions were weak. We will show that this suggestion is misleading by displaying a Lagrangian model in which the Goldberger–Treiman relation holds even if the strong interactions are weak. (Indeed, because of the remarks at the end of the last section, this is the only domain in which, for a Lagrangian field theory, we can be sure it holds.)

However, first I will review the familiar method for obtaining currents from a Lagrangian. A local Lagrangian field theory is determined by a Lagrange density, $\mathscr{L}$, which is a function of a set of fields ϕ^a and their first derivatives $\partial_\mu\phi^a$. The dynamics of the theory is given by Hamilton's principle,

$$\delta \int \mathscr{L}\, d^4x = 0, \tag{5.2}$$

for variations of the fields which vanish at infinity.

Let us consider an infinitesimal transformation of the fields of the form

$$\delta\phi^a = F^a\,\delta\lambda, \tag{5.3}$$

where the Fs are arbitrary functions of the ϕs at the point x, and $\delta\lambda$ is an infinitesimal constant. Let us assume that under this transformation

Associated with these transformations is an isotriplet of axial-vector currents:

$$\mathscr{A}_\mu = \partial_\mu \boldsymbol{\phi} - \mathrm{i}\frac{g}{2M}\bar{\psi}\gamma_\mu\gamma_5\boldsymbol{\tau}\psi. \tag{5.13}$$

The only part of $\mathscr{L}$ not invariant under (5.12) is the pion mass term; therefore,

$$\partial^\mu \mathscr{A}_\mu = -m_\pi^2 \boldsymbol{\phi}. \tag{5.14}$$

This current, as it stands, is not a suitable candidate for A_μ; its one-nucleon matrix elements do not have the right value. Therefore, we define

$$\mathbf{A}_\mu = \frac{2Mg_A}{g}\mathscr{A}_\mu. \tag{5.15}$$

and take A_μ to be the positively-charged component of this triplet.

I leave it to you to verify that in this model (1) all three statements of PCAC given in the preceding section are true, and (2) the Goldberger–Treiman relation is valid. (Always, of course, working only to lowest order in g.)

Problem. Suppose, instead of choosing (5.15) for the axial current, we choose

$$\mathbf{A}_\mu + \alpha\partial_\mu\boldsymbol{\phi},$$

with α some undetermined constant. Is the Goldberger–Treiman relation still valid? If so, why? If not, why not? Answer the question using all three versions of PCAC, if possible.

6 Adler's rule for the emission of one soft pion

We will now develop a formalism, due to S. Adler, for calculating the matrix element of any hadronic process of the form

$$i \to f + \pi, \tag{6.1}$$

where i and f are any hadronic states, in terms of that for the process

$$i \to f. \tag{6.2}$$

For example, we will be able to relate pion production,

$$\mathrm{N} + \mathrm{N} \to \mathrm{N} + \mathrm{N} + \pi, \tag{6.3}$$

to nucleon–nucleon scattering

$$\mathrm{N} + \mathrm{N} \to \mathrm{N} + \mathrm{N}. \tag{6.4}$$

Our method will be to obtain an *exact* formula for the process (6.1), with the pion off the mass shell and with the pion four-momentum, k, close to zero. We will then extrapolate this expression to the mass shell. It is clear

Adler's rule. In any strong-interaction process, to calculate the matrix element for the emission of one soft pion, take the matrix element for the process without the pion, and sum all the terms obtained by attaching the pion to each of the external lines, using gradient coupling.

In other words, lowest-order perturbation in the gradient-coupling theory is exact for the emission of one soft pion.

7 Current commutators

Suppose we attempt to extend the formalism of the previous section to processes involving two soft pions. Clearly, in this case we would have to analyze objects like

$$T\langle f|\partial^\mu A_\mu^a(x)\partial^\nu A_\nu^b(y)|i\rangle.$$

But now when we attempt to pull out the differential operators we obtain not only a double divergence of a matrix element of two currents, but also an equal time commutator from the derivative of the time-ordering operator. Thus, to go further, we need some information about the equal-time commutator.

7.1 *Vector–vector commutators*

Let me review the CVC hypothesis of Feynman and Gell-Mann. Recall our observation that, experimentally, g_V is very close to one. Why should this be so? Feynman and Gell-Mann offered the following explanation; suppose that the weak current for nucleons is of exactly the same structure as that for leptons. Then, the bare value of g_V, the number that occurs in the interaction Lagrangian, would be one, as would the bare value of g_A. Further suppose that the vector current is conserved. (CVC = conserved vector current). Then in this case it can be shown that the g_V is not renormalized by the strong interactions. (For a proof, see Appendix 2). Now, currents conserved by the strong interactions are not that easy to find; the only one known with the right quantum numbers is the isospin current. Therefore, Feynman and Gell-Mann further postulated that V_μ is part of an isotriplet of currents, V_μ^a, and that V_μ^a is proportional to the isospin current I_μ^a; that is to say

$$V_\mu^a = \alpha I_\mu^a, \tag{7.1}$$

with α a constant. This is a stronger hypothesis than conservation, but it has been checked in the famous weak-magnetism experiment, and is now generally accepted.

Originally, α was taken to be two; that is to say, the coupling of the vector current to nucleons was assumed to be exactly equal to the coupling to leptons. The observed small deviation of g_V from one was then ascribed to electromagnetic corrections. However, ever since the famous SU(3) analysis of the weak currents by Cabibbo, it has been realized that this assumption is based on a too-naive formulation of universality; Cabibbo theory predicts

$$\alpha = 2g_V, \tag{7.2}$$

where g_V is identified with the cosine of the Cabibbo angle.

Now, in most simple models of the strong interactions (e.g. Yukawa-type Lagrangians), the equal-time commutators of the fourth components of the isospin currents have the same algebraic structure as those of the associated charges; i.e.

$$[I_0^a(\mathbf{x}, 0), I_0^b(\mathbf{y}, 0)] = i\varepsilon_{abc} I_0^c(\mathbf{x}, 0)\delta^3(\mathbf{x}-\mathbf{y}). \tag{7.3}$$

In more complicated models, there may be additional terms on the right, proportional to gradients of delta-functions, which vanish when we integrate the currents to make the charges. For simplicity, we will ignore the possible occurrence of these terms. However, as you may readily check, none of the calculations we will do will depend on the assumption of their absence.

From the preceding equations, we find

$$[V_0^a(\mathbf{x}, 0), V_0^b(\mathbf{y}, 0)] = 2ig_V\varepsilon_{abc} V_0^c(\mathbf{x}, 0)\delta^3(\mathbf{x}-\mathbf{y}). \tag{7.4}$$

7.2 *Vector–axial commutators*

These follow directly from the identification of V_μ^a with the isospin current, and the statement that A_μ^a is an isotriplet:

$$[V_0^a(\mathbf{x}, 0), A_0^b(\mathbf{y}, 0)] = 2ig_V\varepsilon_{abc} A_0^c(\mathbf{x}, 0)\delta^3(\mathbf{x}-\mathbf{y}). \tag{7.5}$$

7.3 *Axial–axial commutators*

There is no direct experimental check on these objects (other than the applications we are going to discuss). However, there are some general theoretical arguments that make the commutators we are going to assume particularly attractive. I will sketch two lines of argument.

(1) **Chirality principle.** Suppose we follow the original suggestion of Feynman and Gell-Mann and introduce parity-violation into the weak interactions only through the use of the projection matrix $1+i\gamma_5$ in the definition of the currents. This means that the parity-transformed currents, $V_\mu - A_\mu$, only involve the matrix $1-i\gamma_5$. These are orthogonal projection matrices; therefore

$$[(V_0^a(\mathbf{x}, 0) + A_0^a(\mathbf{x}, 0)), (V_0^b(\mathbf{y}, 0) - A_0^b(\mathbf{y}, 0))] = 0. \tag{7.6}$$

Now let us abstract Eq. (7.6) from the Feynman–Gell-Mann theory and adopt it as a general rule. (I will call this 'the chirality principle'.) Then we deduce that

$$[A_0^a(\mathbf{x}, 0), A_0^b(\mathbf{y}, 0)] = 2ig_V \varepsilon_{abc} V_0^c(\mathbf{x}, 0)\delta^3(\mathbf{x}-\mathbf{y}). \tag{7.7}$$

(2) *Universality principle.* Roughly, universality is the statement that all the weak interactions have the same strength. Originally, when only nuclear beta-decay was known, this was formulated as the requirement that the baryon–lepton coupling have the same strength as the lepton–lepton coupling. However, with the current plethora of hadrons, it is difficult to unambiguously generalize this statement. Should one require that all hadronic coupling constants be equal? Or should the sum of the squares be equal to the lepton constant squared? Anyway, how does one compare coupling constant for particles of different spin? To avoid these difficulties, Gell-Mann suggested the following definition of universality: 'The algebra generated by repeated equal-time commutation of the fourth components of the total weak-current with its adjoint must be the same as the corresponding algebra generated from the lepton currents alone.'

This definition has several obvious advantages: (1) if the only hadrons were nucleons, it would lead to the old definition of universality; (2) it is independent of the details of the hadron spectrum and the structure of the strong interactions; (3) within any particular model of the strong interactions, it fixes the relative scale of the hadron and lepton currents, as well as the relative scale of the baryon and boson parts of the hadron current.

It can be shown that this definition of universality, together with the Cabibbo theory, leads to the commutators we have written down. However, since the argument involves considerable use of SU(3), I will not give it here.

I should complete this section by giving some examples of Lagrangian models in which these commutators hold and which also obey PCAC. (Note that the gradient-coupling model will not do; in it, the fourth components of axial currents commute.)* Fortunately, I am spared this labor; any of the Lagrangians discussed by Professor Zumino in his lectures will do the job.†

8 The Weinberg–Tomozawa formula and the Adler–Weisberger relation

In this section we will derive a formula for the *s*-wave threshold scattering length in the elastic scattering of a pion off any hadron target.

* They are canonical momentum densities.

† See *Hadrons and Their Interactions* (Academic Press, New York and London, 1968).

Appendix 1. Notational conventions

1 We use the metric $g_{\mu\nu}$ for which

$$g_{00} = -g_{11} = -g_{22} = -g_{33} = 1.$$

2 The differential operator ∂_μ is defined as $\partial/\partial x^\mu$; likewise

$$\Box^2 = \partial^\mu \partial_\mu = \partial_0^2 - \nabla^2,$$

and the Klein–Gordon equation is

$$(\Box^2 + m^2)\phi = 0.$$

2 The γ-matrices are defined by

$$\gamma_\mu \gamma_\nu + \gamma_\nu \gamma_\mu = 2g_{\mu\nu}.$$

Also, for any vector a,

$$\not{a} = a_\mu \gamma^\mu.$$

The Dirac equation is

$$(i\not{\partial} - m)\psi = 0$$

4 The propagators for spin 0 and spin 1/2 fields are

$$i(p^2 - m^2)^{-1} \quad \text{and} \quad i(\not{p} - m)^{-1},$$

respectively.

5 In a matrix element, every particle carries with it a kinematic factor of $(2\pi)^{-3/2}(2E)^{-1/2}$. To shorten equations, the product of these factors is frequently denoted simply by (K.F.).

Appendix 2. No-renormalization theorem.

In this appendix we wish to show that if the vector current is proportional to the isospin current, the value of g_V does not depend on the strong interactions. In other words, g_V is not renormalized by the strong interactions.

Let us define the total isospin, **I**, by

$$I^a = \int d^3x I_0^a(\mathbf{x}, 0).$$

If we neglect electromagnetism, **I** is conserved; it commutes with the Hamiltonian and turns states of a given energy into states of the same energy. Likewise, from its definition, it commutes with the momentum, and turns states of a given momentum into states of the same momentum. Therefore, **I** must turn one-nucleon states at rest into one-nucleon states at rest. Now the components of **I** are a set of operators closed under commutation; in mathematical language they form a Lie algebra, one that is, in fact, the same as the angular-momentum Lie algebra. Therefore, the one-nucleon states at rest must form the basis for a two-dimensional representation of this algebra.

However, we know from ordinary angular-momentum theory that there is only one such representation, the spin-$\frac{1}{2}$ representation, unique up to a similarity transformation. We fix the similarity transformation by defining the neutron and proton states to be eigenstates of I_z, and choosing the relative phase such that the matrix elements of the raising and lowering operators are real. Thus, the matrix elements of **I** between one-nucleon states at rest are uniquely determined by the isospin algebra; they can not depend on any strong-interaction parameters.

But now we are almost home. For, by Eq. (7.1)

$$I^a = \alpha \int d^3x\, V_0^a(\mathbf{x}, 0),$$

and the right-hand side of this equation, evaluated between one-nucleon states at rest, is clearly linearly related to g_V, since the momentum transfer, k, is zero. Thus, g_V can not depend on the strong interactions, and we have proved the theorem.

Appendix 3. Threshold *S*-matrix and threshold scattering lengths

In this appendix, we will consider a two-particle scattering process with initial momenta (p, q) and final momenta (p', q'), and relate the value of the S-matrix near threshold to the s-wave scattering length.

Near threshold

$$\langle p', q'|(S-1)|p, q\rangle = A\delta^4(p' + q' - p - q).$$

with A some constant. Let us introduce a new basis, labeled by the total and relative momenta, P and k.

$$\langle P', k'|(S-1)|P, k\rangle = A\delta(E-E')\delta^3(\mathbf{P}-\mathbf{P}').$$

If we go to the center-of-mass frame (i.e. restrict ourselves to states with $P=0$) we may write the center of mass S-matrix as

$$\langle \mathbf{k}'|(S-1)|\mathbf{k}\rangle = A\delta(E-E').$$

Note that for small **k**,

$$E = \frac{\mathbf{k}^2}{2m_1} + \frac{\mathbf{k}^2}{2m_2} = \frac{\mathbf{k}^2}{2\mu},$$

where m_1 and m_2 are the masses of the two particles, and μ is the reduced mass. The s-wave state, $|k\rangle$, is defined by

$$|k\rangle = \frac{1}{4\pi k}\int d^3\mathbf{k}\, \delta(k - |\mathbf{k}|)|\mathbf{k}\rangle.$$

3

Dilatations

(1971)

1 Introduction

It is an old idea in particle physics that, in some sense, at sufficiently high energies the masses of the elementary particles should become unimportant. In recent years this somewhat vague hope has acquired a more definite form in the theory of scale transformations, or dilatations. These are transformations that would be exact invariances of the world if all elementary particle masses (more generally, all dimensionful couplings) vanished. The hope is that, by studying these approximate symmetries in Lagrangian field theories, we can gain knowledge about how scale transformations behave in the real world, and learn something about those kinematic realms where the effects of masses are indeed unimportant, or at least simply calculable. This is a hope that has been fulfilled for broken chiral symmetries, by the study of models such as the sigma-model.

The purpose of these lectures is to report on the progress of such investigations. As we shall see, there is still much that is obscure. In particular, the connection between the sorts of things I will be talking about and the famous experimental scaling of deep inelastic electroproduction remains to be unravelled. Nevertheless, what has been done so far has already yielded some remarkable theoretical surprises.

Section 2 is a description of the formal Lagrangian theory of scale transformations. Some of the Ward identities that express broken scale invariance are derived, models of Nambu–Goldstone scale-symmetry breaking are discussed, and the connection between scale invariance and conformal invariance is explained. These last two topics will not be returned to in the remainder of these lectures, and a reader who is only interested in the main line of the argument may skip the Sections (2.3 and 2.4) that discuss them.

Section 3 is an investigation of broken scale invariance in renormalized perturbation theory. We find that the Ward identities derived in the preceding section are lies (in the polite current parlance, they contain anomalies), and we find the true equations that replace them, the Callan–Symanzik equations.

Section 4 explains how, despite the Callan–Symanzik anomalies (in fact, because of them) scale invariance may still be regained in appropriate kinematic regions. This astonishing magic trick is due to Ken Wilson.

My own contributions to this field have been minor. I have learned most of what I know from conversations with Curtis Callan, John Ellis, Roman Jackiw, Kurt Symanzik, Kenneth Wilson, and Bruno Zumino. I would like to express both my gratitude to them and my hope that I have not distorted their ideas too badly.

2 The formal theory of broken scale invariance

2.1 *Symmetries, currents, and Ward identities*

We will deal in these lectures with Lagrangian field theories:[1] that is to say, theories involving a set of fields, which we will assemble into a big vector, ϕ, and whose dynamics are determined by a function of the fields and their first derivatives, $\mathscr{L}(\phi, \partial_\mu\phi)$, called the Lagrangian, via Hamilton's principle:

$$\delta I \equiv \delta \int d^4x\, \mathscr{L} = 0, \tag{2.1}$$

for solutions of the equations of motion. I is called the action integral.

Let us consider some infinitesimal transformation of the fields:

$$\phi \to \phi + \delta\phi. \tag{2.2}$$

(We will always suppress the infinitesimal parameter that should properly multiply $\delta\phi$.) Let us suppose that under this transformation

$$\delta I = \int d^4x\, \Delta, \tag{2.3}$$

where Δ is some function of the fields and their derivatives. (Eq. (2.3) is supposed to be true for general fields, not just for solutions of the equations of motion.) If Δ vanishes, the transformation has no effect on the dynamics and is called a symmetry.

Then one can show from the general formalism of Lagrangian field theory,[2] without any assumptions other than those stated that:

(a) It is always possible to define an object, j^μ, called 'the current associated with the transformation' such that

$$\partial_\mu j^\mu = \Delta. \tag{2.4}$$

Do not be misled by the notation into assuming that the current is always theories,[5] it is possible to define an energy–momentum tensor, $\theta^{\mu\nu}$, such that

$$s^\mu = x_\nu \theta^{\mu\nu}. \tag{2.18}$$

This should not be too much of a surprise. After all, the conserved currents associated with other geometrical transformations, such as translations and Lorentz transformations, are also written in terms of the energy-momentum tensor. Also, Eq. (2.18) implies that exact scale invariance is equivalent to the vanishing of the trace of the energy-momentum tensor:

$$\partial^\mu s_\mu = \theta^\mu_\mu = 0. \tag{2.19}$$

This should strike a familiar chord, if you remember that for free electromagnetism (the prototype of a scale-invariant field theory), the energy-momentum tensor is traceless. (This is why, in scalar theories of gravitation, there is no bending of light by the sun; there is nothing for the scalar graviton to couple to.)

On the other hand, Eq. (2.19) is far from self-evident. The energy-momentum tensor $\theta_{\mu\nu}$ is not, in general, the conventional symmetric energy–momentum tensor of Belinfante; it is another conserved symmetric tensor which differs from the Belinfante tensor by extra terms which do not affect the construction of the total four-momentum nor the Lorentz generators, but which are important in the scale current. Also, (2.19) is not valid for a general field theory; a theory allows the construction of an energy–momentum tensor obeying (2.19) only if a certain condition is met. I will first state this (obscure) condition and then attempt to clarify its meaning.

Let $\Sigma^{\mu\nu}$ be the spin matrix, that matrix that occurs in the transformation law of the fields under infinitesimal Lorentz transformations,

$$\delta^{\mu\nu}\phi = [x^\mu \partial^\nu - x^\nu \partial^\mu + \Sigma^{\mu\nu}]\phi.$$

Then, we can find an energy–momentum tensor such that (2.19) is valid if and only if

$$\frac{\partial \mathscr{L}}{\partial(\partial_\nu \phi)} \cdot [g^{\mu\nu} d + \Sigma^{\mu\nu}]\phi = \partial_\nu \sigma^{\mu\nu}, \tag{2.20}$$

where $\sigma^{\mu\nu}$ is some tensor function of the fields and their derivatives. It is easy to check that all renormalizable field theories, indeed all theories of the interactions of fields of spin $\leqslant 1$ where the derivative interactions are of the same form as in renormalizable theories, satisfy (2.21).

This is, as promised, obscure. To understand what is going on, let us consider the case of exact scale invariance, Eq. (2.10). In this case, a little computation reveals a surprise: we can construct from the energy–

2.4 *Hidden scale invariance*

It has long been known that there are two ways a symmetry that commutes with the Lorentz group (an internal symmetry or a chiral symmetry) can be realized in a quantum field theory:

(1) The symmetry may be manifest. The ground state of the theory (the vacuum) is invariant under the symmetry group, particles arrange themselves into degenerate multiplets corresponding to irreducible representations of the group, and S-matrix elements are invariant. If a small term is added to the Lagrangian to break the symmetry, the masses of the particles in the multiplets split, and the symmetry of S-matrix elements becomes only approximate. This seems to be the case in nature for isospin.

(2) The symmetry may be hidden. (This case, first investigated by Nambu and Goldstone, is usually called 'spontaneously broken symmetry', but this terminology is deceptive; the symmetry is not really broken – the currents associated with the symmetry are still conserved and the Ward identities are still valid – it is just that the consequences of the symmetry are less obvious than in the other case.) The ground state of the theory (the vacuum) is not invariant and particles do not arrange themselves into multiplets: however, there do appear a set of special particles, the 'Goldstone bosons'. These are massless spinless mesons, scalar or pseudoscalar depending on whether the symmetry current is vector or axial. Instead of a symmetric S-matrix, we obtain low-energy theorems, statements relating an S-matrix element for a process involving one zero-energy, zero-momentum Goldstone boson to an S-matrix element with the boson absent. If a small term is added to the Lagrangian to break the symmetry, the Goldstone bosons acquire a small mass and the low-energy theorems require an off-mass-shell extrapolation, become only approximate on the mass shell. This seems to be the case in nature for the chiral transformations of $SU(2) \times SU(2)$; the would-be Goldstone bosons are the pions.

A cheap way to get some feeling for symmetries realized in the Nambu–Goldstone manner is by the study of 'phenomenological Lagrangians'. These are Lagrangian field theories that embody the Goldstone phenomenon, at least in lowest order of perturbation theory. Frequently these models are hard to take seriously as real physical theories; for example, the non-linear chiral Lagrangians, which have played such a prominent role in investigations of current algebra, are non-renormalizable, so that it is impossible to proceed beyond lowest-order perturbation theory in an unambiguous way. Nevertheless, they have at least two uses: they can offer clues to general theorems that one might try to establish by more careful methods, and they can offer counter-examples to wrong argu-

3 The death of scale invariance

3.1 *Some definitions and technical details*

Let us return to the low-energy theorems discussed in Sect. 2.1. For simplicity, I will restrict myself to the theory of a single self-interacting scalar meson,

$$\mathscr{L}=\frac{1}{2}\,\partial_\mu\phi\partial^\mu\phi-\frac{\lambda_0}{4!}\,\phi^4-\frac{1}{2}\,\mu_0^2\phi^2, \tag{3.1}$$

but I will try to conduct the discussion in such a way that the generalization to more complicated theories, such as the meson–nucleon theory of Sect. 2.2, is obvious.

Let us denote the one-particle-irreducible[8] renormalized Green's function with n external lines by

$$\Gamma^{(n)}(p_1\,\ldots\,p_n)$$

where the ps are the momenta carried by the external lines, oriented so they all go inward. (The ps are, of course, not all independent; their sum must be zero.) Likewise, let us denote the one-particle-irreducible Green's function with n external meson lines and one insertion of Δ, the divergence of the scale current (in this case, $\mu_0^2\phi^2$) by

$$\Gamma_\Delta^{(n)}(k;\,p_1\,\ldots\,p_n),$$

where the ps are defined as before, and k is the momentum carried by the insertion. Then it is a trivial exercise in Fourier transforms to see that in this case the low-energy theorem (2.6) becomes

$$\left(\sum_{r=1}^{n-1}p_r\cdot\frac{\partial}{\partial p_r}+nd-4\right)\Gamma^{(n)}(p_1\,\ldots\,p_n)=-\mathrm{i}\Gamma_\Delta^{(n)}(0;\,p_1\,\ldots\,p_n). \tag{3.2}$$

where d is the scale dimension of the scalar field. The analysis of the preceding section has told us that d is one, of course, but I would like to suppress this information momentarily, both to make it easy for you to see the generalization to Fermi fields, for which d is $\frac{3}{2}$, and for another reason, which will become clear shortly.

Equation (3.2) is fairly easy to understand. The first two terms on the left are just the transformation law of the fields, Eq. (2.9) written in momentum space; the sum only runs over $n-1$ momenta because only $n-1$ of the momenta are independent. The factor of four appears because in passing from the Fourier transform of a T^*-product to the conventionally defined $\Gamma^{(n)}$, we must factor out a four-dimensional δ-function, which leaves the four behind as it passes through the differential operator. The right-hand side of the equation is just the left-hand side of (2.6).

Fig. 1

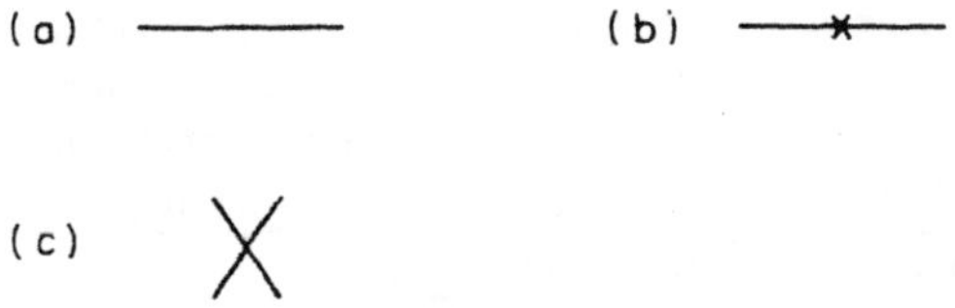

Just to check that we have not made a sign error, let us check (3.2) in lowest-order perturbation theory. To zeroth order, $\Gamma^{(2)}$ is given by Fig. 1(a), and $\Gamma_\Delta^{(2)}$ is given by Fig. 1(b), where the cross denotes the mass insertion. Thus Eq. (3.2) becomes

$$\left(p\cdot\frac{\partial}{\partial p}-2\right)(-\mathrm{i})(p^2-\mu^2)=-\mathrm{i}(2\mu^2), \tag{3.3}$$

which is correct. Likewise, to first order, $\Gamma^{(4)}$ is given by Fig. 1(c), and there is no contribution to $\Gamma_\Delta^{(4)}$. Thus Eq. (3.2) becomes

$$\sum_{i=1}^{3}\left(p_i\cdot\frac{\partial}{\partial p_i}\right)(-\mathrm{i}\lambda)=0 \tag{3.4}$$

which is again correct.

We can write (3.2) in another form, which will be useful in the sequel. Let us trade the *p*s for a set of variables consisting of

$$s=\sum_{r=1}^{n} p_r^2, \tag{3.5}$$

and the dimensionless kinematic variables $p_i\cdot p_j/s$. Then ordinary dimensional analysis tells us that $\Gamma^{(n)}$ is of the form

$$\Gamma^{(n)}=s^{(4-n)/2}F^{(n)}\left(\frac{s}{\mu^2},\lambda,\frac{p_i\cdot p_j}{s}\right), \tag{3.6}$$

where μ is the renormalized mass, and λ is the renormalized coupling constant. (Remember, λ is dimensionless.) From this it is easy to see that (3.2) can equivalently be written as

$$\left[\mu\frac{\partial}{\partial\mu}+n(1-d)\right]\Gamma^{(n)}(p_1\ldots p_n)=\mathrm{i}\Gamma_\Delta^{(n)}(0;p_1\ldots p_n). \tag{3.7}$$

3.2 *A disaster in the deep Euclidean region*

The Euclidean region is that region in multi-particle momentum space in which all four-momenta are Euclidean; that is to say, they all have real space parts and imaginary time parts. The deep Euclidean region is that part of the Euclidean region in which the magnitude of *s* (defined by Eq. (3.5)) gets very large, while the dimensionless variables $p_i \cdot p_j/s$ stay fixed; furthermore, no partial sum of the *p*s is zero. The deep Euclidean region is the maximally unphysical limit in which to study Green's functions: all external lines are far off the mass shell, and, furthermore, no matter how the diagram is cut in two, the momentum transferred between the two halves is far off the mass shell.

There are famous bounds on the behaviour of Feynman amplitudes in the deep Euclidean region, first established by Weinberg.[9] For the functions that appear in Eq. (3.7), these Weinberg bounds say that $\Gamma^{(n)}$ grows no faster than $s^{(4-n)/2}$, times a polynomial in $\ln(s/\mu^2)$, to any finite order in renormalized perturbation theory. (The coefficients in the polynomial are, in general, functions of the coupling constant and of the dimensionless variables $p_i \cdot p_j/s$. Also, the order of the polynomial grows with the order of perturbation theory.) Likewise, $\Gamma_\Delta^{(n)}$ grows no faster than $s^{(2-n)/2}$, again times a polynomial in $\ln(s/\mu^2)$. Crudely, the reason why $\Gamma_\Delta^{(n)}$ grows less rapidly than $\Gamma^{(n)}$ is that, in the deep Euclidean region, all internal momenta are getting large; adding a mass insertion adds an internal propagator, which knocks out one power of *s*.

We can now combine this with the broken-scale-invariance low-energy theorems, Eq. (3.7), to get a much more powerful statement about the asymptotic behaviour of the Green's functions than is given by the Weinberg bounds alone. For the Weinberg bounds tell us that in the deep Euclidean region, the right-hand side of (3.7) is negligible compared to the individual terms on the left-hand side. Thus, we can neglect it, and obtain an equation for the asymptotic form of $\Gamma^{(n)}$,

$$\left[\mu\frac{\partial}{\partial\mu}+n(1-d)\right]\Gamma_{as}^{(n)}=0, \tag{3.8}$$

where the subscript indicates the asymptotic form in the deep Euclidean region. Or, since *d* is one,

$$\mu\frac{\partial}{\partial\mu}\Gamma_{as}^{(n)}=0. \tag{3.9}$$

That is to say, there are no logarithmic factors in every order of renormalized perturbation theory. This is indeed a powerful statement; unfortunately it is also a false one; anyone who has ever done any Feynman calculation involving a closed loop knows that the logarithms are in fact present. Therefore, Eq. (3.7), from which we deduced the false statement (3.9), must itself be false. In the current technical language the Ward identities of broken scale invariance must contain anomalies. Phrased more straightforwardly, *the entire theoretical structure of Sect. 2 is a lie!*

For the self-energy operator, for example, the solution to Eq. (3.8) becomes

$$\Gamma_{as}^{(2)} \sim s(s/\mu^2)^{1-d}. \tag{3.11}$$

Expanding the exponent and discarding terms of order g^4 and higher, we find

$$\Gamma_{as}^{(2)} \sim s - \frac{g^2}{8\pi^2}\, s \ln(s/\mu^2). \tag{3.12}$$

This is the correct asymptotic behaviour, to this order in perturbation theory. Similar remarks apply to the nucleon field; here d must be changed from its naive value ($\frac{3}{2}$) to

$$d = 3/2 + \frac{g^2}{32\pi^2}. \tag{3.13}$$

(As it happens, these changes in d not only fix up (3.7), but also take care of all the anomalies in the Ward identities for arbitrary momentum transfers.)

This phenomenon is frequently described by saying that the fields acquire anomalous dimensions. This is a slightly misleading way of putting things, for it tempts us to confuse scale dimensions and dimensions in the sense of dimensional analysis, and to think that something counter to common sense has occurred. It is only the scale dimensions of the fields that have changed as a result of the interactions; the dimensions in the sense of dimensional analysis remain firmly fixed at one and $\frac{3}{2}$. I emphasize again that there is no logical connection between these two sets of numbers; the (discredited) analysis of Sect. 2 is the only thing that ever led us to believe they were equal.

It would be very pleasant if this phenomenon persisted in higher orders of perturbation theory, if all anomalies could be absorbed into a redefinition of the scale dimension. If this were the case, then we would have a correct theory of broken scale invariance with almost as much predictive power as the false theory of Sect. 2; the only price would be the introduction of a new, dynamically determined parameter for each field in the theory, its anomalous dimension.

Unfortunately, this is not what happens. To see this, let us return to the theory of a self-interacting meson field, Eq. (3.1), and assume that Eq. (3.8) is valid in all orders of perturbation theory, except that d is not one. Then it is easy to see that

$$\mu \frac{\partial}{\partial \mu} [\Gamma_{as}^{(4)}/(\Gamma_{as}^{(2)})^2] = 0. \tag{3.14}$$

That is to say, although both the numerator and denominator of this expression have factors of $\ln(s/\mu^2)$ in their asymptotic expansions, the

Fig. 2

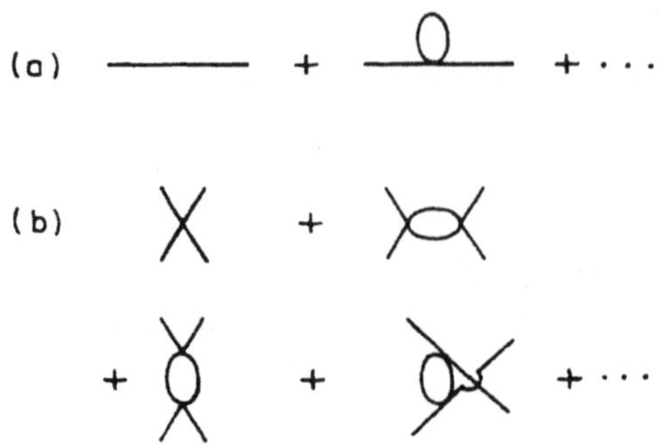

logarithms cancel in the ratio, order by order in perturbation theory. Let us check this prediction by calculating the ratio to order λ^2. The relevant diagrams for $\Gamma^{(2)}$ are shown in Fig. 2(a); the second diagram is just a constant, cancelled by mass renormalization. Thus, we obtain, in the deep Euclidean region,

$$\Gamma^{(2)}_{as} \sim s + O(\lambda^2). \tag{3.15}$$

Figure 2(b) shows the relevant diagrams for $\Gamma^{(4)}$. Every child knows that the three second-order diagrams grow logarithmically for high s; thus we obtain

$$\Gamma^{(4)}_{as} \sim \lambda + a\lambda^2 \ln(s/\mu^2) + b\lambda^2. \tag{3.16}$$

where a and b are constants. (Actually, b is a function of the dimensionless kinematic variables, but, for our purposes, that is as good as being a constant.) Putting this together we find

$$\Gamma^{(4)}_{as}/(\Gamma^{(2)}_{as})^2 \sim \lambda s^{-2} + a\lambda^2 s^{-2} \ln(s/\mu^2) + b\lambda^2 s^{-2} + O(\lambda^3). \tag{3.17}$$

This is in contradiction to our prediction; thus there must be further anomalies in addition to anomalous dimensions.

3.4 *The last anomalies: the Callan–Symanzik equations*

We are lost in a dark wood. Whenever we look at a more complicated situation, investigate higher orders of perturbation theory, new anomalies appear. Happily, there exists a way to order this chaos, discovered by Curtis Callan and Kurt Symanzik,[11] who independently found the true equations that replace the false low-energy theorems (3.7). These equations are surprisingly compact. For the theory of a self-interacting meson field, for example, they involve only two dynamically determined parameters. This is only one more than there would be if the only anomalies were anomalous dimensions.[12] I will now derive the Callan–Symanzik equations for this theory.

The simplest place to begin is in unrenormalized perturbation theory: here one computes the unrenormalized Green's functions (which I will distinguish by a subscript u) in a power series in the bare charge, with the bare mass held fixed. I remind you of the relation between the renormalized Green's functions, which we have been using until now, and the unrenormalized ones

$$\Gamma^{(n)}(p_1 \ldots p_n) = (Z_3)^{n/2}\Gamma_{\mathrm{u}}^{(n)}(p_1 \ldots p_n). \tag{3.18}$$

where Z_3 is the wave-function renormalization constant. Of course, unlike renormalized perturbation theory, unrenormalized perturbation theory is full of divergences. To control these, we will imagine that we have put a cut-off in the theory in some way, say by modifying the Feynman propagator in the standard manner. Once we re-express things in terms of the physical mass and coupling constant, and perform the multiplicative renormalization (3.18), everything is cutoff-independent in the limit of high cutoff; this is just the statement that (3.1) is a renormalizable Lagrangian. I will denote the unrenormalized Green's function with one mass ($\mu_0^2\phi^2$) insertion by $\Gamma_{\mathrm{u}\Delta}^{(n)}(k; p_1 \ldots p_n)$. Then it is easy to see, just by looking at diagrams, that

$$\mathrm{i}\Gamma_{\mathrm{u}\Delta}^{(n)}(0; p_1 \ldots p_n) = \mu_0 \frac{\partial}{\partial \mu_0} \Gamma_{\mathrm{u}}^{(n)}(p_1 \ldots p_n). \tag{3.19}$$

This is the unrenormalized version of the low-energy theorem (3.7). As it stands, it is useless, since the differential operator cannot be turned into a scaling operator; the unrenormalized Green's functions depend on another parameter besides μ_0 with the dimensions of a mass, the cutoff.

Simple power counting[13] shows that if we define

$$\Gamma_{\Delta}^{(n)}(k; p_1 \ldots p_n) = Z(Z_3)^{n/2}\Gamma_{\mathrm{u}\Delta}^{(n)}(k; p_1 \ldots p_n), \tag{3.20}$$

then, for an appropriate (cutoff-dependent) choice of the constant Z, we can make the left-hand side of (3.20) cutoff-independent in the limit of high cutoff. This condition leaves Z undetermined up to a finite (cutoff-independent) multiplicative factor. We will choose this factor later, to make our final equations look as simple as possible.

Putting all this together, we find the consequence of (3.19) for the renormalized Green's functions,

$$\mathrm{i}\Gamma_{\Delta}^{(n)}(0, \ldots) = Z\mu_0 \frac{\partial}{\partial \mu_0} \Gamma^{(n)}(\ldots) - \frac{n}{2} Z\mu_0 \frac{\partial \ln Z_3}{\partial \mu_0} \Gamma^{(n)}(\ldots). \tag{3.21}$$

We have not yet finished, since we have the derivative of a renormalized Green's function with respect to a bare mass, and a renormalized Green's function is cutoff-independent only when expressed in terms of renormalized masses and coupling constants. This is easily taken care of with the aid of the chain rule of differentiation:

$$\left[\left(Z\mu_0 \frac{\partial \mu}{\partial \mu_0}\right)\frac{\partial}{\partial \mu}+\left(Z\mu_0 \frac{\partial \lambda}{\partial \mu_0}\right)\frac{\partial}{\partial \lambda}-\frac{n}{2}\left(Z\mu_0 \frac{\partial \ln Z_3}{\partial \mu_0}\right)\right]\cdot\Gamma^{(n)}(\ldots)$$

$$= i\Gamma_{\Delta}^{(n)}(0, \ldots). \tag{3.22}$$

where all derivatives are taken with fixed cutoff. The only terms in these equations which retain a possible trace of cutoff-dependence are the three terms in parentheses. These, however, are constants, independent of n and the momenta. Therefore, it is easy to show, by evaluating the equations at three independent points and solving for these terms as functions of cutoff-independent quantities, that they are also cutoff-independent, in the limit of high cutoff. (An especially simple choice is the three points at which the renormalized mass, renormalized charge, and scale of the renormalized field are defined.)

We now choose Z (until now undetermined up to a finite multiplicative constant) such that

$$Z\mu_0 \frac{\partial \mu}{\partial \mu_0} = \mu, \tag{3.23}$$

and define

$$\beta = Z\mu_0 \frac{\partial \lambda}{\partial \mu_0}, \tag{3.24}$$

and

$$\gamma = \frac{1}{2} Z\mu_0 \frac{\partial \ln Z_3}{\partial \mu_0}. \tag{3.25}$$

(Dimensional analysis shows that these functions can only depend on λ.) We thus obtain

$$\left[\mu \frac{\partial}{\partial \mu} + \beta(\lambda)\frac{\partial}{\partial \lambda} - n\gamma(\lambda)\right]\Gamma^{(n)}(\ldots) = i\Gamma_{\Delta}^{(n)}(0, \ldots). \tag{3.26}$$

These are the Callan–Symanzik equations.

Some remarks should be made about these equations:

(1) As promised, these are the equations that replace the naive low-energy theorem (3.7). If β were zero, (3.25) would be identical to (3.7), except that the scale dimension, d, would be anomalous:

$$d = 1 + \gamma. \tag{3.27}$$

Unfortunately, as we have already seen, β is not zero; there are further anomalies beyond anomalous dimensions. A direct computation[10] shows that

$$\beta=\frac{3\lambda^2}{16\pi^2}+O(\lambda^3). \tag{3.28}$$

(It will be important in the sequel that this first non-zero term in the perturbation expansion of β is positive.)

(2) The generalization to more complicated field theories is obvious: For every dimensionless coupling constant, there is a β-like term, and, for every field, there is a γ-like term. Thus, for example, in meson–nucleon theory, if we denote a Green's function with n nucleon fields, n antinucleon fields, and m meson fields, by $\Gamma^{(n,n,m)}$, the Callan–Symanzik equations are

$$\left[\mu\frac{\partial}{\partial\mu}+m\frac{\partial}{\partial m}+\beta_1\frac{\partial}{\partial\lambda}+\beta_2\frac{\partial}{\partial g}-2n\gamma_1-m\gamma_2\right]\Gamma^{(n,n,m)}(\ldots)$$
$$=\mathrm{i}\Gamma_\Delta^{(n,n,m)}(0;\ldots)$$

where the βs and the γs are functions of λ, g and μ/m.

(3) Likewise, if we attempt to study Green's functions for objects other than canonical fields (for example, conserved or partially-conserved currents), this will, in general, change the structure of the γ-terms, which refer to a particular field. The β-terms, though, which do not make reference to a choice of fields, but only to the underlying dynamics, will remain unchanged.

(4) Unlike the low-energy theorems of current algebra, which they so closely resemble, the Callan–Symanzik equations are practically useless for low-energy phenomenology. It is the β-terms that make the difference. A current-algebra low-energy theorem is useful because it expresses one Green's function in terms of another in a way that does not depend on strong-interaction dynamics. The Callan–Symanzik equations, on the other hand, express one Green's function in terms of another *and* its derivatives with respect to coupling constants. If you know how to compute *these*, you have already solved the strong-interaction dynamics, and there is no reason for you to be piddling around with low-energy theorems.

(5) Finally, there remains the possibility that β has a zero, and that, for some reason (the bootstrap?) the real value of λ is at this zero. (This is a speculation of Kenneth Wilson, among others). If this is the case, we would regain the naive theory of scale invariance (with anomalous dimensions). There is very little to be said for or against this possibility from a study of perturbation theory alone.

4 The resurrection of scale invariance

4.1 *The renormalization group equations and their solution*

By the arguments we used in Sect. 3.2, in the deep Euclidean region, we can neglect the right-hand side of the Callan–Symanzik equations. Thus, for the case of a self-interacting meson field, we obtain the following differential equations for the asymptotic forms of the Green's functions:

$$\left[\mu \frac{\partial}{\partial \mu} + \beta(\lambda) \frac{\partial}{\partial \lambda} - n\gamma(\lambda)\right] \Gamma^{(n)}_{as} = 0. \tag{4.1}$$

These equations are old friends; they are the equations associated with the so-called renormalization group, first devised by Gell-Mann and Low to study quantum electrodynamics, and later extended to general renormalizable field theories.[14]

The renormalization group Eq. (4.1) can be derived much more simply than the Callan–Symanzik equations. As I stated earlier, the deep Euclidean region is the maximally unphysical region; all external lines and all momentum transfers are far from the mass shell. One's first thought would be that in this regime all memory of the actual meson mass would be lost, and all Green's functions would be independent of the mass. However, this is too naive: the objects we are studying are Green's functions for renormalized fields, expressed as functions of the renormalized coupling constant; both the normalization of the field and the value of the renormalized coupling constant are defined on the mass shell; these quantities remember the mass shell no matter how far we flee into the deep Euclidean region. Therefore, the correct statement is that all memory of the actual value of the mass is lost, except for that which is contained in the scale of the fields and the value of λ. In other words, in the deep Euclidean region, a small change in the mass can always be compensated for by an appropriate small change in λ and an appropriate small rescaling of the fields. The equations (4.1) are just the mathematical expression of this statement.

This argument seems to me to owe very little to asymptotic estimates derived from perturbation theory; therefore I will assume from now on that the renormalization group equations are valid independent of perturbation theory, and try to use them to get information about the asymptotic forms of Green's functions in a non-perturbative way.

The renormalization group equations are fairly easy to solve. To motivate the solution, let us consider a similar equation with a direct dynamical interpretation:

$$\frac{\partial \rho}{\partial t} + v(x) \frac{\partial \rho}{\partial x} = L(x)\rho, \tag{4.2}$$

4

Renormalization and symmetry: a review for non-specialists

(1971)

1 Introduction

I suppose that as good a way as any of explaining the contents of this lecture is to explain the title. By 'renormalization' I mean the removal of infinities for Feynman amplitudes, in perturbation theory, for Lagrangian field theories with polynomial interactions. In particular non-perturbative renormalization (the work of Jaffe, Glimm, etc.) is outside the scope of this lecture, as are the properties of non-polynomial interactions (the work of Efimov, Salam, Lehmann, etc.). By 'renormalization and symmetry' I mean that we will be concerned not only with the renormalization of scattering amplitudes, but also with the renormalization of the matrix elements of conserved and partially conserved currents. In particular, we will discuss some fairly recent results of Symanzik, Benjamin Lee, Preparata, Weisberger, and others. By 'a review for non-specialists' I mean that I hope that this talk will be intelligible to people who can do nothing more complicated than remove the divergences from the self-energy of the electron.

Since renormalization theory has a well-deserved reputation for complexity, it is obvious that I will be able to do all this in a single lecture only by cheating. To be precise, I will explain a very powerful theorem due to Klaus Hepp, but not prove it (this is the cheat); then I will show how a wide variety of results can be obtained from this master theorem by elementary methods.[1]

2 Bogoliubov's method and Hepp's theorem

For simplicity, we will restrict ourselves to field theories involving spin-zero and spin-one-half fields only, which we will call Bose and Fermi fields, respectively. We will write the Lagrangian for such a theory in the

5

Secret symmetry: an introduction to spontaneous symmetry breakdown and gauge fields

(1973)

1 Introduction

Here are some long-standing problems in particle theory:

(1) How can we understand the hierarchical structure of the fundamental interactions? Are the strong, medium strong (i.e. SU(3)-breaking), electromagnetic, and weak interactions truly independent, or is there some principle that establishes connections between them?

(2) How can we construct a renormalizable theory of the weak interactions, one which reproduces the low-energy successes of the Fermi theory but predicts finite higher-order corrections?

(3) How can we construct a theory of electromagnetic interactions in which electromagnetic mass differences within isotopic multiplets are finite?

(4) How can we reconcile Bjorken scaling in deep inelastic electroproduction with quantum field theory? The SLAC–MIT experiments seem to be telling us that the light-cone singularities in the product of two currents are canonical in structure; ordinary perturbation theory, on the other hand, tells us that the canonical structure is spoiled by logarithmic factors, which get worse and worse as we go to higher and higher orders in the perturbation expansion. Are there any theories of the strong interactions for which we can tame the logarithms, sum them up and show they are harmless?

Enormous progress has been made on all of these problems in the last few years. There now exists a large family of models of the weak and electromagnetic interactions that solve the second and third problem, and we have discovered a somewhat smaller family of models of the strong interactions that solve the fourth problem. As we shall see, the structure of these models is such that we are beginning to get ideas about the solution to the (very deep) first problem; connections are beginning to appear in its time component does not annihilate the vacuum state, then the theory necessarily contains a massless spinless meson, with the same internal-symmetry and parity properties as the time component of the current.[5]

At first glance, Goldstone's theorem seems to be a killing blow to the idea that spontaneous breakdown (at least of continuous symmetries) is at work in the real world, for there is not a smidgen of experimental evidence for the existence of massless spinless mesons. However, there is one loophole: there do exist perfectly respectable field theories which do not obey the usual axioms. These are gauge field theories, of which quantum electrodynamics is the most familiar. There is no gauge in which quantum electrodynamics obeys all the axioms simultaneously; if we quantize in a covariant gauge, the theory contains states of negative norm, associated with the longitudinal photons; if we quantize in a gauge in which the theory has only states of positive norm, such as radiation gauge, the theory is not covariant. We will now investigate this loophole in more detail.

2.4 *The Higgs phenomenon in the Abelian model*

I will begin by reviewing the minimal-coupling prescription of ordinary quantum electrodynamics, and its connection with gauge invariance. Let ϕ be a set of fields (not necessarily real and spinless), with dynamics determined by a Lagrange density, $\mathscr{L}(\phi, \partial_\mu\phi)$. Let $\mathscr{L}$ be invariant under a one-parameter group of transformations,

$$\phi \to e^{iQ\omega}\phi, \tag{2.25}$$

where Q is a Hermitian matrix, called the charge matrix. (Conventionally, a set of complex basis fields of definite charge is chosen, so that Q is diagonal. However, for our purposes, it will be more convenient to choose a real set of fields, so that iQ is a real antisymmetric matrix, like the Ts in Eq. (2.20).) The associated infinitesimal transformation is

$$\delta\phi = iQ\phi\delta\omega. \tag{2.26}$$

Now let us consider transformations of the same form as Eq. (2.26), but with $\delta\omega$ space-time dependent (gauge transformations). Our theory is not invariant under these transformations, since

$$\delta(\partial_\mu\phi) = iQ(\partial_\mu\phi)\delta\omega + iQ\phi\partial_\mu(\delta\omega), \tag{2.27}$$

and the second term spoils the invariance. We can take care of this, though, by enlarging the theory and introducing a new field, A_μ, the gauge field, that transforms according to

$$\delta A_\mu = -\frac{1}{e}\partial_\mu(\delta\omega), \tag{2.28}$$

Hence, if we choose

$$\xi = \alpha g, \tag{5.47}$$

we can cancel the annoying cross terms. Note that the interactions in this Lagrange density are still of renormalizable type, dimension less than or equal to four.

6 Asymptotic freedom

6.1 *Operator products and deep inelastic electroproduction*

The topic we are now going to discuss seems, at first glance, to have very little to do with the previous lectures. It is a topic in strong-interaction physics, that of reconciling the apparent scaling in the SLAC–MIT electroproduction experiments with the predictions of quantum field theory. I will begin by summarizing very briefly the standard lore on this problem.[27]

(1) The electroproduction experiments at Stanford measure the total cross sections for the process

$$\text{electron} + \text{nucleon} \to \text{electron} + \text{anything},$$

which is, of course, the same thing as

$$\text{virtual photon} + \text{nucleon} \to \text{anything}.$$

The process is therefore described by two kinematic variables: q^2, the mass of the virtual photon, a negative number, and E, the energy of the virtual photon in the lab frame. It is convenient to trade E for the dimensionless variable

$$x = -q^2/2mE, \tag{6.1}$$

where m is the nucleon mass. Elementary kinematics restricts x to be between zero and one. The nucleons in the experiment are unpolarized, while the virtual photons can be either transverse or longitudinal; thus the cross-section can be described in terms of two dimensionless invariants, $F_i(q^2, x)$, where i is 1 or 2. The Fs are called structure functions; the details of their definitions will not be relevant to our immediate purposes.

As $-q^2$ increases, the Fs quite rapidly lose their dependence on q^2; by $q^2 = -(2\ \text{GeV})^2$, the Fs appear to be functions of x alone, within experimental error. This phenomenon is called Bjorken scaling. There are two schools of thought on Bjorken scaling. (1) Bjorken scaling is a true asymptotic phenomenon. It will persist even if the range of q^2 is increased greatly. (2) SLAC energies are too small for us to believe that we are really in the asymptotic region. Bjorken scaling is some sort of low-energy epiphenomenon, and has nothing to do with true high-energy limits.

6

Classical lumps and their quantum descendants

(1975)

1 Introduction

A stone thrown into a still body of water makes ripples that spread out and eventually die away. The stone disturbs the water, gives it energy, but, even if we ignore friction, this energy tends in the course of time to spread out over the water. If we imagine the water to be infinite in extent then, if we wait long enough, at no point is the water appreciably different from its state before the stone was cast. The disturbance dissipates.

This concept of dissipation can be generalized beyond the hydrodynamic example I have given. Consider a classical field theory, with energy density Θ_{00}, such that Θ_{00} is always greater than or equal to zero and is everywhere zero for the ground state(s) of the theory. Then we will say that a solution of the classical equations of motion is dissipative if

$$\lim_{t\to\infty} \max_{\mathbf{x}} \Theta_{00}(\mathbf{x}, t) = 0. \tag{1.1}$$

Most of the simple field theories with which we are familiar have the property that all of their non-singular solutions of finite total energy are dissipative. This is the case for Maxwell's equations, the Klein–Gordon equation, etc. However, there are some perfectly sensible classical field theories that possess non-singular non-dissipative solutions of finite energy. Among these are some (though not all) of the spontaneously broken gauge theories that have risen to such prominence in recent years.

In the most striking and simplest case, and the one that will occupy most of our attention, these solutions may be time-independent, lumps of energy holding themselves together by their own self-interaction. There are also more complicated kinds of non-dissipative solutions, solutions that are periodic in time or have even more exotic time dependences, oscillating or quivering lumps. Also, if our theory is Lorentz-invariant (or Galileo-invariant), if we can construct stationary lumps, we can also

7

The uses of instantons

(1977)

1 Introduction

In the last two years there have been astonishing developments in quantum field theory. We have obtained control over problems previously believed to be of insuperable difficulty and we have obtained deep and surprising (at least to me) insights into the structure of the leading candidate for the field theory of the strong interactions, quantum chromodynamics. These goodies have come from a family of computational methods that are the subject of these lectures.

These methods are all based on semiclassical approximations, and, before I can go further, I must tell you what this means in the context of quantum field theory.

To be definite, let us consider the theory of a single scalar field in four-dimensional Minkowski space, with dynamics defined by the Lagrangian density

$$\mathscr{L} = \tfrac{1}{2}\partial_\mu\phi\partial^\mu\phi - \tfrac{1}{2}m^2\phi^2 - g^2\phi^4. \tag{1.1}$$

For classical physics, g is an irrelevant parameter. The easiest way to see this is to define

$$\phi' = g\phi. \tag{1.2}$$

In terms of ϕ',

$$\mathscr{L} = \frac{1}{g^2}\left(\tfrac{1}{2}\partial_\mu\phi'\partial^\mu\phi' - \tfrac{1}{2}m^2\phi'^2 - \phi'^4\right). \tag{1.3}$$

Thus, g does not appear in the field equations; if one can solve the theory for any positive g, one can solve it for any other positive g; g is irrelevant. Another way of seeing the same thing is to observe that, in classical physics, g is a dimensionful parameter and can always be scaled to one.

Of course, g *is* relevant in quantum physics. The reason is that quantum large sphere, Eq. (3.31), and one in terms of a volume integral over all four-space, Eq. (3.47).

How much of this depends on the gauge group being SU(2)? Firstly, if the gauge group is U(1), it is easy to see that every mapping of S^3 into U(1) is continuously deformable into the trivial mapping (all of S^3 mapped into a single point). *Thus, for an Abelian gauge field theory, there is no analog of the winding number.* Secondly, for a general simple Lie group, G, there is a remarkable theorem due to Raoul Bott[14] that states that any continuous mapping of S^3 into G can be continuously deformed into a mapping into an SU(2) subgroup of G. *Thus, everything we have discovered for* SU(2) *is true for an arbitrary simple Lie group; in particular, it is true for* SU(n). I stress that 'everything' means *everything*. In particular, not a single numerical factor in the integral formulas for the winding number needs alteration, so long as we choose the normalization of the Cartan inner product appropriately (as we have). Finally, since a general compact Lie group is locally the direct product of an Abelian group and a string of simple groups, *for a general gauge field theory, there is an independent winding number for every simple factor group.*

3.3 *Many vacua*

We have learned a lot about classical gauge field theories; now it is time to confront the quantum theory. In principle, the Euclidean functional integral tells how to go from the classical theory to the quantum theory. As I explained in Sect. 2, we can use the functional integral to study the energy eigenstates of the theory; also, by adding appropriate source terms to the Hamiltonian (equivalently, to the Euclidean action) and then differentiating with respect to the sources at the end of the computation, we can study the expectation values of strings of operators, Euclidean Green's functions. However, for gauge field theories, there is a famous complication; to make the functional integral well-defined, we must impose a gauge-fixing condition.[15]

I will choose to work in axial gauge, $A_3=0$. I have several reasons for this choice. (1) It is possible to show[16] that every non-singular gauge field configuration can be put in axial gauge by a non-singular gauge transformation. It is by no means clear whether this is true for covariant gauges, for example. (2) In axial gauge the functional integral is directly equivalent to a canonical formulation of the theory;[17] there is no need of the ghost terms that occur in covariant gauges, or of the subsidiary conditions on the space of states that are needed in such gauges as $A_0=0$. (3) Most of the treatment in the literature of the phenomena we are about to discuss is in the gauge $A_0=0$. It is nice to show explicitly that the answers do not complicated by nonlinear terms in the expression for the energy density as a function of the field.

One aspect of these nonlinear complications is of physical import: Eq. (4.14) is periodic in q with period e. This is explicable if the theory contains charged particles of charge e. If this is the case, there is a process that can change the charge on our condenser plates by $\pm e$: a particle–antiparticle pair can materialize in the region between the plates, and the particle can fly to one plate and the antiparticle to the other. This process will occur whenever it is energetically favorable. For sufficiently large L', this is equivalent to saying that it will occur whenever it lowers the energy density, because the energetic cost of making a pair is independent of L', and the energetic gain of lowering the energy density is proportional to L'. Thus qs that are equal modulo e lead to identical physics; no matter which one you start out with, pairs are made until the charge on the plates reaches its optimum value, the one that gives minimum energy density.

What if we were to do a parallel computation in a four-dimensional gauge field theory, with non-Abelian external charges? Would we also obtain a force independent of separation? Alas, we would not. There is an L' in Eq. (4.15) because there is an $L'T'$ in Eq. (4.14), that is to say, because even an instanton deep within the loop has a non-negligible effect on the loop integral. This is precisely what does not happen in four dimensions. At large distances from an instanton, A_μ is $g\partial_\mu g^{-1}$, plus terms that fall off far too rapidly to affect the loop integral. However, the loop integral is gauge-invariant, and we can always gauge-transform g such that it is constant everywhere except within a small cone emerging from the instanton perpendicular to the plane of the loop. Whatever confines quarks, it is not instantons.

5 't Hooft's solution of the U(1) problem

5.1 *The mystery of the missing meson*

The U(1) problem is an apparent contradiction between two pieces of accepted wisdom. One is wisdom of the 1970s, that hadronic physics is quantum chromodynamics. The other is wisdom of the 1960s, that hadronic physics is approximately invariant under chiral SU(2)⊗SU(2). Let me remind you of the meaning of these two propositions.

Quantum chromodynamics is a field theory whose dynamical variables are an octet of SU(3) gauge fields and a family of SU(3) triplet Dirac bispinor fields, called quarks. In Minkowski space, the Lagrangian density is

$$\mathscr{L} = -\frac{1}{4g^2}(F_{\mu\nu}, F^{\mu\nu}) + \sum_f \bar{\psi}_f(\mathrm{i}D_\mu\gamma^\mu - m_f)\psi_f, \tag{5.1}$$

where f, called the flavour index, labels the various triplets. The usual exact and approximate symmetries of hadron physics (charge, isospin, Gell-Mann's SU(3), etc.) act only on the flavour indices; all physical hadrons are supposed to be singlets under the gauge group. (This last statement is sometimes called quark confinement; it is still far from proved, although there are some suggestive arguments.) ψ_1 and ψ_2 form an isodoublet, the non-strange quarks; ψ_3 is the strange quark; ψ_4 is the charmed quark; there may or may not be additional flavors.

Chiral SU(2)⊗SU(2) is the group generated by the strangeness-conserving weak-interaction currents and their parity transforms. Its diagonal subgroup is conventional isospin. This group is very close to being an exact symmetry of the strong interactions; it is a much better symmetry than SU(3) and roughly as good a symmetry as isospin. However, were this symmetry to be exact, only the isospin subgroup would be a manifest symmetry; the remainder of the group would be a Nambu–Goldstone symmetry, with three massless Goldstone bosons, the pions. The smallness of the pion mass (on a hadronic mass scale) is a measure of the goodness of the symmetry. This is the picture that stands in back of all the stunningly successful soft-pion computations of the mid 1960s.

Now for the apparent contradiction. In quantum chromodynamics, the limit of perfect SU(2)⊗SU(2) symmetry is the limit in which the non-strange quarks are massless. In this limit, the Lagrangian (4.1) obviously has a further chiral U(1) symmetry; it is invariant under

$$\psi_f \to \mathrm{e}^{-\mathrm{i}\alpha\gamma_5}\psi_f, \quad (f = 1, 2) \tag{5.2}$$

where α is a real number. The associated conserved current is

$$j^5_\mu = \sum_{f=1}^{2} \bar{\psi}_f\gamma_\mu\gamma_5\psi_f. \tag{5.3}$$

I emphasize that the appearance of this additional chiral symmetry is very special to quantum chromodynamics; for example, the σ model has no such additional symmetry in the chiral limit.

Now, either this additional symmetry is manifest or it is spontaneously broken. If it were manifest, all non-massless hadrons would occur in parity doublets. This is not the case; thus it must be spontaneously broken. But if it is spontaneously broken, Goldstone's theorem tells us there must be an associated isoscalar pseudoscalar Goldstone boson. This is the U(1) problem: *what happened to the fourth Goldstone boson?*

One's first thought is that the missing meson is the eta, but this is wrong. The chiral U(1) symmetry is broken by the same mass term that breaks chiral SU(2)⊗SU(2), and thus the fourth Goldstone boson should have roughly the same mass as the pions. The eta is far too heavy. This can be made more precise: using conventional soft-pion methods, Weinberg[28] has shown that a U(1) Goldstone boson must have a mass less than $\sqrt{3}m_\pi$. The eta grossly disobeys this inequality. Also, if we consider the approximation in which the strange quark mass also vanishes, and in which we have perfect chiral SU(3)⊗SU(3) symmetry, the eta takes its place with the pions in an octet of Goldstone bosons. But in this limit we still have an additional U(1) symmetry and we still have a missing meson.

(This should be all that I need to say about the eta. However, there is some confusion abroad on this point, and thus I emphasize that *there is no connection between the eta and the U(1) problem*. The eta is a red herring; it is just another hadron; it is no more a relic of a U(1) Goldstone boson than is the N**.)

It may seem that I have posed an insoluble problem; this is because I have lied to you. In fact, j^5_μ is not a conserved current; it is afflicted with the famous Adler–Bell–Jackiw anomaly.[29] In the limit of N massless quarks,

$$\partial^\mu j^5_\mu = \frac{N}{32\pi^2}\,\varepsilon^{\mu\nu\lambda\sigma}(F_{\mu\nu},\,F_{\lambda\sigma}). \tag{5.4}$$

(Note the similarity between the right-hand side of this equation and the Pontryagin density. This will be important to us later.)

You might think that this is the end of the story; if the current is not conserved, there is no U(1) symmetry to worry about. Alas, life is not so simple. In Sect. 3, we showed that the Euclidean counterpart of the right-hand side of Eq. (5.4) could be written as the divergence of a (gauge-variant) function of A_μ and $F_{\mu\nu}$. It is easy to see that the same construction works in Minkowski space. Thus, if we define

$$J^5_\mu = j^5_\mu - \frac{N}{16\pi^2}\,\varepsilon_{\mu\nu\lambda\sigma}(A^\nu,\,F^{\lambda\sigma} - \tfrac{2}{3}A^\lambda A^\sigma), \tag{5.5}$$

this current is gauge-variant but conserved.

If we work in a covariant gauge (and why should we not?), the added term commutes with the quark fields at equal times. Thus we can derive, for Green's functions made of one J^5_μ and a string of gauge-invariant quark multilinears, chiral U(1) Ward identities of the usual form. And since these are of the usual form, they lead to the usual conclusion: chiral U(1) is a symmetry; either Green's functions made of quark multilinears alone are U(1) symmetric, or there are Goldstone poles in Green's functions for one J_μ^5 and a string of quark multilinears.

Is there no way out? Well, there is one. The Hilbert space of a gauge field theory quantized in a covariant gauge is notoriously full of negative-norm timelike photons and similar gauge phantoms, states that never couple to gauge-invariant operators. Could it be that the Goldstone boson is such a phantom? No, this is not possible; the formulation of the question is wrong. If the Goldstone boson does not couple at all to gauge-invariant operators, it cannot produce a pole in a Green's function for one J_μ^5 and a string of gauge-invariant operators.

The proper formulation of the question was found by Kogut and Susskind,[30] who had the bright idea of looking at the Schwinger model, massless spinor electrodynamics, in 1 + 1 dimensions in a covariant gauge. The Schwinger model is an exactly soluble theory that has properties very close to those we have been discussing. In particular, there is a gauge-invariant axial current with an anomalous divergence and a gauge-variant conserved axial current, and, most important, there is chiral symmetry breakdown without Goldstone poles in gauge-invariant Green's functions. What Kogut and Susskind found in the covariant-gauge Schwinger model were two free massless fields, ϕ_+ and ϕ_-. ϕ_+ creates quanta of positive norm and has the usual propagator; ϕ_- creates quanta of negative norm and has minus the usual propagator. (Remember, a covariant gauge is full of negative-norm states from the very beginning.) All gauge-invariant quantities couple to the sum of these fields, $\phi_+ + \phi_-$; this has zero propagator and produces no singularities. Thus gauge-invariant Green's functions are free of Goldstone poles. However, the gauge-variant conserved current couples to the gradient of the difference, $\partial_\mu(\phi_+ - \phi_-)$. Thus, when one considers a Green's function for one gauge-variant current and a string of gauge-invariant fields, the relative minus sign in the coupling cancels the relative minus sign in the propagators, and Goldstone poles appear where they should. This set-up is called a Goldstone dipole. (The terminology is a bit misleading, because there are only single poles in Green's functions, but I shall stick with it anyway.)

Thus according to Kogut and Susskind, the proper formulation of our question is, is the U(1) symmetry of quantum chromodynamics spontaneously broken via a Goldstone dipole? You might think that this is a question that could be asked seriously only by a field theorist driven mad by spending too many years in too few dimensions. Nevertheless, as 't Hooft[9] brilliantly showed, the answer is yes. The remainder of this section is an explanation of his computation.

5.2 *Prèliminaries: Euclidean Fermi fields*

Before we can treat quantum chromodynamics by functional integration, we must know how to integrate over Euclidean Fermi fields. This section is a description of the theory of such integration, with all mathematical fine points ruthlessly suppressed.[31] I will develop the theory by defining Fermi integration as a 'natural' generalization of Bose integration. At the end, I will justify my definitions by showing that they lead to formulae equivalent to those obtained by conventional canonical quantization.

Let us begin by defining our integration variables. For Bose theories, we integrate over *c*-number Euclidean fields. These are objects that commute with each other at arbitrary separations; they can be thought of as the classical (vanishing $\hbar$) limit of quantum Bose fields. This suggests that the proper variables for a Fermi theory should be classical Fermi fields, objects which *anticommute* with each other at arbitrary separations. Thus, for example, for the theory of a single Dirac field, we would expect our integration variables to be two Euclidean bispinors, $\bar{\psi}$ and ψ, obeying

$$\{\psi(x), \psi(y)\} = \{\bar{\psi}(x), \bar{\psi}(y)\} = \{\psi(x), \bar{\psi}(y)\} = 0, \tag{5.6}$$

for all Euclidean points x and y.

The last of these relations is crucial, for it implies that $\bar{\psi}$ can not be in any sense the adjoint of ψ times some matrix. For if this were so, the last relation (multiplied by the inverse matrix) would state that the sum of two positive semi-definite objects, $\psi\psi^\dagger$ and $\psi^\dagger\psi$, was zero. This would only be possible if ψ vanished, not a happy situation for a prospective integration variable. Thus if we are to have any hope of founding a sensible integration theory, we must treat ψ and $\bar{\psi}$ as *totally independent variables.*

This independence is the main novelty of Euclidean Fermi fields; the rest of the construction is straightforward. We define the Euclidean γ-matrices to be four Hermitian matrices obeying

$$\{\gamma_\mu, \gamma_\nu\} = 2\delta_{\mu\nu}. \tag{5.7}$$

We use these to define the O(4) transformation law for ψ in the usual way, and define $\bar{\psi}$ to transform like the adjoint of ψ. We define γ_5, a Hermitian matrix, by

$$\gamma_5 = \gamma_1\gamma_2\gamma_3\gamma_4. \tag{5.8}$$

Thus, $\bar{\psi}\psi$ is a scalar, $\bar{\psi}\gamma_5\psi$ a pseudoscalar, $\bar{\psi}\gamma_\mu\psi$ a vector, etc.

The Euclidean action for a free Dirac field is

$$S = -\int d^4x\bar{\psi}(i\partial_\mu\gamma_\mu - im)\psi. \tag{5.9}$$

The minus sign is pure convention; we could always absorb it into ψ if we wanted to. (Remember, we are free to transform ψ without touching $\bar{\psi}$.) The i in front of the mass term is not conventional. It is there to insure that the Euclidean propagator is proportional to $(\not{p}+im)/(p^2+m^2)$; if it were not for the i, we would have tachyon poles. If m vanishes, Eq. (5.9) is invariant under chiral transformations,

$$\psi \to e^{-i\alpha\gamma_5}\psi, \ \bar{\psi} \to \bar{\psi} e^{-i\alpha\gamma_5}. \tag{5.10}$$

The quark part of the Euclidean action for quantum chromodynamics is obtained from Eq. (5.9) by replacing ordinary derivatives by covariant derivatives.

So much for the integrand; now for the integration. For Bose fields, we defined functional integration as iterated integration over ordinary numbers. Therefore, let us begin by defining integration for a function of a single anticommuting quantity, a. (Of course, for a single quantity, the anticommutation algebra degenerates to a single equation, $a^2 = 0$.)

We want to define

$$\int da\, f(a), \tag{5.11}$$

for an arbitrary function, f. We want this to have the usual linearity property: the integral of a linear combination of two functions should be the linear combination of the integrals. In addition, we would like the integral to be translation-invariant

$$\int da\, f(a+b) = \int da\, f(a), \tag{5.12}$$

where b is an arbitrary anticommuting quantity. I will now show that these conditions determine the integral, up to a normalization factor.

The reason is that there are only two linearly independent functions of a, 1 and a; all higher powers vanish. We will choose our normalization such that

$$\int da\, a = 1. \tag{5.13}$$

From this, and Eq. (5.12),

$$\int da\, 1 = 0. \tag{5.14}$$

For functions of many anticommuting variables, we define multiple integrals as iterated single integrals. Thus, for example, a complete integration table for the four linearly independent functions of two anticommuting variables, a and $\bar{a}$, is

Note that this is the inverse of the answer we would have obtained had we done the identical integral with ψ and $\bar{\psi}$ complex Bose fields.

I will now show that Eq. (5.22) is the correct answer, that it is identical to the normal field-theoretic expression for the vacuum-to-vacuum transition amplitude in a theory of a quantized Dirac field interacting with external c-number fields. In this theory, this amplitude is the sum of all Feynman graphs with no external Fermi lines. This in turn is the exponential of the sum of all connected (that is to say, one-loop) graphs. Now, if ψ were a Bose field, we know that the amplitude would be the inverse determinant, because we trust functional integration for Bose fields. But the only effect of replacing bosons by fermions is to multiply the one-loop graphs by minus one. This inverts the exponential of the one-loop graphs, that is to say, it turns the inverse determinant into the determinant.

In any theory in which the Fermi fields enter the action at most bilinearly, we can always integrate over the Fermi fields, using Eq. (5.23), before we integrate over the Bose fields. In diagrammatic language, we can always sum the Fermi loops before we integrate over virtual bosons. Thus, because our definition of Fermi integration gives the right answer for a Dirac field in an external c-number field, it also gives the right answer for a Dirac field interacting with a quantum Bose field. In particular, it gives the right answer for quantum chromodynamics.

5.3 *Preliminaries: chiral Ward identities*

In this section is a discussion of the chiral Ward identities for a theory of a set of quantum Dirac fields interacting with c-number gauge fields. In the sequel, we shall use these identities in several different cases; thus it is useful to have them written down in their most general form, at hand when we need them.

Let ψ be a set of Euclidean Dirac fields, assembled into a big vector, which transforms according to some representation of SU(n), not necessarily irreducible, generated by a set of matrices, T^a. Let us define the constant C by

$$\operatorname{Tr} T^a T^b = -C\delta^{ab}. \tag{5.24}$$

Thus, for example, for a set of N fields each transforming according to the n-dimensional representation of SU(n),

$$C = N/2. \tag{5.25}$$

We wish to study the theory of these fields interacting with given c-number gauge fields,

$$S = -\mathrm{i}\int \mathrm{d}^4x\, \bar{\psi}(\gamma_\mu D_\mu - M)\psi, \tag{5.26}$$

parts; the theory contains no massless particles that could give a non-vanishing surface term. Also, on the right we can use

$$\int d^4y(F, \tilde{F}) = 32\pi^2 \nu. \tag{3.47}$$

Thus we obtain

$$\begin{aligned} 2\left\langle \int d^4y\, \bar{\psi} M \gamma_5 \psi(y) \phi^{(1)}(x_1) \ldots \phi^{(m)}(x_m) \right\rangle^A \\ + \frac{\partial}{\partial \alpha} \langle \phi^{(1)}(x_1) \ldots \phi^{(m)}(x_m) \rangle^A \\ = -4\mathrm{i}C\nu \langle \phi^{(1)}(x_1) \ldots \phi^{(m)}(x_m) \rangle^A. \end{aligned} \tag{5.31}$$

Now all our artillery is at the ready; we can begin our assault on quantum chromodynamics.

5.4 *QCD (baby version)*

I will begin by analyzing a baby version of quantum chromodynamics, in which the gauge group is SU(2), and in which there is only a single isodoublet quark, of mass zero. In equations,

$$S = \int d^4x \left[\frac{1}{4g^2} (F, F) - \mathrm{i}\bar{\psi} D_\mu \gamma_\mu \psi \right]. \tag{5.32}$$

After we have worked out the baby theory, we will go on to the real thing.

Most of the analysis of Sect. 3 is essentially unaltered by the presence of a quark. In particular, all of our old instanton solutions are still solutions of the Euclidean equations of motion (with the quark fields set equal to zero). Thus we still have all the θ-vacua, and formulae like

$$E(\theta)/V = -2K \cos\theta\, e^{-S_0}, \tag{3.56}$$

and

$$\langle \theta | (F, \tilde{F}) | \theta \rangle = -64\pi^2 \mathrm{i} K e^{-S_0} \sin\theta, \tag{3.59}$$

remain unaltered. The only effect of the quarks is to insert into the definition of K a term proportional to

$$\det\left[\frac{\mathrm{i}\not{D}}{\mathrm{i}\not{\partial}}\right] = \det\left[\frac{\mathrm{i}(\partial_\mu + A_\mu)\gamma_\mu}{\mathrm{i}\partial_\mu \gamma_\mu}\right], \tag{5.33}$$

where A_μ is the field of an instanton.

This is a trifling alteration, but it is a tremendous trifle, for, as we shall see, $\mathrm{i}\not{D}$ has a vanishing eigenvalue. Thus the determinant vanishes, as does $E(\theta)/V$ and $\langle \theta | (F, \tilde{F}) | \theta \rangle$!

The vanishing eigenvalue can be demonstrated either by a short explicit computation or by a long indirect argument. I will choose the

8

1/*N*

(1979)

1 Introduction

More variables usually means greater complexity, but not always. There exist families of field theories with symmetry group SO(N) (or SU(N)) that become simpler as N becomes larger. More precisely, the solutions to these theories possess an expansion in powers of $1/N$. This expansion is the subject of these lectures.

There are two reasons to study the $1/N$ expansion.

(1) It can be used to analyze model field theories. This is important. Most of us have a good intuition for the phenomena of classical mechanics. We were not born with this intuition; we developed it toiling over problems involving rigid spheres that roll without slipping and similar extreme but instructive simplifications of reality. One reason we have such a poor intuition for the phenomena of quantum field theory is that there are so few simple examples; essentially all we have to play with is perturbation theory and a handful of soluble models. The $1/N$ expansion enables us to enlarge this set.

In Section 2 I develop the $1/N$ expansion for ϕ^4 theory and apply it to two-dimensional models with similar combinatoric structures, the Gross–Neveu model and the $\mathbb{C}P^{N-1}$ model. These models display (in the leading $1/N$ approximation) such interesting phenomena as asymptotic freedom, dynamical symmetry breaking, dimensional transmutation, and non-perturbative confinement; they are worth studying.

(2) It is possible that the $1/N$ expansion, with N the number of colors, might fruitfully be applied to quantum chromodynamics. In the real world, N is 3, so an expansion in powers of $1/N$ may not seem like such a good idea. This objection is without force, as is shown by the following wisecrack by Ed Witten:

$$\frac{e^2}{4\pi}=\frac{1}{137}\Leftrightarrow e=0.30.$$

momentum space, is

$$-\frac{iN}{4\pi}[g_{\mu\nu}p^2 - p^\mu p^\nu]\int_0^1 \frac{dx(1-2x)^2}{\sigma_0^2 - p^2x(1-x) - i\varepsilon}. \qquad (2.59)$$

If we are only interested in long-range forces, that is to say, in small momenta, we may neglect the p^2 in the integrand. We thus obtain

$$-\frac{iN}{12\pi\sigma_0^2}[g_{\mu\nu}p^2 - p_\mu p_\nu]. \qquad (2.60)$$

This corresponds to a term in the effective action of the form

$$S_{\text{eff}} = -\frac{N}{48\pi\sigma_0^2}\int d^2x(\partial_\mu A_\nu - \partial_\nu A_\mu)^2. \qquad (2.61)$$

Aside from a trivial normalization, this is the action for the free electromagnetic field. This is the most astonishing feature of the model; a genuine gauge field has been dynamically generated, produced as a result of radiative corrections in a theory that perturbatively has only short-range interactions. The gauge field now produces a linear potential that confines the *z*s. Not only does the theory have (bosonic) quarks, it has confined quarks.

There is much more that can be said about the $\mathbb{C}P^{N-1}$ model. For example, the classical theory admits instantons (just as in chromodynamics), and there is no infrared cutoff on instanton sizes (again as in chromodynamics). Thus the model can be used as a laboratory for instanton physics (=arena for bloody controversies). Unfortunately, I do not have the time to go into any of this here, and must once again refer you to the literature.[6]

3 Adjoint representations, or, chromodynamics

3.1 *The double-line representation and the dominance of planar graphs*[7] (*)

The $1/N$ expansion is vastly more difficult for SU(N) gauge theories than for any of the theories of Section 2. The source of the difficulty has nothing to do with the traditional problems of chromodynamics, the intricacies of gauge invariance or the uncontrollable infrared divergences. It is just that we have to deal with fields that transform according to the adjoint representation rather than the vector representation, objects that carry two group indices rather than one. ϕ^4 theory is just as difficult if ϕ is in the adjoint representation.

The dynamical variables of the theory are a set of Dirac fields, ψ^a, and a set of gauge fields, $A^a_{\mu b}$, where a and b run from 1 to N. The Dirac fields can be thought of as elements of an SU(N) column vector, the gauge

(*) The large-N expansion for QCD was invented by G. 't Hooft.

intervening SU(N) matrix made of $F_{\mu\nu}$s and covariant derivatives.) With no loss of generality we can study the states made by a single product,

$$Q(x) = B'_1(x)B'_2(x), \tag{3.16}$$

where, for purposes of this argument, I have restored explicit space-time dependence. For simplicity, I will assume we have chosen the flavors of these operators so that B_1, B_2, and Q all have vanishing vacuum expectation values. Then,

$$\begin{aligned}\langle Q^\dagger(x)Q(y)\rangle = {} & \langle B'^\dagger_1(x)B'_1(y)\rangle\langle B'^\dagger_2(x)B'_2(y)\rangle \\ & + \langle B'^\dagger_1(x)B'_2(y)\rangle\langle B'^\dagger_2(x)B'_1(y)\rangle \\ & + \langle B'^\dagger_2(x)B'^\dagger_2(x)B'_1(y)B'_2(y)\rangle_C.\end{aligned} \tag{3.17}$$

The first two terms on the right are of order one, while the third is of order $1/N$, and thus should be dropped in the large-N limit. But the first two terms simply describe the independent propagation of two mesons from x to y. In the large-N limit, quadrilinears make meson pairs and nothing else.

3.3 The 't Hooft model[8]

The arguments of the preceding section have been powerful but abstract; it would be nice to have a concrete example in which we could see them at work. Such an example is provided by the 't Hooft model, large-N chromodynamics in two space-time dimensions. The model is almost exactly soluble; the simplest Green's functions can be found in closed form (in an appropriate gauge), and, although the computation of the particle spectrum requires numerical analysis, it is of a sort that can be carried out on a pocket calculator. The model also serves to eliminate a worrisome possibility, that the arguments of Section 3.2 are internally inconsistent, that confinement cannot exist for arbitrarily large N.

Of course, it is no surprise to find confinement in a two-dimensional gauge theory. As a warm-up for the 't Hooft model, let me remind you how confinement occurs in an even simpler theory, two-dimensional quantum electrodynamics. This theory is defined by

$$\mathscr{L} = \tfrac{1}{2}(F_{01})^2 + \bar{\psi}(i\partial_\mu\gamma^\mu - eA_\mu\gamma^\mu - m)\psi, \tag{3.18}$$

where

$$F_{01} = \partial_0 A_1 - \partial_1 A_0. \tag{3.19}$$

As always in the analysis of a gauge theory, the first step is to pick a gauge. I will choose axial gauge,

$$A_1 = 0. \tag{3.20}$$

Euler characteristic is a topological invariant. (2) If our polygonal surface is topologically equivalent to a sphere with B holes cut out of it and H handles stuck on to it, then

$$\chi = 2 - 2H - B. \tag{A.1.2}$$

Proof of (1). One can convince oneself that there are only three fundamental ways one can change a polygonal surface continuously. (i) One can distort the surface without changing either F, E, or V. Of course, this does not change χ. (ii) One can shrink an edge to a point. This eliminates one edge, merges two vertices, causing a net loss of one vertex, and does not change the number of faces. Thus χ does not change. The same argument applies to the reverse process. (iii) One can shrink a face to a point. If the face is a polygon with n sides, this procedure eliminates one face, n edges, and $n-1$ vertices. Again χ does not change. The same argument applies to the reverse process. Processes like conversion of a face to an edge can be obtained as (iii) followed by the reverse of (ii) and do not require independent analysis.

Proof of (2). The argument goes in stages. First we prove the equation for a sphere, then for a sphere with holes, and finally for a sphere with holes and handles. (i) A sphere can be constructed by taking two n-sided polygons and identifying their perimeters. We thus obtain a surface with n edges, n vertices, and two faces, for which

$$\chi = 2. \tag{A.1.3}$$

(ii) If we have a polygonal surface that is topologically equivalent to a sphere, we cut holes out of it by removing faces. Thus, each hole reduces χ by one, and

$$\chi = 2 - B. \tag{A.1.4}$$

(iii) To make a handle, we cut two holes that are both n-sided polygons (reducing χ by two) and then identify the perimeters of the two polygons (reducing both E and V by n and not changing χ at all). Thus,

$$\chi = 2 - 2H - B. \tag{A.1.5}$$

Appendix 2. The 't Hooft equations

This appendix gives the computations promised in Section 3.3. I will begin with the quark self-energy. We need a preliminary identity:

$$\int \frac{dx}{x \pm i\varepsilon} = \int dx \left[\frac{P}{x} \mp i\pi\delta(x) \right] = \mp i\pi. \tag{A.2.1}$$

As an example, let me compute the power of $1/N$ associated with the leading connected vacuum-to-vacuum graphs containing a correction gluon. We must add the correction gluon to graphs that contain quark loops. The leading graphs are those in which the correction gluon connects two disconnected graphs each of which contains a quark loop; these are $O(N \times N \times 1/N^2) = O(1)$. However, these graphs are trivial; they serve only to exactly cancel the corresponding gluon-exchange graphs in the uncorrected theory. The leading non-trivial graphs are those in which a correction gluon is added to a connected graph containing a quark loop; these are $O(N \times 1/N^2) = O(1/N)$.

Notes and references

1. To my knowledge, the first to observe that ϕ^4 theory became simple for large N was K. Wilson, *Phys. Rev.* **D7**, 2911 (1973). Wilson drew on ideas developed in statistical mechanics, especially Stanley's work on the spherical model (H. E. Stanley, *Phys. Rev.* **176**, 718 (1968)). The auxiliary-field method used here was also developed in statistical mechanics (but not in the context of large-N expansions), by R. L. Statonovich, *Doklady Akad. Nauk. S.S.S.R.* **115**, 1097 (1957). The treatment of ϕ^4 theory given here follows that of S. Coleman, R. Jackiw, and H. D. Politzer, *Phys. Rev.* **D10**, 2491 (1974). Some points left confused in this paper are clarified in L. Abbott, J. Kang, and H. Schnitzer, *Phys. Rev.* **D13**, 2212 (1976).
2. D. Gross and A. Neveu, *Phys. Rev.* **D10**, 3235 (1974).
3. This note is for the cognoscenti only; it is written in shorthand. You may know that you can define an effective potential, much like V, in theories with fundamental scalar fields, and that, in such theories, V can be interpreted as an energy density for general arguments, not just stationary points. This is important; for example, it implies that if V is unbounded below the theory is sick, no matter how nicely behaved V is at its stationary points. *There is no such interpretation of the effective potential for composite fields.* For example, in the case at hand, if we add a source term to the Lagrange density,

$$\mathscr{L} \to \mathscr{L} + J(x)\sigma,$$

this corresponds to adding a term to the Hamiltonian density,

$$\mathscr{H} \to \mathscr{H} - J\phi^a\phi^a + \frac{g_0}{2N} J^2.$$

The J^2 term has no analog for a fundamental scalar field, and destroys the standard energy arguments (except at stationary points of V, where it vanishes).
4. The model was devised by H. Eichenherr, *Nucl. Phys.* **B146**, 215 (1978) and V. Golo and A. Perelomov, *Phys. Lett.* **79B**, 112 (1978). The $1/N$ expansion is worked out in A. D'Adda, M. Lüscher, and P. Di Vecchia, *Nucl. Phys.* **B146**, 63 (1978), and **B152**, 125 (1979).
5. This is a bit too slick. After all, in electrodynamics, gauge invariance could also be thought of as a mere reflection of the presence of a redundant variable, the longitudinal part of the vector potential. The real difference is slightly more subtle. In electrodynamics, it is not possible to eliminate the redundant variables and still have a local theory, as is shown by the Bohm–Aharonov effect; in the case at hand, it is possible to do just this, as is shown by our

derivation of the model. As we shall see, this distinction will disappear when we sum up the radiative corrections.

6. E. Witten, *Nucl. Phys.* **B149**, 285 (1979); A. Jevicki, *Phys. Rev.* **D20**, 3331 (1979); I. Affleck, *Phys. Lett.* **92B**, 149 (1980); *Nucl. Phys.* **B162**, 461 (1980); **B171**, 420 (1980).
7. (This note covers both this subsection and the next.) The large-N expansion for chromodynamics was invented by G. 't Hooft, *Nucl. Phys.* **B72**, 461 (1974). There are numerous parallels and connections with the topological expansions of S-matrix theory; see G. Veneziano, *Nucl. Phys.* **B117**, 519 (1976) and G. Chew and C. Rosenzweig, *Phys. Rep.* **41C**, 263 (1978).
8. G. 't Hooft, *Nucl. Phys.* **B75**, 461 (1974). This paper has spawned a large literature, with two branches. (1) Papers which investigate the model in more detail (for example, by computing form factors) and/or use the model to gain insight into four-dimensional chromodynamics. Some examples: C. G. Callan, N. Coote, and D. J. Gross, *Phys. Rev.* **D13**, 1649 (1976); M. B. Einhorn, *Phys. Rev.* **D14**, 3451 (1976); R. Brower, J. Ellis, M. Schmidt, and J. Weis, *Nucl. Phys.* **B128**, 131, 175 (1977). (2) Papers which attempt to clean up 't Hooft's original derivations and/or to find inconsistencies in the model. To my mind, the paper of this kind that does things best and settles all the problems is I. Bars and M. B. Green, *Phys. Rev.* **D17**, 537 (1978).
9. E. Witten, *Nucl. Phys.* **B160**, 57 (1979).
10. For a review of lumps, see Chapter 6 in this volume.
11. The proposal was made in a lecture given at Harvard in the Spring of 1979. E. Witten in *Recent Developments in Gauge Theories*, 1979 Cargese Lectures, edited by G. 't Hooft *et al.* Plenum (1980).
12. E. Brezin, C. Itzykson, G. Parisi, and J. B. Zuber, *Comm. Math. Phys.* **59**, 35 (1978).

THIRD PART

Here are some photos of Sid in Erice: in 1968, 1969, 1970 and 1973. The one in 1975, taken by Sid, shows Ken Wilson dancing. Three pictures in 1977 show Sid with the Erice Gold Medal for "Best Lecturer". The Mayor of Erice, Antonio Gianquinto, the Minister of the Sicilian Government, Hon. Domenico Cangelosi, and the Rector of the Polytechnic University of Milan, Luigi Dadda, were present at the Ceremony.

1968

1969

From left: Sid Coleman, Giacomo Morpurgo, Howard J. Schnitzer, Robert Salmeron, R.L. Cool, Shelly Glashow, Giuliano Preparata and Marvin Leonard Golberger.

1969

From left: Giacomo Morpurgo, Shelly Glashow and Sid Coleman.

1970

1973

1975

Photo taken by Sid Coleman, showing Ken Wilson dancing in Erice.
In the middle of the photo the tall fellow is Michel Gourdin;
on his left Gerardus 't Hooft and Betteke 't Hooft.

1977
Sid Coleman with the Erice Gold Medal for "Best Lecturer".

1977

The Mayor of Erice, Antonio Gianquinto, congratulates Sid Coleman for his "Gold Medal".

1977

From left: The Minister of the Sicilian Government Domenico Cangelosi, A.Z. and the Rector of the Polytechnic University of Milan Professor Luigi Dadda, congratulating Sid Coleman for the "Gold Medal".

1979
Sidney Coleman at Erice.

"ETTORE MAJORANA PRIZE – ERICE –
SCIENCE FOR PEACE"

TO

SIDNEY COLEMAN

For his superb series of Lectures at the International School of Subnuclear Physics in Erice and for his exceptional contribution in spreading the "Erice geist", worldwide.

Antonino Zichichi

2006

The Diploma of the "Ettore Majorana Prize" awarded to Sid Coleman in 2006.

Every year a special scholarship has been established to honour
the memory of Sid Coleman.
This scholarship is awarded to the "Best Student"
of the Subnuclear Physics School.

Sidney Coleman

Best Student's Award

REFERENCES

[1] S. Coleman
"*Aspect of Symmetry - Selected Erice Lectures*", Cambridge University Press (1985).

[2] A. Zichichi
"*Antiparticles and Antimatter: the basic difference*", p. 94-100 in Proceedings of Karlsruher Nuklidkarte, Commemoration of the 50th Anniversary; G. Pfennig, C. Normand, J. Magill and T. Fanghänel (eds). Publishing editor by G. Weber, Germany, (2008).

[3] L. Maiani
in Proceedings of the Symposium "*The Discovery of Nuclear Antimatter*", to celebrate the 30th anniversary of the discovery of Nuclear Antimatter. Conference Proceedings 53, Italian Physical Society, Bologna, Italy (1995).

[4] *The Origin of the third family*
by C.S. Wu, T.D. Lee, N. Cabibbo, V.F. Weisskopf, S.C.C. Ting, C. Villi, M. Conversi, A. Petermann, B.H. Wiik and G. Wolf, on the XXX Anniversary of the Proposal to search for the Third Lepton at ADONE; O. Barnabei, L. Maiani, R.A. Ricci and F. Roversi Monaco (eds), Rome (1997); and World Scientific (1998).

[5] A. Zichichi
"*Subnuclear Physics - The first fifty years*", a joint publication by the University and the Academy of Sciences of Bologna, Italy (1998); World Scientific Series in 20th Century Physics, Vol. 24, two editions (2000–2001).

CHAIRMAN: A. ZICHICHI

Scientific Secretaries: M. Goebel, A. Kaminska

DISCUSSION

REMEMBERING SIDNEY COLEMAN

- Susskind:

Before I start, I must talk to Sidney for a moment, so please excuse me. [talking to the ceiling] Sidney! Sidney, can you hear me? I have to apologize, because I am extremely jet lagged and I would like to do it really well... But Nino [Zichichi] wouldn't let me do it this morning. What are you saying? You are saying that's no excuse, OK, blame it on Nino.

Sidney was a marvelously colorful person. Those of you who didn't know him have to just look at his photograph to see what a colorful person he was. And as you might expect, he was a fantastic story teller and a master of telling jokes. He had a huge repertoire of stories, but there was something I always marveled at, something incredible. He would tell those stories over and over and over again, exactly the same way, to exactly the same people. He would polish them each time a little bit. And yet he was always funny. It was like listening to a favorite recipe or a favorite piece of music. When Sidney said he wanted to tell this or that story it was like saying "I am going to tell joke nr 117". And everybody would come around saying "Sidney is going to tell joke 117, wow!". He'd tell it and it was really, really funny for the 117th time.

Sidney was of course also capable of incredibly spontaneous humor. You know how you always say to yourself "If only I had been quick enough, I would have said this or that". Sidney was always quick enough. He always had an instantaneous and funny response to almost anything. I will give one example. I don't remember jokes very well, but I do remember this one story extremely well. Sidney, I and Steven Hawking and I don't remember who else were invited during the Jerusalem winter school—I think the year was 1988—to appear on Israeli TV. Sidney was not as well dressed as usual because El Al airlines had lost his luggage. The first person to be interviewed was Steven. The interviewer said "Professor Hawking, are we close to having a complete theory?". Steven responded, in his special way of speaking "Yes, I think by year 2000 we will have an absolute theory of everything" Whereupon the interviewer turned to Sidney and said, "Professor Coleman, do you agree with that?" Without a second's thought, Sidney responded "I don't even think by the year 2000 El Al will find my luggage". In any case Sidney was a marvelous wit and I think that there was something about him that we will miss for years to come. There will never be anyone to take his place. John [Schwarz], why don't you quickly say what you told me a few minutes ago about a real Sidney remark?

- Schwarz:

I didn't know Sidney terribly well but I remember when he was struggling with his diabetes and was starting to pay attention to his health, he said it wasn't until after he turned 50 that he realized that his mind had a body attached to it.

- Susskind:

Wait a second: Sidney is trying to say something... Sidney is upset, he is very upset. OK, Sidney, I'll tell them. Nino, Sidney is very upset with what is going on. He says that Power Point is ruining theoretical physics. We have to get back to the blackboard, so that decent theoretical physics can be done. Yes Sidney, they were applauding you, Nino was nodding his head.

- *Ferrara:*

Sidney Coleman was not very sympathetic with Supersymmetry, String Theory, all this new development. However he gave an actual contribution to Supersymmetry. He was responsible for the invention of the name gravitino. The originally proposed word was rejected by Phys. Rev. and it was emitrion.

ERICE LECTURES 2008

Gerard 't Hooft

Institute for Theoretical Physics
Utrecht University

and

Spinoza Institute
Postbox 80.195
3508 TD Utrecht, the Netherlands

e-mail: g.thooft@uu.nl
internet: http://www.phys.uu.nl/~thooft/

Opening Lecture: Remembering Sidney Coleman

1 Opening Lecture: REMEMBERING SIDNEY COLEMAN

The Amsterdam International EPS Conference[1], June 30 - July 6, 1971, was for me the first occasion to present my theories about the renormalizability of non-Abelian gauge theories with spontaneous symmetry breakdown. My then advisor, Martinus Veltman, had given me the opportunity to explain this in 10 minutes to the audience. "Let me introduce you to two American gangsters", he said to me, after the lecture was over. I was puzzled by the remark at that time, but later realized that this was Veltmans way to talk about his adversaries in physics. Anyway, the two gentlemen in question turned out to be Sheldon Glashow and Sidney Coleman. Talking to these two people, made me realize that these must be the smartest gangsters on the planet. Sidney immediately saw the most essential ingredients of the new theory. He later explained that Quantum Field Theory had looked to him as an ugly monster until exactly that moment, when it received a kiss by our work, turning into a beautiful and wealthy prince.

I met Sidney many times since, and I vividly remember the lectures he gave in the 1975 Erice School[2]. At that school, not only the best student and best secretary were chosen, but also best lecturer. This was Sidney Coleman, who, at the farewell party, also came as the most fancy dressed participant, in his fluorescent pink jacket. His lectures had been about topology in QFT. Previously, when renormalizing the theory, we had always limited ourselves to perturbation expansion, where the topology is trivial.

Sidney would illustrate vividly what topology is, using the cord of his microphone, by which he was connected to the loudspeaker system. Winding the cord around his neck he explained the concept of winding. The students, worried as they were that their lecturer might strangle himself, would never forget the definition of winding number.

It was October 1988 when, during one of my later encounters at Harvard, I visited Sidney at his home, with my two little children. There, we would play the game of Charades; he was very good at it. My children loved it.

As emphasized by others here, Sidney was a great story teller: One of his stories was about how he rescued mankind from a pending planetary disaster, while sitting at a bar, with his friend Carl Sagan. With Sagan, Sidney shared his love for science-fiction. Behind a good glass of whiskey, or two, Sagan showed a problem to Sidney: Suppose that there exist alien, totally unknown, life forms on Mars, or, for that matter, on the Moon as well. The Moon would soon be visited by astronauts. What should space agencies do to avoid the danger that these life forms would infect the Earth, and destroy mankind? How do we avoid contamination of one planet by another in general, when space flight becomes routine?

This became a vivid discussion. But a month later, Sidney had almost forgotten this conversation, so that he was highly surprised when a manuscript fell in his mailbox: "Spacecraft Sterilization Standards and Contamination of Mars", by S. Coleman and C. Sagan. The norm for the possibility of cross contamination from one celestial body to another should be less than 0.1%. Further, astronauts should be put in quarantine, and the chances of survival for any individual organism of any life form from one planet to any other should be less than one in 10,000.

The paper was submitted for publication[3]. Now, this was the only existing publication on this topic, so, NASA decided to follow the advice. The Committee of Space Research, COSPAR, determined that astronauts returning from the Moon would have to go into quarantine for a certain amount of time. The "Lunar Receiving Laboratory" (LRL) was erected in Houston, Texas. Upon their return from the Moon, many astronauts spent boring hours there. Fortunately, dangerous life forms were never found.

References

[1] IProceedings of the Amsterdam International Conference on Elementary Particles, held at Amsterdam, The Netherlands, June 30 - July 6, 1971 (North-Holland Pub. Co., 1972)

[2] International School of Subnuclear Physics, Director A. Zichichi, 13th Course: New Phenomena in Subnuclear Physics 11 July - 1 August 1975.

[3] S. Coleman and C. Sagan, *Spacecraft Sterilization Standards and Contamination of Mars*, Journal of Astronautics and Aeronautics **3**, 22 (1965).

CHAIRMAN: G. 't HOOFT

Scientific Secretaries: P. Fedosenko, A. Slobodeniuk

DISCUSSION I

- *Yeranyan:*

Today in the morning lecture you said that defects (strings) can create some composite objects. Is it possible in this picture to have also decomposition of these objects? If yes, which kind of conditions we need for this?

- *'t Hooft:*

What can happen is that two strings may collide such that a third string may seem to emerge. However, the associated holonomy is found not always to describe a string. If the angle by which they collide is not 90 degrees then this is not a string. In that case we assume that two strings emerge instead of one, so that, here, it could be said that a single string decomposes into two strings moving away from one another. So indeed then the answer is yes, this is an example of a decomposing string. I'm not sure whether this answers your question.

- *Yeranyan:*

Is it possible after some time for two strings to decompose into a single string?

- *'t Hooft:*

This is not so easy. They must align exactly so that the two strings can merge into one. You must remember that the surrounding space-time has to remain flat. This is conceivable, but remember that you can't have excitations of the strings because they would emit gravitational waves and that is not in line with the rules that have been set.

- *Yeranyan:*

But there is some probability that this happens?

- *'t Hooft:*

Not in my present picture. In this picture what you are allowed to do with the strings is very limited. The strings must be straight – we don't allow any excitations. If you would allow for excitations you could say, "Well, after a collision these two strings don't leave a new one stretching between them as was described. Instead, they simply generate an excitation that then spreads along the strings." This would be what conventional string theorists would have come forward with. But then the surrounding space-time could not be flat. And remember that this would generate all curvature terms in space-time as dictated by Einstein's gravity equations, and this wouldn't be allowed according to the rules I have set. This is a very important constraint that makes it much harder to figure out which possibilities there are after a collision.

- *Shakirov:*

In your talk you have mentioned the (3N-4) – parameter space of possible solutions to the collision problem. But it seems that if there is an actual solution with N intermediate strings, then arbitrarily large N will also do. So my question is: what fundamental principle fixes N in this situation? Are triple collisions possible?

- *'t Hooft:*

In principle N could be anything and what fixes that is what you decide your model to be. The model defines the rules that apply when two defects hit each other. After the collision a new defect structure arises; which defect structure this is, you are free to choose. I haven't yet specified the model so I could choose anything I like – I can take this N to be anything I like. One must impose that the surrounding holonomies should describe locally flat space, and we assume that there should be a legal solution for all cases that may arise. If you take N too small then we might not have a solution.. Maybe under some circumstances there won't be any solution no matter what N is. In that case I have to try something else, I could even think of having strings cross-lining in different directions, or do something altogether different: think of other kinds of defects. This has to be sorted out, it hasn't been worked out yet and I haven't yet been able to find a completely solid prescription which covers every eventuality. That is very difficult, so that hasn't been done, but once you are able to do that you must decide: which N will I choose? That choice is reminiscent of the choice of the matter Lagrangian. In principle I could choose anything I like; one choice may mimic the natural world better than others, we just try everything; that same freedom is present in this model.

- *Shakirov:*

And are triple collisions possible?

- *'t Hooft:*

In the continuum case the triple collision will have measure zero because there will always be a very tiny mismatch, but in the crystalline version of this gravity model, the situation is again different. Then, I have only integral numbers so some integers might accidentally match and so you might have three strings hitting exactly at the same point. This is then conceivable; again, then you have to make some solid prescription as to what happens under those circumstances.

- *Bruno:*

Why are the instantons so very important for the mixing of pseudoscalar mesons?

- *'t Hooft:*

Well, first of all this is a question about instantons, but I will answer it now. Pseudoscalars break the symmetry we call 'chiral symmetry'; this is a spontaneous symmetry breaking, which would normally lead to massless (or very light) pseudoscalar particles. Instantons, however, break the symmetry explicitly, and so they can generate mass. The breaking is only in the U(1) sector of the chiral symmetry, and this is why the mass is only generated in the singlet channels of these states. The singlets are singled out. For SU(2), this is the eta particle, for SU(3) it is eta-prime. Consequently, the pseudoscalars mix along these lines. This was not understood until the instantons came along.

In the scalars you don't see the instanton effects so clearly, but there is a strong enhancement of an effective six quark interaction. Normally in QCD you assume those interactions not to be present or to be strongly suppressed because basically quarks just interact indirectly only with each other by the exchange of gluons. But now it seems there is a very strongly enhanced effective six quark interaction, so the question is what does the six quark interaction do, and we find that it gives the kind of splittings between the two quark and the four quark meson states, which is precisely the kind suggested by observations. So, that is why we think that instantons also play a role in the scalar meson sector, scalar two quark mesons or scalar tetraquark mesons. They cause mixing, and this solves a problem we had in purely phenomenological QCD when trying to understand the tetraquarks. There was a mystery about the single two quark mesons as to why they seem to be as light as a uū or dbar-d quark pair, while the decay properties suggest it to be sbar-s, which one would have thought to be heavier than uū or dbar-d or ubar-d. So the tetraquarks were thought of as a possible solution. Now we say there is a strong mixture between the two quark channel and the four quark channel and the mixture is precisely the kind of interaction induced by instantons, and therefore instantons explain away the difficulties we had in QCD.

- *Nicotri:*

Where does field theory enter in your description of gravity? Is it possible to put quantum field theory in the space you have described? If yes, what happens to it?

- *'t Hooft:*

Field theory as we know is quantum field theory. Well, we have not been successful in quantizing our theory at all, so the question of adding quantized fields may be premature. Classical field theory, will not be allowed in the intervening continuum because that's really flat, so that we cannot allow for any sources of energy or momentum density there. What I have in mind is that the defect lines represent matter, and in fact represent all dynamical degrees of freedom: matter, including the gravitons and whatever we have. So these are now string-like objects and not particles. Now I imagine that at a later stage of the development of this theory, numerous collisions cause the string-like defects effectively to fluctuate – they come and go from the vacuum, and so they might lead to vacuum fluctuations. Then there are excitations in these fluctuations. The minimal vacuum fluctuations then get excited and those would be models for the elementary particles at this later stage, for instance by assuming that you use the renormalization group equations to go from the Planck scale to the standard model scale – and there you invoke the language of a quantum field theory to describe the excitations. So quantum field theory comes in at a distance scale far away from the Planck scale. Thus, you apply quantum field theory as an effective way to describe the renormalization group procedure to go from short distances to long distances. That's where common field theory comes in, and not at the Planck scale. The Planck scale itself would only see the degrees of freedom that I described as defects.

- *Carrasco:*

Many fundamental approaches to building quantum gravity seem to start by piecing together flat classical pieces of space-time and then looking for some way to quantize. I would expect a purely quantum informatic world where any 4D basis like position of momentum are highly suspect – with classical space-time emerging much closer to our scale then the Planck scale. Maybe you can comment on the preference for the sort of reverse engineering.

- *'t Hooft:*

I am not sure I understood the question. Most procedures of quantized gravity start with Einstein questions which are differentiable and so they refer to smooth manifolds but have a nontrivial metric, as a metric depends in a nontrivial way on space and time coordinates so you get an action generated by curvature, you get a Riemann tensor to describe this curvature and together these form the degrees of freedom obeying Einstein equations. Then you replace the Poisson brackets by commutators , which would be the conventional way of quantizing gravity. Now I argue - well, we have seen what that leads to, there are difficulties In the form of divergences. Let's start instead with an extreme "top-down" procedure where I start asking myself first what could possibly happen at a scale small compared to the Planck scale, assuming that that is a well-defined concept at all. According to string theory, you might probably have a problem defining the notion of a "short distance", well, I just define it to be a long distance divided by a very large number and then I get a definition of a short distance. It may not mean much physically, but at least I can think of what it would have meant; I would introduce such continuum coordinates. My claim is that if I would reach distance scales short compared to Planck scale that way, I am entering an unphysical domain where it is no longer useful to talk about events taking place at all. What does this mean? Well, I remove all events there, I remove all particles there, so that what you are left with is only flat space-time, totally flat, not even Riemann curvature - it is a completely trivial flat Cartesian coordinate frame. The Cartesian coordinate frame means nothing physically because nothing is happening in that coordinate frame it all. We use this only semantically, to use a language to describe what goes on there. The physically relevant things are then the defects in this flat metric and I choose defects containing the least amount of information. Straight defect lines really represent practically zero information, the only information they represent is where the defect line is, how is it oriented and what is its velocity. Those are just a bunch of real numbers, typically five or ten real numbers, so that we have a finite number of degrees of freedom in every region of spacetime. Much like what you get with a black hole. The entropy of a black hole is restricted; it is limited, amounting to one point, or one bit of information per unit of area. This is the kind of theory that I want to have. Bits of information, scarcely spread over the spacetime. And what I do is one way of achieving this.

- *Susskind:*

I think the question was a little bit different, I think a question was: why do you start with a classical theory and then quantize it?

- *'t Hooft:*

Yes, that is a good question. You may have started right away with the quantum version of a theory. What is the quantum version of a defect line? First of all, these are string-like objects and are not particles, so there you have a difficulty – how do you quantize directly a string, not string excitations which are usually quantized in string theory. String excitations are again fields, which are quantized more or less conventionally, but we have strings ranging from minus infinity to plus infinity, unless they are bounded by or joined with some other strings. Quantizing that is quite a different story, so my procedure is that I am not going to quantize that in the conventional way but in an unconventional way. As I am advocating some time now, you might think of quantum mechanics in a way very different from the usual. I bypass the step of going from Poisson brackets to Dirac brackets, I directly go from the classical theory to the Hilbert space. I want to define directly the Hilbert space. These things which are defined here could each define basis elements of a Hilbert space. Then I say, the way they evolve is how this basis of Hilbert space evolves. That is the way I would like to quantize the theory. That is, in some sense, more down-to-earth, maybe that is how you need to discuss Planck length physics. In the quantization you can't really use the Poisson brackets anymore, you may have to quantize it the way I described. But that means, when we start from the classical theory and quantize, we basically do not change the theory into a quantum theory, but rather change the language which we use to talk about the theory. And that is an equally important modification of the theory, as I have claimed for some time. And I believe that's the way in which this model could be turned into a quantum model. So "quantizing" the theory, I say, is to turn a theory into a quantum theory by turning the language which I use to talk about it into a quantum language. This is all really needed to quantize a theory. The rest is only mathematics.

- *Johansson:*

So you say in this proposal of gravity as defects in a crystal there are no gravitons, and presumably then no gravity waves. Does this mean that this theory differs from classical general relativity at the macroscopic scale?

- *'t Hooft:*

It deviates from the classical gravity, in some sort, in a minimal way, because I still have Einstein's equations. Now, the Ricci tensor vanishes everywhere except the defect lines. So on the lines where I have the Ricci tensor, I can identify that with the energy-momentum tensor. And thus I have Einstein's equations. It is not that crazy from this point of view. The material sources are nothing but the defect lines.

- *Johansson:*

But presumably there are some effective gravity waves in your framework?

- *'t Hooft:*

So, in my framework I indeed have no obvious gravity waves. Physically speaking, the gravity wave is a bunch of gravitons moving around. Gravitons are particles. In my theory now, those particles are replaced by defect lines. And that's what a gravity wave is – even a gravitational field around the Earth is a bunch of defect lines in this, otherwise perfect, crystal.

- *Johansson:*

But you do predict that gravity waves can be physically measured?

- *'t Hooft:*

I do not want to change conventional physics. Of course, gravity waves are predicted by ordinary Einstein's equations, without any quantum mechanics. I do not want to make a wrong prediction. And the same thing for the Standard Model – of course, Standard Model is a great theory. I do not want to change it. My theory is completely different, it concerns physics at the Planck scale. And how it connects to the real world, is going to be a very complicated problem. Yes, I do want to get the real world, as you all believe it is today, I do want to get it out. Gravity waves are a feature of large-scale physics, and they should be observed as usually expected.

- *Stavenga:*

Can you bypass the difficulty of the consistency conditions by relaxing the condition of the string condition to be real? Imaginary string tensors would perhaps allow for more space to solve the consistency conditions.

- *'t Hooft:*

If the string tension would become imaginary, I wouldn't have Minkovski space anymore.

- *Stavenga:*

But you could have had a simpler theory instead.

- *'t Hooft:*

But you have to remember that I have used SL(2,C) just as a mathematical trick. Replacing SO(3,1) matrices by SL(2,C) matrices is just for mathematical convenience. In terms of the matrix elements of SL(2,C) the trace of the Lorentz transformation is complex. But in terms of the SO(3,1) matrices it is still a real number. However, I find an element of the Lorentz group which does not correspond to a moving string. So the claim is that, the subset of Lorentz transformations which describe the moving string is five-dimensional. A static string has a three-dimensional space of three rotations, and when you move it, you can only move in the transverse direction, that is two more. All together, you have a five-dimensional space of string holonomies. That's also five dimensional in the SO(3,1) notation of the Lorentz group. However, all the SO(3,1) matrices form a six-dimensional space. So you see, you get a one dimensional constraint, while, in terms of SO(3,1) the trace is always a real number. The constraint is more complicated to express in SO(3,1), so in order to save time I do calculations in the SL(2,C) notation. This way you can see that chosing a complex string constant would not solve the problem.

- *Ciuffoli:*

Is there the presence of a sort of instant interaction? So, for example, if there are two parallel strings – can one feel the presence of the other?

- *'t Hooft:*

No, there is no tidal force. We here on Earth feel the presence of the Moon, because the Moon generates tides on the Earth – high tide and low tide on the water. In two dimensions plus time, particles do not generate tidal forces. Two planets, no matter how close they are one from another, they do not disturb the life on each other's surface. Our strings, provided that they are flat strings, i.e, straight strings – if there were wiggles on the strings, then these wiggles would emit gravitational waves and they would affect one each other in this way – if they are straight, as I demand in this theory, then there is no direct effect of one string on the other. There is a topological effect – one string can rotate around the other, after less than three hundred sixty degrees. But that is, in a sense, a global effect. And that is also present in two plus one dimensions. Every particle has a conical singularity around it, so the other particle can go around it by rotating less than three hundred sixty degrees. But that is global, not local. So two strings do not affect each other at all, no matter how close they are to each other. Only when they try to get through each other, only then they affect each other.

- *Tziveloglou:*

You have presented to us, in the two plus one dimensional case, the setup of a problem. Then you try to run a problem, where you let a set of particles evolve. You obtain a specific result – a big crunch. I would like to ask, given the optimism that for the string like defects if we increase the number of strings that are produced after collisions you might get something consistent, can you speculate about what will a similar program give us in three plus one dimensions? A specific result which you would expect from this theory, in general?

- *'t Hooft:*

This program was very simple in 2 + 1 dimensions. As usual, we start with a Big Bang, with particles moving outwards (in two dimensions plus one time). They go around each other many times, and it is difficult to draw of course, because these are all topologically nontrivial spaces. These particles always end up in a Big Crunch, in a very complicated way. That is a kind of game you can play. Then you ask, can I do the same in three plus one dimensions?

- *Tziveloglou:*

It seems that when you deal with strings and multiple collisions between strings, you might get a more and more dense configuration of strings, attached to each other, and maybe in an eventual case something like a curved or full of defects spacetime or something like that.

- *'t Hooft:*

This is a picture, imagine a bunch of straight strings, each surrounded by a defect angle and each moving. Then what happens? Well, they cross each other. Whenever they cross, they branch, and they make segments of new strings. That does not seem ever to stop, that gives you a more and more dense situation, smaller and smaller string segments which are then glued together by joints. There is no reason why it should stop, splitting things further, etc. This means that you are right, a simple configuration may evolve into something extremely complicated.

- *Tziveloglou:*

Do you understand what kind of spacetime would emerge from this process?

- *'t Hooft:*

It looks pretty hopeless. There are two problems now. The first problem is what happens at spatial infinity, because these strings all extend to infinity, so what is the boundary condition in space here. Another question is: what happens as time evolves – will this not end up in a gigantic fractal, of some sort. And I suspect yes. In other words, you get shorter and shorter string segments, so that this thing might not behave in an interesting way. This is why I said – OK, this was chapter one, chapter two is to go to crystalline gravity. I am going to make more restrictions here, I am not going to the entire Poincare group – the entire Lorentz group plus translations in space and time – but, I am going to use a discrete subset. You can think of this restriction as of introducing a lattice in space and time, where the lattice length is of order of the Planck length. Then we might again be in business, because there is a shortest length. In spite of this, the crystal might generate more and more defects, but pure crystal regions may also evolve, and maybe this thing could get to some sort of equilibrium. That would be my model, and we have a better chance here that it evolves in a physically more interesting way. That is a hope for the future, to get a decent model out of this.

- *Maier:*

Developing this theory seems to be quite involved. What is your opinion, how long will it take you until you are able to calculate observables?

- *'t Hooft:*

I need a few more graduate students to do the hard work. Apart from that, of course, we need imagination. This is just an attempt, and it may or may not succeed. But if you do not try, you will never succeed. If you do not shoot, you always miss. As you see, I am shooting here.

From Extra-dimensions: Multiple brane scenarios and their contenders

Ignatios Antoniadis*

Department of Physics, CERN - Theory Division, CH-1211 Geneva 23, Switzerland.

February 24, 2009

Abstract

- Motivations, the problem of mass hierarchy, main BSM proposals
- Strings, branes and extra dimensions
- Models of intersecting branes

Contents

*On leave from CPHT (UMR du CNRS 7644, Ecole Polytechnique, 91128 Palaiseau, France

1 Departure from the Standard Model

1.1 Why Beyond the Standard Model?

Almost all research in theoretical high energy physics of the last thirty years has been concentrated on the quest for the theory that will replace the Standard Model (SM) as the proper description of nature at the energies beyond the TeV scale. Thus, it is important to know if this quest is justified or not, given that the SM is a very successful theory that has offered to us some of the most striking agreements between experimental data and theoretical predictions. We will start by briefly examining the reasons for leaving behind such a successful SM. Actually there is a variety of such arguments, both theoretical and experimental, that lead to the undoubtable conclusion that the Standard Model should be only an effective theory of a more fundamental one.

We briefly mention some of the most important arguments. Firstly, the Standard Model does not include gravity. It says absolutely nothing about one of the four fundamental forces of nature. Another is the popular "mass hierarchy" problem. Within the SM, the mass of the higgs particle is extremely sensitive to any new physics at higher energies and its natural value is of order of the Planck mass, if the SM is valid up to that scale. This is several orders of magnitude higher than the electroweak scale implied by experiment. The way to cure this discrepancy within the SM requires an incredible fine tuning of parameters. Also, the SM does not explain why the charges of elementary particles are quantized. Other equally important reasons include that the SM does not describe the dark matter or the dark energy of the universe. It does not explain the observed neutrino masses and oscillations and does not predict any gauge coupling unification, suggested by experiments.

Here, we describe some of these arguments in more detail, starting from gravity. If we follow the intuition that the fundamental theory should describe within the same quantum framework all forces of nature, we need to extend the Standard Model in a way that it consistently includes this force. However this is particularly difficult for a variety of reasons. The most serious is that gravity cannot be quantized consistently as a field theory. Another one is that the new theory has to provide a natural explanation for the apparent weakness of gravity compared to the other three fundamental interactions.

The gravitational force between two particles of mass m is

$$F_{\text{grav}} = G_N \frac{m^2}{r^2}, \qquad G_N^{-1/2} = M_{\text{Planck}} = 10^{19}\ GeV \tag{1}$$

If we compare this with the electric force

$$F_{\text{el}} = \frac{e^2}{r^2} \tag{2}$$

for, say the proton, we get

$$\frac{F_{\text{grav}}}{F_{\text{el}}} = \frac{G_N m_{\text{proton}}^2}{e^2} \simeq 10^{-40} \tag{3}$$

So gravity is an extremely weak force at low energies (small masses - large distances) even compared to the weak force as is shown in the table below.

force	range	intensity of 2 protons	intensity at 10^{-16} cm
Gravitation	∞	10^{-38}	10^{-30}
Electromagnetic	∞	10^{-2}	10^{-2}
Weak (radioactivity β)	10^{-15} cm	10^{-5}	10^{-2}
Strong (nuclear forces)	10^{-12} cm	1	10^{-1}

The distance at which gravitation becomes comparable to the other interactions is 10^{-33} cm, the Planck length. The energy that corresponds to this length is of order the Planck mass and is $\sim 10^{15}\times$ the LHC energy! Thus, another great difficulty is that even if we manage to construct such a model, its phenomenological verification will be extremely difficult.

Next, we describe the mass hierarchy problem. The mass of the higgs particle is very sensitive to high energy physics. For example, the main one loop radiative corrections from the virtual exchange of top quarks and the higgs particle itself are

$$\mu_{\text{eff}}^2 = \mu_{\text{bare}}^2 + \left(\frac{\lambda}{8\pi^2} - \frac{3\lambda_t^2}{8\pi^2}\right)\Lambda^2 + \cdots \tag{4}$$

where λ is the higgs quartic coupling, λ_t is the top Yukawa coupling and Λ is the ultraviolet (UV) cutoff, set by the scale where new physics appears. From the masses of all other particles and for reasons that we will explain later, we know that the higgs mass should be at the weak scale. Requiring the validity of the Standard Model at energies $\Lambda \gg \mathcal{O}(100)$GeV imposes an "unnatural" order by order fine tuning between the bare mass parameter μ_{bare}^2 in the Lagrangian and the radiative corrections.

For example, for $\Lambda \sim \mathcal{O}(M_{\text{Planck}}) \sim 10^{19}$ GeV and a loop factor of the order of 10^{-2}, we get $\mu_{\text{1-loop}}^2 \sim 10^{-2}\times 10^{38} = \pm 10^{36}$ (GeV)^2. Thus, we need $\mu_{\text{bare}}^2 \sim \mp 10^{36}$ (GeV)^2+10^4 (GeV)^2. An incredible adjustment is then required, at the level of 1 part per 10^{32}: $\mu_{\text{bare}}^2/\mu_{\text{1-loop}}^2 = -1\pm 10^{-32}$. Even more, at the next and all higher orders, a new adjustment is required. The correction will be of the order of $\mathcal{O}(100)$ GeV only beyond 17 loops: $(10^{-2})^N \times 10^{38} \leq 10^4$ $\Rightarrow N \geq 17$ loops. If we want to avoid these unnatural adjustments we need $10^{-2}\Lambda^2 \leq 10^4$ (GeV)^2 $\Rightarrow \Lambda \leq 1$ TeV. The resolution of the mass hierarchy problem should reside within the energy reach of LHC!

Another open question is the charge quantization: all observed color singlet states have integer electric charges. The symmetry group of the SM is $SU(3) \times SU(2)_L \times U(1)_Y$ and the electric charge given by: $Q = T_3 + Y$. The representations of the symmetry group of all fields are:

$$\begin{aligned} q &= (3,2)_{1/6} & \qquad q = \begin{pmatrix} u_{2/3} \\ d_{-1/3} \end{pmatrix} \\ u^c &= (\bar{3},1)_{-2/3} \\ d^c &= (\bar{3},1)_{1/3} \\ \ell &= (1,2)_{-1/2} & \qquad \ell = \begin{pmatrix} \nu_0 \\ e_{-1} \end{pmatrix} \\ e^c &= (1,1)_1 \end{aligned}$$

where the subscripts denote the hypercharges in the left column and the electric charges in the right one. However, the SM does not tell us anything about this choice of the hypercharges, that guarantees the observed electric charge quantization.

Also, the experimental indication of gauge coupling unification is a feature of nature that asks for an explanation. The renormalization group evolution of the three SM gauge couplings $\alpha_i = \frac{g_i^2}{4\pi}$ is given by:

$$\frac{d\alpha_i}{d\ln Q} = -\frac{b_i}{2\pi}\alpha_i^2 \quad \Rightarrow \quad \alpha_i^{-1}(Q) = \alpha_i^{-1}(Q_0) - \frac{b_i}{2\pi}\ln\frac{Q}{Q_0} \tag{5}$$

where Q is the energy scale and b_i are the one-loop beta-function coefficients. The extrapolation of the low energy experimental data at high energies under the 'desert' assumption indicates an approximate unification of all couplings at energies of order $M_{GUT} \simeq 10^{15}-10^{16}$GeV. By doing a more precise analysis however, using the current experimental precision, on finds that the Standard Model fails to predict such a unification.

One more reason for going beyond the Standard Model is the existence of dark matter. Astronomical observations tell us that the ordinary baryonic matter is only a tiny fraction of the energy of the universe. There are observations that point towards the fact that a kind of non-luminous matter is out there and that is actually much more abundant than baryonic matter, consisting of around 25% of the total energy density of the universe. A natural explanation is that dark matter consists of a new kind of particles that are stable, massive at the electroweak scale and weakly interacting.

We could mention many other important reasons that drive us beyond the SM, like neutrino masses, dark energy and so on, but we will stop here. Even this brief and incomplete presentation shows that there are a lot of fundamental questions that the SM cannot address so the search for a new fundamental theory is definitely justified.

Modifying the SM though is not as easy as it might seem. If we try to identify the scale Λ of new physics by looking at what energies there are deviations from the SM predictions by observations, we will be disappointed. With the exception of dark matter and the higgs mass, all experimental indications suggest that Λ is very high. For example, the "see-saw" mechanism for the neutrino masses gives a very large scale: $m_\nu \sim \frac{v^2}{M} \sim 10^{-2}$ eV $\Rightarrow M \sim 10^{13}$ GeV, where v is the vacuum expectation value (VEV) of the SM higgs. Similarly, the scale of gauge coupling unification is even higher: $M_{\rm GUT} \sim 10^{16}$ GeV.

Even more, modifying the Standard Model in a way that solves the hierarchy problem is highly restricted by numerous experiments. Any new interactions could violate various Standard Model predictions that have been experimentally tested with good accuracy, such as lepton and baryon number conservation, absence of flavor changing neutral currents (FCNC), CP violation etc. For example, theories with baryon number violating operators like

$$\frac{1}{\Lambda_B^2}\left(\bar{q}\gamma^\mu u^c\right)\left(\bar{\ell}\gamma_\mu d^c\right) \tag{6}$$

can lead to proton decay:

$$p = [uud] \longrightarrow (e^+[\bar{d}d]) = e^+\pi^0, \qquad uu \longrightarrow e^+\bar{d} \tag{7}$$

The bound from the Superkamiokande experiment on the proton lifetime of $\tau_p \geq 2.6 \times 10^{33}$ years highly restricts the allowed scale: $\Lambda_B \geq 10^{15}$ GeV.

Similarly for the lepton number; in this case, the lowest dimensional operator that violates lepton number has dimension five:

$$\frac{1}{\Lambda_L}\left(\ell H^*\right)^2 \tag{8}$$

where H is the Higgs field. This operator leads in particular to Majorana neutrino masses. However, as we said before, neutrino masses are very low, $m_\nu = \dfrac{v^2}{\Lambda_L} \leq 0.1$ eV. With a VEV for the higgs at $\mathcal{O}(100)$ GeV, the scale Λ_L is also constrained to very high values: $\Lambda_L \geq 10^{14}$ GeV. Finally, FCNC operators like

$$\frac{1}{\Lambda_F^2}\left(\bar{d}_L\gamma_\mu s_L\right)^2 \tag{9}$$

are restricted by the measured value of $K_0 - \overline{K}_0$ mixing to be $\Lambda_F \geq 10^6$ GeV.

It is very challenging to construct a model that provides us with a solution to the theoretical problems mentioned above and at the same time to comply with all experimental requirements.

1.2 Directions Beyond the Standard Model

One way to encapsulate in the low energy regime the effects of a more fundamental theory is to use effective operators. These can be produced by integrating out heavy states in the fundamental theory. However, we can always write down generic effective operators even when that theory is unknown. We just require Lorentz symmetry and gauge invariance of the low energy effective action. These terms need not be renormalizable because by definition, the validity of an effective theory holds only up to the scale where the high energy physics appears.

Here we show an example of integrating out a heavy gauge boson of mass M

$$-\frac{1}{4}F_{\mu\nu}^2 - \frac{1}{2}M^2 A_\mu^2 + g\bar{\psi}\not{A}\psi \quad \Rightarrow \quad \frac{g^2}{M^2}\left(\bar{\psi}\gamma^\mu\psi\right)^2 \tag{10}$$

It gives rise to a six-dimensional non-renormalizable four-fermion effective operator. This is for instance the case in the popular Fermi theory, obtained by integrating out at low energies $E << m_W$ the heavy $SU(2)$ gauge bosons $W^{\pm}$:

$$\mathcal{L}_{\text{eff}} = G_F \left(\bar{q}_L \gamma_\mu q_L + \bar{\ell}_L \gamma_\mu \ell_L \right)^2 + \cdots \tag{11}$$

with $G_F = g_2^2/m_W^2$.

Generally, effective field theory parametrizes our ignorance of the high energy physics by introducing local effective operators O_n^i of dim $(4+n)$

$$\mathcal{L}_{\text{eff}} = \mathcal{L}_{\text{SM}} + \sum_i \frac{c_n^i}{M^n} O_n^i \tag{12}$$

for $E \ll M$; M is the new physics scale where the heavy states become dynamical. If this is not too far away from the electroweak scale, the lowest dimensional (in powers of M) operators O_n^i can significantly affect the low energy physics. They can shift the mass of the higgs or violate symmetries of the Standard Model such as baryon number, lepton number, induce FCNC and important CP violation, etc. as we mentioned previously.

Effective field theory offers a systematic but blind way to explore the physics beyond the Standard Model. The alternative is to propose a totally new theory and study its consequences. There are several such proposals. They all cure some major weaknesses of the Standard Model but they also have their own drawbacks. Keeping the mass hierarchy as guidance, the following proposals exist so far:

1. Compositeness

2. Symmetry:
   - Supersymmetry
   - Higgs as pseudo-goldstone boson, little Higgs
   - conformal
   - Higgs as component of a higher-dimensional gauge field

3. Low UV cutoff:
   - low scale gravity $\Rightarrow$ · large extra dimensions
     · warped extra dimensions
   - DGP localized gravity
   - low string scale $\Rightarrow$ · low scale gravity
     · ultra weak string coupling
   - large N degrees of freedom
   - higgsless models

4. Live with the hierarchy:
   - landscape of vacua, environmental selection
   - split supersymmetry

We will briefly describe only the first two and later we will focus on strings and extra dimensions.

The first proposal for physics beyond the SM is "compositeness". This idea postulates the existence of new strong dynamics at the energy range of TeV. The higgs field which breaks the electroweak symmetry is then a bound state of fermion bilinears just like the pions that break chiral symmetry in QCD. A concrete model realizing this idea is called "technicolor". However, most technicolor models have conflicts with electroweak precision measurements, leading in particular to the appearance of FCNC's, and are generally disfavored.

Another idea is to introduce a new symmetry. This provides a natural way to keep the mass of the higgs small and solve the mass hierarchy. After all, there are various examples where symmetries have been used to naturally obtain small masses for fermions and bosons. In the case of fermions, it is the chiral symmetry as it can be seen in the following simple Lagrangian

$$\mathcal{L}_F = i\bar{\psi}_L \not{D} \psi_L + i\bar{\psi}_R \not{D} \psi_R + m\left(\bar{\psi}_L \psi_R + \bar{\psi}_R \psi_L\right) \tag{13}$$

When $m = 0$ there is an enhanced symmetry $\psi_L \to e^{i\theta_1}\psi_L; \quad \psi_R \to e^{i\theta_2}\psi_R$ which breaks when the mass terms are included. Thus, imposing this symmetry on our model, one forbids any mass terms in the Lagrangian. Moreover, in their presence, since chiral symmetry is "softly" broken, radiative corrections behave like $\delta m \propto g^2 m$ where g is the gauge coupling (e.g. in QED with $\psi \equiv$ electron and $g \equiv e$). An alternative reason to understand why there is no $g^2\Lambda$ term, is because in a relativistic quantum field theory, any linear divergence cancels between the electron and positron contributions.

The corresponding symmetry for vector bosons is the gauge invariance. Consider for instance a massive abelian vector field:

$$\mathcal{L}_V = -\frac{1}{4}F_{\mu\nu}^2 - \frac{1}{2}m^2 A_\mu^2 \tag{14}$$

We see again that $m = 0$ respects $A_\mu \to A_\mu + \partial_\mu \omega$. This is for instance how the gauge invariance of QED forbids any mass for the photon.

Is there a symmetry that can protect scalar masses, too? We can introduce for example a shift symmetry, implying that the Lagrangian should depend only through derivatives of the scalar field:

$$\phi \to \phi + c \quad \Rightarrow \quad \mathcal{L}\left(\partial_\mu \phi\right) = \frac{1}{2}(\partial_\mu \phi)^2 + \frac{1}{\Lambda^4}\left[(\partial_\mu \phi)^2\right]^2 + \cdots \tag{15}$$

This is the case for a Goldstone boson without potential and in particular it does not hold when a quartic coupling is included. Scenarios like the "Little Higgs" take advantage of this property and treat the higgs as a pseudo-goldstone boson. In this case, the shift symmetry is broken by new gauge interactions, that generate quartic higgs interactions at higher orders.

Another possibility is to use scale invariance:

$$x^\mu \to a x^\mu \quad \varphi_d \to a^{-d}\varphi(ax) \tag{16}$$

where d is the conformal dimension fo the field φ_d and $d = 1$ for a scalar field. This symmetry allows a quartic coupling but in a renormalizable theory is hardly broken by radiative corrections. In order to take advantage of this symmetry, the Standard Model should be embedded in a conformally invariant theory at the TeV scale.

Finally, we can use a different strategy by postulating a new symmetry that connects scalars to fermions or gauge fields:

$$\delta\phi = \xi\psi \quad \text{or} \quad \delta\phi = \epsilon_\mu A^\mu \tag{17}$$

Then the chiral or gauge invariance that protects the fermion or the gauge boson masses should also protect the masses of their scalar partners. This idea leads us to supersymmetry or to theories with extra dimensions, correspondingly. Indeed, the second case leads to the identification of the higgs field with an internal component of a higher-dimensional gauge field. In the following we will concentrate on strings and extra dimensions.

2 Strings and extra dimensions

The appropriate and most convenient framework for low energy supersymmetry and grand unification is the perturbative heterotic string. Indeed, in this theory, gravity and gauge interactions have the same origin, as massless modes of the closed heterotic string, and they are unified at the string scale M_s. As a result, the Planck mass M_P is predicted to be proportional to M_s:

$$M_P = M_s/g\,, \tag{18}$$

where g is the gauge coupling. In the simplest constructions all gauge couplings are the same at the string scale, given by the four-dimensional (4d) string coupling, and thus no grand unified group is needed for unification. In our conventions $\alpha_{\rm GUT} = g^2 \simeq 0.04$, leading to a discrepancy between the string and grand unification scale $M_{\rm GUT}$ by almost two orders of magnitude. Explaining this gap introduces in general new parameters or a new scale, and the predictive power is essentially lost. This is the main defect of this framework, which remains though an open and interesting possibility [1].

The other other perturbative framework that has been studied extensively in the more recent years is type I string theory with D-branes. Unlike in the heterotic string, gauge and gravitational interactions have now different origin. The latter are described again by closed strings, while the former emerge as excitations of open strings with endpoints confined on D-branes [2]. This leads to a braneworld description of our universe, which should be localized on a hypersurface, i.e. a membrane extended in p spatial dimensions, called p-brane (see Fig. 1). Closed strings propagate in all nine dimensions of string theory: in those extended along the p-brane, called parallel, as well as in the transverse ones. On the contrary, open strings are attached on the p-brane. Obviously, our p-brane world must have at least the three known dimensions of space. But it may contain more: the extra $d_\parallel = p - 3$ parallel dimensions must have a finite size, in order to be unobservable at present energies, and can be as large as TeV$^{-1} \sim 10^{-18}$ m [3]. On the other hand, transverse dimensions interact with us only gravitationally and experimental bounds are much weaker: their size should be less than about 0.1 mm [4]. In the following, I review the main properties and experimental signatures of low string scale models [5, 6].

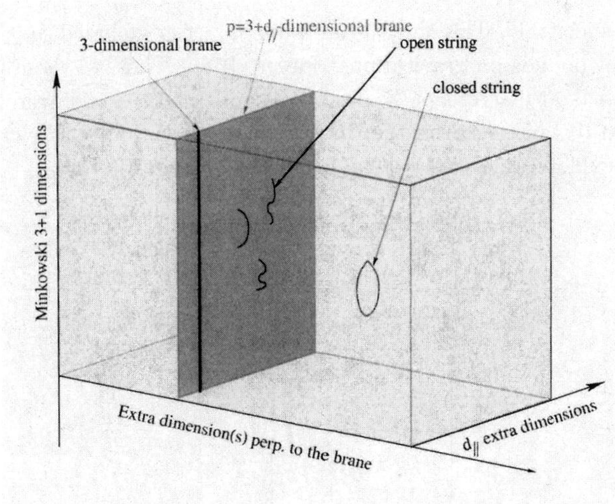

Figure 1: In the type I string framework, our Universe contains, besides the three known spatial dimensions (denoted by a single blue line), some extra dimensions ($d_{\parallel} = p - 3$) parallel to our world p-brane (green plane) where endpoints of open strings are confined, as well as some transverse dimensions (yellow space) where only gravity described by closed strings can propagate.

2.1 Framework of low scale strings

In type I theory, the different origin of gauge and gravitational interactions implies that the relation between the Planck and string scales is not linear as (18) of the heterotic string. The requirement that string theory should be weakly coupled, constrain the size of all parallel dimensions to be of order of the string length, while transverse dimensions remain unrestricted. Assuming an isotropic transverse space of $n = 9 - p$ compact dimensions of common radius $R_{\perp}$, one finds:

$$M_P^2 = \frac{1}{g^4} M_s^{2+n} R_{\perp}^n \,, \qquad g_s \simeq g^2 \,. \tag{19}$$

where g_s is the string coupling. It follows that the type I string scale can be chosen hierarchically smaller than the Planck mass [7, 5] at the expense of introducing extra large transverse dimensions felt only by gravity, while keeping the string coupling small [5]. The weakness of 4d gravity compared to gauge interactions (ratio M_W/M_P) is then attributed to the largeness of the transverse space $R_{\perp}$ compared to the string length $l_s = M_s^{-1}$.

An important property of these models is that gravity becomes effectively $(4 + n)$-dimensional with a strength comparable to those of gauge interactions at the string scale. The first relation of Eq. (19) can be understood as a consequence of the $(4+n)$-dimensional Gauss law for gravity, with

$$M_*^{(4+n)} = M_s^{2+n}/g^4 \tag{20}$$

the effective scale of gravity in $4 + n$ dimensions. Taking $M_s \simeq 1$ TeV, one finds a size for the extra dimensions $R_{\perp}$ varying from 10^8 km, .1 mm, down to a Fermi for $n = 1, 2$, or 6 large dimensions, respectively. This shows that while $n = 1$ is excluded, $n \geq 2$ is allowed by present experimental bounds on gravitational forces [4, 8]. Thus, in these models, gravity appears to us very weak at macroscopic scales because its intensity is spread in the "hidden" extra dimensions. At distances shorter than $R_\perp$, it should deviate from Newton's law, which may be possible to explore in laboratory experiments (see Fig. 2).

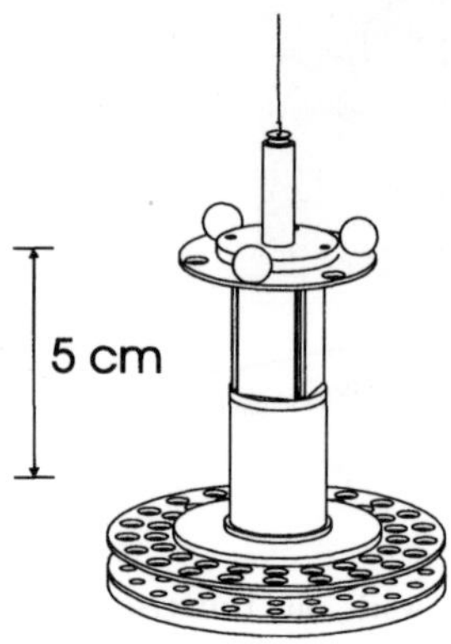

Figure 2: Torsion pendulum that tested Newton's law at 55 μm.

The main experimental implications of TeV scale strings in particle accelerators are of three types, in correspondence with the three different sectors that are generally present: (i) new compactified parallel dimensions, (ii) new extra large transverse dimensions and low scale quantum gravity, and (iii) genuine string and quantum gravity effects. On the other hand, there exist interesting implications in non accelerator table-top experiments due to the exchange of gravitons or other possible states living in the bulk.

3 Experimental implications in accelerators

3.1 World-brane extra dimensions

In this case $RM_s \gtrsim 1$, and the associated compactification scale $R_\parallel^{-1}$ would be the first scale of new physics that should be found increasing the beam energy [3, 9, 10]. There are several reasons for the existence of such dimensions. It is a logical possibility, since out of the six extra dimensions of string theory only two are needed for lowering the string scale, and thus the effective p-brane of our world has in general $d_\parallel \equiv p - 3 \leq 4$. Moreover, they can be used to address several physical problems in braneworld models, such as obtaining different SM gauge couplings, explaining fermion mass hierarchies due to different localization points of quarks and leptons in the extra dimensions, providing calculable mechanisms of supersymmetry breaking, etc.

The main consequence is the existence of Kaluza-Klein (KK) excitations for all SM particles that propagate along the extra parallel dimensions. Their masses are given by:

$$M_m^2 = M_0^2 + \frac{m^2}{R_\parallel^2} \quad ; \quad m = 0, \pm 1, \pm 2, \ldots \tag{21}$$

where we used $d_\parallel = 1$, and M_0 is the higher dimensional mass. The zero-mode $m = 0$ is identified with the 4d state, while the higher modes have the same quantum numbers with the lowest one, except for their mass given in (21). There are two types of experimental signatures of such dimensions [9, 11, 12]: (i) virtual exchange of KK excitations, leading to deviations in cross-sections compared to the SM prediction, that can be used to extract bounds on the compactification scale; (ii) direct production of KK modes.

On general grounds, there can be two different kinds of models with qualitatively different signatures depending on the localization properties of matter fermion fields. If the latter are localized in 3d brane intersections, they do not have excitations and KK momentum is not conserved because of the breaking of translation invariance in the extra dimension(s). KK modes of gauge bosons are then singly produced giving rise to generally strong bounds on the compactification scale and new resonances that can be observed in experiments. Otherwise, they can be produced only in pairs due to the KK momentum conservation, making the bounds weaker but the resonances difficult to observe.

When the internal momentum is conserved, the interaction vertex involving KK modes has the same 4d tree-level gauge coupling. On the other hand, their couplings to localized matter have an exponential form factor suppressing the interactions of heavy modes. This form factor can be viewed as the fact that the branes intersection has a finite thickness. For instance, the coupling of the KK excitations of gauge fields $A^\mu(x,y) = \sum_m A_m^\mu \exp i\frac{my}{R_\parallel}$ to the charge density $j_\mu(x)$ of massless localized fermions is described by the effective action [13]:

$$\int d^4x \sum_m e^{-\ln 16 \frac{m^2 l_s^2}{2R_\parallel^2}} j_\mu(x)\, A_m^\mu(x)\,. \tag{22}$$

After Fourier transform in position space, it becomes:

$$\int d^4x\, dy \, \frac{1}{(2\pi \ln 16)^2} e^{-\frac{y^2 M_s^2}{2\ln 16}} j_\mu(x)\, A^\mu(x,y)\,, \tag{23}$$

from which we see that localized fermions form a Gaussian distribution of charge with a width $\sigma = \sqrt{\ln 16}\, l_s \sim 1.66\, l_s$.

To simplify the analysis, let us consider first the case $d_\parallel = 1$ where some of the gauge fields arise from an effective 4-brane, while fermions are localized states on brane intersections. Since the corresponding gauge couplings are reduced by the size of the large dimension $R_\parallel M_s$ compared to the others, one can account for the ratio of the weak to strong interactions strengths if the $SU(2)$ brane extends along the extra dimension, while $SU(3)$ does not. As a result, there are 3 distinct cases to study [12], denoted by (t,l,l), (t,l,t) and (t,t,l), where the three positions in the brackets correspond to the three SM gauge group factors $SU(3) \times SU(2) \times U(1)$ and those with l (longitudinal) feel the extra dimension, while those with t (transverse) do not.

In the (t,l,l) case, there are KK excitations of $SU(2) \times U(1)$ gauge bosons: $W_{\pm}^{(m)}$, $\gamma^{(m)}$ and $Z^{(m)}$. Performing a χ^2 fit of the electroweak observables, one finds that if the Higgs is a bulk state (l), $R_{\parallel}^{-1} \gtrsim 3.5$ TeV [14]. This implies that LHC can produce at most the first KK mode. Different choices for localization of matter and Higgs fields lead to bounds, lying in the range $1-5$ TeV [14].

In addition to virtual effects, KK excitations can be produced on-shell at LHC as new resonances [11] (see Fig. 3). There are two different channels, neutral Drell–Yan processes

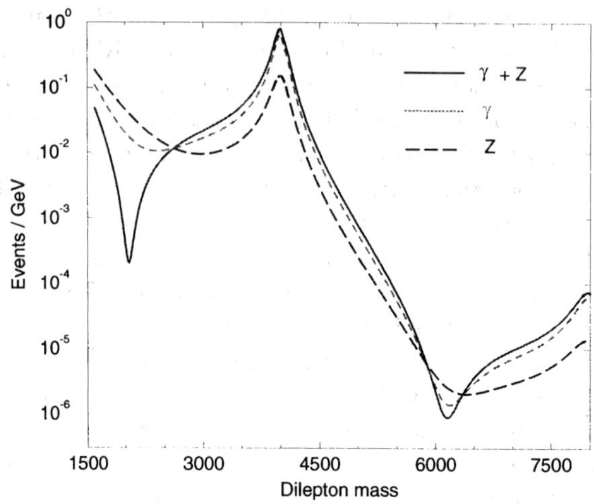

Figure 3: Production of the first KK modes of the photon and of the Z boson at LHC, decaying to electron-positron pairs. The number of expected events is plotted as a function of the energy of the pair in GeV. From highest to lowest: excitation of $\gamma + Z$, γ and Z.

$pp \to l^+l^-X$ and the charged channel $l^{\pm}\nu$, corresponding to the production of the KK modes $\gamma^{(1)}, Z^{(1)}$ and $W_{\pm}^{(1)}$, respectively. The discovery limits are about 6 TeV, while the exclusion bounds 15 TeV. An interesting observation in the case of $\gamma^{(1)} + Z^{(1)}$ is that interferences can lead to a "dip" just before the resonance. There are some ways to distinguish the corresponding signals from other possible origin of new physics, such as models with new gauge bosons. In fact, in the (t,l,l) and (t,l,t) cases, one expects two resonances located practically at the same mass value. This property is not shared by most of other new gauge boson models. Moreover, the heights and widths of the resonances are directly related to those of SM gauge bosons in the corresponding channels.

In the (t,l,t) case, only the $SU(2)$ factor feels the extra dimension and the limits set by the KK states of $W^{\pm}$ remain the same. On the other hand, in the (t,t,l) case where only $U(1)_Y$ feels the extra dimension, the limits are weaker and the exclusion bound is around 8 TeV. In addition to these simple possibilities, brane constructions lead often to cases where part of $U(1)_Y$ is t and part is l. If $SU(2)$ is l the limits come again from $W^{\pm}$, while if it is t

Table 1: Limits on $R_\perp$ in mm.

Experiment	$n=2$	$n=4$	$n=6$
Collider bounds			
LEP 2	5×10^{-1}	2×10^{-8}	7×10^{-11}
Tevatron	5×10^{-1}	10^{-8}	4×10^{-11}
LHC	4×10^{-3}	6×10^{-10}	3×10^{-12}
NLC	10^{-2}	10^{-9}	6×10^{-12}
Present non-collider bounds			
SN1987A	3×10^{-4}	10^{-8}	6×10^{-10}
COMPTEL	5×10^{-5}	-	-

then it will be difficult to distinguish this case from a generic extra $U(1)'$. A good statistics would be needed to see the deviation in the tail of the resonance as being due to effects additional to those of a generic $U(1)'$ resonance. Finally, in the case of two or more parallel dimensions, the sum in the exchange of the KK modes diverges in the limit $R_\parallel M_s >> 1$ and needs to be regularized using the form factor (22). Cross-sections become bigger yielding stronger bounds, while resonances are closer implying that more of them could be reached by LHC.

On the other hand, if all SM particles propagate in the extra dimension (called universal)[1], KK modes can only be produced in pairs and the lower bound on the compactification scale becomes weaker, of order of 300-500 GeV. Moreover, no resonances can be observed at LHC, so that this scenario appears very similar to low energy supersymmetry. In fact, KK parity can even play the role of R-parity, implying that the lightest KK mode is stable and can be a dark matter candidate in analogy to the LSP [15].

3.2 Extra large transverse dimensions

The main experimental signal is gravitational radiation in the bulk from any physical process on the world-brane. In fact, the very existence of branes breaks translation invariance in the transverse dimensions and gravitons can be emitted from the brane into the bulk. During a collision of center of mass energy $\sqrt{s}$, there are $\sim (\sqrt{s}R_\perp)^n$ KK excitations of gravitons with tiny masses, that can be emitted. Each of these states looks from the 4d point of view as a massive, quasi-stable, extremely weakly coupled (s/M_P^2 suppressed) particle that escapes from the detector. The total effect is a missing-energy cross-section roughly of order:

$$\frac{(\sqrt{s}R_\perp)^n}{M_P^2} \sim \frac{1}{s}\left(\frac{\sqrt{s}}{M_s}\right)^{n+2}. \tag{24}$$

Explicit computation of these effects leads to the bounds given in Table 1. However, larger radii are allowed if one relaxes the assumption of isotropy, by taking for instance two large dimensions with different radii.

[1]Although interesting, this scenario seems difficult to be realized, since 4d chirality requires non-trivial action of orbifold twists with localized chiral states at the fixed points.

Fig. 4 shows the cross-section for graviton emission in the bulk, corresponding to the process $pp \to jet + graviton$ at LHC, together with the SM background [16]. For a given

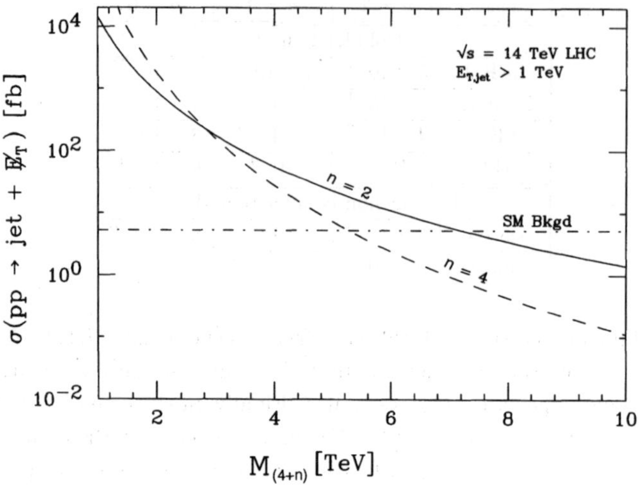

Figure 4: Missing energy due to graviton emission at LHC, as a function of the higher-dimensional gravity scale M_*, produced together with a hadronic jet. The expected cross-section is shown for $n = 2$ and $n = 4$ extra dimensions, together with the SM background.

value of M_s, the cross-section for graviton emission decreases with the number of large transverse dimensions, in contrast to the case of parallel dimensions. The reason is that gravity becomes weaker if there are more dimensions because there is more space for the gravitational field to escape. There is a particular energy and angular distribution of the produced gravitons that arise from the distribution in mass of KK states of spin-2. This can be contrasted to other sources of missing energy and might be a smoking gun for the extra dimensional nature of such a signal.

In Table 1, there are also included astrophysical and cosmological bounds. Astrophysical bounds [17, 18] arise from the requirement that the radiation of gravitons should not carry on too much of the gravitational binding energy released during core collapse of supernovae. In fact, the measurements of Kamiokande and IMB for SN1987A suggest that the main channel is neutrino fluxes. The best cosmological bound [19] is obtained from requiring that decay of bulk gravitons to photons do not generate a spike in the energy spectrum of the photon background measured by the COMPTEL instrument. Bulk gravitons are expected to be produced just before nucleosynthesis due to thermal radiation from the brane. The limits assume that the temperature was at most 1 MeV as nucleosynthesis begins, and become stronger if temperature is increased.

3.3 String effects

At low energies, the interaction of light (string) states is described by an effective field theory. Their exchange generates in particular four-fermion operators that can be used to extract independent bounds on the string scale. In analogy with the bounds on longitudinal extra dimensions, there are two cases depending on the localization properties of matter fermions. If they come from open strings with both ends on the same stack of branes, exchange of massive open string modes gives rise to dimension eight effective operators, involving four fermions and two space-time derivatives [20, 13]. The corresponding bounds on the string scale are then around 500 GeV. On the other hand, if matter fermions are localized on non-trivial brane intersections, one obtains dimension six four-fermion operators and the bounds become stronger: $M_s \gtrsim 2 - 3$ TeV [13, 6]. At energies higher than the string scale, new spectacular phenomena are expected to occur, related to string physics and quantum gravity effects, such as possible micro-black hole production [21, 22, 23]. Particle accelerators would then become the best tools for studying quantum gravity and string theory.

Direct production of string resonances in hadron colliders leads generically to a universal deviation from Standard Model in jet distribution [24]. In particular, the first Regge excitation of the gluon has spin 2 and a width an order of magnitude lower than the string scale, leading to a characteristic peak in dijet production; similarly, the first excitations of quarks have spin 3/2. The dijet cross-section is shown in Fig. 5 for LHC energies. The

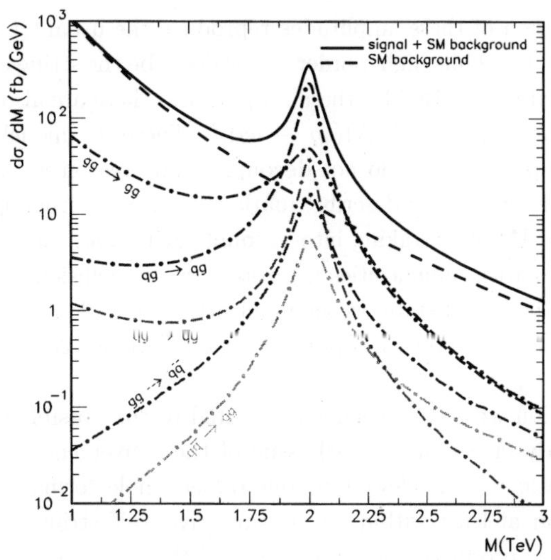

Figure 5: Production of the first Regge excitations at LHC in the dijet channel, for a string scale $M_s = 2$ TeV. The cross-section is plotted as a function of the dijet invariant mass M.

reason for the universal behavior is that tree N-point open superstring amplitudes involving at most two fermions and gluons are completely model independent from the details of the compactification, including the number of supersymmetries that are left unbroken in four dimensions (even if all are broken). Such tree-level amplitudes do not receive contributions from KK, string winding or closed string graviton modes, but are given as a universal sum over exchanges of open string Regge excitations lying in Regge trajectories with masses

$$M_n^2 = M_s^2 n \quad ; \quad n = 0, 1, \ldots \tag{25}$$

and maximal spin $n+1$.

The relevant partonic cross-sections for dijet production, involving at most two quarks are $|\mathcal{M}(gg \to gg)|^2$, $|\mathcal{M}(gg \to q\bar{q})|^2$, $|\mathcal{M}(q\bar{q} \to gg)|^2$ and $|\mathcal{M}(qg \to qg)|^2$, where g and q denote gluons and quarks, respectively. They can be obtained from the following two independent processes, up to crossing symmetries [25]:

$$\begin{aligned}
|\mathcal{M}(gg \to gg)|^2 &= g^4 \left(\frac{1}{s^2} + \frac{1}{t^2} + \frac{1}{u^2}\right) \times \left[\frac{9}{4}\left(s^2V_s^2 + t^2V_t^2 + u^2V_u^2\right) - \frac{1}{3}\left(sV_s + tV_t + uV_u\right)^2\right] \\
|\mathcal{M}(gg \to q\bar{q})|^2 &= g^4 \frac{t^2+u^2}{s^2}\left[\frac{1}{6}\frac{1}{tu}(tV_t + uV_u)^2 - \frac{3}{8}V_tV_u\right],
\end{aligned} \tag{26}$$

where s, t, u are the usual Mandelstam variables in string mass units and

$$V_s = -\frac{tu}{s}\, B(t,u) = 1 - \frac{2}{3}\pi^2\, tu + \ldots \qquad V_t : s \leftrightarrow t \qquad V_u : s \leftrightarrow u\,. \tag{27}$$

The low energy expansion of these amplitudes reproduce the usual QCD expressions (upon inserting the color factors), while higher order terms describe the string corrections due to the exchange of Regge excitations. Besides these amplitudes, there are also those involving four quarks, such as $|\mathcal{M}(q\bar{q} \to q\bar{q})|^2$ and $|\mathcal{M}(qq \to qq)|^2$. These are model dependent, because the details of the compactification do not decouple. The reason is that they involve four vertices containing twist fields that describe quark states arising from open strings stretched in brane intersections. However, taking into account QCD color factors, their contribution is suppressed because parton luminosities in proton-proton collisions at TeV energies favor at least one gluon in the initial state, as shown clearly in Fig. 6. As a result, the dominant contribution comes from the model independent cross-sections (26) leading to the effect of Fig. 5.

We finish this section with some comments related to the possible micro-black hole production. Independently on the unresolved issue of the convergence of string perturbation theory in the kinematic region relevant to micro-black hole formation, there is a simple argument showing that at least within the perturbative TeV string framework, the energy threshold for black hole production is far above the LHC reach. Indeed, a string size black hole has a horizon radius $r_H \sim 1$ in string units, while the d-dimensional Newton's constant behaves as $G_N \sim g_s^2$. It follows that the mass of a d-dimensional black hole is [26]:

$$M_{\rm BH} \sim r_H^{d/2-1}/G_N \simeq 1/g_s^2\,. \tag{28}$$

Thus, for a weakly coupled theory, this energy threshold is much higher than the string and the higher dimensional Planck scales M_s and M_* of eq. (20). Comparing this energy threshold

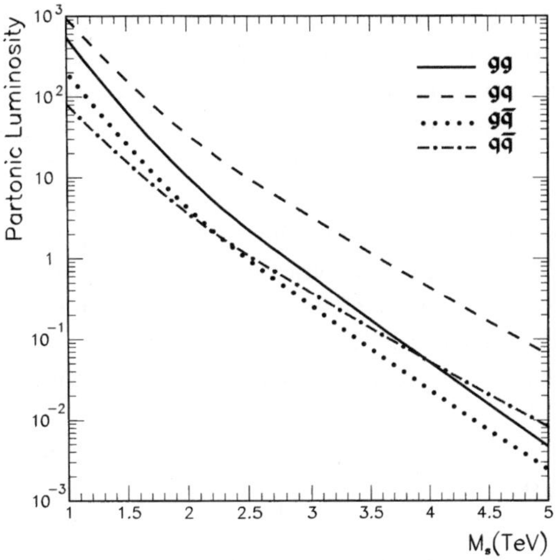

Figure 6: Comparison of parton luminosities in pp collisions at TeV energies.

with the mass of Regge excitations (25), one finds $n \sim 1/g_s^4$ which is actually compatible with the relation one obtains by identifying the black hole entropy $S_{\rm BH} \sim 1/G_N \sim 1/g_s^2$ with the perturbative string entropy $S_{\rm string} \sim \sqrt{n}$. Using now relation (19), and the value of the Standard Model gauge couplings $g_s \simeq g^2 \sim 0.1$, one finds that the energy threshold $M_{\rm BH}$ of micro-black hole production is about four orders of magnitude higher than the string scale, implying that one would produce 10^4 string states before reaching $M_{\rm BH}$.

4 Supersymmetry in the bulk and short range forces

4.1 Sub millimotor forces

Besides the spectacular predictions in accelerators, there are also modifications of gravitation in the sub-millimeter range, which can be tested in "table-top" experiments that measure gravity at short distances. There are three categories of such predictions:

(i) Deviations from the Newton's law $1/r^2$ behavior to $1/r^{2+n}$, which can be observable for $n = 2$ large transverse dimensions of sub-millimeter size. This case is particularly attractive on theoretical grounds because of the logarithmic sensitivity of SM couplings on the size of transverse space [27], that allows to determine the hierarchy [28].

(ii) New scalar forces in the sub-millimeter range, related to the mechanism of supersymmetry breaking, and mediated by light scalar fields φ with masses [29, 5]:

$$m_\varphi \simeq \frac{m_{susy}^2}{M_P} \simeq 10^{-4} - 10^{-6}\ {\rm eV}\,, \tag{29}$$

for a supersymmetry breaking scale $m_{susy} \simeq 1 - 10$ TeV. They correspond to Compton wavelengths of 1 mm to 10 μm. m_{susy} can be either $1/R_\parallel$ if supersymmetry is broken by compactification [29], or the string scale if it is broken "maximally" on our world-brane [5]. A universal attractive scalar force is mediated by the radion modulus $\varphi \equiv M_P \ln R$, with R the radius of the longitudinal or transverse dimension(s). In the former case, the result (29) follows from the behavior of the vacuum energy density $\Lambda \sim 1/R_\parallel^4$ for large $R_\parallel$ (up to logarithmic corrections). In the latter, supersymmetry is broken primarily on the brane, and thus its transmission to the bulk is gravitationally suppressed, leading to (29). For $n = 2$, there may be an enhancement factor of the radion mass by $\ln R_\perp M_s \simeq 30$ decreasing its wavelength by an order of magnitude [28].

The coupling of the radius modulus to matter relative to gravity can be easily computed and is given by:

$$\sqrt{\alpha_\varphi} = \frac{1}{M}\frac{\partial M}{\partial \varphi} \ ; \quad \alpha_\varphi = \begin{cases} \frac{\partial \ln \Lambda_{\rm QCD}}{\partial \ln R} \simeq \frac{1}{3} \ \text{ for } R_\parallel \\ \\ \frac{2n}{n+2} = 1 - 1.5 \text{ for } R_\perp \end{cases} \tag{30}$$

where M denotes a generic physical mass. In the longitudinal case, the coupling arises dominantly through the radius dependence of the QCD gauge coupling [29], while in the case of transverse dimension, it can be deduced from the rescaling of the metric which changes the string to the Einstein frame and depends slightly on the bulk dimensionality ($\alpha = 1 - 1.5$ for $n = 2 - 6$) [28]. Such a force can be tested in microgravity experiments and should be contrasted with the change of Newton's law due the presence of extra dimensions that is observable only for $n = 2$ [4, 8]. The resulting bounds from an analysis of the radion effects are [30]:

$$M_* \gtrsim 6\,\text{TeV}\,. \tag{31}$$

In principle there can be other light moduli which couple with even larger strengths. For example the dilaton, whose VEV determines the string coupling, if it does not acquire large mass from some dynamical supersymmetric mechanism, can lead to a force of strength 2000 times bigger than gravity [31].

(iii) Non universal repulsive forces much stronger than gravity, mediated by possible abelian gauge fields in the bulk [17, 32]. Such fields acquire tiny masses of the order of M_s^2/M_P, as in (29), due to brane localized anomalies [32]. Although their gauge coupling is infinitesimally small, $g_A \sim M_s/M_P \simeq 10^{-16}$, it is still bigger that the gravitational coupling E/M_P for typical energies $E \sim 1$ GeV, and the strength of the new force would be $10^6 - 10^8$ stronger than gravity. This is an interesting region which will be soon explored in micro-gravity experiments (see Fig. 7). Note that in this case supernova constraints impose that there should be at least four large extra dimensions in the bulk [17].

In Fig. 7 we depict the actual information from previous, present and upcoming experiments [8, 28]. The solid lines indicate the present limits from the experiments indicated. The excluded regions lie above these solid lines. Measuring gravitational strength forces at short distances is challenging. The horizontal lines correspond to theoretical predictions, in particular for the graviton in the case $n = 2$ and for the radion in the transverse case. These limits are compared to those obtained from particle accelerator experiments in Ta-

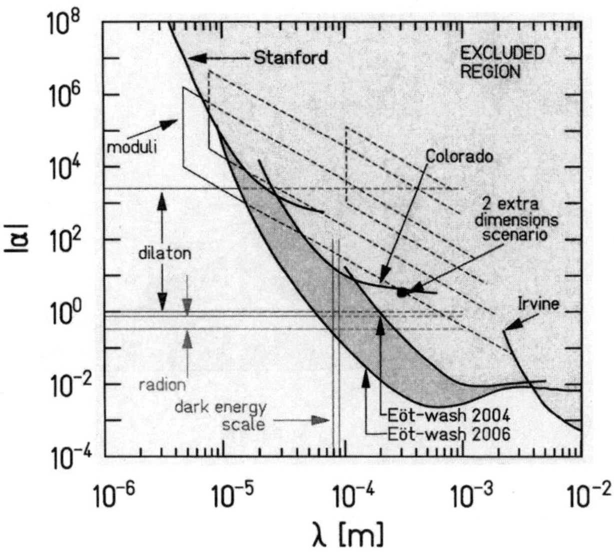

Figure 7: Present limits on new short-range forces (yellow regions), as a function of their range λ and their strength relative to gravity α. The limits are compared to new forces mediated by the graviton in the case of two large extra dimensions, and by the radion.

ble 1. Finally, in Figs. 8 and 9, we display recent improved bounds for new forces at very short distances by focusing on the left hand side of Fig. 7, near the origin [8].

5 Electroweak symmetry breaking

Non-supersymmetric TeV strings offer also a framework to realize gauge symmetry breaking radiatively. Indeed, from the effective field theory point of view, one expects quadratically divergent one-loop contributions to the masses of scalar fields. The divergences are cut off by M_s and if the corrections are negative, they can induce electroweak symmetry breaking and explain the mild hierarchy between the weak and a string scale at a few TeV, in terms of a loop factor [33]. More precisely, in the minimal case of one Higgs doublet H, the scalar potential is:

$$V = \lambda(H^\dagger H)^2 + \mu^2(H^\dagger H)\,, \tag{32}$$

where λ arises at tree-level. Moreover, in any model where the Higgs field comes from an open string with both ends fixed on the same brane stack, it is given by an appropriate truncation of a supersymmetric theory. Within the minimal spectrum of the SM, $\lambda = (g_2^2 + g'^2)/8$, with g_2 and g' the $SU(2)$ and $U(1)_Y$ gauge couplings. On the other hand, μ^2 is generated at one

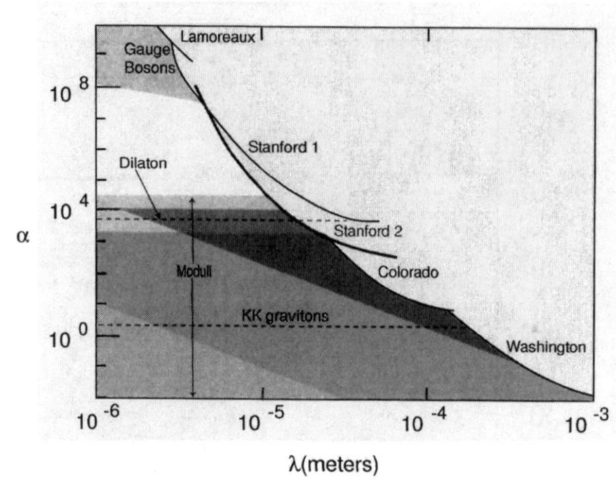

Figure 8: Bounds on non-Newtonian forces in the range 6-20 μm (see S. J. Smullin et al. [8]).

loop:

$$\mu^2 = -\varepsilon^2\, g^2\, M_s^2\,, \tag{33}$$

where ε is a loop factor that can be estimated from a toy model computation and varies in the region $\epsilon \sim 10^{-1} - 10^{-3}$.

Indeed, consider for illustration a simple case where the whole one-loop effective potential of a scalar field can be computed. We assume for instance one extra dimension compactified on a circle of radius $R > 1$ (in string units). An interesting situation is provided by a class of models where a non-vanishing VEV for a scalar (Higgs) field ϕ results in shifting the mass of each KK excitation by a constant $a(\phi)$:

$$M_m^2 = \left(\frac{m + a(\phi)}{R}\right)^2\,, \tag{34}$$

with m the KK integer momentum number. Such mass shifts arise for instance in the presence of a Wilson line, $a = q\oint \frac{dy}{2\pi} gA$, where A is the internal component of a gauge field with gauge coupling g, and q is the charge of a given state under the corresponding generator. A straightforward computation shows that the ϕ-dependent part of the one-loop effective potential is given by [34]:

$$V_{eff} = -Tr(-)^F\,\frac{R}{32\,\pi^{3/2}}\,\sum_n e^{2\pi i n a}\int_0^\infty dl\; l^{3/2} f_s(l)\; e^{-\pi^2 n^2 R^2 l} \tag{35}$$

where $F = 0, 1$ for bosons and fermions, respectively. We have included a regulating function $f_s(l)$ which contains for example the effects of string oscillators. To understand its role we will consider the two limits $R >> 1$ and $R << 1$. In the first case only the $l \to 0$ region contributes to the integral. This means that the effective potential receives sizable contributions only from the infrared (field theory) degrees of freedom. In this limit we would

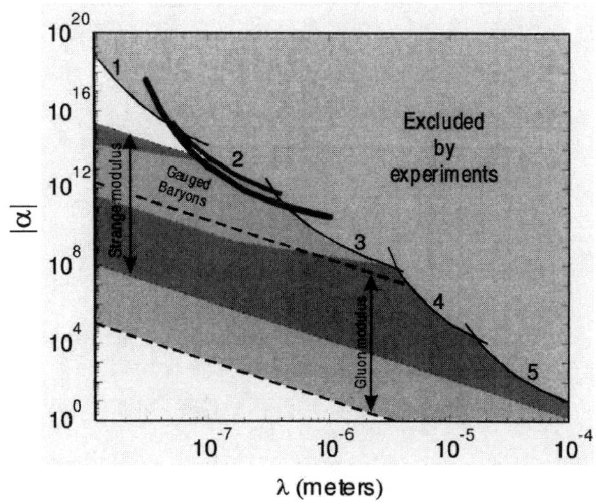

Figure 9: Bounds on non-Newtonian forces in the range of 10-200 nm (see R. S. Decca et al. in Ref. [8]). Curves 4 and 5 correspond to Stanford and Colorado experiments, respectively, of Fig. 8 (see also J C. Long and J. C. Price of Ref. [8]).

have $f_s(l) \to 1$. For example, in the string model considered in [33]:

$$f_s(l) = \left[\frac{1}{4l}\frac{\theta_2}{\eta^3}(il + \frac{1}{2})\right]^4 \to 1 \qquad \text{for} \qquad l \to 0, \tag{36}$$

and the field theory result is finite and can be explicitly computed. As a result of the Taylor expansion around $a = 0$, we are able to extract the one-loop contribution to the coefficient of the term of the potential quadratic in the Higgs field. It is given by a loop factor times the compactification scale [34]. One thus obtains $\mu^2 \sim g^2/R^2$ up to a proportionality constant which is calculable in the effective field theory. On the other hand, if we consider $R \to 0$, which by T-duality corresponds to taking the extra dimension as transverse and very large, the one-loop effective potential receives contributions from the whole tower of string oscillators as appearing in $f_s(l)$, leading to squared masses given by a loop factor times M_s^2, according to eq. (33).

More precisely, from the expression (35), one finds:

$$\varepsilon^2(R) = \frac{1}{2\pi^2}\int_0^\infty \frac{dl}{(2\,l)^{5/2}}\frac{\theta_2^4}{4\eta^{12}}\left(il + \frac{1}{2}\right) R^3 \sum_n n^2 e^{-2\pi n^2 R^2 l}\,, \tag{37}$$

which is plotted in Fig. 10. For the asymptotic value $R \to 0$ (corresponding upon T-duality to a large transverse dimension of radius $1/R$), $\varepsilon(0) \simeq 0.14$, and the effective cut-off for the mass term is M_s, as can be seen from Eq. (33). At large R, $\mu^2(R)$ falls off as $1/R^2$, which is the effective cut-off in the limit $R \to \infty$, as we argued above, in agreement with field theory results in the presence of a compactified extra dimension [35, 29]. In fact, in the limit

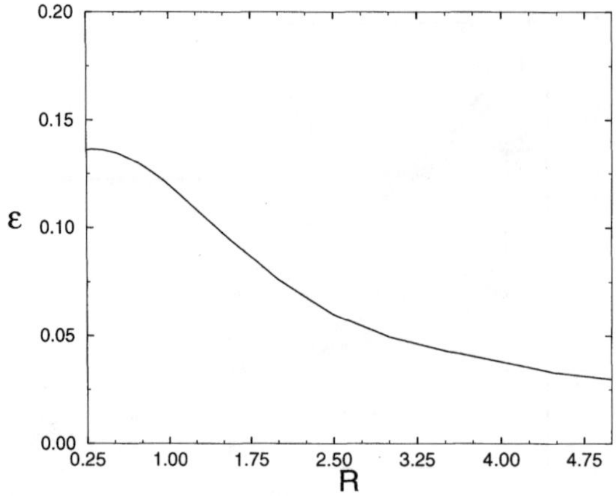

Figure 10: The coefficient ε of the one loop Higgs mass (33).

$R \to \infty$, an analytic approximation to $\varepsilon(R)$ gives:

$$\varepsilon(R) \simeq \frac{\varepsilon_\infty}{M_s R}, \qquad \varepsilon_\infty^2 = \frac{3\,\zeta(5)}{4\,\pi^4} \simeq 0.008\,. \tag{38}$$

The potential (32) has the usual minimum, given by the VEV of the neutral component of the Higgs doublet $v = \sqrt{-\mu^2/\lambda}$. Using the relation of v with the Z gauge boson mass, $M_Z^2 = (g_2^2 + g'^2)v^2/4$, and the expression of the quartic coupling λ, one obtains for the Higgs mass a prediction which is the Minimal Supersymmetric Standard Model (MSSM) value for $\tan\beta \to \infty$ and $m_A \to \infty$: $m_H = M_Z$. The tree level Higgs mass is known to receive important radiative corrections from the top-quark sector and rises to values around 120 GeV. Furthermore, from (33), one can compute M_s in terms of the Higgs mass $m_H^2 = -2\mu^2$:

$$M_s = \frac{m_H}{\sqrt{2}\,g\varepsilon}, \tag{39}$$

yielding naturally values in the TeV range.

6 Standard Model on D-branes

The gauge group closest to the Standard Model one can easily obtain with D-branes is $U(3) \times U(2) \times U(1)$. The first factor arises from three coincident "color" D-branes. An open string with one end on them is a triplet under $SU(3)$ and carries the same $U(1)$ charge for all three components. Thus, the $U(1)$ factor of $U(3)$ has to be identified with *gauged* baryon number. Similarly, $U(2)$ arises from two coincident "weak" D-branes and the corresponding abelian factor is identified with *gauged* weak-doublet number. Finally, an extra $U(1)$ D-brane is necessary in order to accommodate the Standard Model without breaking the baryon number [36]. In principle this $U(1)$ brane can be chosen to be independent of the other two collections with its own gauge coupling. To improve the predictability of the model, we choose to put it on top of either the color or the weak D-branes [37]. In either case, the model has two independent gauge couplings g_3 and g_2 corresponding, respectively, to the gauge groups $U(3)$ and $U(2)$. The $U(1)$ gauge coupling g_1 is equal to either g_3 or g_2.

Let us denote by Q_3, Q_2 and Q_1 the three $U(1)$ charges of $U(3) \times U(2) \times U(1)$, in a self explanatory notation. Under $SU(3) \times SU(2) \times U(1)_3 \times U(1)_2 \times U(1)_1$, the members of a family of quarks and leptons have the following quantum numbers:

$$\begin{array}{ll} Q & (\mathbf{3},\mathbf{2};1,w,0)_{1/6} \\ u^c & (\bar{\mathbf{3}},\mathbf{1};-1,0,x)_{-2/3} \\ d^c & (\bar{\mathbf{3}},\mathbf{1};-1,0,y)_{1/3} \\ L & (\mathbf{1},\mathbf{2};0,1,z)_{-1/2} \\ l^c & (\mathbf{1},\mathbf{1};0,0,1)_{1} \end{array} \tag{40}$$

The values of the $U(1)$ charges x, y, z, w will be fixed below so that they lead to the right hypercharges, shown for completeness as subscripts.

It turns out that there are two possible ways of embedding the Standard Model particle spectrum on these stacks of branes [36], which are shown pictorially in Fig. 11. The quark

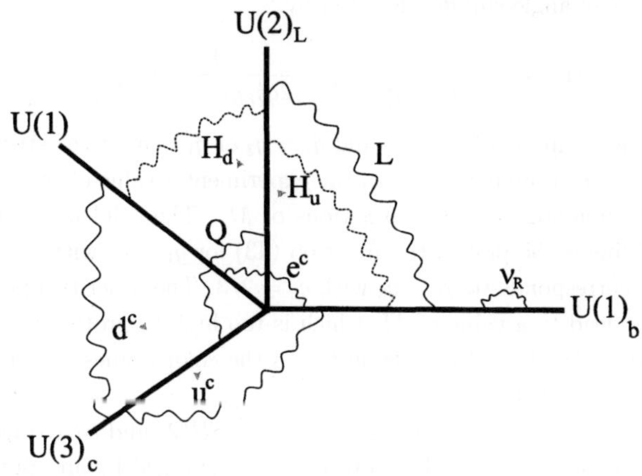

Figure 11: A minimal Standard Model embedding on D-branes.

doublet Q corresponds necessarily to a massless excitation of an open string with its two ends on the two different collections of branes (color and weak). As seen from the figure, a fourth brane stack is needed for a complete embedding, which is chosen to be a $U(1)_b$ extended in the bulk. This is welcome since one can accommodate right handed neutrinos as open string states on the bulk with sufficiently small Yukawa couplings suppressed by the large volume of the bulk [38]. The two models are obtained by an exchange of the up and down antiquarks, u^c and d^c, which correspond to open strings with one end on the color branes and the other either on the $U(1)$ brane, or on the $U(1)_b$ in the bulk. The lepton doublet L arises from an open string stretched between the weak branes and $U(1)_b$, while the antilepton l^c corresponds to a string with one end on the $U(1)$ brane and the other in the bulk. For completeness, we also show the two possible Higgs states H_u and H_d that are both necessary in order to give tree-level masses to all quarks and leptons of the heaviest generation.

6.1 Hypercharge embedding and the weak angle

The weak hypercharge Y is a linear combination of the three $U(1)$'s:

$$Y = Q_1 + \frac{1}{2}Q_2 + c_3 Q_3 \quad ; \quad c_3 = -1/3 \text{ or } 2/3\,, \tag{41}$$

where Q_N denotes the $U(1)$ generator of $U(N)$ normalized so that the fundamental representation of $SU(N)$ has unit charge. The corresponding $U(1)$ charges appearing in eq. (40) are $x = -1$ or 0, $y = 0$ or 1, $z = -1$, and $w = 1$ or -1, for $c_3 = -1/3$ or $2/3$, respectively. The hypercharge coupling g_Y is given by [2]:

$$\frac{1}{g_Y^2} = \frac{2}{g_1^2} + \frac{4c_2^2}{g_2^2} + \frac{6c_3^2}{g_3^2}\,. \tag{42}$$

It follows that the weak angle $\sin^2\theta_W$, is given by:

$$\sin^2\theta_W \equiv \frac{g_Y^2}{g_2^2 + g_Y^2} = \frac{1}{2 + 2g_2^2/g_1^2 + 6c_3^2 g_2^2/g_3^2}\,, \tag{43}$$

where g_N is the gauge coupling of $SU(N)$ and $g_1 = g_2$ or $g_1 = g_3$ at the string scale. In order to compare the theoretical predictions with the experimental value of $\sin^2\theta_W$ at M_s, we plot in Fig. 12 the corresponding curves as functions of M_s. The solid line is the experimental curve. The dashed line is the plot of the function (43) for $g_1 = g_2$ with $c_3 = -1/3$ while the dotted-dashed line corresponds to $g_1 = g_3$ with $c_3 = 2/3$. The other two possibilities are not shown because they lead to a value of M_s which is too high to protect the hierarchy. Thus, the second case, where the $U(1)$ brane is on top of the color branes, is compatible with low energy data for $M_s \sim 6 - 8$ TeV and $g_s \simeq 0.9$.

From Eq. (43) and Fig. 12, we find the ratio of the $SU(2)$ and $SU(3)$ gauge couplings at the string scale to be $\alpha_2/\alpha_3 \sim 0.4$. This ratio can be arranged by an appropriate choice of the relevant moduli. For instance, one may choose the color and $U(1)$ branes to be D3 branes while the weak branes to be D7 branes. Then, the ratio of couplings above can be explained by choosing the volume of the four compact dimensions of the seven branes to be $V_4 = 2.5$ in string units. This being larger than one is consistent with the picture above. Moreover it predicts an interesting spectrum of KK states for the Standard model, different from the naive choices that have appeared hitherto: the only Standard Model particles that have KK descendants are the W bosons as well as the hypercharge gauge boson. However, since the hypercharge is a linear combination of the three $U(1)$'s, the massive $U(1)$ KK gauge bosons do not couple to the hypercharge but to the weak doublet number.

[2]The gauge couplings $g_{2,3}$ are determined at the tree-level by the string coupling and other moduli, like radii of longitudinal dimensions. In higher orders, they also receive string threshold corrections.

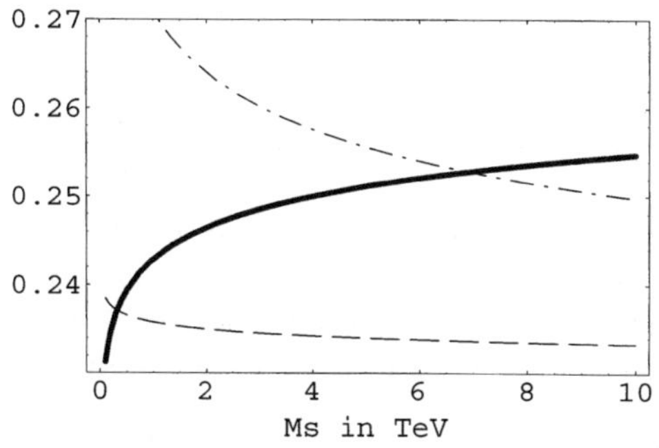

Figure 12: The experimental value of $\sin^2\theta_W$ (thick curve), and the theoretical predictions (43).

6.2 The fate of $U(1)$'s, proton stability and neutrino masses

It is easy to see that the remaining three $U(1)$ combinations orthogonal to Y are anomalous. In particular there are mixed anomalies with the $SU(2)$ and $SU(3)$ gauge groups of the Standard Model. These anomalies are cancelled by three axions coming from the closed string RR (Ramond) sector, via the standard Green-Schwarz mechanism [39]. The mixed anomalies with the non-anomalous hypercharge are also cancelled by dimension five Chern-Simmons type of interactions [36]. An important property of the above Green-Schwarz anomaly cancellation mechanism is that the anomalous $U(1)$ gauge bosons acquire masses leaving behind the corresponding global symmetries. This is in contrast to what would had happened in the case of an ordinary Higgs mechanism. These global symmetries remain exact to all orders in type I string perturbation theory around the orientifold vacuum. This follows from the topological nature of Chan-Paton charges in all string amplitudes. On the other hand, one expects non-perturbative violation of global symmetries and consequently exponentially small in the string coupling, as long as the vacuum stays at the orientifold point. Thus, all $U(1)$ charges are conserved and since Q_3 is the baryon number, proton stability is guaranteed.

Another linear combination of the $U(1)$'s is the lepton number. Lepton number conservation is important for the extra dimensional neutrino mass suppression mechanism described above, that can be destabilized by the presence of a large Majorana neutrino mass term. Such a term can be generated by the lepton-number violating dimension five effective operator $LLHH$ that leads, in the case of TeV string scale models, to a Majorana mass of the order of a few GeV. Even if we manage to eliminate this operator in some particular model, higher order operators would also give unacceptably large contributions, as we focus on models in which the ratio between the Higgs vacuum expectation value and the string scale is just of order $\mathcal{O}(1/10)$. The best way to protect tiny neutrino masses from such contributions is to impose lepton number conservation.

A bulk neutrino propagating in $4+n$ dimensions can be decomposed in a series of 4d

KK excitations denoted collectively by $\{m\}$:

$$S_{kin} = R_\perp^n \int d^4x \sum_{\{m\}} \left\{ \bar{\nu}_{Rm} \not{\partial} \nu_{Rm} + \bar{\nu}_{Rm}^c \not{\partial} \nu_{Rm}^c + \frac{m}{R_\perp} \nu_{Rm} \nu_{Rm}^c + c.c. \right\}, \tag{44}$$

where ν_R and ν_R^c are the two Weyl components of the Dirac spinor and for simplicity we considered a common compactification radius $R_\perp$. On the other hand, there is a localized interaction of ν_R with the Higgs field and the lepton doublet, which leads to mass terms between the left-handed neutrino and the KK states ν_{Rm}, upon the Higgs VEV v:

$$S_{int} = g_s \int d^4x H(x) L(x) \nu_R(x, y=0) \quad \to \quad \frac{g_s v}{R_\perp^{n/2}} \sum_m \nu_L \nu_{Rm}, \tag{45}$$

in strings units. Since the mass mixing $g_s v / R_\perp^{n/2}$ is much smaller than the KK mass $1/R_\perp$, it can be neglected for all the excitations except for the zero-mode ν_{R0}, which gets a Dirac mass with the left-handed neutrino

$$m_\nu \simeq \frac{g_s v}{R_\perp^{n/2}} \simeq v \frac{M_s}{M_p} \simeq 10^{-3} - 10^{-2} \text{ eV}, \tag{46}$$

for $M_s \simeq 1 - 10$ TeV, where the relation (19) was used. In principle, with one bulk neutrino, one could try to explain both solar and atmospheric neutrino oscillations using also its first KK excitation. However, the later behaves like a sterile neutrino which is now excluded experimentally. Therefore, one has to introduce three bulk species (at least two) ν_R^i in order to explain neutrino oscillations in a 'traditional way', using their zero-modes ν_{R0}^i [40]. The main difference with the usual seesaw mechanism is the Dirac nature of neutrino masses, which remains an open possibility to be tested experimentally.

Acknowledgments

Work supported in part by the European Commission under the ERC Advanced Grant 226371.

References

[1] Fore a review, see e.g. K. R. Dienes, Phys. Rept. **287** (1997) 447 [arXiv:hep-th/9602045]; and references therein.

[2] C. Angelantonj and A. Sagnotti, *Phys. Rept.* **371** (2002) 1 [Erratum-ibid. **376** (2003) 339] [arXiv:hep-th/0204089].

[3] I. Antoniadis, *Phys. Lett.* B **246** (1990) 377.

[4] D. J. Kapner, T. S. Cook, E. G. Adelberger, J. H. Gundlach, B. R. Heckel, C. D. Hoyle and H. E. Swanson, Phys. Rev. Lett. **98** (2007) 021101.

[5] N. Arkani-Hamed, S. Dimopoulos and G. R. Dvali, *Phys. Lett.* B **429** (1998) 263 [arXiv:hep-ph/9803315]; I. Antoniadis, N. Arkani-Hamed, S. Dimopoulos and G. R. Dvali, *Phys. Lett.* B **436** (1998) 257 [arXiv:hep-ph/9804398].

[6] For a review see e.g. I. Antoniadis, *Prepared for NATO Advanced Study Institute and EC Summer School on Progress in String, Field and Particle Theory, Cargese, Corsica, France (2002)*; and references therein.

[7] J. D. Lykken, *Phys. Rev.* D **54** (1996) 3693 [arXiv:hep-th/9603133].

[8] J. C. Long and J. C. Price, *Comptes Rendus Physique* **4** (2003) 337; R. S. Decca, D. Lopez, H. B. Chan, E. Fischbach, D. E. Krause and C. R. Jamell, *Phys. Rev. Lett.* **94** (2005) 240401; R. S. Decca et al., arXiv:0706.3283 [hep-ph]; S. J. Smullin, A. A. Geraci, D. M. Weld, J. Chiaverini, S. Holmes and A. Kapitulnik, arXiv:hep-ph/0508204; H. Abele, S. Haeßler and A. Westphal, in 271th WE-Heraeus-Seminar, Bad Honnef (2002).

[9] I. Antoniadis and K. Benakli, *Phys. Lett.* B **326** (1994) 69.

[10] K. R. Dienes, E. Dudas and T. Gherghetta, Phys. Lett. B **436** (1998) 55 [arXiv:hep-ph/9803466]; Nucl. Phys. B **537** (1999) 47 [arXiv:hep-ph/9806292].

[11] I. Antoniadis, K. Benakli and M. Quirós, *Phys. Lett.* B **331** (1994) 313 and *Phys. Lett.* B **460** (1999) 176; P. Nath, Y. Yamada and M. Yamaguchi, *Phys. Lett.* B **466** (1999) 100 T. G. Rizzo and J. D. Wells, *Phys. Rev.* D **61** (2000) 016007; T. G. Rizzo, *Phys. Rev.* D **61** (2000) 055005; A. De Rujula, A. Donini, M. B. Gavela and S. Rigolin, *Phys. Lett.* B **482** (2000) 195;

[12] E. Accomando, I. Antoniadis and K. Benakli, *Nucl. Phys.* B **579** (2000) 3.

[13] I. Antoniadis, K. Benakli and A. Laugier, *JHEP* **0105** (2001) 044.

[14] P. Nath and M. Yamaguchi, *Phys. Rev.* D **60** (1999) 116004; *Phys. Rev.* D **60** (1999) 116006; M. Masip and A. Pomarol, *Phys. Rev.* D **60** (1999) 096005; W. J. Marciano, *Phys. Rev.* D **60** (1999) 093006; A. Strumia, *Phys. Lett.* B **466** (1999) 107; R. Casalbuoni, S. De Curtis, D. Dominici and R. Gatto, *Phys. Lett.* B **462** (1999) 48; C. D. Carone, *Phys. Rev.* D **61** (2000) 015008; A. Delgado, A. Pomarol and M. Quirós, *JHEP* **1** (2000) 30.

[15] G. Servant and T. M. P. Tait, *Nucl. Phys.* B **650** (2003) 391.

[16] G. F. Giudice, R. Rattazzi and J. D. Wells, *Nucl. Phys.* B **544** (1999) 3; E. A. Mirabelli, M. Perelstein and M. E. Peskin, *Phys. Rev. Lett.* **82** (1999) 2236; T. Han, J. D. Lykken and R. Zhang, *Phys. Rev.* D **59** (1999) 105006; K. Cheung and W.-Y. Keung, *Phys. Rev.* D **60** (1999) 112003; C. Balázs *et al.*, *Phys. Rev. Lett.* **83** (1999) 2112; L3 Collaboration (M. Acciarri *et al.*), *Phys. Lett.* B **464** (1999) 135 and **470** (1999) 281: J. L. Hewett, *Phys. Rev. Lett.* **82** (1999) 4765.

[17] N. Arkani-Hamed, S. Dimopoulos and G. Dvali, *Phys. Rev.* D **59** (1999) 086004.

[18] S. Cullen and M. Perelstein, *Phys. Rev. Lett.* **83** (1999) 268; V. Barger, T. Han, C. Kao and R. J. Zhang, *Phys. Lett.* B **461** (1999) 34.

[19] K. Benakli and S. Davidson, *Phys. Rev.* D **60** (1999) 025004; L. J. Hall and D. Smith, *Phys. Rev.* D **60** (1999) 085008.

[20] E. Dudas and J. Mourad, Nucl. Phys. B **575** (2000) 3 [arXiv:hep-th/9911019]; S. Cullen, M. Perelstein and M. E. Peskin, *Phys. Rev.* D **62** (2000) 055012; D. Bourilkov, *Phys. Rev.* D **62** (2000) 076005; L3 Collaboration (M. Acciarri *et al.*), *Phys. Lett.* B **489** (2000) 81.

[21] P. C. Argyres, S. Dimopoulos and J. March-Russell, Phys. Lett. B **441** (1998) 96 [arXiv:hep-th/9808138]; T. Banks and W. Fischler, arXiv:hep-th/9906038.

[22] S. B. Giddings and S. Thomas, *Phys. Rev.* D **65** (2002) 056010 [arXiv:hep-ph/0106219]; S. Dimopoulos and G. Landsberg, *Phys. Rev. Lett.* **87** (2001) 161602 [arXiv:hep-ph/0106295].

[23] P. Meade and L. Randall, arXiv:0708.3017 [hep-ph].

[24] L. A. Anchordoqui, H. Goldberg, D. Lust, S. Nawata, S. Stieberger and T. R. Taylor, Phys. Rev. Lett. **101** (2008) 241803 [arXiv:0808.0497 [hep-ph]].

[25] D. Lust, S. Stieberger and T. R. Taylor, Nucl. Phys. B **808** (2009) 1 [arXiv:0807.3333 [hep-th]].

[26] G. T. Horowitz and J. Polchinski, Phys. Rev. D **55** (1997) 6189 [arXiv:hep-th/9612146].

[27] I. Antoniadis, C. Bachas, *Phys. Lett.* B **450** (1999) 83.

[28] I. Antoniadis, K. Benakli, A. Laugier and T. Maillard, *Nucl. Phys.* B **662** (2003) 40 [arXiv:hep-ph/0211409].

[29] I. Antoniadis, S. Dimopoulos and G. Dvali, *Nucl. Phys.* B **516** (1998) 70; S. Ferrara, C. Kounnas and F. Zwirner, *Nucl. Phys.* B **429** (1994) 589.

[30] E. G. Adelberger, B. R. Heckel, S. Hoedl, C. D. Hoyle, D. J. Kapner and A. Upadhye, Phys. Rev. Lett. **98** (2007) 131104.

[31] T. R. Taylor and G. Veneziano, *Phys. Lett.* B **213** (1988) 450.

[32] I. Antoniadis, E. Kiritsis and J. Rizos, *Nucl. Phys.* B **637** (2002) 92.

[33] I. Antoniadis, K. Benakli and M. Quirós, *Nucl. Phys.* B **583** (2000) 35.

[34] I. Antoniadis, K. Benakli and M. Quiros, *New Jour. Phys.* **3** (2001) 20.

[35] I. Antoniadis, C. Muñoz and M. Quirós, *Nucl. Phys.* B **397** (1993) 515; I. Antoniadis and M. Quirós, *Phys. Lett.* B **392** (1997) 61; A. Pomarol and M. Quirós, *Phys. Lett.* B **438** (1998) 225; I. Antoniadis, S. Dimopoulos, A. Pomarol and M. Quirós, *Nucl. Phys.* B **544** (1999) 503; A. Delgado, A. Pomarol and M. Quirós, *Phys. Rev.* D **60** (1999) 095008; R. Barbieri, L. J. Hall and Y. Nomura, *Phys. Rev.* D **63** (2001) 105007.

[36] I. Antoniadis, E. Kiritsis and T. N. Tomaras, *Phys. Lett.* B **486** (2000) 186; I. Antoniadis, E. Kiritsis, J. Rizos and T. N. Tomaras, *Nucl. Phys.* B **660** (2003) 81.

[37] G. Shiu and S.-H. H. Tye, *Phys. Rev.* D **58** (1998) 106007; Z. Kakushadze and S.-H. H. Tye, *Nucl. Phys.* B **548** (1999) 180; L. E. Ibáñez, C. Muñoz and S. Rigolin, *Nucl. Phys.* B **553** (1999) 43.

[38] K. R. Dienes, E. Dudas and T. Gherghetta, *Nucl. Phys.* B **557** (1999) 25 [arXiv:hep-ph/9811428]; N. Arkani-Hamed, S. Dimopoulos, G. R. Dvali and J. March-Russell, *Phys. Rev.* D **65** (2002) 024032 [arXiv:hep-ph/9811448]; G. R. Dvali and A. Y. Smirnov, *Nucl. Phys.* B **563** (1999) 63.

[39] A. Sagnotti, *Phys. Lett.* B **294** (1992) 196; L. E. Ibáñez, R. Rabadán and A. M. Uranga, *Nucl. Phys.* B **542** (1999) 112; E. Poppitz, *Nucl. Phys.* B **542** (1999) 31.

[40] H. Davoudiasl, P. Langacker and M. Perelstein, Phys. Rev. D **65** (2002) 105015 [arXiv:hep-ph/0201128].

CHAIRMAN: I. ANTONIADIS

Scientific Secretaries: G. Moraitis, E. Müller

DISCUSSION I

- *Nicotri:*

You have shown several types of models, some inspired by or directly related to String Theory and other more phenomenological models. Some of the models you have shown appear to be non-renormalisable. Is this true? If so, can we trust the results obtained through them? We solve some problems but we lose renormalisability.

- *Antoniadis:*

The field theory models with extra dimensions are not renormalisable so they require some cutoff which should be nearby, and this is provided by String Theory. That's the reason why in my discussion I have always used String theory as the underlying theory.

- *Shakirov:*

My question is about the Kaluza-Klein dark matter candidate. You have mentioned that in models with universal extra dimensions the lightest Kaluza-Klein excitation is stable due to conservation of Kaluza-Klein parity and thus provides a good example of dark matter candidate. But what Standard Model particle is responsible for that? Is it the Kaluza-Klein neutrino or is it something else?

- *Antoniadis:*

It is an excitation of the photon.

- *Baudis:*

If I can comment, you can also have the Kaluza-Klein partner of the Z or others, depending on the number of dimensions. So there are several candidates that have been studied but the neutrino has been excluded for instance.

- *Shakirov:*

So it is model dependent?

- *Antoniadis:*

As in supersymmetry there are several models but the photon is the most natural lightest Kaluza-Klein excitation from the point of view that it remains light under radiative corrections, as the photino in SUSY. But in a complete model you can have combinations of the KK photon with other particles.

- *Baudis:*

You have mentioned the two extreme models of large extra dimensions, one in which only gravity propagates into the extra dimension and one – universal extra dimensions – in which all the Standard Model particles are allowed to propagate into the extra dimension. These cases have been studied, there are predictions for direct and indirect detection in accelerators etc. However, you said that the truth lies somewhere in between. Can you please elaborate on this? Are there any predictions from these more realistic models (for colliders, direct and indirect dark matter searches)?

- *Antoniadis:*

Tomorrow I will show you a particular embedding of the Standard Model in this context which I think will make this point clear. The main idea is that when you try to construct the Standard Model from, say, a String Theory with branes then for every factor of the Standard Model you associate a stack of branes: three coincident branes describing the strong interactions, two branes describing the weak interactions and the hypercharge can have component in another brane. Every stack of these branes points into different directions in the extra dimensions. In this context a universal extra dimension common to all brane stacks is very unlikely. As an example, you want to explain for instance why the weak coupling is two times smaller than the strong coupling. The easiest way to explain it is to say that the strong branes are three-branes while the weak branes extend into an extra dimension whose size is twice the string size.

- *Baudis:*

But you still have conservation of momentum in these extra dimensions.

- *Antoniadis:*

It depends on how the quarks and leptons arise. Consider for instance the case where the weak interaction branes extend into an extra dimension while the strong ones do not. Then, since the quark doublet corresponds to an open string that is stretched between the strong and the weak branes, it is always localized in this dimension. But the lepton doublet could only see the weak branes and would therefore propagate into the extra dimension. So depending on the particular model, you can have particles that propagate into extra dimensions and others that are localised. That's what I meant by saying a realistic model is usually somewhere in between the two extreme cases I discussed.

- *Ni:*

How can one distinguish between extra dimensions and supersymmetry if missing energy is detected at the LHC?

- *Antoniadis:*

One possibility is to study the angular distribution in order to infer the spin of the particle that you do not detect and produces missing energy.

- *Ni:*

So the decay products and the spin are different?

- *Antoniadis:*

Yes, and the angular distribution depends on the spin.

- *Johansson:*

Can you comment on what happens to the GUT scale if the Planck scale is brought down to the TeV level? Does it follow or does it stay where it is now?

- *Antoniadis:*

There is no GUT scale and in general no unification in this framework. Tomorrow I will mention one example where you can still make a prediction for $\sin^2(\theta_W)$ but it has nothing to do with ordinary unification.

- *Nath:*

I have a short comment. One of the interesting results from extra dimensions is the modification of Newton's inverse square law which goes like $1/x^{2+n}$ where n is an integer. There is another mechanism which was investigated by Haim Goldberg and myself where we considered the implications of exchange of a scale invariant hidden sector. Here one finds that Newton's law is modified to $1/x^{2+\delta}$ where δ could be fractional. So it appears possible to discriminate between the two models.

- *Antoniadis:*

Are the particles in the hidden sector massless?

- *Nath:*

Yes, they form a continuum, the propagator goes like $1/(p^2)^\delta$ so they are conformal. So in other words, once an experiment finds a result, it is possible using detailed analysis to discriminate between those two models.

- *Antoniadis:*

As far as I know, in the experimental analysis the deviation [from Newton's law] is parametrised by a Yukawa potential. But of course, once they find a deviation the discrimination you described should be possible.

- *Akiba:*

I am confused by the slide where you write that the size of the extra dimensions is smaller than 45microns [slide with experimental apparatus used by Adelberger et al. (2006)]. On this slide you write that the dark energy scale is 85microns, what does this mean?

- *Antoniadis:*

If one translates the dark energy into the cosmological constant this corresponds to an energy of the order of 0.001eV. This energy scale corresponds to a length of 85 micrometers. It is the first time that a microgravity experiment is performed at distances smaller than the dark energy length scale. There are actually some theoretical works suggesting that gravity may change behavior at distances shorter than this length scale.

- *Akiba:*

So this 85 microns is based on this relation. But the experiment [by Adelberger et al.] shows that the size of the extra dimensions must be smaller than 45microns, so is this a contradiction?

- *Antoniadis:*

No, once you have a cosmological constant in the theory there is another scale besides Planck: gravity has two length scales, one is the Planck length which is 10^{-33}cm and the other is the dark energy length which is 85 micrometers. It is the first time that Newton's law is measured at distances shorter than the dark energy length.

- *t'Hooft:*

Those length scales are only orders of magnitude, not absolute numbers. It does not make sense to give them to two decimal places.

- *Tannenbaum:*

Why isn't it ruled out?

- *Antoniadis:*

Ruled out what?

- *Akiba:*

Don't those two numbers contradict each other?

- *Antoniadis:*

No, they don't contradict each other, it is just that there have been some theories that predicted that gravity maybe behaves completely differently at distances shorter than the dark energy length scale. This is the first measurement that tells you that gravity continues to have the same behaviour for such distances.

- *Akiba:*

Thank you, I have another question about the non-existence of the GUT scale. If there are extra dimensions, is it just an accident that the coupling constants appear to become equal at 10^{16} GeV?

- *Antoniadis:*

Yes.

- *Dubynskiy:*

You mentioned microgravity experiments which could test the validity of Newton gravity at short distances. Are these experiments based on techniques of experimental particle physics or do they use classical techniques? What are the best experimental results at the moment? Are there any experiments other from the one you described above?

- *Antoniadis:*

The best measurement of the gravitational force in the laboratory is done by this experiment. On the other hand, the main experimental test of the idea of low scale gravity should come from accelerators.

- *Dubynskiy:*

On your previous slides you said that for n=2 extra dimensions their size should be about a 0.1millimeter and now you are talking about micrometers. Is the case n=2 excluded by these experiments or not?

- *Antoniadis:*

No, it is not excluded. For n=2 the fundamental scale of higher dimensional gravity should be around 10 TeV.

CHAIRMAN: I. ANTONIADIS

Scientific Secretaries: C. S. Park, I. Samsonov

DISCUSSION II

- Schwarz:

You mentioned one D-brane construction that gives the GUT-model. Can you get a Yukawa coupling for the up-type quarks?

- Antoniadis:

Not for this model. It is generated non-perturbatively though.

- Shakirov:

It seems to me that this topic was not raised in this school so far. I would like to ask you a question about the probability of mini black holes occurring at LHC. There are some people who think that this will destroy the Earth. What kind of estimation can you give?

- Antoniadis:

The probability of such process is practically zero.

- 't Hooft:

The question of black holes is by itself a well-posed question. There is high expectation that mini black holes occur. We all believe that if they appear they will have to decay rapidly. But, do you have any idea about mini black holes?

- Antoniadis:

First of all, as you said, mini black holes have short lifetime. On the other hand, as I mentioned, LHC would not reach the energy to produce black holes because there are already experimental constraints on the string scale. Since the theory is perturbative, the black hole mass is quite high. Finally, there is the question that the calculation of black hole production is based on classical general relativity. This approximation to my knowledge is not quite justified because it falls into a regime in which string perturbation theory diverges.

- 't Hooft:

In principle, it is conceivable that the Planck length is indeed in the TeV domain, which would mean that you might even think of collisions beyond the Planck energy if the Planck energy is TeV, and you might suspect that the collision of particles would give mini black holes, which would rapidly decay emitting Hawking radiation. But that speculation itself, though most of the things are far-fetched, is not that ridiculous. If you accept all these things, you might as well believe this. In principle, black holes defined this way are not totally excluded if, as you say, the Planck length is TeV.

- Susskind:

I do not think that the lightest black hole is of Planck scale mass. I think it is of Planck mass divided by closed string coupling constant squared.

- Antoniadis:

Yes. It is the string scale divided by the coupling constant squared.

- *Susskind:*

How does it compare with the Planck scale? It is much higher, isn't it?

- *Antoniadis:*

Yes. It is higher than the Planck scale.

- *Susskind:*

But it is very much higher in this theory. How much?

- *Antoniadis:*

It can be two orders of magnitude higher.

- *Susskind:*

Only two orders of magnitude.

- *Bettini:*

Apart from the theoretical speculation that you can do, there are observational facts. It's very clear. We have cosmic rays. Every single second hundred thousands of particles hit the Earth. These are the protons and nuclei which are more energetic than any LHC collisions in the center of mass.

- *'t Hooft:*

That was the counter-argument that the people bring forward against the anxiety that the black holes appear at LHC. Cosmic rays make much more black holes which all rapidly decay and they are just exotic bound states and that's it. We have nothing spectacular of this sort. It would be just another form of matter which is unstable like an unstable particle. This is the argument that this form of matter would not be produced very copiously in the energetic cosmic rays. It would be very interesting physics if in this extremely far-fetched scenario something like this would happen.

- *Johansson:*

I have a question regarding Lüst-Steiberger-Taylor formula. You do two expansions: one in the loop parameter g_{YM} or g_{string}, and another in α', the string Regge slope parameter. You keep the first order in g_{YM}, but then include all orders in α'. What is the reason for assuming that $\alpha' \gg g_{YM}$? Is there any good reason for that?

- *Antoniadis:*

We just set up the string coupling to be small for phenomenological reasons.

- *Johansson:*

But is g_{YM} of the same order as the QCD coupling constant?

- *Antoniadis:*

The string coupling is of order $\alpha = g_{YM}^2/4\pi$ which is of order one tenth. Therefore the corrections are small.

- *Dubinsky:*

Many people who are doing supersymmetry argue that it allows unifying gauge couplings. They refer to the result of precise experimental measurement which in their opinion favor gauge unification. On the contrary, you said that the large extra dimensions require that unification has to be dropped. What is your interpretation of the experimental data on the gauge unification? Which idea is more natural, to unify the forces or to have extra dimensions?

- *Antoniadis:*

As I said in my talk, in this framework gauge coupling unification is not natural. Although there are some theoretical proposals, there is no satisfactory explanation within this context. In my opinion, in this framework the apparent unification of coupling constants, by extrapolating the low-energy data, should be considered rather accidental. There might be other explanations for the value of the weak angle, as the example that I mentioned in my talk, but these are independent from conventional unification. By the way, in models with intersecting branes unification is not very natural, even when the string scale is high, for several reasons: on the one hand every stack of branes has its own gauge coupling, while on the other hand there are many extras states in the spectrum beyond those of the MSSM. To my knowledge, there is no satisfactory model with unification in this context.

- *Tannenbaum:*

I have a comment about the black holes. Even if the energy of an LHC collision were turned into a black hole, there would not be enough mass to attract anything except at the Planck length scale. My neighbor is only a few centimeters from me, but exerts very little force on me. So I doubt that a microscopic black hole would do any damage or be noticed. However, there is a better disaster scenario in heavy ion collisions. The vacuum of the universe could be changed at the interaction point to the zero energy state, and then there is enough potential energy to collapse the universe. That is why I like the cosmic rays argument within the Moon or other counterexamples of a disaster rather than a theoretical argument.

- *'t Hooft:*

First of all, it would be a black hole with just the mass equal to the energy that you plug in. It can have the mass of one nucleus and then it would rapidly decay through the Hawking radiation. There are some crazy speculations that it would hit the wall of the detector or the rock around it and it would start eating the matter at the very small scale and growing. It violates the physics of Hawking radiation because we know from the textbooks that it has tremendously high Hawking temperature. Because we know that it would emit, and would not eat, but vomit Hawking radiation and decrease the size. You really have to modify the physics beyond Hawking radiation if you want to change this scenario to get that the black hole start eating instead of radiating away things. In the extremely unlike case that all we think and understand the Hawking radiation is completely wrong, only then the black hole can start accreting matter. But even then you would have the difficulty because such black holes would be copiously produced already in cosmic rays. Nothing has been seen of any of this even in the scale of billions of years. And that is another indirect indication that all this is completely wrong. A black hole which actually accretes is not made by any machine like LHC and that is the end of it. There are many people who think that LHC might be a dangerous machine, but those are people who have read too many science fiction stories.

- *Tannenbaum:*

But I have another comment. My favorite disaster scenario at RHIC is not black holes. I thought there was enough that may have to do with anything. But in heavy ion collisions, if you change the vacuum of the universe for one small amount to the stable vacuum, namely nothing, there were enough potential energy from where we are than the whole world falls to the lower level. I thought that was much more serious because you can imagine that the potential energy is there.

- *'t Hooft:*

Yes. Indeed, if the quark-gluon plasma will be another stable form of matter which could lower the effective vacuum state we are in, yes, that is another science fiction scenario. But that could much easier be contradicted by saying that then the cosmic rays would abound much earlier, and since it affects the vacuum, there was no problem about Lorenz transformation. In case of black holes, the counter-argument is that they are produced so fast by cosmic rays that just go through the Earth and the Earth has been spared that way. But those arguments would also easily be contradicted, so even that is not a realistic scenario.

- *Tannenbaum:*

So, I can accept the shifting of the vacuum of the universe did not happen because the moon is still there. But you seem to have an argument based on theory.

- *'t Hooft:*

The game they wanted to play is to put as little as possible theory, because as soon as you do that, it is easy to say this can possibly happen. So people say that let's only look the observation, and say that not only all the stars at infinity are still there as if nothing of the sort ever happens. But we can do a more rigorous analysis and that there is no way that the cosmic rays will produce much more energetic particles than we ever will be able to do at CERN. They are all around us. You can even have head-on collisions, but the cosmic rays now and then would produce much more black holes whatever dangerous things you can think of, then CERN can ever do. So, one can use those arguments, and there has been extensive study. There is a paper by Allison and coauthors, a paper by Giddings, who made extensive study of all possible cases and scenarios you can think of, and they all basically say that whatever they are, CERN does not hurt humanity.

- *Nikiforov:*

It looks as if it is unrealistic to produce a black hole which would consume and absorb our world. But is it still possible that any object in LHC like black hole could damage the detector or burn the beam pipe?

- *'t Hooft:*

Even that is hardly imaginable. The detector could be destroyable by all sorts of reasons. The magnet or superconductor might burn accidentally. We can imagine all sorts of accidents. Things like that can happen and that is bad enough. But I don't think any dangerous particle will destroy the detectors. The detector might itself be also the reason, but not the particle that goes through it.

- *Nath:*

If that were to happen, it would be a fantastic event. It would be the first experimental confirmation of the Hawking radiation in the laboratory.

Predicted Signals at the LHC from Technicolor: Erice Lecture

Adam Martin*

Department of Physics, Yale University, New Haven CT, 06520

December 15, 2008

Abstract

In this lecture I give an introduction to technicolor and extended technicolor theories. I discuss the issues models of dynamical electroweak symmetry breaking struggle with and propose how non QCD-like dynamics, such as a 'walking' or nearly-conformal technicolor coupling, can invalidate many conventional arguments against technicolor. I provide a short discussion of AdS/CFT-inspired extra-dimensional (Higgsless) models and their similarities and differences with older 4D technicolor models. To conclude, a summary of some possible LHC signatures is given.

Introduction

Electroweak symmetry breaking is observed in nature: the weak bosons $W^{\pm}, Z^0$ are quite massive while the photon is massless. To remain consistent with gauge invariance, Lorentz invariance and unitarity, massive gauge bosons can only come about through spontaneously broken symmetry and the Higgs mechanism. Therefore, our electroweak theory is incomplete without some apparatus for spontaneous electroweak symmetry breaking (EWSB). One way to see this inconsistency is in the scattering amplitude of longitudinally polarized massive gauge bosons. These amplitudes violate unitarity at energy $\mathcal{O}(\text{TeV})$ unless new contributions are added. Although we know spontaneous symmetry breaking must occur, the exact nature of its breaking is unknown. I believe this is the most pressing question in High Energy Physics today and one that the LHC will hopefully answer. Until we have the answer, all possibilities should be explored and their similarities and differences fully fleshed out.

The setup of this lecture is the following: After briefly motivating extensions of the standard model and dynamical symmetry breaking (Section 1), I give an introduction to classic, rescaled-QCD technicolor (Section 2) and its extension Extended Technicolor (Section 3). The hurdles which theories of dynamical electroweak symmetry breaking face are then laid out in Section 4. Many of these hurdles are based on the assumption that technicolor behaves just like QCD, and in Section 5 I illustrate how one example of non QCD-like behavior – a

*adam.martin@yale.edu

nearly conformal or 'walking' technicolor coupling constant – can dramatically improve the situation. Other technicolor reviews which cover similar information are Ref. [1, 2, 3, 4, 5, 6]. In Section 6 I investigate modeling dynamical electroweak symmetry breaking using an extra dimension. Finally, collider signals for two different phenomenological models of technicolor are explored at length in Section 7.

1 Why do we need the Higgs?: Separating Fact and Fiction

The simplest candidate for the Higgs mechanism is a single Higgs doublet whose potential is arranged by hand to have a minimum at non-zero vacuum expectation value. Although adequate, the simple Higgs scalar is theoretically unsatisfying. The mass parameters in the potential are sensitive to the highest momentum in the theory, Λ_{UV}. In order to preserve a hierarchy between Λ_{UV} and the weak scale Λ_{EW}, an incredible and 'unnatural' degree of tuning is necessary. A second problem with the simple Higgs theory is triviality. Triviality is the statement that the cutoff of the theory cannot be taken arbitrarily high without the theory becoming free [3, 7, 8]. Therefore the simple Higgs doublet model must have a finite cutoff. The straightforward way to see this is to look at the renomalization group equation for the Higgs quartic coupling.

To rectify these unsatisfactory aspects of the Higgs theory, we believe new physics is necessary. The most advocated extension is supersymmetry, in which the naturalness problems is improved by adding a symmetry between bosons and fermions. As a result, each standard model (SM) particle has a superpartner with opposite statistics yet equal couplings. Supersymmetry accomplishes EWSB while remaining weakly coupled, making a plethora of calculations – spectrum, cross sections, decay rates, etc. all doable. This is certainly a nice feature, but it is by no means a requirement. Additionally, even the minimal viable model, the MSSM, leads to an improved unification of the gauge couplings and a dark matter candidate. While these 'free' MSSM aspects are theoretically interesting, they are not unique, nor are they necessary for EWSB. Moreover, fundamental spin-0 fields – which are a key ingredient in supersymmetry – have never been observed in nature.

1.1 Dynamical Symmetry Breaking: Higgs mechanism without the Higgs

The phase transitions analogous to EWSB which have been observed happen dynamically. Two obvious precedents are chiral symmetry breaking in QCD and superconductivity.

In QCD, in the limit of massless u, d quarks, the theory has an $SU(2)_L \otimes SU(2)_R \otimes U(1)_B$ global symmetry, where L, R refer to rotations among the left-handed, right-handed chirality quarks. At low energy, to a good approximation, the states fall into representations of $SU(2)_V$ – where both left-handed and right-handed fermions feel the same rotation, rather than the full symmetry. What has happened to the axial combination of the rotations? This symmetry is not manifest in the spectrum, thus it must be broken. However, since the theory only consists of massless quarks and gluons, we are forced to conclude that the symmetry is broken as a result of the strong interaction itself. A simple way to present this, consistent with the observed symmetry-breaking pattern, is to assume the bilinear fermion operator $q_L\bar{q}_R$ has obtained a nonzero expectation value. This is often referred to as the fermions *forming a condensate.*

Dynamical symmetry breaking is also at work in superconductivity. Though the situation is quite different – the system of interest is non-relativistic and the interactions are weak – the properties of superconductors were explained by introducing a charge -2 composite of electrons [9]. Therefore, as in QCD, the symmetry breaking is explained not by fundamental scalar fields, but by new 'high-energy' dynamics.

In addition to having precedents in nature, dynamical symmetry breaking via asymptotically free gauge dynamics also explains large scale hierarchies. Given a UV scale and a coupling at that high scale, an exponentially smaller scale is automatically generated – without fine tuning any parameters! This is the mechanism of dimensional transmutation.

$$\Lambda_{IR} \sim \Lambda_{UV} \exp\left(\frac{-8\pi^2}{b_0 g^2(\Lambda_{UV})}\right) \tag{1}$$

The basic goal of technicolor [10, 11] theories is to break electroweak symmetry through some new asymptotically-free gauge interaction. We explore the basic setup in the next Section. Once dynamical electroweak symmetry breaking has been arranged, dimensional transmutation naturally explains why Λ_{EW} is exponentially smaller than Λ_{UV}. No other extension of the standard model can explain the gauge hierarchy as simply or as naturally! However, just as in QCD, dynamical electroweak symmetry breaking necessarily leads us to strong coupling and incalculability.

2 Classic Technicolor

Classic technicolor is heavily inspired by QCD. It was invented at a time when the heaviest fermion was the bottom quark, whose mass of ~ 5 GeV is much less than the weak scale.

The building blocks of technicolor are a set of massless fermions (technifermions) which feel a new non-abelian gauge interaction (technicolor) with new force-carrying fields (technigluons). We assume for simplicity that the technifermions are charged in the fundamental representation, though other representations can also be used. The left-handed components of the technifermions form electroweak doublets, while the right-handed components are electroweak singlets. Both left-handed and right-handed technifermions carry hypercharge. These charge assignments are summarized below:

$$T = \begin{pmatrix} U_L \\ D_L \end{pmatrix} \in (N_{TC}, 2)_{Y_L}$$

$$U_R \in (N_{TC}, 1)_{Y_U}\ , \quad D_R \in (N_{TC}, 1)_{Y_D}$$

under $(SU(N_{TC}), SU(2)_w)_{U(1)_Y}$, where $SU(N_{TC})$ is the technicolor strong interaction. Ignoring electroweak interactions, the technicolor sector has a global symmetry $SU(N_F)_L \otimes SU(N_F)_R$ where N_F is the number of techniflavors.

We assume the technicolor interaction becomes strong at $\sim \Lambda_{EW}$, causing the formation of a techni-condensate:

$$\langle U\bar{U}\rangle = \langle D\bar{D}\rangle = 4\pi F_T^3, \text{where } \langle U\bar{U}\rangle = 2\langle U_L U_R^\dagger\rangle. \tag{2}$$

As the left-handed technifermions carry electroweak quantum numbers while the right-handed technifermions do not, the formation of the techni-condensate breaks electroweak symmetry.

If we temporarily shut off the gauge couplings, the pattern of global symmetry-breaking in the technicolor sector is $SU(N_F)_L \otimes SU(N_F)_R$ down to the diagonal combination $SU(N_F)_D$ [1]. The fact that the low-energy theory contains a residual (or 'custodial') $SU(N_F)_D$ symmetry in the limit of zero gauge couplings is extremely important. Turning on the gauge interactions, this 'custodial' symmetry forces the $W^\pm$ and Z^0 masses have the correct pattern (up to loop-level electroweak corrections):

$$m_W^2 \sim \frac{g^2 \frac{N_F}{2} F_T^2}{4} = m_Z^2 \cos^2\theta, \tag{3}$$

where $N_F/2$ is the number of electroweak doublet technifermions. For a technicolor theory with more than one electroweak doublet (2 techniflavors), the symmetry-breaking produces more Goldstone bosons than can be eaten by the gauge bosons. These extra states are pseudoscalars and are massless at tree level. These states are generally called *techni-pions* and we will return to the interactions which give them mass shortly.

Without the ability to calculate at strong coupling, the simplest way to estimate the properties of an EW-scale strong interaction is to scale what we know from QCD. These estimates are often accompanied by additional scalings to account for a different number of technicolors or techniflavors as compared to QCD. However, all these scalings are naive in that they have no first-principles motivation. We have already drawn the connection between the pions in QCD and the techni-pions – including the longitudinal $W^\pm, Z$ of technicolor. We can extend the analogy further, postulating spin-1 vector and axial isovector mesons ρ_T, a_T, and isosinglets ω_T. The corresponding QCD states are the ρ, a_1, and ω. In QCD the spin-1 states are narrow, making them easier to find experimentally, however we expect technicolor would also lead to spin-0 and spin-2 states as well. For a naive estimate the masses and widths, we simply rescale the QCD values by factors of $\frac{F_T}{f_\pi}$, $\frac{N_{TC}}{3}$ and $\frac{N_F}{2}$ [5].

$$M_{\rho_T} \sim \sqrt{\frac{3}{N_{TC}}} \times 2\ \text{TeV}, \quad \Gamma(\rho_T \to W_L W_L) \sim 500\Big(\frac{3}{N_{TC}}\Big)^{3/2}\ \text{GeV}. \tag{4}$$

Given the many new states – techni-vectors, techni-pions, etc. in a technicolor model, how will they be produced and observed at colliders like the LHC? So far we have not described any interaction between the technicolor states and the standard model quarks and leptons, thus new technicolor states must be produced through their interactions with gauge bosons. The simplest mechanism, known as vector meson dominance, is shown on the left below: A SM gauge boson, $\gamma, W^\pm, Z$ is produced from the collision of two fermions. This gauge boson

[1] This often appears in the literature as $SU(2N_D)_L \otimes SU(2N_D)_R \to SU(2N_D)_D$ where N_D labels the number of electroweak doublets

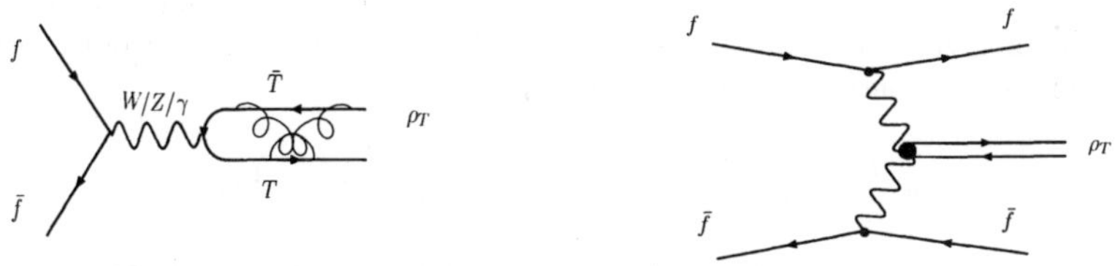

feels the effects of technifermions, thus can mix into a bound state of the technifermions, leading to an effective $f\bar{f} \rightarrow \rho_T$ interaction. This is identical to the mechanism used to describe QCD vector meson production in e^+e^- collisions. Another production mechanism, shown in the right-hand side of the above figure, is to produce the techniresonances by the fusion of gauge bosons which have been emitted by the colliding quarks. This mechanism is known as vector-boson fusion. Some early technicolor phenomenology was explored in Refs. [12, 13, 14].

3 Extending Technicolor

While the chiral symmetry breaking caused by technicolor provides the mechanism to give the $W^{\pm}, Z^0$ bosons mass, the technicolor interaction alone does not provide way to give mass to the standard model fermions. This crucial job is easily accomplished by the SM Higgs. However, to maintain a natural explanation of the weak scale (Eq.(1)) we cannot use scalars and we must get fermion masses using gauge dynamics alone.

The most-studied method for accomplishing this feat is known as Extended Technicolor [15]. The basic postulate of ETC is that there is a new gauge interaction – named extended technicolor – under which both fermions and technifermions transform. Putting SM fermions and technifermions into the same representations, a fermion can turn into a technifermion by emitting an ETC gauge boson. Since we see no other light gauge bosons (other than the SM gauge bosons), the ETC symmetry must be broken at some very high scale, giving the corresponding ETC gauge boson masses of the same order of magnitude. Ideally the breaking of ETC is also done dynamically, otherwise we have merely replaced the Higgs hierarchy problem for the ETC-hierarchy problem. Given the ambiguity about the structure and the severe constraints and functions ETC must fulfill, an exact model of ETC has so far been beyond our grasp. We must be content to play with the effective theory which results from integrating out the massive ETC bosons, leaving the full theory for the future. This is a bit of a cop-out, but no more than is necessary in SUSY or any other extension of the standard model.

Integrating out the massive ETC gauge bosons, we are left with a set of dimension six operators at the ETC energy scale [6]:

$$g^2_{ETC}\Big(\alpha_{ab}\frac{(\bar{T}\gamma_\mu t^a T)(\bar{T}\gamma^\mu t^b T)}{M^2_{ETC}}\Big|_{ETC} + \beta_{\alpha\beta}\frac{(\bar{T}\gamma_\mu t^a T)(\bar{q}\gamma^\mu t^b q)}{M^2_{ETC}}\Big|_{ETC} + \gamma_{\alpha\beta}\frac{(\bar{q}\gamma_\mu t^a q)(\bar{q}\gamma^\mu t^b q)}{M^2_{ETC}}\Big|_{ETC}\Big), \tag{5}$$

where t^a, t^b label the ETC generators, g_{ETC} is the ETC gauge coupling, and the '$|_{ETC}$' indicate that these operators are generated at the matching scale M_{ETC}. The field content of the operators indicates what role they play. The α_{ab} term contains only techni-fields and is responsible for giving mass to the techni-pions. The β_{ab} term contains both technicolor and SM fermions, thus once Fierz rearranged leads to:

$$\beta_{ab}\frac{g_{ETC}^2(\bar{T}_b T_b)(\bar{q}_a q_a)}{M_{ETC}^2} \longrightarrow \left(\frac{\beta_{ab} g_{ETC}^2 \langle\bar{T}T\rangle|_{ETC}}{M_{ETC}^2}\right)\bar{q}q\,, \tag{6}$$

and hence to quark and lepton masses. The remaining term with coefficient γ_{ab} contains only SM fields and will lead to potentially dangerous FCNC interactions.

While the techni-bilinear $\langle\bar{T}T\rangle$ is fixed at the electroweak scale by G_F the fermion mass estimates above depend on its value at the ETC scale. The two techni-bilinear values are related by the renormalization group equations (RGE). Ignoring all interactions except for the strong $SU(N_{TC})$, the RGE for $\langle\bar{T}T\rangle$ is [2]:

$$\langle\bar{T}T\rangle = \langle\bar{T}T\rangle \exp\left(\int_{\Lambda_{TC}}^{M_{ETC}} \frac{d\mu}{\mu}\gamma_m(\mu)\right)\,, \tag{7}$$

where γ_m is the anomalous dimension of $\langle\bar{T}T\rangle$. As technicolor is a strongly coupled theory, we cannot calculate this exactly, however we can get some idea by using QCD as an analogy. In QCD and at lowest order in perturbation theory, $\gamma_m \sim \mathcal{O}(\alpha) \ll 1$. If these two rather drastic approximations – QCD-like behavior and the adequacy of lowest order perturbation theory – can be extended to technicolor, then the anomalous dimension is small and can be dropped. In that extreme case,

$$\langle\bar{T}T\rangle\Big|_{ETC} \approx \langle\bar{T}T\rangle\Big|_{TC} \equiv 4\pi F_T^3\,. \tag{8}$$

Working with this QCD-inspired, small anomalous dimension assumption, we can plug in the techni-condensate into the β_{ab} term to get an estimate of the SM-fermion masses it leads to:

$$m_q, m_\ell \sim \frac{g_{ETC}^2}{M_{ETC}^2}\langle\bar{T}T\rangle|_{ETC} \sim \frac{g_{ETC}^2}{M_{ETC}^2}(4\pi F_T^3). \tag{9}$$

4 Obstacles to Overcome

Flavor-Changing Neutral Currents (FCNC)

Even at its inception it was recognized that ETC would be tightly constrained by flavor-changing neutral currents (FCNC) [15]. These FCNC come from the γ_{ab} terms of (5), and are most stringently constrained by $\Delta S = 2$ interactions in the Kaon system. The ETC-mediated effective interaction which contributes to $\Delta S = 2$ is [5]:

$$\frac{g_{ETC}^2\theta_{sd}^2}{M_{ETC}^2}(\bar{s}\Gamma_\mu d)(\bar{s}\Gamma^\mu d) + h.c.,$$

where Γ_μ is some Dirac matrix structure, and the quark flavors s, d have been selected by the ETC generators. We have included a mixing angle θ_{ds} which is presumably $\mathcal{O}(1)$. The

[2] Conventionally, following naive dimension analysis, $\Lambda_{TC} \cong 4\pi F_T$

$K - \bar{K}$ mass difference is constrained to be $\Delta m_K < 3.5 \times 10^{-12}$ MeV, which we can turn into a constraint on the ETC interactions [5]:

$$\frac{M_{ETC}}{g_{ETC}\sqrt{Re(\theta_{ds}^2)}} \geq 1.3 \times 10^3 \text{ TeV}. \tag{10}$$

The constraints on the imaginary part of the $K - \bar{K}$ mass difference, ϵ_K are tighter by three orders of magnitude, implying [5]

$$\frac{M_{ETC}}{g_{ETC}\sqrt{Im(\theta_{ds}^2)}} \geq 1.6 \times 10^4 \text{ TeV}. \tag{11}$$

Using ETC-scale parameters which pass the FCNC constraints, the resulting quark and lepton masses are forced to be very small,

$$m_q, m_l \sim \frac{0.1 \text{ MeV}}{\left(\frac{N_F}{2}\right)^{3/2} |\theta_{ds}|^2}. \tag{12}$$

Unless we are willing to fine-tune the mixing angles, $\mathcal{O}$(GeV) quark are masses difficult to achieve, and the top mass looks impossible.

Technicolor and Precision Electroweak

While FCNC are a problem of ETC and not the fundamental technicolor, that does not make technicolor problem-free. Technifermions, by definition and necessity, communicate with the EW gauge bosons. The properties of the electroweak sector have been measured to extreme accuracy at experiments such as LEP and SLD and thus the a priori strong contributions to electroweak observables are tightly constrained [16, 17]. Assuming the contributions of

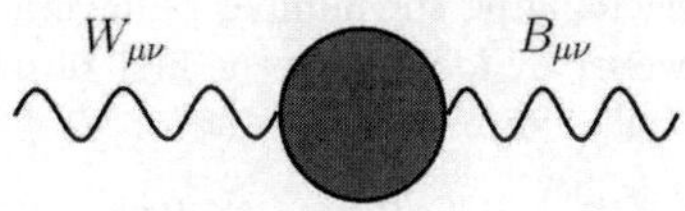

Figure 1: A typical technicolor contribution to precision electroweak quantities. The blob represents an incalculable quagmire of technifermions and technigluons.

new physics are felt only in the EW gauge sector – an assumption which is certainly valid up to ETC scale effects in classic technicolor – all deviations in the EW sector from tree level SM can be parameterized in terms of three quantities called S, T, and U. Precision measurements, taken together, restrict the allowed region of S,T, and U.

Deviations from SM tree level values come from two sources: loop level SM effects and tree-level new physics effects. The former contain divergences which are cut off by the Higgs mass, while the latter do not. However, to constrain new physics, one typically subtracts the SM-loop contribution from the extracted value of S,T,U. To maintain a finite observable the new physics contribution must inherit a piece which depends on the Higgs mass – whatever is necessary to cancel the SM loop part. Unfortunately, this often-confusing procedure results in an odd, residual scale dependence in the new physics contribution.

As we are unable to calculate the blobs in figure (1), we must make some assumption. The assumption made in the classic paper by Peskin and Takeuchi [16] is that the vector and axial current correlation functions which define S are each saturated by a single, narrow resonance. Rescaling with colors and flavors, the result is:

$$S = -4\pi \frac{d}{dq^2}\Big(\Pi_{VV}(q^2) - \Pi_{AA}(q^2)\Big)|_{q^2=0}$$

$$\text{single resonance} : S_{TC} = 4\pi\Big(1 + \frac{M^2_{\rho_T}}{M_{a_T}}\Big)\frac{F_T^2}{M^2_{\rho_T}} \cong 0.25\frac{N_C}{3}\frac{N_F}{2}, \tag{13}$$

where $M_{\rho_T}(M_{a_T})$ is the mass of the first vector (axial) techni-resonance. This value is difficult to reconcile with the current PEW best fit: $S = -0.10 \pm 0.10$ [18] (setting reference Higgs mass to 117 GeV – the limits are more negative if we increase the reference scale). QCD-like technicolor, especially with large values for N_C (large N_F too, though this is a subtler issue as we will see in the next section) appear to be ruled out. However, there is no first-principles reason why the current correlation functions in *all* strongly coupled theories are dominated by a single resonance. Such saturation does happen in QCD, but there is no reason to believe it is generic.

5 Non QCD-like Technicolor Dynamics: 'Walking Technicolor'

A common assumption in the last section was that the technicolor dynamics could be adequately represented by rescaling QCD. While QCD is a useful (and our only full) probe of strong dynamics, there is no reason to believe it is typical. However, playing devil's advocate, one may ask: what argument is there that a strongly coupled theory can be very different from QCD? Consider changing the number of flavors in QCD. At 2 or 3 flavors we have QCD as we know it, however by looking at the first terms in the beta function we get the sense that this QCD-like behavior cannot persist for all N_F. Setting $N_C = 3$:

$$\beta(\alpha_T) = -2\beta_0\frac{\alpha_T^2}{4\pi} - 2\beta_1\frac{\alpha_T^3}{(4\pi)^2} + \cdots \tag{14}$$

$$\text{for fundamentals} : \beta_0 = 11 - \frac{4}{3}\frac{N_F}{2}, \qquad \beta_1 = 102 - \frac{38}{3}N_F. \tag{15}$$

Eventually, at $N_F \approx 16$, the β_0 coefficient becomes negative indicating a loss of asymptotic freedom. A non-asymptotically free theory is not what we want for technicolor, however at intermediate values of N_F the theory may have drastically different properties than in 2 or 3 flavor QCD. For example, at $N_F \sim 12$, one can balance the β_0 and β_1 terms, leading to a zero in the beta function to order α_T^4. The value of α_T where this vanishing occurs, α_T^* is known as an interacting fixed point, since a vanishing beta function means α_T^* stays nearly fixed over wide range of scales [3]. This behavior is clearly distinct from a QCD-like beta function.

[3]For 12 flavor QCD, the value one gets for the interacting fixed point coupling using perturbation theory is $\alpha_T^* = \frac{6\pi}{25}$.

While the complete details of a theory with slowly running, yet strong coupling are, so far, incalculable, rescaling QCD is not necessarily a good approximation. Attempts to study the behavior of gauge theories with different numbers of fermion flavors using lattice techniques, are currently underway [19, 20, 21, 22, 23]. The idea of a slowly evolving technicolor coupling is known as Walking Technicolor [24, 25, 26]. In the next section we work out some of the ramifications of a walking coupling on the low-energy theory.

Walking and FCNC:

The entire line of logic that lead to the FCNC problem started with the assumption that the anomalous dimension of the technifermion condensate was small, $\mathcal{O}(\alpha_{TC})$ in analogy with QCD. Relaxing this assumption, we can ask what happens if the anomalous dimension is large, order 1. In that case, the condensate at the ETC scale, the scale relevant for determine fermion masses, is not the same as the TC scale condensate.

$$\langle\bar{T}T\rangle_{ETC} = \langle\bar{T}T\rangle_{TC}\exp\Big(\int_{\Lambda_{TC}}^{M_{ETC}}\frac{d\mu}{\mu}(1+\mathcal{O}(\alpha_T))\Big) \sim \langle\bar{T}T\rangle_{TC}\frac{M_{ETC}}{\Lambda_{TC}}, \tag{16}$$

$$\text{thus} \quad m_q, m_\ell \lesssim \frac{g_{ETC}^2}{M_{ETC}^2}(4\pi F_T^3)\Big(\frac{M_{ETC}}{\Lambda_{TC}}\Big) \sim \mathcal{O}(\text{few GeV})\,. \tag{17}$$

While this result isn't perfect, second generation fermion masses and possibly the b mass now look tractable. The tecnipion masses, caused by the α_{ab} term of Eq.(5) are also enhanced by walking,

$$m_{\pi_T} \sim \frac{g_{ETC}\langle\bar{T}T\rangle_{TC}}{F_T\Lambda_{TC}}. \tag{18}$$

One consequence of enhanced techni-pion masses which we will visit later is that the decays of technivector mesons to techni-pions, $\rho_T \to \pi_T\pi_T$, etc. may be kinematically closed.

While the enhancement one gets from a large anomalous dimension definitely eases the tension between FCNC and reasonable quark masses, how does one actually calculate the anomalous dimension in a strongly coupled theory? One technique employed early on to calculate the anomalous dimension in walking TC theories is the rainbow approximation of the Schwinger-Dyson equation (SDE) for the technifermion propagator [27, 28, 29]. The SDE are equations of motion for operators and extend beyond perturbation theory. However, to get a simple expression, typically one truncates the perturbation expansion at lowest order, while simultaneously neglecting the effects of the technicolor interaction on the technigluon propagator. With these assumptions, one indeed finds the anomalous dimension for the technifermion bilinear to be large.

$$\gamma_{m,SDE} = 1 - \sqrt{1 - \frac{3C_2(r)\alpha(\mu)}{\pi}} \to 1 \text{ as } \alpha(\mu) \to \frac{\pi}{3C_2(r)} \equiv \alpha_c, \tag{19}$$

where α_c is interpreted as the critical coupling necessary for chiral symmetry breakdown. This expression does reduce to the conventional form at small coupling, however the assumptions involved are severe. Attempts to improve the SDE analysis, both analytically and numerically have been attempted, with some evidence that the lowest order analysis is stable [30, 31, 32]. While the exact value of the anomalous dimension in Eq. (19) may be debatable, it is clear that large anomalous dimensions are possible in strongly coupled theories and can significantly improve the FCNC situation.

Walking and the S parameter

The effects of walking on the precision electroweak parameters are more speculative. At the very least, the QCD-based assumptions must be scrapped and we have to resign ourselves to the incalculability of generic strong interactions [33]. Said another way, the saturation of the correlation functions involved in S,T,U is unjustified – one may expect a large, slowly varying coupling to result in a whole tower of (closely spaced) resonances, all contributing to low-energy observables.

There has been some evidence that the near-conformal behavior of walking technicolor can also lead to a naturally reduced S-parameter [32, 34, 35, 36]. The primary argument is that near-conformal behavior is accompanied by a near parity doubling of the resonance spectrum. Thus the techni-rho and techni-a_T are nearly degenerate, as are their higher resonances. Simply from the definition of S (Eq. (4)), one can see that a degenerate spectrum leads to small deviations in the SM electroweak sector [4]. The arguments for a degenerate resonance spectrum are based on numerical SDE results and by simple models of the effect of intermediate energy scales on S.

Walking and the top mass

Based on SDE calculations, the maximum anomalous dimension a fermion bilinear can attain is $\gamma_m = 1$. However, even with maximal enhancement from walking, the large top quark mass is impossible to achieve with ETC scales consistent with Eq. (10, 11). In order to accommodate the top mass, additional ingredients must be added to technicolor (this is why, in my opinion, it is the biggest problem with technicolor theories). Several mechanisms have been suggested to ameliorate this problem:

- Tumbling [38, 39] technicolor theories: This idea suggests that there are different ETC scales for each generation: Λ_i for generation i, with $\Lambda_1 \gg \Lambda_2 \gg \Lambda_3$. By separating the scales, we can keep Λ_1 consistent with Eq.(10, 11), while lowering Λ_3 enough to generate realistic top masses. Unfortunately, it is difficult to build models which achieve multiple scales using gauge dynamics alone [40, 41].
- Topcolor-Assisted technicolor [42]: Another mechanism to enhance the top mass is to give the third generation separate, stronger $SU(3)_2$ and $U(1)_2$ gauge interactions. The $SU(3)_2$ (and $U(1)_2$) of the third generation and the $SU(3)_1, (U(1)_1)$ of the light generations are spontaneously broken at a scale ~ 5 TeV down to conventional color $SU(3)$ and hypercharge, leaving an octet of massive colored gauge bosons and a massive Z'. The heavy gauge bosons couple strongly to the third generation. If the coupling is strong enough, the dimension 6 four-fermion operator generated by heavy gauge boson exchange can cause a dynamical mass to form. The formalism which demonstrates

[4]For simple resonance models with zero or negative S see Ref. [37].

the formation of a dynamical mass resums the four-fermion interactions to all orders, and is due to Nambu and Jona-Lasinio [43, 44]. The exchange of the extra $U(1)$ gauge boson is arranged to be attractive for the top quark but repulsive for the bottom quark – so we get an enhanced top mass without a large bottom mass. This usually requires either a strong $U(1)_2$ coupling, leading to a Landau pole below the ETC scale, or an extreme tuning in the coupling of the third generation's $SU(3)$ coupling.

- Larger anomalous dimension: The limit $\gamma \lesssim 1$ is purely a result of the SDE approach. Forgetting this approximation, the actual limit on the anomalous dimension comes from unitarity and is $\gamma \lesssim 2$. Depending on the other ETC parameters, anomalous dimensions of $\gamma \sim 1.5$ are sufficient to generate the top mass with a consistent ETC scale, so perhaps the SDE is just an insufficient tool. Models aimed at $\gamma > 1$ have been proposed, under the name 'Strong ETC' [45], or, more recently, Conformal Technicolor [46].

Walking at low N_C with different fermion representations

In the above efforts in walking technicolor, we have focused on keeping the fermions in the fundamental representation while increasing their number. This is not the only approach. In fact, a whole interesting class of models have recently been constructed which utilize a small $SU(N_{TC})$ technicolor gauge group, and a small number of technifermions in a large technicolor representation [47, 48, 49, 50]. These theories have a minimal value of the perturbative S parameter, which can be reduced either by walking dynamics (see above) or by contributions from a new 'lepton' doublet. Depending on the N_{TC} and N_F, the existence of such 'leptons' is necessary to cancel the global $SU(2)_w$ Witten anomaly, however the hierarchy between the lepton and neutrino within the additional doublet necessary for a negative S is left as a requirement of the ETC theory. Within this class of theories there has been considerable effort to map the conformal window – the ranges of N_F, N_C for various technifermion representations where the theory is believed to exhibit walking behavior [51, 52]. There are also lattice efforts underway using different fermion representations [20, 53].

6 New Tools for Technicolor: Extra-dimensional Models

While walking technicolor offers a potential escape from FCNC and precision electroweak constraints, many results are simply incalculable. While incalculable models are difficult to rule out, they have fairly limited predictive power.

Recently much progress has been made describing strongly coupled gauge theories in d dimensions by using weakly coupled 'dual' gravity theories in $d+1$ dimensions. This duality has been most rigorously proven between specific supersymmetric conformal $4D$ theory and a particular string theory sitting in an 5d AdS background. This duality, known as AdS/CFT [54, 55], has inspired a new effort in modelling strong dynamics based on using weakly coupled 5D models. The general thinking is that we don't necessarily need a complete duality to gain some new insight; perhaps the symmetry pattern and separation of scales present in AdS is sufficient to gain some insight into the behavior of important operators like $\langle \bar{T}T \rangle$.

The starting point for an AdS model of strong interactions is a slice of AdS space: an interval in the fifth coordinate, z between two branes. The geometry in the interval is

$$ds^2 = \frac{\ell_0^2}{z^2}(\eta_{\mu\nu}dx^\mu dx^\nu - dz^2), \; z \in (\ell_0, \ell_1) \tag{20}$$

Without the presence of the branes, this geometry is invariant under scalings in z. Interpreting the fifth dimension as the energy/renormalization scale, the 5D model describes a theory which is scale invariant between two momentum scales, $\Lambda_{UV} = \frac{1}{\ell_0}$ and $\Lambda_{IR} = \frac{1}{\ell_1}$. The UV scale is the cutoff of the theory, while the IR scale is where conformal invariance is spontaneously broken. The AdS geometry alone models the 4D near conformality, but we also need to include chiral symmetry. This is done by including gauge fields in the bulk of AdS; to model a 4D theory with $SU(2)_L \otimes SU(2)_R$ chiral symmetry we introduce $SU(2)_L \otimes SU(2)_R$ 5D gauge fields propagating in the interval [5]. The strength of the 5D interactions is set by the 5D gauge coupling g_5. To model chiral symmetry breaking we can include a bulk scalar field – a 5D Higgs – which interacts with the gauge fields and has a nonzero vev. By choosing the bulk Higgs couplings correctly, when the Higgs breaks chiral symmetry (down to $SU(2)_V$) it also breaks electroweak symmetry in the observed pattern. In the 4D interpretation, the bulk Higgs corresponds to the $\langle \bar{T}T \rangle$ condensate. Alternatively, it is possible to break chiral and electroweak symmetry in these models using boundary conditions alone. Models without a bulk Higgs are known as 'Higgsless' models [56, 57, 58].

The next step in the 5D model is to reduce it to a 4D theory. This is done by the usual technique of KK decomposition, i.e. by writing each 5D field as a sum over 4D fields times a profile in the fifth dimension. The profiles are solved for using the classical equations of motion accompanied by boundary conditions (which are necessary since we are working in a finite extra dimension). By restricting to classical EOM, we are making the assumption that the theory is weakly coupled and that quantum effects $\mathcal{O}(g_5^2)$ are negligible. A small 5D couplings corresponds to a 4D theory with a large number of colors (large-N_C).

With the profiles determined, the fifth dimension can be integrated over and we are left with a series of 4D fields. Within each KK tower, all fields have the same quantum numbers but different 5D profiles and thus different mass. The lightest fields – known as 'zero-modes' – are interpreted as the SM gauge fields $W^\pm, Z, \gamma$. The subsequent replicas of the SM fields are now the resonances of the strongly interacting theory.

Using a 5D description we have modeled a strongly interaction theory, yet the observables we get at the end are different than what we get from 4D models. In a 4D model, we know the microscopic degrees of freedom – the technifermion lagrangian, but we know nothing about the resonance masses and interactions. In 5D the opposite is true; we have the masses and the couplings of the resonances, but the only clues we have about the fundamental theory are the chiral symmetry and large-N_C.

With the resonance masses and couplings determined, we can calculate observables like S. Unfortunately, in the simplest AdS-5D models we find S is large and positive [57, 59, 60].

[5]We put $SU(2)_L \otimes SU(2)_R$ in the bulk rather than $SU(2)_L \otimes U(1)_Y$ because we want our low energy theory (after chiral symmetry breaking) to contain a custodial symmetry.

By examining the spectrum, we can get some idea as to why, after all this 5D work, we are back to square one. The spectrum – the mass ordering of different parity excitations of the simplest AdS models is very similar to QCD, and the correlation functions involved in S are indeed dominated by the lowest resonances. While it is certainly a useful tool, 5D modeling needs to be extended beyond simple AdS in order to provide useful insight into phenomenologically viable technicolor.

7 Technicolor Phenomenology in the LHC Era

As we have seen, model building for strongly coupled models of EWSB is extremely difficult. The simplest approaches, either based on QCD-rescaling or 5D-AdS models do not yield viable models. In order to proceed, phenomenologists have adopted a new approach of effective lagrangians. Rather than attempt to write a complete high-energy theory, instead write down a consistent set of interesting interactions, imposing phenomenological constraints to limit the number and size of interactions. In the next section we explore two example phenomenological technicolor models and their phenomenology.

Low-Scale Technicolor

To have consistent fermion masses without excessive FCNC, technicolor must be a walking theory, and therefore we need a lot of technimatter (matter drives the β_0 coefficient to smaller values). This technimatter can either be many fermions in the fundamental representation, or technifermions in multiple representations. Because each technidoublet contributes to the effective EWSB scale (Eq. (3)), there is a relation between the F_T scale of each technifermion and $v = 246$ GeV. For N_F technifermions in the same representation, the relation is $v^2 = NF_T^2$. If the technifermions are in multiple represenstations the relation is generalized to $v^2 = \sum_i N_i F_{T,i}^2$, where we sum over all represenatations [6]. In either case, lots of technimatter implies a small F_T scale. The class of walking theories which contains a low scale are known as Low-Scale Technicolor (LSTC) scenarios.

In a Low-Scale Technicolor scenario the resonances associated with the low scale will be light $\sim \mathcal{O}(200 - 500$ GeV). Assuming these resonances can be treated in isolation (that is, without considering the other resonances in the theory), the couplings of the low-scale resonances to fermions and gauge bosons are suppressed by the ratio of F_i scales and the ratio of masses,

$$g_{ff\rho_T} \sim g_{ffW} \frac{F_1}{v} \frac{m_W}{M_{\rho_T}}. \tag{21}$$

This allows light resonances without violating direct detection bounds from LEP and the Tevatron. Additionally, lots of technimatter implies a large chiral symmetry in LSTC models and therefore a lot of techni-pions. These techni-pions can interact with the SM gauge bosons, fermions, and the TC resonances. Because of walking effects, the techni-pion mass

[6] The compositeness scale dictated by (Eq. (1)) depends on the representation so different representations have different F_T, with the largest technifermion representation having the largest F_T scale. The simplest multi-representation model is Two-Scale technicolor [61], which postulates only two technicolor scales, F_1 and F_2.

is enhanced relative to the spin-1 techni-mesons (ρ_T, a_T). As a result, decay modes like $\rho_T \to \pi_T^+\pi_T^-, a_T \to 3\pi_T$ are likely to be closed and the resonances are forced to decay into SM gauge bosons. The couplings to SM gauge bosons are small (see Eq. (21), therefore the TC resonances in this scenario are strikingly narrow, $\Gamma_\rho \sim \mathcal{O}(.5\ \text{GeV})$ [62]. The final assumption of LSTC is that the ρ_T and a_T are nearly degenerate [63]. This assumption is based on the idea, as explained in section (5) that the S parameter is reduced in walking theories due to spectrum degeneracy.

The simplest and most explored LSTC scenario, known as the technicolor 'straw-man' model postulates $\mathcal{O}(9)$ doublets, so $F_T \sim \frac{v}{3}$. Several Tevatron studies of the 'straw-man' model have been performed in the past, focusing on the modes $p\bar{p} \to \rho_T \to \pi_T W$ [64, 65, 66, 67, 68]. Due to the excessive $\bar{t}t$ background $\rho_T \to \pi_T W$ is impossible at the LHC. Instead, the best modes for discovery are $\rho_T \to WZ \to 3\ell + \nu, a_T \to \gamma W \to \ell\nu\gamma$ and $\omega_T \to \gamma Z \to \ell\ell\gamma$. These decay modes have multiple photons and leptons, thus milder backgrounds and better mass resolution. The discovery potential in these three modes at the LHC was investigated recently in Ref. [69]. The signals are striking – all of these resonances could be discovered within the first few fb^{-1} of LHC running.

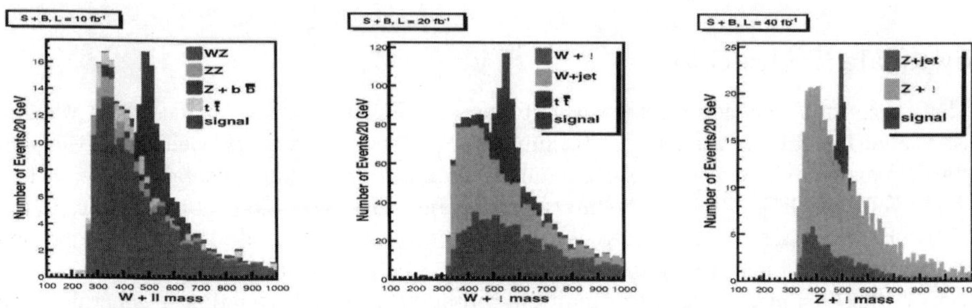

Figure 2: LSTC signal plus SM background in the $WZ, W\gamma$, and $Z\gamma$ channels. The signal in each channel is dominated by a single resonance: ρ_T in WZ, a_T in $W\gamma$ and ω_T in $Z\gamma$. These resonances could be discovered within the first few fb^{-1} of LHC running. The neutrino momentum in the WZ and $W\gamma$ channels is reconstructed by the reqiurement that $M_{(p_\ell + p_\nu)} = M_W$. The details of the analysis leading to these plots can be found in Ref. [69].

Resonance searches in the modes $3\ell + \nu, \ell\nu\gamma$, and $\ell\ell\gamma$ have received little attention, partially because they have no SUSY analogue. In the MSSM one could imagine searches for $H^\pm$ in the $3\ell + \nu$, or $\ell\nu\gamma$ channels, however, the $H^\pm W^\pm Z$, $H^\pm\gamma W^\pm$ interactions vanish at tree level in the MSSM (or in any other two-Higgs doublet model).

Holographic Technicolor

Until recently, all modern technicolor collider studies were done in the LSTC context [7]. While these are important studies, it is also important to explore other effective theories and frameworks of EW-scale strong interactions.

To study a wider array of technicolor models and access phenomenology not present in the LSTC, a variant of the Higgsless setup, known as Holographic Technicolor (HTC) was developed in Ref. [70, 71] . The setup of HTC is identical to the Higgsless setup reviewed in section (6), but with one important difference – the assumption that all bulk fields feel the same 5D geometry has been dropped. Instead, HTC assumes a more general action

$$\mathcal{L} = -\frac{1}{g_5^2}\int d^5x\ \omega_V(z)(F_{V,NM}F_V^{NM}) + \omega_A(z)(F_{A,NM}F_A^{NM}) \tag{22}$$

$$\omega_{V,A}(z) = \frac{\ell_0}{z}\exp o_{V,A}\Big(\frac{z}{\ell_1}\Big)^4, \quad o_{V,A} \lesssim 0$$

The extra parameters in HTC relative to traditional Higgsless are the geometry parameters $o_{V,A}$. Nonzero values for these parameters indicate a deviation from AdS in the IR which grows stronger with increasing z. The exact power of 4 is unimportant; a different power of z can be compensated with a different $o_{V,A}$ and will lead to the same physics. In practice $\omega_V \neq \omega_A$ allows us to dial the properties of the axial resonances independently of the vector resonances. This is not possible in LSTC, where a techni-parity symmetry is assumed. Actually, assuming a techni-parity symmetry is artificial since the electroweak interactions *automatically* break parity. By divorcing vector from axial, we can study a much wider subset of technicolor scenarios. A few benchmark scenarios which focused on $M_{\rho_T} \cong M_{a_T}$ for precision electroweak considerations [70], were studied in [72, 73]. The key difference between Ref. [72, 73] and LSTC is that both the ρ_T and a_T can couple to $WZ, W\gamma$, resulting in the distinct two-peak feature seen below. In LSTC, the $a_T WZ$ and $\rho_T WZ$ couplings are suppressed by techniparity. More complicated processes, such as vector boson fusion $pp \to W/Z + jj$, which probes the resonance effects on unitarizing gauge boson scattering, and associated production $pp \to W^* \to W/Z + \rho_T$ have also been studied. These processes, which involve more final state particles and elaborate cuts to remove SM backgrounds, often require high luminosity ($100 - 300$ fb^{-1}) for discovery.

8 Conclusions

The goal of the LHC is to unravel the mechanism behind electroweak symmetry breaking. We should be open to all options and not just those with pretty, but irrelevant aspects. Technicolor is the *only* completely natural solution to the gauge hierarchy problem, however it involves strong coupling. I view this as an interesting, compelling challenge rather than a deterrent. There is, as yet, no evidence for Technicolor. However, as with supersymmetry or most other schemes for beyond the SM physics – technicolor cannot be ruled out. To ameliorate the severe flavor constraints, modern technicolor is believed to have a walking

[7] The reason so many studies exist for the 'straw-man' technicolor model is that it is currently the only technicolor model implemented in the ubiquitous Monte Carlo program PYTHIA.

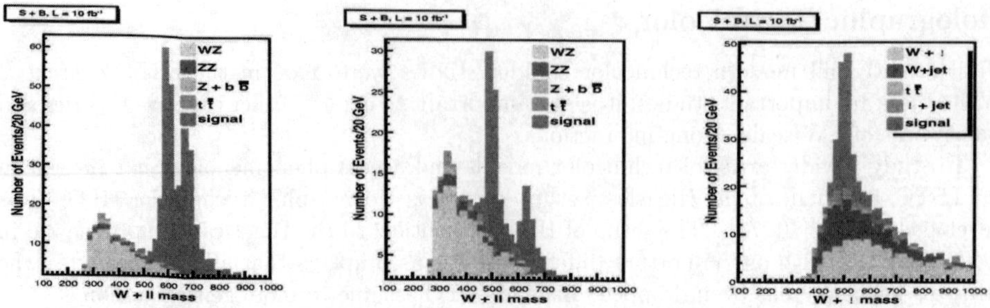

Figure 3: Possible resonance signals in the WZ channel (left two figures), or $W\gamma$ channel (rightmost figure) generated within the HTC framework. By dialing the geometry parameters o_V, o_A we recreate a scenario where *both* the two lightest charged resonance KK states couple to WZ or $W\gamma$, leading to the striking two-peak feature. For the details of the parameter choices in these plots see Ref. [72]

(nearly conformal) coupling over a wide range of energies. This departure from familiar strong interactions makes model building challenging, and phenomenologists have turned to effective theories for collider studies. While not fundamental, effective theories of strong interactions describe a host of striking features which are well within the grasp of the LHC.

Acknowledgements

I would like to thank the organizers Professors Antonino Zichichi and Gerard 't Hooft for the invitation to attend and lecture at this school. I would also like to thank Prof. Kenneth Lane for recommending me to the organizers and for his guidance in preparing the lecture and notes. AM is supported by DOE grant DE-FG02-92ER-40704.

References

[1] K. D. Lane, "An Introduction to technicolor," hep-ph/9401324.

[2] K. D. Lane, "Technicolor," hep-ph/9501249.

[3] R. S. Chivukula, "Models of electroweak symmetry breaking," hep-ph/9803219.

[4] K. D. Lane, "Technicolor 2000," hep-ph/0007304.

[5] K. Lane, "Two lectures on technicolor," hep-ph/0202255.

[6] C. T. Hill and E. H. Simmons, "Strong dynamics and electroweak symmetry breaking," *Phys. Rept.* **381** (2003) 235–402, hep-ph/0203079.

[7] K. G. Wilson, "Renormalization group and critical phenomena. 2. Phase space cell analysis of critical behavior," *Phys. Rev.* **B4** (1971) 3184–3205.

[8] K. G. Wilson and J. B. Kogut, "The Renormalization group and the epsilon expansion," *Phys. Rept.* **12** (1974) 75–200.

[9] J. Bardeen, L. N. Cooper, and J. R. Schrieffer, "Theory of superconductivity," *Phys. Rev.* **108** (1957) 1175–1204.

[10] S. Weinberg, "Implications of Dynamical Symmetry Breaking: an addendum," *Phys. Rev.* **D19** (1979) 1277–1280.

[11] L. Susskind, "Dynamics of Spontaneous Symmetry Breaking in the Weinberg-Salam Theory," *Phys. Rev.* **D20** (1979) 2619–2625.

[12] E. Eichten, I. Hinchliffe, K. D. Lane, and C. Quigg, "Super Collider Physics," *Rev. Mod. Phys.* **56** (1984) 579–707.

[13] J. Bagger *et. al.*, "The Strongly interacting W W system: Gold plated modes," *Phys. Rev.* **D49** (1994) 1246–1264, hep-ph/9306256.

[14] J. Bagger *et. al.*, "CERN LHC analysis of the strongly interacting W W system: Gold plated modes," *Phys. Rev.* **D52** (1995) 3878–3889, hep-ph/9504426.

[15] E. Eichten and K. D. Lane, "Dynamical Breaking of Weak Interaction Symmetries," *Phys. Lett.* **B90** (1980) 125–130.

[16] M. E. Peskin and T. Takeuchi, "A new constraint on a strongly interacting Higgs sector," *Phys. Rev. Lett.* **65** (1990) 964–967.

[17] M. Golden and L. Randall, "Radiative corrections to electroweak parameters in technicolor theories," *Nucl. Phys.* **B361** (1991) 3–23.

[18] **Particle Data Group** Collaboration, C. Amsler *et. al.*, "Review of particle physics," *Phys. Lett.* **B667** (2008) 1.

[19] T. Appelquist, G. T. Fleming, and E. T. Neil, "Lattice Study of the Conformal Window in QCD-like Theories," *Phys. Rev. Lett.* **100** (2008) 171607, 0712.0609.

[20] S. Catterall and F. Sannino, "Minimal walking on the lattice," *Phys. Rev.* **D76** (2007) 034504, 0705.1664.

[21] A. Deuzeman, M. P. Lombardo, and E. Pallante, "Hunting for the Conformal Window," 0810.3117.

[22] Z. Fodor, K. Holland, J. Kuti, D. Nogradi, and C. Schroeder, "Probing technicolor theories with staggered fermions," 0809.4890.

[23] Z. Fodor, K. Holland, J. Kuti, D. Nogradi, and C. Schroeder, "Nearly conformal electroweak sector with chiral fermions," 0809.4888.

[24] B. Holdom, "Raising the sideways scale," *Phys. Rev.* **D24** (1981) 1441.

[25] T. W. Appelquist, D. Karabali, and L. C. R. Wijewardhana, "Chiral Hierarchies and the Flavor Changing Neutral Current Problem in Technicolor," *Phys. Rev. Lett.* **57** (1986) 957.

[26] K. Yamawaki, M. Bando, and K.-i. Matumoto, "Scale invariant technicolor model and a technidilaton," *Phys. Rev. Lett.* **56** (1986) 1335.

[27] H. Pagels, "Departures from Chiral Symmetry: A Review," *Phys. Rept.* **16** (1975) 219.

[28] R. Fukuda and T. Kugo, "Schwinger-Dyson Equation for Massless Vector Theory and Absence of Fermion Pole," *Nucl. Phys.* **B117** (1976) 250.

[29] K. Higashijima, "Dynamical Chiral Symmetry Breaking," *Phys. Rev.* **D29** (1984) 1228.

[30] T. Appelquist, K. D. Lane, and U. Mahanta, "ON THE LADDER APPROXIMATION FOR SPONTANEOUS CHIRAL SYMMETRY BREAKING," *Phys. Rev. Lett.* **61** (1988) 1553.

[31] A. G. Cohen and H. Georgi, "WALKING BEYOND THE RAINBOW," *Nucl. Phys.* **B314** (1989) 7.

[32] M. Kurachi and R. Shrock, "Study of the change from walking to non-walking behavior in a vectorial gauge theory as a function of N(f)," *JHEP* **12** (2006) 034, hep-ph/0605290.

[33] K. D. Lane, "Technicolor and precision tests of the electroweak interactions," hep-ph/9409304.

[34] T. Appelquist and F. Sannino, "The physical spectrum of conformal SU(N) gauge theories," *Phys. Rev.* **D59** (1999) 067702, hep-ph/9806409.

[35] T. Appelquist, P. S. Rodrigues da Silva, and F. Sannino, "Enhanced global symmetries and the chiral phase transition," *Phys. Rev.* **D60** (1999) 116007, hep-ph/9906555.

[36] M. Kurachi and R. Shrock, "Behavior of the S parameter in the crossover region between walking and QCD-like regimes of an SU(N) gauge theory," *Phys. Rev.* **D74** (2006) 056003, hep-ph/0607231.

[37] M. Knecht and E. de Rafael, "Patterns of spontaneous chiral symmetry breaking in the large N(c) limit of QCD-like theories," *Phys. Lett.* **B424** (1998) 335–342, hep-ph/9712457.

[38] S. Raby, S. Dimopoulos, and L. Susskind, "Tumbling Gauge Theories," *Nucl. Phys.* **B169** (1980) 373.

[39] Y. Kikukawa and N. Kitazawa, "Tumbling and technicolor theory," *Phys. Rev.* **D46** (1992) 3117–3122.

[40] T. Appelquist, M. Piai, and R. Shrock, "Fermion masses and mixing in extended technicolor models," *Phys. Rev.* **D69** (2004) 015002, hep-ph/0308061.

[41] T. Appelquist, N. D. Christensen, M. Piai, and R. Shrock, "Flavor-changing processes in extended technicolor," *Phys. Rev.* **D70** (2004) 093010, hep-ph/0409035.

[42] C. T. Hill, "Topcolor assisted technicolor," *Phys. Lett.* **B345** (1995) 483–489, hep-ph/9411426.

[43] Y. Nambu and G. Jona-Lasinio, "Dynamical model of elementary particles based on an analogy with superconductivity. I," *Phys. Rev.* **122** (1961) 345–358.

[44] Y. Nambu and G. Jona-Lasinio, "Dynamical model of elementary particles based on an analogy with superconductivity. II," *Phys. Rev.* **124** (1961) 246–254.

[45] R. S. Chivukula, A. G. Cohen, and K. D. Lane, "ASPECTS OF DYNAMICAL ELECTROWEAK SYMMETRY BREAKING," *Nucl. Phys.* **B343** (1990) 554–570.

[46] M. A. Luty and T. Okui, "Conformal technicolor," *JHEP* **09** (2006) 070, hep-ph/0409274.

[47] F. Sannino and K. Tuominen, "Techniorientifold," *Phys. Rev.* **D71** (2005) 051901, hep-ph/0405209.

[48] N. Evans and F. Sannino, "Minimal walking technicolour, the top mass and precision electroweak measurements," hep-ph/0512080.

[49] D. D. Dietrich, F. Sannino, and K. Tuominen, "Light composite Higgs and precision electroweak measurements on the Z resonance: An update," *Phys. Rev.* **D73** (2006) 037701, hep-ph/0510217.

[50] R. Foadi, M. T. Frandsen, T. A. Ryttov, and F. Sannino, "Minimal Walking Technicolor: Set Up for Collider Physics," *Phys. Rev.* **D76** (2007) 055005, 0706.1696.

[51] D. D. Dietrich and F. Sannino, "Walking in the SU(N)," *Phys. Rev.* **D75** (2007) 085018, hep-ph/0611341.

[52] T. A. Ryttov and F. Sannino, "Conformal Windows of SU(N) Gauge Theories, Higher Dimensional Representations and The Size of The Unparticle World," *Phys. Rev.* **D76** (2007) 105004, 0707.3166.

[53] T. DeGrand, Y. Shamir, and B. Svetitsky, "Exploring the phase diagram of sextet QCD," 0809.2953.

[54] S. S. Gubser, I. R. Klebanov, and A. M. Polyakov, "Gauge theory correlators from non-critical string theory," *Phys. Lett.* **B428** (1998) 105–114, hep-th/9802109.

[55] E. Witten, "Anti-de Sitter space and holography," *Adv. Theor. Math. Phys.* **2** (1998) 253–291, hep-th/9802150.

[56] C. Csaki, C. Grojean, H. Murayama, L. Pilo, and J. Terning, "Gauge theories on an interval: Unitarity without a Higgs," *Phys. Rev.* **D69** (2004) 055006, hep-ph/0305237.

[57] C. Csaki, C. Grojean, L. Pilo, and J. Terning, "Towards a realistic model of Higgsless electroweak symmetry breaking," *Phys. Rev. Lett.* **92** (2004) 101802, hep-ph/0308038.

[58] G. Cacciapaglia, C. Csaki, C. Grojean, and J. Terning, "Curing the ills of Higgsless models: The S parameter and unitarity," *Phys. Rev.* **D71** (2005) 035015, hep-ph/0409126.

[59] K. Agashe, A. Delgado, M. J. May, and R. Sundrum, "RS1, custodial isospin and precision tests," *JHEP* **08** (2003) 050, hep-ph/0308036.

[60] K. Agashe, C. Csaki, C. Grojean, and M. Reece, "The S-parameter in holographic technicolor models," *JHEP* **12** (2007) 003, 0704.1821.

[61] K. D. Lane and E. Eichten, "Two Scale Technicolor," *Phys. Lett.* **B222** (1989) 274.

[62] K. D. Lane and M. V. Ramana, "Walking technicolor signatures at hadron colliders," *Phys. Rev.* **D44** (1991) 2678–2700.

[63] E. Eichten and K. Lane, "Low-scale technicolor at the Tevatron and LHC," 0706.2339.

[64] E. Eichten and K. D. Lane, "Low-scale technicolor at the Tevatron," *Phys. Lett.* **B388** (1996) 803–807, hep-ph/9607213.

[65] E. Eichten, K. D. Lane, and J. Womersley, "Finding low-scale technicolor at hadron colliders," *Phys. Lett.* **B405** (1997) 305–311, hep-ph/9704455.

[66] K. D. Lane, "Technihadron production and decay in low-scale technicolor," *Phys. Rev.* **D60** (1999) 075007, hep-ph/9903369.

[67] K. Lane and S. Mrenna, "The collider phenomenology of technihadrons in the technicolor Straw Man Model," *Phys. Rev.* **D67** (2003) 115011, hep-ph/0210299.

[68] K. Lane, "Search for low-scale technicolor at the Tevatron," hep-ph/0605119.

[69] G. Brooijmans *et. al.*, "New Physics at the LHC: A Les Houches Report. Physics at Tev Colliders 2007 – New Physics Working Group," 0802.3715.

[70] J. Hirn and V. Sanz, "A negative S parameter from holographic technicolor," *Phys. Rev. Lett.* **97** (2006) 121803, hep-ph/0606086.

[71] J. Hirn and V. Sanz, "The fifth dimension as an analogue computer for strong interactions at the LHC," hep-ph/0612239.

[72] J. Hirn, A. Martin, and V. Sanz, "Benchmarks for new strong interactions at the LHC," *JHEP* **05** (2008) 084, 0712.3783.

[73] J. Hirn, A. Martin, and V. Sanz, "Describing viable technicolor scenarios," *Phys. Rev.* **D78** (2008) 075026, 0807.2465.

CHAIRMAN: A. MARTIN

Scientific Secretaries: P. Kopnin, A. Tochin

DISCUSSION

- *Nicotri:*

You have said that in Technicolor the coupling constant is running slowly. How is the running (or "walking") of the coupling constant encoded in the holographic representation of this theory? Do you assume that it does not run or is it expressed in some part of the theory?

- *Martin:*

In the holographic five-dimensional versions of Technicolor theories the coupling constant really doesn't enter at all. Essentially the fifth dimension is just a tool for making an effective theory of the resonances and the electroweak symmetry breaking, so the dynamics of any coupling, associated with that, doesn't really show up.

- *Goebel:*

Related to the S-T plot from Erler Langacker. The S and T parameters are computed with a fixed Higgs mass. Why do you compare these parameters with a higgsless model?

- *Martin:*

Usually you have to specify what the Higgs mass you're using when you're calculating S, since the calculation, as you've said, these blobs here depend on it. With no Higgs you have to instead supply something else; either the first resonance mass, or a cut-off of the theory, some scale present in the Technicolor model. Here, I think, 1 TeV was used, though I don't quite remember.

- *Dubynskiy:*

I have a very short question. When you introduced the basic concept of Technicolor you drew an analogy between QCD and TC, thus introducing for pions -- technipions, for the rho-meson – technimeson etc. But why did you introduce also the gauge bosons for the real pions?

- *Martin:*

When condensation occurs and Technicolor breaking happens, I get a whole bunch of Nambu-Goldstone bosons. This is exactly what happens if you took away all quark masses and gauge interactions in the chiral Lagrangian; you would just have three degrees of freedom, the pions π. Here (in Technicolor) we have a whole bunch more Nambu-Goldstone bosons. Technicolor contains some gauge interactions, the Standard Model as part of that symmetry. Once the (global) symmetry breaking happens, three of what would be just the Goldstone bosons are eaten by the W and the Z: this is the Higgs mechanism in a Technicolor theory. The rest of the guys that aren't eaten are still pseudoscalars, and they're still in the spectrum. Those are the technipions. The π_T eaten by the W and the Z essentially become the longitudinal degrees of freedom.

- *Dubynskiy:*

Are they composed from techniquarks?

- *Martin:*

Yes.

- *Kaminska:*

Is there a natural way to place the electroweak symmetry breaking mechanism in the history of the Universe and explain why, how and when it happened?

- *Martin:*

You want to know the cosmological implications and how this would happen in the Universe? Well, Technicolor is really just a model of electroweak energies. What happenes from then on, we don't even dare to say. It's hard enough to get it to work for the energies that we do know. But yes, if confirmed, there should be immediate work in that area, because, I'm sure, there are plenty of interesting questions and answers that we can get.

- *Marshall:*

Towards the end of your talk you showed a handful of plots with invariant mass. There is a nice signal over background. I'm wondering if that went through any detector simulation at all?

- *Martin:*

It went through the three detector simulations available to a phenomenologist like me, just known as the "Pretty Good Simulator" or PGS. We tuned it (PGS) a little bit according to some recommendations from ATLAS collaborators that we had during this LHC school (Les Houches BSM 2007), the proceedings of which are in the archive, arXiv:0802.3715 [hep-ph].

- *Marshall:*

A comment to you and any other phenomenologist who has his or her favorite model. There is a standard event format now, HEPMC, which was created by ATLAS and was adopted by LCG and some other groups, and if you can translate your events into HEPMC, it's straightforward to run them through the ATLAS detector simulation, see what comes out with real resolutions and real detector effects.

- *Martin:*

Can I, as a phenomenologist, publish something using your ATLAS simulator? We (Les Houches 2007 collaboration) have had problems with that before.

- *Marshall:*

The right way to do this, I guess, is to join up with the Exotics group, and you can be a collaborator with the ATLAS Exotics group. They'll need some help interpreting the data, I think.

- *Susskind:*

In the original Technicolor idea, you could pretty much call it higgsless, this was just QCD scaled up by a factor of a thousand. Then there would be a broad resonance, there would be the scaled up version of the sigma, which is not much of a resonance, there would be this broad washed out thing that you would not see as a narrow Higgs boson. I believe the idea in that sense of a higgsless theory is ruled out, that in any kind of theory the has to be a narrow resonance, whose mass is probably not much bigger than 170 GeV. Take this as a question: in all of your theories that you're talking about □ "walking", "crawling", "dying" Technicolor, whatever, □ is there generally a narrow resonance that looks like the Higgs boson?

- *Martin:*

A narrow, spin zero, scalar – no.

- *Susskind:*

Then there's something wrong with what you're saying.

- *Martin:*

Why do you assume that it is ruled out?

- *Susskind:*

I think it's ruled out by high precision electroweak things. I think we know this pretty tight connection between the width of the resonance or the width of the Higgs and the mass of the Higgs, and I think we know that the mass of the Higgs cannot be pushed up much above about 170 GeV, at which point it's fairly narrow.

- *Martin:*

But these analyses are based on using calculations of a Higgs plus corrections in perturbation theory, so they assume that your New Physics at that scale is weakly coupled…

- *Susskind:*

So are you saying that there is not going to be a Higgs boson at the LHC? In these theories?

- *Martin:*

I would assume it (the Higgs) was like you would say in original Technicolor: there would be a wide sigma-like resonance that would be difficult to discover.

- *Susskind:*

I think that's ruled out.

- *Martin:*

No, we can talk about this later, I disagree.

- *Ni:*

I am wondering if there are any viable WIMP-like or axion-like particles as Dark Matter candidates in the Technicolor theory?

- *Martin:*

That is a good question and one that is difficult to answer without any explicit model of what the matter content is, its charge under this Technicolor theory. People have made models where they assumed there are some technibaryons, the technibaryon number would have to be protected and of course we have to assign charges so that it's not violated by anomalies, and they have done some studies. The biggest problem is that with this strongly coupled theory people cannot calculate the annihilation cross-section. No one has any idea how to figure out the relic densities and they can't figure out exactly how these things are going to annihilate. There have been some studies, I know, say Chivukula and his collaborators, on technibaryon Dark Matter in the early nineties and recently there is a Technicolor group based in Denmark with Francesco Sannino and they have recently written a paper about technibaryon Dark Matter. I'm not an expert on that, so I would refer you to those papers.

- *Kopnin:*

The techniquarks transform under the SU(2)-left group of the Standard Model. How does this affect the unification of the couplings? Is that considered?

- *Martin:*

We couldn't care less. There are a whole bunch of states that are coming in, maybe they are even charged under color or something like that, so the unification is definitely not something that Technicolor cares about at all. People have thought about it and have made models where you can arrange things to work, but it's certainly not a driving force behind the model.

- *'t Hooft:*

Could you elaborate a bit more on the five-dimensional picture?

- *Martin:*

For me any of these 5D theories are just basically an effective theory, they're non-renormalizable, they're going to break down at some scale, it's just an effective tool that's used to calculate resonances and how they interact. If you take the simplest theory where they have basically just the AdS metric for the gauge fields, you can calculate S and you can find that it's too big. Since we don't know, what we'll see at the LHC, we tried to be more general, this (*referring to work* arXiv: 0712.3783 [hep-ph]) is just a parametrization with a few different dials that will allow us to generate a different spectrum then we can get usually.

- *'t Hooft:*

It seems to be quite opposite to your theories of "walking technicolor" where things scale slowly. In a five-dimensional theory nothing scales slowly because such a theory is non-renormalizable.

- *Martin:*

In a 5D theory this presence of the infrared brane is what breaks the conformal invariance, that's where everything happens all at once, so what this (*referring to different metrics*) is actually doing, is introducing a little bit of symmetry breaking that extends into higher energies or in the extra dimension. So, it's something that is still on the same sphere. Things are happening all at one scale, there is a sort of a slow effect this time in the form of the z-dependent function here that incorporates that part.

- *Susskind:*

You're missing the point that the 5-dimensional gravity theory is the 4-dimensional gauge theory. This whole thing is using the duality between the conformal field theories or approximately conformal field theories and 5-dimensional ...

-*'t Hooft:*

That's what he is doing, but it sounds different.

- *Susskind:*

That's what the point is. I don't know if that's what he is doing or not, but that is what people do.

- *Martin:*

What we are doing is basically saying that tools we have so far to describe the model, don't work. We need another tool. This is just an attempt to do that.

- *Susskind:*

But this is presumably a tool for studying a 4-dimensional quantum field theory by using a 5-dimensional gravity dual.

- *Martin:*

We don't need this actual duality. The LHC is not going to give us precision data. We are using this as a nice way to set up our effective theory, so I don't need any of the higher mathematics of the AdS/CFT duality, it's just a model.

- *Susskind:*

But the 5-dimensional field theory is inconsistent, it's not renormalizable and so forth. So what might be behind and be the completion of it, be the real theory, would be some 4-dimensional gauge theory being described through AdS/CFT-type duality.

- *Martin:*

Absolutely, I just don't know that 4D theory.

- *'t Hooft:*

After having a nice little bit of confusion, this was just the last question and the end of today's exciting discussion section. I thank the speaker once again.

The One-Parameter Model at LHC

James Maxin[a], Eric Mayes[b], D.V. Nanopoulos[a,b,c 1]

[a]*Department of Physics, Texas A&M University, College Station, TX 77843-4242, USA*

[b]*Astroparticle Physics Group, Houston Advanced Research Center (HARC), Mitchell Campus, Woodlands, TX 77381, USA*

[c]*Academy of Athens, Division of Natural Sciences, 28 Panepistimiou Avenue, Athens 10679, Greece*

Abstract

No-scale supergravity is a framework where it is possible to naturally explain radiative electroweak symmetry breaking and correlate it with the effective SUSY breaking scale. Many string compactifications have a classical no-scale structure, resulting in a one-parameter model (OPM) for the supersymmetry breaking soft terms, which results in a highly constrained subset of mSUGRA. We investigate the allowed supersymmetry parameter space for a generic one-parameter model taking into account the most recent experimental constraints. We also survey the possible signatures which may be observable at the Large Hadron Collider (LHC). Finally, we compare collider signatures of OPM to those from a model with non-universal soft terms, in particular those of an intersecting $D6$-brane model.

[1] Lecturer in this school

I. INTRODUCTION

No-scale supergravity (nSUGRA) [1] is a framework where it is possible to naturally explain radiative electroweak symmetry breaking (REWSB) [2–4] and correlate it with the gravitino mass, or more generally, the effective SUSY breaking scale. In the simplest no-scale models, the gravitino mass $m_{3/2}$ remains undetermined at the classical level, and is instead fixed by radiative corrections to be near the electroweak scale [1]. Thus, in this framework, we find that the scale of supersymmetry breaking is correlated with the electroweak scale [1]. Another striking feature of nSUGRA is that the cosmological constant vanishes at tree-level. Although it is presently known that the cosmological constant is in fact non-zero, its very small value is still consistent with the no-scale framework with small corrections. Furthermore, it is well-known that the Kähler moduli of Type I, IIB orientifold, and heterotic string compactifications have a classical no-scale structure [5–7]. It has been shown that in Type IIB orientifold compactifications, this type of supersymmetry breaking corresponds to turning on RR and NS fluxes, which are generically present in order to cancel tadpoles as well as to stabilize closed-string moduli. Indeed, this is the case in the so-called large volume models [8] [9]. This combined with the generic appearance of the no-scale structure across many string compactifications leads to the idea that supersymmetry breaking is moduli dominated.

The most studied model of supersymmetry breaking is minimal supergravity (mSUGRA), which arises from adopting the simplest ansatz for the Kähler metric, treating all chiral superfields symmetrically. In this framework, $\mathcal{N} = 1$ supergravity is broken in a hidden sector which is communicated to the observable sector through gravitational interactions. Such models are characterized by the following parameters: a universal scalar mass m_0, a universal gaugino mass $m_{1/2}$, the Higgsino mixing μ-parameter, the Higgs bilinear B-parameter, a universal trilinear coupling A_0, and $\tan\beta$. One then determines the B and $|\mu|$ parameters by the minimization of the Higgs potential triggering REWSB, with the sign of μ remaining undetermined. Thus, we are left with only four parameters. Although, this is one of the most generic frameworks that can be adopted, many string compactifications typically yield expressions for the soft terms which are even more constrained, due to the no-scale structure which emerges naturally in these theories assuming that supersymmetry breaking is dominated by the Kähler moduli and/or dilaton. In particular, in such nSUGRA models, we will generically have $m_0 = m_0(m_{1/2})$ and $A = A(m_{1/2})$. This reduces the number of free parameters compared to mSUGRA down to two, $m_{1/2}$ and tanβ. In fact, adopting a strict no-scale framework, one can also fix the B-parameter as $B = B(m_{1/2})$, and thus we are led to a *one-parameter* model where all of the soft terms may be fixed in terms of $m_{1/2}$.

In this framework, we can identify two extreme cases. The first of these is referred to as the special *dilaton* scenario,

$$m_0 = \frac{1}{\sqrt{3}} m_{1/2}, \qquad A = -m_{1/2}, \qquad B = \frac{2}{\sqrt{3}} m_{1/2}. \tag{1}$$

while the second is referred to as the strict *moduli* scenario,

$$m_0 = 0, \qquad A = 0, \qquad B = 0. \tag{2}$$

For many string compactifications, the soft-terms will have such a form, and in the past such models were studied within the context of the the free-fermionic class of models in particular those with a flipped $SU(5)$ gauge group [10]. Interestingly, soft terms for heterotic M-theory compactifications with moduli dominated supersymmetry breaking have the above forms as special limits [11]. In addition, the so-called large-volume models have been studied extensively [8] [9] in recent years and the generic soft terms for this framework have been calculated in [12], and have a form nearly identical to the special dilaton scenario. These models involve Type IIB compactifications where the moduli are stabilized by fluxes and quantum corrections to the Kähler potential generate an exponentially large volume. This exponentially large volume may lower the string scale to an intermediate level which can be in the range $m_s \sim 10^{3-15}$ GeV. More generally, no-scale moduli-dominant scenarios of SUSY breaking are favored by F-theory [13], as are models with a flipped SU(5) gauge group [14] [15].

II. ALLOWED PARAMETER SPACE OF OPM

A one-parameter model of the above form has been much studied in the past [16–18]. However, the last such analysis was performed some years ago. In the intervening time, the experimental constraints on SUSY models have been updated. In addition, the experimental determination of the top quark mass has become considerably more precise in recent years. Finally, with the advent of the Supercritical String Cosmology (SSC), regions of the parameter space which would not ordinarily satisfy the WMAP constraints on the relic density are allowed due to a diluting factor of $\mathcal{O}(10)$ [19]. The allowed parameter space for the models defined by Eqs. 1 and Eqs. 2 was investigated in [20] using updated experimental constraints. For the strict moduli scenario, it is found that there are no spectra which satisfy all constraints. However, solutions may potentially be found when non-leading order corrections to the no-scale model are taken into account. For a detailed study concerning these corrections to the no-scale model, see [21].

For the dilaton scenario, a small region of the parameter space which satisfies all constraints is found. The parameter space which results in a relic density satisfying the WMAP limits is shown in Fig. 1. If the relic neutralino LSP comprises a sub-dominant component of the dark matter, we should not impose the lower bound on the WMAP limits. Thus, the parameter space for the four cases considered are 1) $0.1109 \leq \Omega_{\chi^o}h^2 \leq 0.1177$, 2) $0.095 \leq \Omega_{\chi^o}h^2 \leq 0.129$, 3) $0 \leq \Omega_{\chi^o}h^2 \leq 0.1177$, and 4) $0 \leq \Omega_{\chi^o}h^2 \leq 0.129$. Note that the parameter space shown in Fig. 1 is allowed by all constraints, except those regions noted. Fig. 2 also displays the parameter space allowed by all constraints, though the parameter space for the SSC model of the dark matter density is shown in totality.

Imposing all the experimental constraints, we find that the viable WMAP parameter space is in the range $\tan\beta = 35.2$ to $\tan\beta = 38$ as shown in Fig. 1. Extending the dark matter density to the SSC model as shown in Fig. 2, the allowed parameter space expands from $\tan\beta = 10.2$ to $\tan\beta = 38$. We show in Table I the allowed ranges for the CP-even Higgs boson mass satisfying all constraints for each range of $\Omega_{\chi^o}h^2$ within the parameter space. As we see, the range of the Higgs mass is highly constrained in each of the cases. For the superpartner spectra allowed by the constraints, there are only two hierarchal mass patterns of the four lightest sparticles: 1) $\widetilde{\chi}_1^0 < \widetilde{\tau} < \widetilde{e}_R < \widetilde{\chi}_1^\pm$, and 2) $\widetilde{\chi}_1^0 < \widetilde{\tau} < \widetilde{e}_R < \widetilde{\nu}_\tau$. The stau is NLSP in each case. A characteristic of this coannihilation region is the nearly degenerate mass of the lightest neutralino and the stau, which is in fact what we find for these spectra in the WMAP region. The LSP in both the WMAP and SSC regions allowed by the constraints is found to be Bino-like.

It has been shown in [22, 23] that it is possible to obtain mass measurements of the supersymmetric particles in the neutralino-stau coannihilation region by utilizing each final state and parameterizing kinematical observables, such as those described in the previous section, in terms of the SUSY masses. The goal of such an analysis would be to determine the mSUGRA model parameters m_0, $m_{1/2}$, A_0, and $\tan\beta$ since we want to determine the

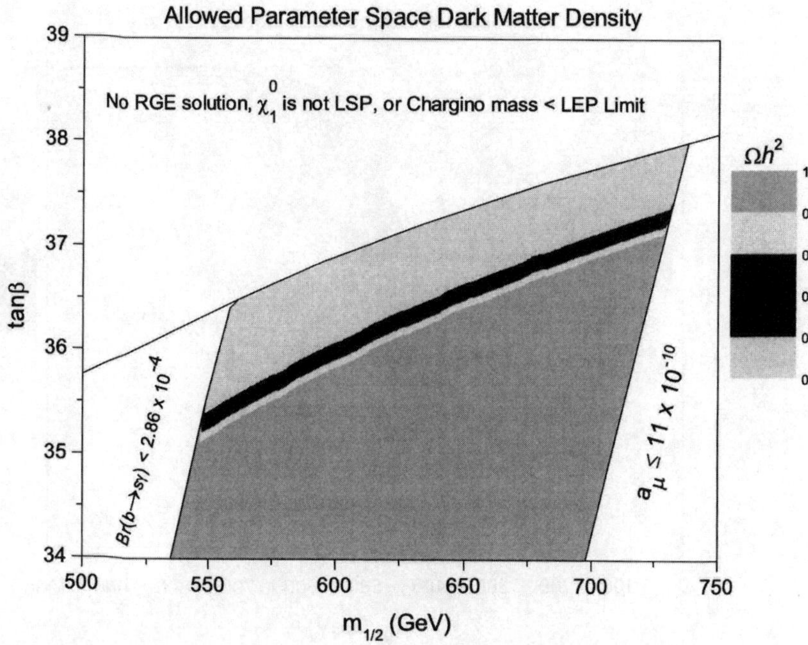

FIG. 1: Parameter space allowed by all constraints. The thin shaded areas constitute the WMAP region. The shaded area below the WMAP region is allowed by all constraints, though $0.129 < \Omega_{\chi^o} h^2 \leq 1.1$. The region on the far right side of the plot is excluded by the $g_\mu - 2$ results. The region on the far left side of the plot is excluded by $Br(b \to s\gamma) < 2.86 \times 10^{-4}$. The remaining area at the top of the plot is excluded for the reasons noted.

dark matter content and the neutralino-proton cross section, while the fifth mSUGRA model parameter, sign(μ), is assumed to be positive, since this is preferred by measurements of the $b \to s\gamma$ branching ratio and the muon $g_\mu - 2$ [24]. To determine the mSUGRA parameters, we need four kinematical observables which are linearly independent functions of those parameters. The determination of the parameters is then accomplished by inverting four such functional relationships. An analysis of the type was discussed in [25].

The OPM parameter space satisfying all constraints which has been found is a subset of the minimal supergravity (mSUGRA) parameter space, whose collider signals has been

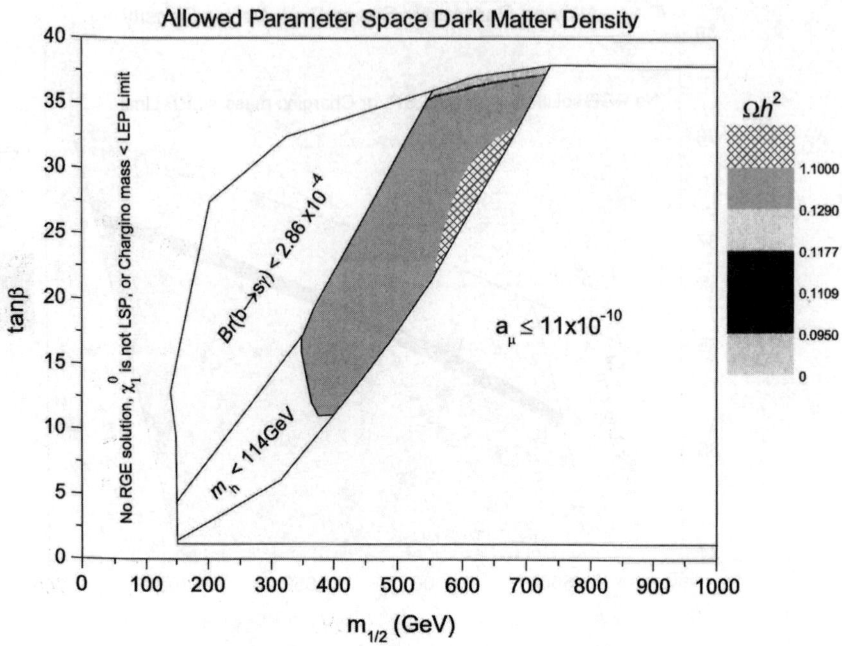

FIG. 2: Parameter space allowed by all constraints for the SSC model. The WMAP region is the small thin region at the top. The shaded area below the WMAP region is allowed by all constraints, though $0.129 < \Omega_{\chi^o} h^2 \leq 1.1$. The region excluded by $m_h < 114$ GeV satisfies all other constraints, including $\Omega_{\chi^o} h^2 \leq 1.1$. The cross-hatched region satisfies all constraints, though $\Omega_{\chi^o} h^2 > 1.1$. The remaining areas are excluded for the reasons noted.

the subject of much study over the years. In particular, the allowed parameter space of OPM satisfying the WMAP constraints falls into the coanhilation regions of mSUGRA. However, since this is a very small subset of the mSUGRA parameter space, the allowed superpartner spectra are somewhat constrained, and thus the possible signals of the model which might be observed at LHC are constrained as well. Once there is experimental data from LHC available, one may perform the analysis discussed in the previous paragraph to determine the mSUGRA model parameters. These may then be compared to the above OPM parameter space allowed by constraints. If the experimentally determined mSUGRA

TABLE I: Allowed ranges of the CP-even Higgs boson mass (in GeV) which satisfy the WMAP and SSC dark matter density limits as well as all other constraints.

$\Omega_{\chi^o} h^2$	m_h (GeV)
$0.1109 \leq \Omega_{\chi^o} h^2 \leq 0.1177$	117.17 - 118.58
$0.095 \leq \Omega_{\chi^o} h^2 \leq 0.129$	117.16 - 118.60
$0 \leq \Omega_{\chi^o} h^2 \leq 0.1177$	117.17 - 118.64
$0 \leq \Omega_{\chi^o} h^2 \leq 0.129$	117.16 - 118.64
$0 \leq \Omega_{\chi^o} h^2 \leq 1.1$	114.00 - 118.64

parameters happen to coincide with the allowed OPM parameter space, then this may provide an important clue to the structure of the underlying theory at high energy scales. One may also focus on the superpartner spectra of OPM and the resulting experimental signatures which should be observed at LHC if we live in a one-parameter model universe. We give a generic discussion of the experimental signatures of OPM in the next section.

In a one-parameter model universe, predominantly squarks and gluinos will be produced at LHC. To discuss the possible phenomenology of the one-parameter model, a possible strategy is to select one spectrum from each of the two regions of the parameter space: WMAP and SSC. We could then analyze the probable channels and resulting signatures at LHC, and can construct observables such as the opposite sign (OS) ditau invariant mass. The anticipated states decaying from the squarks and gluinos involve hadronic jets and tau, so the OS ditau invariant mass may provide some clues leading to discovery at LHC. To examine these probable channels, the cross-sections and branching ratios can then be calculated with PYTHIA [26] and cross-checked with ISAJET [27], using ISASUGRA to calculate the sparticle masses. Many analyses have been completed on the entire mSUGRA parameter space, the stau-neutralino coannihilation region in particular. We do not repeat these analyses, but would focus on this much more constrained region of the mSUGRA parameter space predicted by the one-parameter model. The most constructive analysis of the one-parameter model phenomenology is to study the signatures

- 1 tau lepton, 1 tau and ≥1 b-jet, 2 tau leptons, 1 lepton, 2 leptons and ≥2 jets, 2 jets, 3 jets, 1 b-jet

at LHC. These are also the relevant signals for SSC. A simulation of the number of events

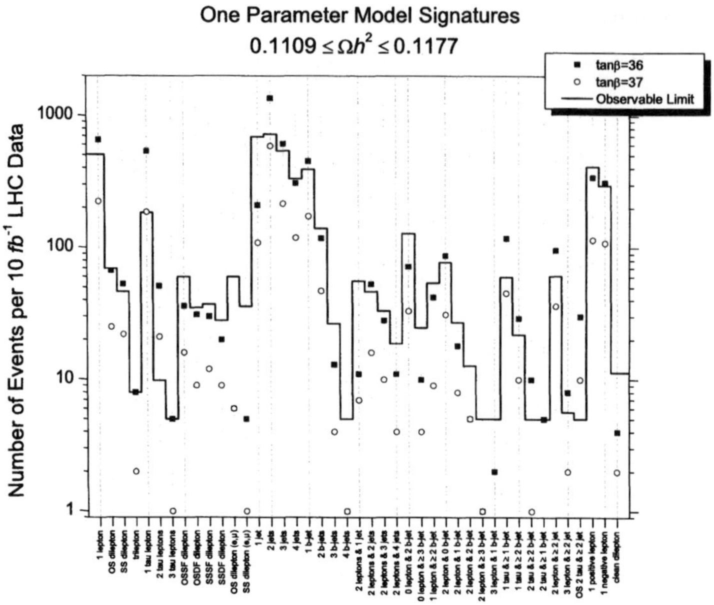

FIG. 3: Signatures of superpartner spectra within WMAP region. Spectra in this plot are listed in Table II and Table III. The solid bold line represents the observable limit computed in accordance with the requirements in Section III due to the Standard Model background. The markers represent the number of events per 10 fb^{-1} LHC data. An additional cut of $P_T^{miss} > 275$ GeV was applied in this plot.

for different signatures is shown in Fig 3 for the non-SSC case. See [20] for a more in-depth discussion.

III. SIGNATURES OF OPM VS. NON-UNIVERSALITY AT LHC

Much of the older work towards constructing semi-realistic string vacua was done in the context of heterotic strings. In particular, many of the most phenomenologically interesting models were those constructed within the free-fermionic formulation [10], and it is really from these types of models that the one-parameter model was first defined [16–18]. As we have mentioned, the same basic structure of the one-parameter model also arises in the context of heterotic M-theory constructions as well as Type IIB orientifold flux compactifications.

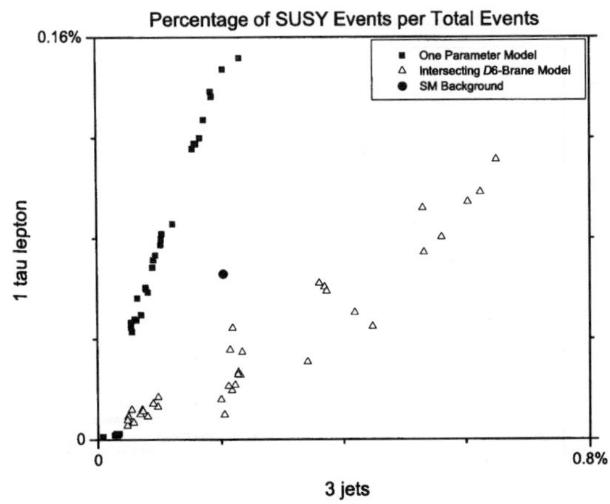

FIG. 4: Percentage of 1 tau lepton vs. 3 jet events per 10 fb^{-1} of integrated luminosity at LHC, for both the intersecting $D6$-brane model and the one-parameter model. The round marker indicates the observable limit due to the Standard Model background. Total events = signal + background.

Besides these types of string vacua, there are also recent constructions involving Type IIA compactifications involving $D6$-branes intersecting at angles. Such models have been the subject of much study in recent years, and we refer the reader to [28, 29] for recent reviews. The soft terms for intersecting $D6$-branes are in general non-universal [30], in contrast in the one-parameter model as well as the standard framework, mSUGRA. It is an interesting question to consider the LHC signatures of the one-parameter model compared to those of a model with non-universal soft terms, such as an intersecting D6-brane model. An analysis of this type has also been performed in [31]

In ref. [32] an explicit example of a supersymmetric intersecting $D6$-brane model in Type IIA string theory was constructed which possesses many desirable phenomenological properties. In particular, the model has three generations of SM fermions and also exhibits automatic gauge coupling unification at tree-level. In addition, it is possible to obtain correct masses and mixings for both up and down-type quarks as well as the tau lepton. The soft supersymmetry breaking terms were also calculated for this model, and it was shown that there are regions within the parameter space which may generate the observed dark matter density and superpartner spectra satisfying all presently known constraints.

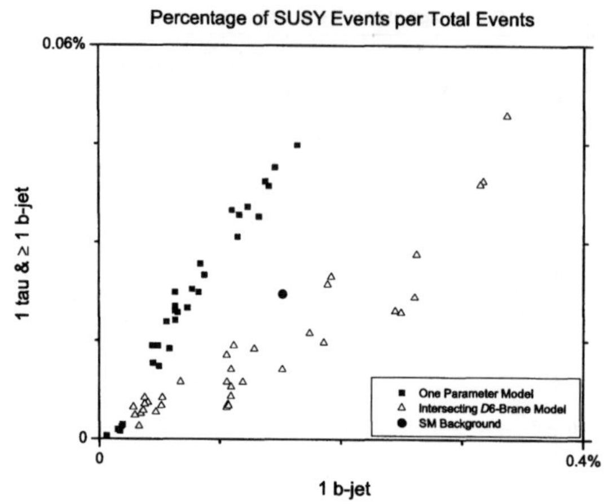

FIG. 5: Percentage of 1 tau and ≥ 1 b-jet vs. 1 b-jet events per 10 fb^{-1} of integrated luminosity at LHC, for both the intersecting $D6$-brane model and the one-parameter model. The round marker indicates the observable limit due to the Standard Model background. Total events = signal + background.

Given the desirable phenomenology of this model, it provides a suitable candidate with which to compare the one-parameter model. In particular, the question that we would like to address is if there are distinguishing characteristics in the collider signatures between this class of string vacua constructed with intersecting $D6$-branes in Type IIA and other string constructions mentioned earlier having soft terms similar to OPM.

As discussed in the previous section, there are signatures at LHC that could provide distinguishing characteristics of the one-parameter model. For the present, we want to investigate if these signatures differ from the intersecting $D6$-brane model. In Figs.4-10 we plot two of these collider signatures against one other for both the one-parameter model and the intersecting $D6$-brane model. For each of the cases shown, there seems to be a clear separation between the one-parameter model and the intersecting $D6$-brane model. All spectra simulated in these figures from both models are within the WMAP region of the allowed parameter space. Namely, thirty-one spectra from the one-parameter model and thirty-five spectra from the intersecting $D6$-brane model were simulated in the event generator. Table II lists the different patterns of mass hierarchies for the one-parameter

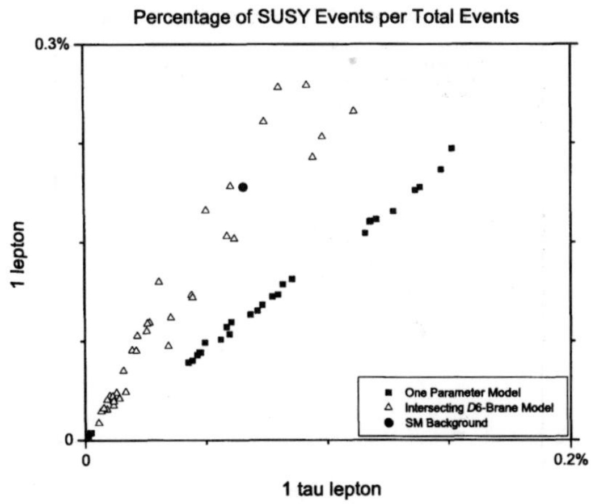

FIG. 6: Percentage of 1 lepton vs. 1 tau lepton events per 10 fb^{-1} of integrated luminosity at LHC, for both the intersecting $D6$-brane model and the one-parameter model. The round marker indicates the observable limit due to the Standard Model background. Total events = signal + background.

model parameter space as the intersecting $D6$-brane model. The mass patterns of the thirty-five superpartner spectra for the intersecting $D6$-brane model simulated in the event generator are the *ID6BraneP1* chargino pattern. This is in contrast to the two different stau patterns of the one-parameter model, *OPMP1* and *OPMP2*. The thirty-one OPM spectra simulated in the event generator were a combination of both the *OPMP1* and *OPMP2* stau mass patterns. The patterns in these charts corroborate the details given in the previous section that the number of tau events as a percentage of overall lepton events for the one-parameter model will be high. In fact, the percentage of tau events is much higher in the one-parameter model than in the intersecting $D6$-brane model. We will perform a complete phenomenological analysis of the intersecting $D6$-brane model in the future, but suffice it to say that by examining the mass pattern for *ID6BraneP1* in Table II, it is evident that the decay of the $\widetilde{\chi}_1^{\pm}$ or $\widetilde{\chi}_2^0$ to a $\widetilde{\tau}_1$ is no longer present due to the $\widetilde{\tau}_1$ now being heavier than the $\widetilde{\chi}_1^{\pm}$ and $\widetilde{\chi}_2^0$. The elimination of this channel alone will reduce the production of tau, as the plots show.

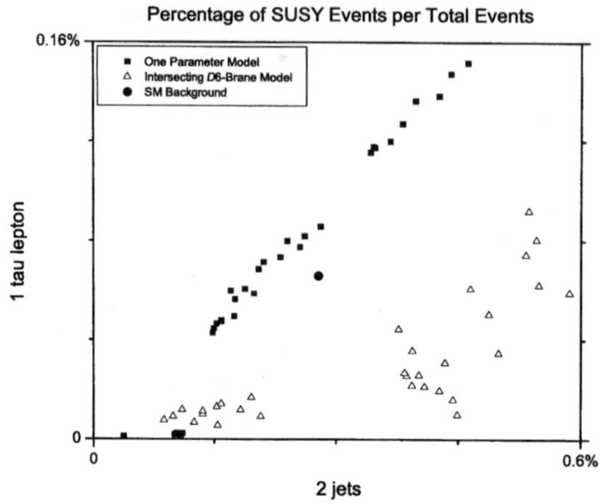

FIG. 7: Percentage of 1 tau lepton vs. 2 jet events per 10 fb^{-1} of integrated luminosity at LHC, for both the intersecting $D6$-brane model and the one-parameter model. The round marker indicates the observable limit due to the Standard Model background. Total events = signal + background.

TABLE II: Mass patterns of spectra allowed by all constraints for the one-parameter model and intersecting $D6$-brane model.

Model	Pattern No.	Pattern Type	Mass Pattern
one-parameter	*OPMP1*	Stau	$\widetilde{\chi}_1^0 < \widetilde{\tau} < \widetilde{e}_R < \widetilde{\chi}_1^{\pm}$
one-parameter	*OPMP2*	Stau	$\widetilde{\chi}_1^0 < \widetilde{\tau} < \widetilde{e}_R < \widetilde{\nu}_\tau$
intersecting $D6$-brane	*ID6BraneP1*	Chargino	$\widetilde{\chi}_1^0 < \widetilde{\chi}_1^{\pm} < \widetilde{\chi}_2^0 < \widetilde{\tau}$
intersecting $D6$-brane	*ID6BraneP2*	Chargino	$\widetilde{\chi}_1^0 < \widetilde{\chi}_1^{\pm} < \widetilde{\tau} < \widetilde{\chi}_2^0$
intersecting $D6$-brane	*ID6BraneP3*	Chargino	$\widetilde{\chi}_1^0 < \widetilde{\chi}_1^{\pm} < \widetilde{\tau} < \widetilde{e}_R$
intersecting $D6$-brane	*ID6BraneP4*	Stau	$\widetilde{\chi}_1^0 < \widetilde{\tau} < \widetilde{\chi}_1^{\pm} < \widetilde{\chi}_2^0$
intersecting $D6$-brane	*ID6BraneP5*	Stau	$\widetilde{\chi}_1^0 < \widetilde{\tau} < \widetilde{e}_R < \widetilde{\chi}_1^{\pm}$

IV. CONCLUSION

An updated survey the allowed parameter space and experimental signatures of the one-parameter model has been discussed. Our motivation for studying this model stems from

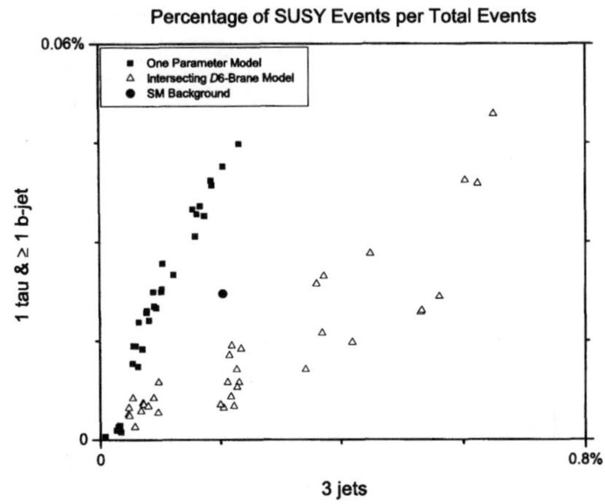

FIG. 8: Percentage of 1 tau and $\geq$ 1 b-jet vs. 3 jet events per 10 fb^{-1} of integrated luminosity at LHC, for both the intersecting $D6$-brane model and the one-parameter model. The round marker indicates the observable limit due to the Standard Model background. Total events = signal + background.

the appearance of the universal soft supersymmetry breaking ansatz across multiple types of string compactifications due to the classical no-scale structure in these models. These include weak coupled and heterotic M-theory vacua, as well as Type IIB flux vacua, in particular the so-called large-volume compactification models. In the strict moduli dominant case, we found that there are no regions of the parameter space which can satisfy all of the constraints, whereas there is a small parameter space allowed for the dilaton scenario. The dominant component of the signatures observable at LHC are tau and hadronic jets. In future work, we plan to study the observable signatures at LHC for this model in somewhat more depth.

The collider signatures of the one-parameter model were compared to a model with non-universal soft terms, in particular an intersecting $D6$-brane model with interesting phenomenological properties. We found that for one particular intersecting $D6$-brane pattern of mass hierarchies, there are possible distinguishing characteristics between these two classes of models. Although there may also be a lot of overlap in the observable signatures of these two models, there are regions of the parameter space of each class which may give strikingly different observable signatures which may be used to distinguish them. Thus, it may be

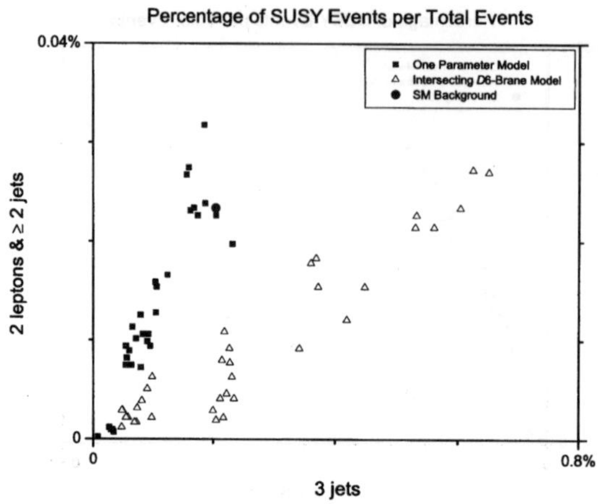

FIG. 9: Percentage of 2 leptons and $\geq$ 2 jets vs. 3 jet events per 10 fb^{-1} of integrated luminosity at LHC, for both the intersecting $D6$-brane model and the one-parameter model. The round marker indicates the observable limit due to the Standard Model background. Total events = signal + background.

possible for LHC to say something about the structure of the underlying theory at high-energies, e.g. universality vs. non-universality. We plan to investigate this possibility more deeply in an upcoming paper.

V. ACKNOWLEDGMENTS

This research was supported in part by the Mitchell-Heep Chair in High Energy Physics (CMC), by the Cambridge-Mitchell Collaboration in Theoretical Cosmology, and by the DOE grant DE-FG03-95-Er-40917. One of us (DVN) would like to thank the co-inventer of the one-parameter model, Antonino Zichichi for his warm hospitality at this school.

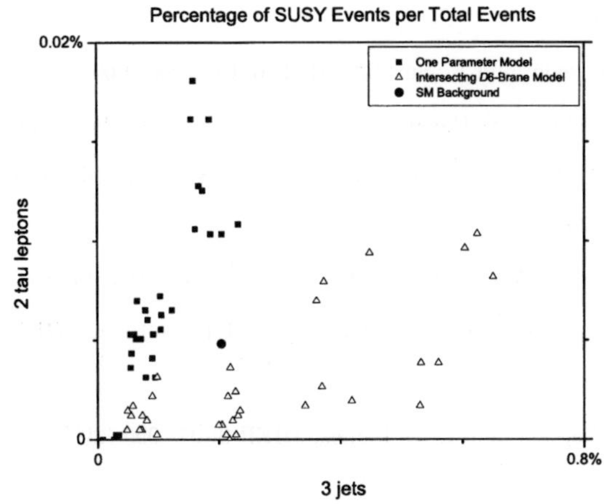

FIG. 10: Percentage of 2 tau leptons vs. 3 jet events per 10 fb^{-1} of integrated luminosity at LHC, for both the intersecting $D6$-brane model and the one-parameter model. The round marker indicates the observable limit due to the Standard Model background. Total events = signal + background.

[1] E. Cremmer, S. Ferrara, C. Kounnas, and D. V. Nanopoulos, Phys. Lett. B **133**, (1983) 61; J. Ellis, A. Lahanas, D. V. Nanopoulos, and K. Tamvakis, Phys. Lett. B **134**, (1984) 429; J. Ellis, C. Kounnas, and D. V. Nanopoulos, Nucl. Phys. B **241**, (1984) 406 and B **247** (1984) 373; For a review see A. Lahanas and D. V. Nanopoulos, Phys. Rep. **145**, (1987) 1

[2] J. R. Ellis, D. V. Nanopoulos and K. Tamvakis, Phys. Lett. B **121**, 123 (1983).

[3] J. R. Ellis, J. S. Hagelin, D. V. Nanopoulos and K. Tamvakis, Phys. Lett. B **125**, 275 (1983).

[4] L. Alvarez-Gaume, J. Polchinski and M. B. Wise, Nucl. Phys. B **221**, 495 (1983).

[5] J. L. Lopez and D. V. Nanopoulos, Int. J. Mod. Phys. A **11**, (1996) 3439-3478. arXiv:hep-ph/9412332v2

[6] V. Kaplunovsky and J. Louis, Phys. Lett. B **306**, (1993) 269. arXiv:hep-th/9303040v1

[7] A. Brignole, L. Ibá$\tilde{n}$ez, and C. Mun$\tilde{o}$z, Nucl. Phys. B **422**, (1994) 125. arXiv:hep-ph/9707209

[8] V. Balasubramanian, P. Berglund, J. P. Conlon and F. Quevedo, JHEP **0503**, (2005) 007.

arXiv:hep-th/0502058

[9] J. P. Conlon, F. Quevedo, and K. Suruliz, JHEP **0508**, (2005) 007. arXiv:hep-th/0505076

[10] I. Antoniadis, J. R. Ellis, J. S. Hagelin and D. V. Nanopoulos, Phys. Lett. B **194** (1987) 231; Phys. Lett. B **205** (1988) 459; Phys. Lett. B **208** (1988) 209 [Addendum-ibid. B **213** (1988) 562]; Phys. Lett. B **231** (1989) 65.

[11] Tianjun Li, Phys. Rev. **D59**, (1999) 107902. arXiv:hep-ph/9804243v1

[12] J. P. Conlon, S. S. Abdussalam, F. Quevedo, and K. Suruliz, JHEP **0701**, (2007) 032. arXiv:hep-th/0610129v2

[13] L. Aparicio, D. G. Cerdeno and L. E. Ibanez, JHEP **0807**, 099 (2008) [arXiv:0805.2943 [hep-ph]].

[14] C. Beasley, J. J. Heckman and C. Vafa, arXiv:0806.0102 [hep-th].

[15] J. Jiang, T. Li, D. V. Nanopoulos and D. Xie, arXiv:0811.2807 [hep-th].

[16] J. L. Lopez, D. V. Nanopoulos and A. Zichichi, Phys. Lett. B **319** (1993) 451 [arXiv:hep-ph/9306226].

[17] J. L. Lopez, D. V. Nanopoulos and A. Zichichi, Int. J. Mod. Phys. A **10**, 4241 (1995) [arXiv:hep-ph/9408345].

[18] J. L. Lopez, D. V. Nanopoulos and A. Zichichi, Phys. Rev. D **52**, 4178 (1995) [arXiv:hep-ph/9502414]; For a review, see `Searching for the Superworld: A Volume in Honor of Antonino Zichichi on the Occasion of the Sixth Centenary Celebrations of the University of Turin, Italy`, Antonino Zichichi, Sergio Ferrara, Rudolf M. Mössbauer, World Scientific (2007), ISBN 9812700188, 9789812700186

[19] D.V. Nanopoulos, talk given at this school, 'How Supercritical String Cosmology Affects LHC'.

[20] J. A. Maxin, V. E. Mayes and D. V. Nanopoulos, arXiv:0809.3200 [hep-ph].

[21] B. Dutta, Y. Mimura and D. V. Nanopoulos, Phys. Lett. B **656**, 199 (2007) [arXiv:0705.4317 [hep-ph]].

[22] R. Arnowitt *et al.*, Phys. Lett. B **639** (2006) 46.

[23] R. Arnowitt *et al.*, Phys. Lett. B **649** (2007) 73.

[24] G. W. Bennett *et al* (Muon g-2 Collaboration), Phys. Rev. Lett. **92**, 161802 (2004). arXiv:hep-ex/0401008

[25] R. Arnowitt, B. Dutta, A. Gurrola, T. Kamon, A. Krislock, D. Toback, Phys. Rev. Lett. **100**,

231802 (2008). arXiv:hep-ph/0802.2968v2

[26] T. Sjostrand, S. Mrenna , and P. Skands, JHEP **0605**, (2006) 026. arXiv:hep-ph/0603175v2

[27] F.E. Paige, S.D. Protopopescu, H. Baer, and X. Tata, arXiv:hep-ph/0312045v1

[28] R. Blumenhagen, M. Cvetic, P. Langacker and G. Shiu, Ann. Rev. Nucl. Part. Sci. **55**, 71 (2005) [arXiv:hep-th/0502005].

[29] R. Blumenhagen, B. Kors, D. Lust and S. Stieberger, Phys. Rept. **445**, 1 (2007) [arXiv:hep-th/0610327].

[30] B. Kors and P. Nath, Nucl. Phys. B **681**, 77 (2004) [arXiv:hep-th/0309167].

[31] D. Feldman, Z. Liu and P. Nath, Phys. Lett. B **662**, 190 (2008) [arXiv:0711.4591 [hep-ph]].

[32] C. M. Chen, T. Li, V. E. Mayes and D. V. Nanopoulos, Phys. Lett. B **665**, 267 (2008) [arXiv:hep-th/0703280]; Phys. Rev. D **77**, 125023 (2008) [arXiv:0711.0396 [hep-ph]].

CHAIRMAN: D. NANOPOULOS

Scientific Secretaries: S. Cassel, A. Maier

DISCUSSION I

- Susskind:

Suppose that there is SUSY. But, it just happens to be a little bit out of range. A factor of five, two in the energy. A factor ten, twenty, even a hundred, maybe even a thousand. Is there a smoking gun, perhaps the Higgs mass or something that would be the best way to guess that it is there and approximately where it is?

- Nanopoulos:

A very good question. Now, as I was showing in the morning, for instance in the Higgs histogram, and this is of course for a class of models where we get into this region with a Higgs mass in the range 114-119 GeV. So first you know very well that there is this rigorous upper bound on the mass of the Higgs to be lower than 135 GeV and separately, most of us believe it would be much less, of course above 114 GeV. So, the Higgs boson, if it would be in this region, would be a good smoking gun. I am not saying that I then would be happy and that would do it. If nature has done it in a way that it is just above the TeV range or whatever it is, this region we are talking about, I will not be very happy.

- Susskind:

It's certainly a bad thing but I mean we would have some best guess.

- Nanopoulos:

Yes. Maybe we have somehow with this hierarchy problem. I would guess that what I would do is set my students to go back through all these calculations to see what the limits are, and how far we could push let's say for a correction to the Higgs mass. That's what I would do.

- Musaev:

On your slides you have mentioned three possibilities of how the fine tuning problem can be solved. One of them was some new dynamics associated with the electroweak symmetry breaking. What does it mean?

- Nanopoulos:

What I have in mind is something like technicolour. This story gives you another cutoff that is not related to SUSY.

- Musaev:

You mentioned such particles as h, A, H^+, H^-. What are these particles?

- Nanopoulos:

As you know, in the SUSY theory, and you can see it immediately from the cancellation of the anomalies, if I have one Higgs doublet, like we have in the Standard Model, I mean non-SUSY Standard Model, then of course we should have its fermionic counterpart. For instance, if I have a Higgs boson, I have a corresponding Higgsino. If I have only one Higgs doublet, then I'm going to produce a Higgsino. Now, if it was *only* one Higgsino doublet then I will have a contribution to the Adler-Bell-Jackiw anomaly. In order to cancel this, you would need another Higgsino doublet. Then I am going to have eight states. From these, three are going to be absorbed by the W^+, W^- and Z^0

and then we are left with five Higgses. The h is the light Higgs. This is close to the Standard Model. Then we have the A as a heavy Higgs and then we have H^+ and H^-, which are the charged Higgs bosons. So these are new Higgs-like states coming out of those. We have very definite reasons to be there. This is the minimal thing. We have no reason at the moment to go any further. So these are the states I am talking about.

- *Reeb:*

You have somewhat motivated your one parameter model, and you and other people before have talked about the mSUGRA model. To me it looks like this is a convenient way of parameterizing the more than 100 free parameters of the MSSM. Is it worthwhile to concentrate so much on these most simple parameterizations of the MSSM, since after all some totally different fundamental model at the Planck scale might be true, which likely gives totally different low-energy phenomenology?

- *Nanopoulos:*

First thing is, you always start with the simplest possible case and you move on from there. But this is not just based on completion, it is based also on SUSY GUTs or even string theory and if you play around with models, then you will see that in *many* of these models, not *all* of these models, you are getting this type of behaviour. There are minimal extensions of these minimal models. For instance, one thing that my friend John Ellis has been pushing for some years now, is that it is possible that in the Higgs sector, the masses of the Higgs and the Higgsino may be a little different from this universality that I am talking about. But what I am saying here: First, we have this kind of simplicity. We start from the simplest and go to the complex and we have models to solve where this thing is coming out. If we did not have any model to solve then I would feel a little bit differently. That is kind of putting it in a nutshell. This is somehow also related in a way with Lenny's [L. Susskind] question. But if we open this kind of worm box, then it can get a bit more complicated.

- *Reeb:*

From this we may conclude, that is does not make much sense to parametrize our low-energy MSSM by such high scale physics, that we do not know anything about?

- *Nanopoulos:*

No, absolutely the contrary. Because, at least in my generation, our conclusion was that the real world is not this world here, but the real world is up there at the Planck scale, and somehow what we see here are the relics of this world. Actually these give us a fantastic handle on the symmetries, on the breaking, all these things and John [Schwarz] I understand spoke about F-theory, Sergio Ferrara spoke about supergravity. The real game all of us play, at least for my generation, we believe that the real stuff is up there and all of this are consequences of this thing. So, and that's why I was showing this transparency with bottom up and top down approach that we try to fix in a way that we could make some sense. That is our main philosophy. Focusing on the model down here, I mean, we know more or less what it is. We put a little bit of neutrino masses here and stuff like that, but that solves many problems. And then we make this kind of super-jump to go to the Planck scale. That is what the philosophy is all about, that we believe – and I think I speak for many people of my age in this room - that the real story finally is the game up there and all of this is giving us handles on how to develop this theory.

- *Baudis:*

In the "one parameter model", how light can the neutralino be? You said it's highly constrained. And what are the predicted neutralino-nucleon cross sections? What are the constraints you get?

- *Nanopoulos:*

Tomorrow, I am going to give you some of these things. I will give you some cross sections tomorrow, when I will speak about this and then you will see.

- *Baudis:*

You also said it is ruled out.

- *Nanopoulos:*

No, what was ruled out was one of these models, the strict no-scale moduli scenario.

- *Baudis:*

But this is what you call one parameter model?

- *Nanopoulos:*

No! These are different examples. Let me just make clear what is happening here. What I call one parameter model it is this kind of thing.

- *Baudis:*

OK, so I understand that was a special case.

- *Nanopoulos:*

Right! You are getting something like that but of course different expressions from the ones that I have here. This is a superstring phenomenological type motivated case which we were considering. But you can take now for the same analysis that I showed you and you can do for your own favourite model. And actually, in this analysis that we did with Nino [A. Zichichi] 15 years ago, we were considering cases more general than that and then the issue there was, we did basically what we have all the data, the same analysis presented here with all the data to give examples of how it is possible. All of the things that we know to be depending only on one parameter for this thing. Let me now tell you something that I didn't have time in the morning to explain – that eventually you see this kind of thing and I believe from string theory or whatever it is eventually going to be a no parameter model. Now you say what he's crazy. No, I'm not because the no-scale structure does not only give you some boundary conditions, but is also giving you a way to connect because of the minimisation of the potential. This potential depends on this specific moduli and then you make EWSB but still you have not finished. You have another parameter that you minimize with respect to this parameter and this is the no-scale condition. And if you use this, then you determine all your parameters in your theory, but we have not dared yet to do it in an explicit way because we would like to learn more about how things are coming out. So, I repeat it again, we use this as an example, and we are analysing a general construction for this type of structure from F-theory models that are going to do the same analysis that I presented here.

- *Fedosenko:*

You told us of ways to protect the Higgs boson mass from large quantum corrections. Could you tell us in some detail or give some examples of little Higgs models and composite models?

- *Nanopoulos:*

What I understand from work I believe by Howard Georgi and others on non-linear realized symmetry, that there you have cancellation of the loops not between fermions and bosons but between fermions and the conclusion for when you get this type of cancellation is that you get different bounds for the Higgs bosons and actually they call it "little Higgs" because they get a light Higgs.

- *Fedosenko:*

In a gauge coupling unification picture including the MSSM, we see the breaking of the lines around one TeV. What is the reason for this breaking?

- *Nanopoulos:*

Now you're getting me in trouble with Nino [A. Zichichi] here. This is because the electroweak scale, and you run down and get different degrees of freedom when you go from the 100 GeV down to ~1 GeV. But now you have to be very careful, actually with all of these things because this is like a cartoon figure. You have to be very careful how you run. When Nino, etc did this thing with the gaugino masses, they showed that they play a very important role in this case. I consider that this is one of the essential points in this game. This is the unification of the coupling constants that you get when you run from 100 GeV to 10^16 GeV. Assuming that nothing is happening in between, these things meet with respect to LEP. This is something that is very interesting but I would like to see other approaches for how we could explain all this kind of thing. It is not enough to explain only one thing, we have to take the whole data that we have in our case, and this is one crucial point in the discussion of the unification of the coupling constants. Now you can say well maybe it's accidental, well OK if it's accidental, it's accidental, and nature is using it to throw us that kind of sun in our eyes, and we do this kind of unification, but this unification means nothing, just accidental. I don't believe so, but you never know.

- *Zichichi:*

I would like to say that in my lecture, I will discuss the importance of the running of the gaugino masses to go down by three orders of magnitude in the threshold.

- *Nagy:*

OK, I dare to ask this question because Professor Zichichi stated at the beginning that there would be no punishment for questions. My question is about the fundamental need for SUSY: Do you know anyone who calculated Standard Model Higgs mass corrections non-perturbatively, e.g. on the lattice? Because for a long time, it seemed to me that the instability of the Higgs mass may be due to some mathematical inconsistency of the perturbative expansion and might not be present if one does not rely on perturbation theory and there would be no natural need for SUSY.

- *Nanopoulos:*

I think a lot of ink has been spent on this kind of thing. I don't think so, I mean many many people have tried many many years. We have several of these people here, we have Lenny [L. Susskind] here who invented technicolour. We have people here on SUSY, including myself. Now a lot of people, they have thought very hard about this and believe me I don't think they would move on with this thing before they have really thought very hard about this problem. So, I think this is a genuine problem. The problem is there, but maybe our philosophy for this kind of thing may change. I'm not a string landscape guy at all, I'm on the opposite side, but I give you an example now, so Lenny [L. Susskind] may use it: let's go back to quantum mechanics for this. Now, when quantum mechanics appeared, some people could not get used to the idea of the scientific principle. It was very very strange that you dare not to have trajectories for the electron say and this was a very hard question at the time for people to absorb this kind of thing. I don't know, maybe we have to completely change our philosophy for how we look at things, and what we are asking from nature to understand. So, certainly the problem is there, but maybe by shifting our philosophy, there is of course experimental data not just the philosophical, but that would be a different story. I would believe if I had a student I would not ask him on try to solve this problem non-perturbatively, I don't mean by strong dynamics, but inside electroweak theory.

- *Dubynskiy:*

You said that in the exact SUSY the bosonic and the fermionic contribution to the vacuum energy cancel each other. But SUSY is badly broken. Could we really advocate SUSY with that argument? The actual problem has never gone and we are still are many orders of magnitude far from there.

- *Nanopoulos:*

SUSY says that the mass of the boson is equal to the mass of the fermion. If you break it, then you can see easily that this thing is the difference of the boson mass squared minus the fermion mass squared. So what you are doing here, is you are using this relation, and I don't want my fixed mass to be above its own number. So you have this nice number here, and that's why it gives you an upper bound. In order to avoid problems, you should call it effective SUSY breaking. There is something to discuss, because you may have a certain new theorem that SUSY breaking occurs at high energies, when it gets down to the Standard Model. This thing that it effectively feels is this kind of difference which is of the order of 1 TeV or whatever is this difference.

- *Arnowitt:*

In the no-scale model, when you stabilize the T_i moduli all the soft breaking parameters are functions of $m_{1/2}$. Are these functional relations universal, or do they depend in how the T moduli are stabilized?

- *Nanopoulos:*

No, stabilization of T is going to be related with what is the vacuum expectation value of T, but not related with the forms of this thing here. But of course, I have to have a complete model, and I think I can use this in a given string compactification. You can play such a kind of game because I have several Kahler moduli. It's not going to be as simple as I presented here. It's not this Ferrara form [-3log(T+T*)]. It's going to be more complicated and then of course I have to do my business there, then get what you're getting from there, that's why if you put different string compactifications, you get different relations. I'm going to use the philosophy what is going to be like that. It's not a universally understood property of string compactification.

- *Carrasco:*

If you had to characterize your D6 brane model with a number of tunable parameters, what number would it be? Are there tunable parameters?

- *Nanopoulos:*

That's a good question. I did this to show you if this thing is in the D6 models. This is now far different the one parameter model. The reason I put it there, the philosophy is to see how we can separate these things here from this one parameter model. I didn't mean to say the D6 model as we have it today is a one parameter model. We hope very soon that this is going to reduce to this level here. But it is just to give us some comparison with something that is not the one parameter model.

- *Carrasco:*

Given unprecedented "engineering" possibilities in F-theory, is it possible to engineer a fit to the one parameter model? What is the scope of the engineering?

- *Nanopoulos:*

Well, I think the answer is yes, and actually there are already papers that have done an aggressive analysis in what we call a modulus dominated breaking scenario who are using F-theory and actually that's what we're parameterising now. You can see there it is getting this type of form that I'm plotting – not the relations, but it will get to this thing, yes. I would like to stress once more the excitement that John [Schwarz] was talking about, that this is a very interesting development of this thing here, and actually if you think about this to be a non-perturbative F-theory it is dual to the IIA for this kind of thing. So I would not be surprised that these things are developing and actually we should be a little bit more clever and use this duality, because certain problems can be solved easier in IIA, and others in IIB.

- *Ni:*

Why are there multiple peaks in the histogram of the Higgs masses?

- *Nanopoulos:*

The Higgs mass histogram is not a statistical plot of the Higgs mass based on a random distribution, but simply a plot of the calculated Higgs masses for the points of the parameter space which survived the constraints. The important thing here is not the shape of the distribution, but the range of Higgs of calculated Higgs masses for the different models, which is in the range 117-119 GeV.

- *Nikiforov:*

In the one parameter model, the difference of the signatures between tanβ=36 and tanβ=37 is just a normalization factor. This means that in order to distinguish between different parameters of the model one needs to have a good control over the luminosity measurement as well as over the acceptance of the detector. Are there any other more solid ways to distinguish between different parameters?

- *Nanopoulos:*

Well, let me say this is I gave you what we have done here. This is our background that we have used. A lot of these things are simulation. I'm not saying that this is how you're going to see it in the LHC, of course we know when LHC comes into action then we go back and use the real data. We are using all the conditions that people are using, and we try to do the best from this consideration. Now of course as time is moving we're going to use a little more background. This is to show to you what types of relation we are going to find. This is really the beginning, at least for my group, of what we try to do for this. I'm not claiming that this is exactly what we're going to find at the LHC. So here is the analysis of what we are doing and that we will find it. But what is interesting is that we have this kind of separation, a lot of different models we have played with and in all of them you have this kind of correlation because that's what someone was talking about before. There is going to be a matter of correlations, they find different signals and then we see if it is extra dimensions or if it is black holes or supersymmetry and all of this. These things are going to be all related. That's what I'm trying to show you here.

- *Ferrara:*

Suppose SUSY is discovered at the LHC. There would be some clear signal which would discriminate between supergravity and gauge mediated breaking, looking at gluino masses or masses of sleptons, squarks.

- *Nanopoulos:*

Oh yes, clearly. The gluino masses. It is very clear, let's get the signal, and you will see in the next few months we will know very clearly which one is which. That will be a very clear case. It is one thing about this scenario with $m_0 = 0$ which causes the gauge-mediated kind of thing, the good thing here is that you are avoiding FCNCs which are very nasty in SUSY theories. If you go to SUSY theories and you are not careful, you are going to have a big problem. As I suggested about 30 years ago with John Ellis, you somehow undo the Glashow-Maiani mechanism of this kind of thing, which is a fantastic thing for the Standard Model. So we have to be very careful, and actually this is one example of this case that it gives you absence. They all come from that gauge interactions. All these quarks they have the same quantum numbers. So there are cancellations there and that's why you are getting this kind of thing "naturally."

How Supercritical String Cosmology Affects LHC

D.V. Nanopoulos

Department of Physics, Texas A&M University, College Station, TX 77843-4242, USA
University of Athens, Physics Department, Nuclear and Particle Physics Section, GR-157 71, Athens, Greece
Astroparticle Physics Group, Houston Advanced Research Center (HARC), Mitchell Campus, Woodlands, TX 77381, USA
Academy of Athens, Division of Natural Sciences, 28 Panepistimiou Avenue, Athens 10679, Greece

Abstract

In supercritical string cosmology (SSC), a time-dependent dilaton leads to a smoothly evolving dark energy and modifies the regions of the mSUGRA parameter space where the observed value of the dark matter relic density may be obtained. In particular, the dilaton dilutes the supersymmetric dark matter density (of neutralinos) by a factor $\mathcal{O}(10)$ and consequently relaxes the allowed parameter mSUGRA parameter space. The final states expected at the LHC in this scenario, consist of Z bosons, Higgs bosons, and/or high energy taus. From this, it is possible to characterize these final states and determine the model parameters. Using these parameters, we determine the dark matter content and the neutralino-proton cross section. All these techniques can also be applied to determine model parameters in SSC models with different SUSY breaking scenarios.

1 Introduction

With the advent of the era of precision cosmology, it has been determined that the energy content of the universe is around 23% dark matter and 73% dark energy [1]. The origin of dark matter can be explained in supersymmetry (SUSY) models, where the lightest SUSY particle, the neutralino (in most SUSY models) [2], is the dark matter candidate. SUSY combined with supergravity grand unification (SUGRA GUT) [3], resolves a number of the problems inherent in the standard model (SM). On the other hand, the origin of dark energy is not well understood. The simplest proposal is to add a cosmological constant in Einstein's equations. However, the curious fact that the dark matter content is comparable to the dark energy content at the present remains a puzzle. Another proposal is that a quintessence scalar field is responsible for dark energy [4]. However, this requires the field to have a very small mass and is not well motivated in particle physics models. In the context of string theory, the dilaton can play the role of dark energy [5, 6]. In this talk, I will investigate experimental signatures of SUSY as consequences of a rolling dilaton in the Q-cosmology scenario [5] which offers an alternative framework that establishes the Supercritical (or non-critical) String Cosmology (or SSC). In the SSC framework, the dark energy has two components: One component arises from the dilaton, ϕ, and the other arises from the Q^2 which is associated with the central charge deficit. Both Q and the dilaton have time dependent pieces. It was shown that the SSC scenario [7] is consistent with the smoothly evolving dark energy at least for the last ten billion years $(0 < z < 1.6)$, in accordance with the very recent observations on supernovae [8]. In addition, the SSC scenario suggest a possible vacuum index of refraction, resulting in a measurable time-lag for high-energy gamma-rays traveling from distance sources [9]. Amazingly, such a time-delay has recently been measured by the MAGIC telescope [10], as well as more recently by HESS and Fermi telescopes [11].

The presence of this time dependent dilaton affects the relic density calculation since it modifies the Boltzman equation in the following way: $\frac{dn}{dt} + 3Hn + \langle\sigma v\rangle(n^2 - n_e q^2) - \dot{\phi}n = 0$ The relic density is given by

$$\Omega h^2 = R \times (\Omega h^2)_0 \tag{1}$$

where $R \sim \exp[\int_{x_0}^{x_f}(\dot{\phi}H^{-1}/x)dx]$ and $(\Omega h^2)_0$ denotes the relic density that is obtained by ordinary cosmology. It is possible to determine R by solving for ϕ from the field equations for this SSC scenario. The value of R is about 0.1 in order to satisfy the recent observation of the evolution of dark energy in the range $0 < z < 1.6$. This new factor changes the profile of dark matter allowed region in SUSY models.

To investigate the SUSY signatures we use the minimal SUGRA (mSUGRA) model and calculate the dark matter content in the context of the SSC framework. We note that the low energy limit of string theory is certainly much more complicated than mSUGRA, and there are many different effective theories, depending on the details of compactification and SUSY breaking [12]. The relevant dark matter phenomenological analyses are highly model dependent [13]. In some cases, such as the orbifold-compactified heterotic models [12], there might be situations in which the couplings of matter with stabilized dilatons lead to non-thermal dark matter, thus leading to completely different phenomenology.

However, the SSC framework is characterized by a non-stabilized dilaton which runs in cosmic time [7]. In this context, it is possible to have thermalization of weakly interacting dark matter, such as the mSUGRA lightest neutralino ($\tilde{\chi}_1^0$) which couples to the dilaton. In this sense, the mSUGRA framework provides a sufficiently non-trivial and generic pilot study of the novel effects the running dilaton has on the abundance of thermal dark matter relics.

The mSUGRA parameters are the universal scalar mass, m_0, the universal gaugino mass, $m_{1/2}$, the universal soft breaking trilinear coupling constant, A_0, the ratio of Higgs expectation values, $\tan\beta$, and the sign of μ, the bilinear Higgs coupling constant. In the case of the standard cosmology, if we concentrate on smaller values of m_0 and $m_{1/2}$, then the stau neutralino ($\tilde{\tau}_1$-$\tilde{\chi}_1^0$) coannihilation region is the only dark matter allowed region (which is also allowed by the $g_\mu - 2$ constraint) [14]. However, due to the presence of the extra factor R, the parameter space in the SSC scenario requires larger values of m_0. This is because a smaller annihilation cross section is required in the presence of the dilaton contribution. The magnitude of m_0 is however, still much smaller than the focus point region. This difference in m_0 will produce new types of signals at the LHC for the SSC model. For example, in the case of the standard cosmology, the allowed region for low m_0 requires the $\tilde{\tau}_1$ and $\tilde{\chi}_1^0$ to have nearly degenerate masses within $\sim$ 10 GeV. This produces low energy τ's in the final states [15, 16, 17]. In contrast, in the SSC model, Z bosons, Higgs bosons or high energy taus appear in the final states. These final states, which we will discuss for this SSC scenario, actually exist in most regions of the SUSY parameter space. Therefore, even without any cosmological motivation, searching for these signals is a worthwhile exercise. Furthermore, even though we have used the SSC as our motivation to probe the signals of the SUGRA model, one can come up with any other cosmological framework where the Boltzmann equation is modified in such a way that the universe is not really overclosed in this wide region of SUGRA paramater space. This analysis is valid for all these scenarios. One can also use other SUSY breaking scenarios in the context of SSC. Our analysis of signals can still be applicable in those new scenarios. However, additional observables may be required in order to determine the model parameters.

The determination of the factor R in Eq. 1 is important since it will tell us whether we satisfy the cosmological observation for the evolution of dark energy for $0 < z < 1.6$. In order to calculate R we need to calculate the relic density precisely at the collider. In this paper, we first show how to analyze and develop appropriate cuts to extract the signals in the newly allowed parameter space in order to determine model parameters. We also construct new observables necessary for the determination of such parameters. Then, using these parameters, we determine the accuracy of the result for the dark matter content. Finally, when the LHC will be operating, the dark matter direct detection experiments also will be probing the SUSY parameter space. The neutralino-proton scattering cross section is different for this newly allowed parameter space compared to the standard cosmology case. Thus we also determine the accuracy of the result for the neutralino proton scattering cross section based on the LHC measurements. This will be very useful when we will combine the data from these direct detection experiments with the LHC data to extract the final model.

2 Parameter Space

The mSUGRA model parameters are already significantly constrained by various experimental results. Most important for limiting the parameter space are: (i) the light Higgs mass bound of $M_{h^0} > 114$ GeV from LEP [18], (ii) the $b \to s\gamma$ branching ratio bound of $1.8 \times 10^{-4} < \mathcal{B}(B \to X_s\gamma) < 4.5 \times 10^{-4}$ (we assume here a relatively broad range, since there are theoretical uncertianties in extracting the branching ratio from the data) [19], (iii) the 2σ bound on the dark matter relic density: $0.095 < \Omega_{\rm CDM}h^2 < 0.1117$ [1], (iv) the bound on the lightest chargino mass of $M_{\tilde{\chi}_1^\pm} > 104$ GeV from LEP [20] and (v) the muon magnetic moment anomaly a_μ, where one gets a 3.3σ deviation from the SM from the experimental result [21, 22, 23]. Assuming the future data confirms the a_μ anomaly, the combined effects of $g_\mu - 2$ and $M_{\tilde{\chi}_1^\pm} > 104$ GeV then only allows $\mu > 0$.

Since the mSUGRA parameters determine the masses of our supersymmetric particles, they also determine the dark matter (or neutralino) relic density. We can find which regions of the parameter space will agree with the dark matter relic density observed by WMAP. The region allowed by WMAP for the standard big bang cosmology is vastly different from the region for the SSC model [7]. The comparison of these two regions is shown in Fig. 1 for the case of $A_0 = 0$ and $\tan\beta = 40$. We see a clear separation between the standard big bang cosmology region and the SSC region. The standard big bang cosmology region is the very narrow region ranging from 350 GeV $\lesssim m_{1/2} \lesssim$ 900 GeV (850 GeV $\lesssim M_{\tilde{g}} \lesssim$ 2000 GeV), and 200 GeV $\lesssim m_0 \lesssim$ 350 GeV. The SSC region is much broader for $m_{1/2} \lesssim$ 800 GeV, and is higher in m_0 which ranges from 400 to 500 GeV.

3 Signals at the LHC

The SSC region of parameter space has some unique characteristics which are distinguishable from the $\tilde{\tau}_1$-$\tilde{\chi}_1^0$ coannihilation region which appears for the lower values of m_0. For example, let us consider the decay chains of the dominant SUSY production mechanism at the LHC, which will produce the squark and gluino, in pairs (e.g., $\tilde{q}\tilde{g}$). In the coannihilation region, the dominant decay chain for the squark ($\tilde{q}_L$) is $\tilde{q}_L \to q\tilde{\chi}_2^0 \to q\tau\tilde{\tau}_1 \to q\tau\tau\tilde{\chi}_1^0$. Here $\tilde{\chi}_2^0$ is the second lightest neutralino. Thus, this region produces τ's, along with jets and missing transverse energy, $\not{E}_{\rm T}$. However, the characteristic decay in the SSC region is $\tilde{\chi}_2^0 \to h^0\tilde{\chi}_1^0$. In this case, we would expect $h^0 \to b\bar{b}$, along with jets and $\not{E}_{\rm T}$. Figure 2 shows the branching ratios for the $\tilde{\chi}_2^0$ decay. As we increase $m_{1/2}$, the branching ratios shift from Higgs dominant decay chains to τ dominant decay chains. However, we will easily distinguish this SSC τ dominated region from the coannihilation region by observing a large mass difference between the $\tilde{\tau}_1$ and $\tilde{\chi}_1^0$. For even lower $m_{1/2}$ values ($m_{1/2} \lesssim$ 350 GeV), the $\tilde{\chi}_2^0 \to Z\tilde{\chi}_1^0$ decay becomes dominant. Typical mass spectra are shown for points in the Higgs boson, Z boson and two τ final state dominated regions in Tables 1, 2, and 3.

We have three possible signals in this model. These signals are the following:

- Higgs + jets + $\not{E}_{\rm T}$
- Z + jets + $\not{E}_{\rm T}$
- 2 τ + jets + $\not{E}_{\rm T}$

A detailed discussion of these signals may be found in [24].

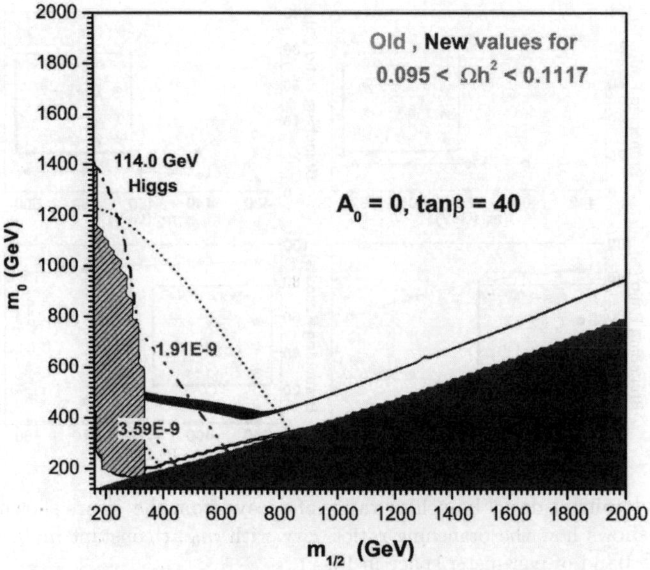

Figure 1: WMAP allowed parameter space for the SSC and standard big bang cosmology shown for $A_0 = 0$ $\tan\beta = 40$. The very thin green (grey) band is where the neutralino relic density calculated by standard big bang cosmology agrees with the WMAP3 limits $0.0950 < \Omega h^2 < 0.1117$. The thicker dark purple band shows the same agreement using the SSC calculation of the relic density. Also shown are the Higgs mass boundary (dashed dotted blue line), muon $g_\mu - 2$ boundaries (dashed and dotted red lines), a hatched cyan region which is excluded by $b \to s\gamma$ experimental bounds, and a lower solid red region where the neutralino is not the lightest supersymmetric particle.

Table 1: SUSY masses (in GeV) and dominant branching ratios for $\tilde\chi_2^0$ for the point $m_0 =$ 471 GeV, $m_{1/2} = 440$ GeV, $\tan\beta = 40$, $A_0 = 0$, and $\mu > 0$. Notice that the $\tilde\chi_2^0 \to \tilde\tau_1\tau$ is kinematically forbidden. For this point, $\Omega_{\tilde\chi_1^0} h^2 = 0.089$ and $\sigma_{p-\tilde\chi_1^0} = 2.42 \times 10^{-9}$ pb. The total production cross section for this point is $\sigma = 1.61$ pb.

$\tilde g$	$\tilde u_L$	$\tilde t_2$	$\tilde b_2$	$\tilde e_L$	$\tilde\tau_2$	$\tilde\chi_2^0$	$\mathcal{B}(\tilde\chi_2^0 \to h^0\tilde\chi_1^0)$ (%)
	$\tilde u_R$	$\tilde t_1$	$\tilde b_1$	$\tilde e_R$	$\tilde\tau_1$	$\tilde\chi_1^0$	$\mathcal{B}(\tilde\chi_2^0 \to Z\tilde\chi_1^0)$ (%)
1041	1044	954	958	557	532	341	86.8
	1017	768	899	500	393	181	13.0

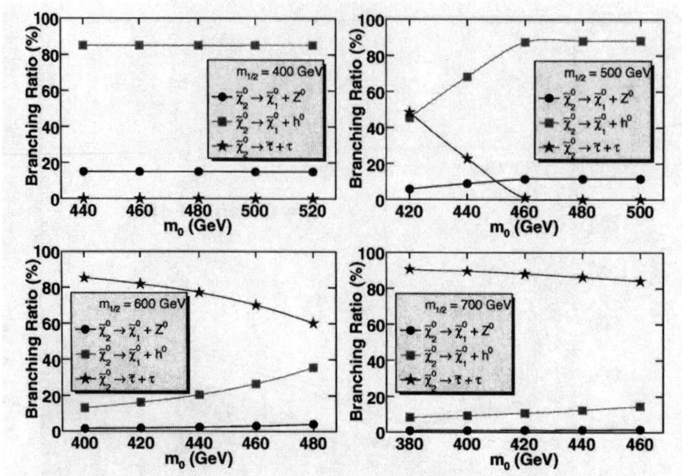

Figure 2: The dominant decay branching ratios of decays from the $\tilde{\chi}_2^0$ are shown here. Each of the four plots shows how the branching ratios vary with m_0 at constant $m_{1/2}$. Together they survey the SSC band of parameter space in Fig. 1.

Table 2: SUSY masses (in GeV) and dominant branching ratios for $\tilde{\chi}_2^0$ for the point $m_0 =$ 471 GeV, $m_{1/2} = 320$ GeV, $\tan\beta = 40$, $A_0 = 0$, and $\mu > 0$. We chose this point to examine despite the fact that it is within the region excluded by $b \to s\gamma$. We did this to examine the behavior of $\tilde{\chi}_2^0 \to Z\tilde{\chi}_1^0$ at its maximal branching ratio. The total production cross section for this point is $\sigma = 7.10$ pb.

$\tilde{g}$	$\tilde{u}_L$ / $\tilde{u}_R$	$\tilde{t}_2$ / $\tilde{t}_1$	$\tilde{b}_2$ / $\tilde{b}_1$	$\tilde{e}_L$ / $\tilde{e}_R$	$\tilde{\tau}_2$ / $\tilde{\tau}_1$	$\tilde{\chi}_2^0$ / $\tilde{\chi}_1^0$	$\mathcal{B}(\tilde{\chi}_2^0 \to h^0\tilde{\chi}_1^0)$ (%) / $\mathcal{B}(\tilde{\chi}_2^0 \to Z\tilde{\chi}_1^0)$ (%)
785	838 / 821	763 / 598	768 / 708	519 / 487	493 / 389	241 / 129	0. / 99.6

Table 3: SUSY masses (in GeV) and dominant branching ratios for $\tilde{\chi}_2^0$ for the point $m_0 =$ 440 GeV, $m_{1/2} = 600$ GeV, $\tan\beta = 40$, $A_0 = 0$, and $\mu > 0$. For this point, $\Omega_{\tilde{\chi}_1^0} h^2 = 0.106$ and $\sigma_{p-\tilde{\chi}_1^0} = 7.19 \times 10^{-10}$ pb. The total production cross section for this point is $\sigma = 0.446$ pb.

$\tilde{g}$	$\tilde{u}_L$ / $\tilde{u}_R$	$\tilde{t}_2$ / $\tilde{t}_1$	$\tilde{b}_2$ / $\tilde{b}_1$	$\tilde{e}_L$ / $\tilde{e}_R$	$\tilde{\tau}_2$ / $\tilde{\tau}_1$	$\tilde{\chi}_2^0$ / $\tilde{\chi}_1^0$	$\mathcal{B}(\tilde{\chi}_2^0 \to h^0\tilde{\chi}_1^0)$ (%) / $\mathcal{B}(\tilde{\chi}_2^0 \to \tau\tilde{\tau}_1)$ (%)
1366	1252	1153	1153	594	574	462	20.5
	1211	957	1094	494	376	249	77.0

4 Determining Model Parameters

It has been shown in [16, 17] that we can obtain mass measurements of the supersymmetric particles in the neutralino-stau coannihilation region by utilizing each final state and parameterizing kinematical observables, such as those described in the previous section, in terms of the SUSY masses. Our goal is to determine the mSUGRA model parameters m_0, $m_{1/2}$, A_0, and $\tan\beta$ since we want to determine the dark matter content and the neutralino-proton cross section. The fifth mSUGRA model parameter, sign(μ), is assumed to be positive, since this is preferred by measurements of the $b \to s\gamma$ branching ratio and the muon $g_\mu - 2$ [21]. To determine the mSUGRA parameters, we will thus need four kinematical observables which are linearly independent functions of those parameters. The determination of the parameters is then accomplished by inverting four such functional relationships.

As discussed above, certain regions of the mSUGRA parameter space might give rise to very different signals. For each region we can combine different observables to determine the four mSUGRA parameters. However, so far, there are not four observables which can be made for each signal described above. Thus we introduce the following additional kinematical observables which are valid in any region: M_{eff}, $M_{\text{eff}}^{(b)}$, and $M_{\text{eff}}^{(2b)}$.

The effective mass, M_{eff}, is defined by

$$M_{\text{eff}} = p_T^{\text{jet1}} + p_T^{\text{jet2}} + p_T^{\text{jet3}} + p_T^{\text{jet4}} + \not{E}_{\text{T}}, \tag{2}$$

where all four leading jets are not b-tagged jets. This combination carries the information of the characteristic SUSY scale. The majority of the p_T of the jets is characteristic of the gluino and first two generation squark decays, and the majority of the $\not{E}_{\text{T}}$ is due to the lightest neutralino escaping the detector. As such, the observable M_{eff} depends only on the mSUGRA parameters m_0 and $m_{1/2}$. This is because the parameters A_0 and $\tan\beta$ only affect the third generation superparticles.

When we construct the effective mass distribution, we use the following cuts[25]:

- At least one jet with $p_T \geq 100$ GeV and an additional three jets $p_T \geq 50$ GeV, where all such jets have $|\eta| \leq 2.5$,
- No isolated leptons with $|\eta| \leq 2.5$,

- $\not{E}_{\rm T} \geq 200$ GeV and $\not{E}_{\rm T} \geq 0.2 \times M_{\rm eff}$,
- Transverse Sphericity, $S_T > 0.2$,
- We reject events where any of the four leading jets is tagged as a b jet[26].

We find the peak value with an iterative fitting technique. First, we fit the distribution iteratively with an asymmetric gaussian function. The purpose of the iterative fit is simply to find the ideal fitting range. Once that is found, we fit with a cubic polynomial to find the peak position. A sample effective mass distribution showing the result of this procedure is shown in Fig. 3. This figure also shows that as $m_{1/2}$ increases, the peak increases.

Two very similar observables can also be constructed. The b effective mass, $M_{\rm eff}^{(b)}$, is defined by

$$M_{\rm eff}^{(b)} = p_T^{\rm jet1(b)} + p_T^{\rm jet2} + p_T^{\rm jet3} + p_T^{\rm jet4} + \not{E}_{\rm T}, \tag{3}$$

and the $2b$ effective mass, $M_{\rm eff}^{(2b)}$, is similarly defined by

$$M_{\rm eff}^{(2b)} = p_T^{\rm jet1(b)} + p_T^{\rm jet2(b)} + p_T^{\rm jet3} + p_T^{\rm jet4} + \not{E}_{\rm T}. \tag{4}$$

Here there are no restrictions on the non-leading jets; they can be either b-tagged or not. By including the leading b jets, which are primarily decay products of the superpartners to the third generation quarks, we include information about the parameters A_0 and $\tan\beta$.

To construct these distributions, we use the very same cuts as we used for $M_{\rm eff}$, with some exceptions. For the $M_{\rm eff}^{(b)}$ distribution, the leading $p_{\rm T}$ jet must be tagged as a b jet, otherwise we reject the event. For the $M_{\rm eff}^{(2b)}$ distribution, the two leading $p_{\rm T}$ jets must both be tagged as b jets. Again, the non-leading jets in $M_{\rm eff}^{(b)}$ and $M_{\rm eff}^{(2b)}$ can be either b-tagged or not. We also use the same fitting algorithm for these distributions as we have used for $M_{\rm eff}$. Sample $M_{\rm eff}^{(b)}$ and $M_{\rm eff}^{(2b)}$ effective mass distributions are shown in Figs. 4 and 5.

Now that we have all the observables we need, we can determine the four mSUGRA parameters in any region. We will now describe examples of this method for a Higgs boson dominated region, a $\tilde{\tau}$ dominated region, and a Z boson dominated region.

Higgs + jets + $\not{E}_{\rm T}$: For the Higgs dominant region, we use the following four observables to determine our mSUGRA parameters:

- Effective Mass: $M_{\rm eff}^{\rm peak} = f_1(m_0, m_{1/2})$
- b Effective Mass: $M_{\rm eff}^{(b)\ \rm peak} = f_2(m_0, m_{1/2}, A_0, \tan\beta)$
- $2b$ Effective Mass: $M_{\rm eff}^{(2b)\ \rm peak} = f_3(m_0, m_{1/2}, A_0, \tan\beta)$
- Higgs plus jet invariant mass: $M_{bbj}^{\rm 2nd,end} = f_4(m_0, m_{1/2})$

These functional forms are determined by examining how each kinematical observable changes while varying one of the mSUGRA parameters. Examples of this are shown in Figs. 6 and 7. To determine our mSUGRA parameters, we invert these functional forms into functions of the mSUGRA parameters in terms of the kinematical observables. Then we can simply plug in the

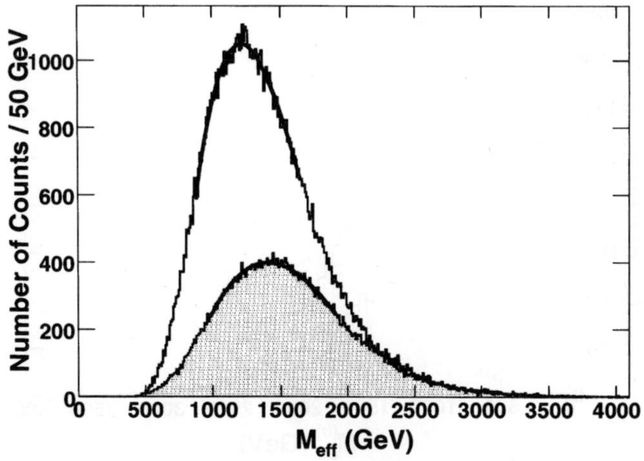

Figure 3: The effective mass distribution reconstructed through PGS in two 500 fb^{-1} mSUGRA samples at $(m_0,\ m_{1/2}) = (471\text{ GeV},\ 400\text{ GeV})$ and $(471\text{ GeV},\ 480\text{ GeV})$ for the black histogram with red(gray) fit and blue(gray, filled) histogram with dark blue(dark gray) fit, respectively. We fix $\tan\beta = 40$, $A_0 = 0$, and $\mu > 0$.

values of the observables into the inverted functions to solve for the mSUGRA parameters. To get the uncertainties of the mSUGRA determinations, we propagate the uncertainties of the measured observables through the inverted functions using a Monte Carlo method. An example of an analysis of this type is shown in [24].

5 Conclusions and Discussion

In this paper, we have studied the mSUGRA final states at the LHC which are motivated by supercritical string cosmology (SSC). In the SSC case, the time dependent dilaton not only contributes to the dark energy but also to the Boltzman equation which determines the dark matter content of the universe. Consequently the dark matter profile in this model is different compared to the standard cosmology. We found that the dark matter allowed region has larger values of m_0 compared to the standard cosmology case. Thus, the final states in the SSC scenario are different from those of the standard cosmology. For example, in the case of standard cosmology for smaller values of m_0 (also allowed by the $g_\mu - 2$ constraint), we have low energy taus in the final state due to the proximity of the stau to the neutralino mass in the stau-neutralino coannihilation region. On the other hand, in the SSC case the final states contain Z bosons, Higgs bosons or high energy taus. In fact these final states dominate in most of the allowed SUGRA parameter space. Therefore, by analyzing the parameter space of the SSC model we actually investigate most regions of the SUGRA parameter space at the LHC. We analyzed the signals involving Higgs + jets + $\not{E}_T$, Z + jets + $\not{E}_T$, and 2τ + jets + $\not{E}_T$ and

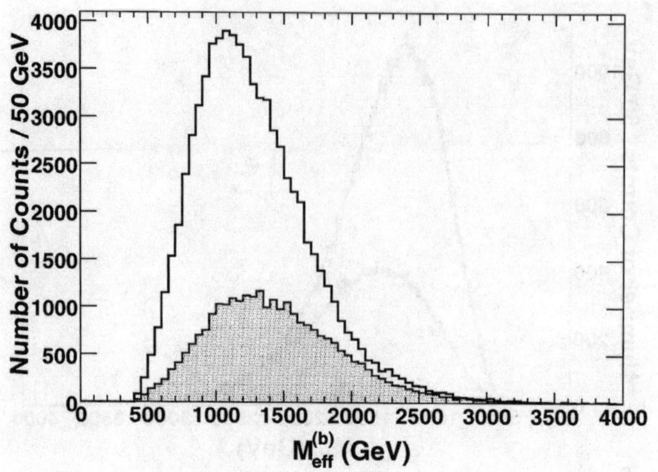

Figure 4: Same as Fig. 3, except that this is the b effective mass distribution and that the peak fits are not shown.

constructed observables such as the endpoints of invariant mass distributions M_{bbj}, M_{Zj}, and $M_{j\tau\tau}$ and the peak point of $M_{\tau\tau}$.

When the LHC will be operating, we will also have results from the dark matter direct detection experiment. We found that the cross section for these experiments can be predicted from the LHC measurements with an accuracy of 195% for 1000 fb^{-1} of data in the Higgs boson dominated region and 21% for 500 fb^{-1} of data in the stau dominated region. This cross section however includes uncertainity due to the form factors.

Acknowledgments

This work of is supported in part by the DOE grant DE-FG02-95ER40917. I would like to thank Professor Antonino Zichichi for his warm hospitality at this school, as well as my coauthors in reference [24] for a fruitful collaboration.

References

[1] WMAP Collaboration, D.N. Spergel *et al.*, Astrophys. J. Suppl. **148** (2003) 175.

[2] H. Goldberg, Phys. Rev. Lett. **50** (1983) 1419; J. Ellis, J. Hagelin, D. Nanopoulos, K. Olive, and M. Srednicki, Nucl. Phys. B **238** (1984) 453.

[3] D. Z. Freedman, P. Van Niewenhuisen, and S. Ferrara, Phys. Rev. D **13** (1976) 3214; S. Deser and B. Zumino, Phys. Lett. B **65** (1976) 369; A.H. Chamseddine, R. Arnowitt, and

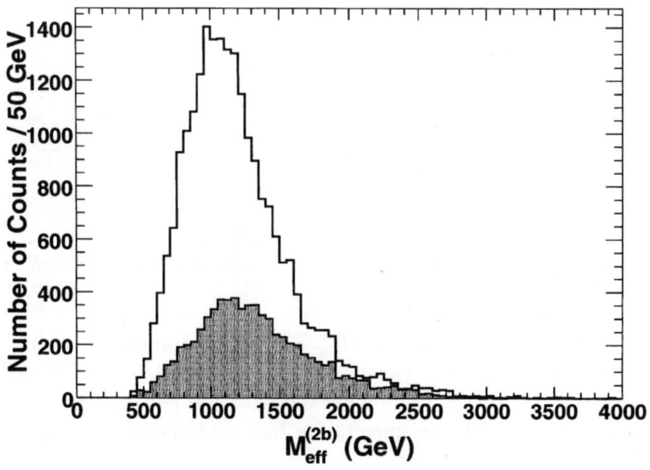

Figure 5: Same as Fig. 4, except that this is the $2b$ effective mass distribution.

P. Nath, Phys. Rev. Lett. **49** (1982) 970; R. Barbieri, S. Ferrara, and C.A. Savoy, Phys. Lett. B **119** (1982) 343; L. Hall, J. Lykken, and S. Weinberg, Phys. Rev. D **27** (1983) 2359; P. Nath, R. Arnowitt, and A.H. Chamseddine, Nucl. Phys. B **227** (1983) 121; For a review, see P. Nilles, Phys. Rep. **100** (1984) 1.

[4] R. R. Caldwell, R. Dave, and P. J. Steinhardt, Phys. Rev. Lett. **80** (1998) 582; A. Hebecker and C. Wetterich, Phys. Rev. Lett. **85** (2000) 3339; M. Gasperini, Phys. Rev. D **64** (2001) 043510; M. Gasperini, F. Piazza, and G. Veneziano, Phys. Rev. D **65** (2001) 023508; L. Amendola and D. Tocchini-Valentini, Phys. Rev. D **64** (2001) 043509.

[5] I. Antoniadis, C. Bachas, J. R. Ellis, and D. V. Nanopoulos, Phys. Lett. B **211** (1988) 393; Nucl. Phys. B **328** (1989) 117; Phys. Lett. B **257** (1991) 278; J. R. Ellis, N.E. Mavromatos, and D. V. Nanopoulos, Lectures given at International Workshop on Recent Advances in the Superworld, Woodlands, TX, 13-16 Apr 1993, Published in Woodlands Superworld 1993:3-26 (QCD161:I966:1993) e-Print: hep-th/9311148; J. R. Ellis, N. E. Mavromatos, and D. V. Nanopoulos, Mod. Phys. Lett. A **10** (1995) 1685; Phys. Lett. B **619** (2005) 17.

[6] M. Gasperini, F. Piazza, and G. Veneziano, Phys. Rev. D **65** (2002) 023508; M. Gasperini and G. Veneziano, Phys. Rept. **373** (2003) 1.

[7] A. B. Lahanas, N. E. Mavromatos, and D. V. Nanopoulos, PMC Phys. A **1** (2007) 2; Phys. Lett. B **649** (2007) 83.

[8] A. G. Riess *et al.*, [arXiv:astro-ph/0611572].

[9] J. R. Ellis, N. E. Mavromatos and D. V. Nanopoulos, Phys. Lett. B **665**, 412 (2008) [arXiv:0804.3566 [hep-th]].

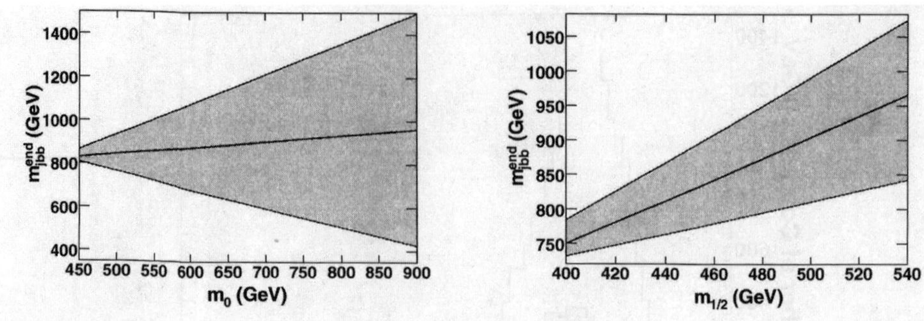

Figure 6: The left plot shows the change in $M_{bbj}^{\text{2nd,end}}$ under variations of m_0 ($m_{1/2} = 440$ GeV) within the Higgs dominant region of parameter space. The right plot shows the same for variations in $m_{1/2}$ ($m_0 = 471$ GeV). Combining the functions plotted results in the functional form $M_{bbj}^{\text{2nd,end}} = f_4(m_0, m_{1/2})$. The 1σ uncertainty bands (dashed lines) represent 500 fb^{-1} of data.

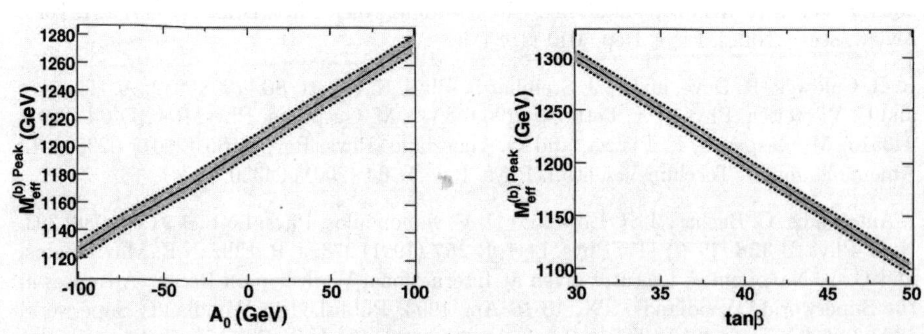

Figure 7: The left plot shows the change in $M_{\text{eff}}^{(b)\,\text{peak}}$ under variations of A_0 within the Higgs dominant region of parameter space ($m_0 = 471$ GeV, $m_{1/2} = 440$ GeV). The right plot shows the same for variations in $\tan\beta$. Combining the functions plotted along with variations in m_0 and $m_{1/2}$ results in the functional form $M_{\text{eff}}^{(b)\,\text{peak}} = f_2(m_0, m_{1/2}, A_0, \tan\beta)$. The 1σ uncertainty bands (dashed lines) represent 1000 fb^{-1} of data.

[10] J. Albert *et al.* [MAGIC Collaboration and Other Contributors Collaboration], Phys. Lett. B **668**, 253 (2008) [arXiv:0708.2889 [astro-ph]].

[11] J. Ellis, N. E. Mavromatos and D. V. Nanopoulos, arXiv:0901.4052 [astro-ph.HE].

[12] See, for instance: P. Binetruy, M. K. Gaillard, and B. D. Nelson, Nucl. Phys. B **604** (2001) 32; P. Binetruy, A. Birkedal-Hansen, Y. Mambrini, and B. D.Nelson, Eur. Phys. J. C **47** (2006) 481.

[13] For a recent review on the predictions of different models see: N. E. Mavromatos, "LHC Physics and Cosmology," in Proc. Lake Louise Winter Institute 2007: Fundamental Interactions, 19-24 Feb 2007, Lake Louise, Alberta, Canada (World Sci. 2008) [arXiv:hep-ph/0708.0134] and references therein.

[14] J. Ellis, K. Olive, Y. Santoso, and V. Spanos, Phys. Lett. B, **565** (2003) 176; R. Arnowitt, B. Dutta, and B. Hu, hep-ph/0310103; H. Baer, C. Balazs, A. Belyaev, T. Krupovnickas, and X. Tata, JHEP **0306** (2003) 054; A. B. Lahanas and D.V. Nanopoulos, Phys. Lett. B, **568** (2003) 55; U. Chattopadhyay, A. Corsetti, and P. Nath, Phys. Rev. D**68** (2003) 035005; E. Baltz and P. Gondolo, JHEP **0410** (2004) 052; A. Djouadi, M. Drees, and J. L. Kneur, JHEP **0603** (2006) 033; G. Belanger, S. Kraml, and A. Pukhov, Phys. Rev. D **72** (2005) 015003.

[15] R. Arnowitt *et al.*, Phys. Rev. Lett. **100** (2008) 231802.

[16] R. Arnowitt *et al.*, Phys. Lett. B **639** (2006) 46.

[17] R. Arnowitt *et al.*, Phys. Lett. B **649** (2007) 73.

[18] ALEPH, DELPHI, L3, OPAL Collaborations, G. Abbiendi *et al.*(The LEP Working Group for Higgs Boson Searches), Phys. Lett. B, **565** (2003) 61.

[19] M. Alam *et al.*, Phys. Rev. Lett. **74** (1995) 2885.

[20] Particle Data Group, S. Eidelman *et al.*, Phys. Lett. B, **592** (2004) 1.

[21] Muon $g-2$ Collaboration, G. Bennett *et al.*, Phys. Rev. Lett. **74** (2004) 161802.

[22] M. Davier, [arXiv:hep-ex/0312065].

[23] K. Hagiwara, A. Martin, D. Nomura, and T. Teubner, Phys. Rev. D **69** (2004) 093003.

[24] B. Dutta, A. Gurrola, T. Kamon, A. Krislock, A. B. Lahanas, N. E. Mavromatos and D. V. Nanopoulos, arXiv:0808.1372 [hep-ph].

[25] I. Hinchliffe *et al.*, Phys. Rev. D **55** (1997) 5520; I. Hinchliffe and F. E. Paige, Phys. Rev. D **61** (2000) 095011.

[26] PGS4 is a parameterized detector simulator. We use version 4 (`http://www.physics.ucdavis.edu/~conway/research/software/pgs/pgs4-general.htm`) in the CMS detector configuration. We assume the τ identification efficiency with $p_T^{\rm vis} > 20$ GeV is 50%, while the probability for a jet being mis-identified as a τ is 1%. The b-jet tagging efficiency in PGS is ∼42% for $E_T > 50$ GeV and $|\eta| < 1.0$, and degrading between $1.0 < |\eta| < 1.5$. The b-tagging fake rate for c and light quarks/gluons is $\sim 9\%$ and 2%, respectively.

CHAIRMAN: D. NANOPOULOS

Scientific Secretaries: R. Patterson, A. Suchecka

DISCUSSION II

- *Aprile:*

What neutralino mass is most favored for direct detection in the case of a 10^{-9} pb spin-independent cross section, which is less favorable than the cross section predicted by the benchmark point in the stau coannihilation region?

- *Nanopoulos:*

In this particular story, we're talking about neutralinos of 185 GeV or something like that.

- *Aprile:*

So higher mass, then...

- *Nanopoulos:*

Yes.

- *Aprile:*

What are the favored signatures for this neutralino mass range at the LHC?

- *Nanopoulos:*

If you go to yesterday's talk and look at the slides that have "SSC" on them, you will see there that I give a table with all the favored signatures.

- *Slobodeniuk:*

The standard cosmology describes our world, but we postulate the number of dimensions. Does string cosmology have a dynamical principle which explains why our universe has four dimensions?

- *Nanopoulos:*

In all these models that I and others are working on, we always start with ten or eleven dimensions, and then we compactify, coming down to four. In conventional or critical string theory, which means strings living in a Minkowski, non-expanding universe there is no answer to why D=4. In a realistic string theory that involves strings in an expanding universe, or curved space-time, that as I mentioned in my lecture was invented by I. Antoniadis, John Ellis and myself, the number of space-time dimensions D becomes a dynamical variable that is determined by the infrared fixed point of the Renormalization Group flow of the world sheet conformal theory central charge.

- *Carrasco:*

So you found this surprising relation between cosmic acceleration and the square of the string coupling that's generic to a particular class of string models describing cosmic acceleration. Presumably there exist other classes of string theory models of cosmic acceleration. Could you briefly comment about distinguishing features of these other models and whether they suggest any connection between string coupling and acceleration?

- *Nanopoulos:*

When I was a kid and working with Glashow, he was telling me if you get a good relation even from the wrong model, you should go on with this relation, which is more or less our attitude here, to be honest with you. A lot of this stuff may not be realized in nature. We were very surprised when we found something like that. So, there is a class of models. Now, of course, the models where you have a fixed dilaton – the "standard" way – then of course you cannot get such a relation, because the way that the string coupling constant gets in is that it goes through the dilaton door. So in order to have this, you have to have this kind of dependence with the dilaton. Now, that does not mean that you necessarily have to go non-critical to get the dilaton doing this thing. But, I like this relation, this is true.

- *Hoehn:*

You mentioned that the dark matter contribution in your model was negligible at the time of nucleosynthesis. Could you comment a bit more on this? Why is this the case? What is the effect of it, and why is it necessary?

- *Nanopoulos:*

It is imperative that during primordial nuclesynthesis the baryonic density is the "Standard Cosmology" one, because it works! What I was referring to was the fact that the mechanism that dilutes by a factor of order 10 the dark matter density is inoperative in the case of the baryonic density and this is exactly what we found by our explicit calculations.

- *Zichichi:*

Do you get Λ_{GUT} out of nothing?

- *Nanopoulos:*

I get it out of the acceleration. I relate this kind of thing with the acceleration.

- *Zichichi:*

But, you get the right value out of...?

- *Nanopoulos:*

Out of the unification.

- *Zichichi:*

Ah, experimental physics wins!

- *Nanopoulos:*

That's right, that's right.

High Scale Physics Connection to LHC Data[1]

Pran Nath

Department of Physics, Northeastern University, Boston, MA 02115-5000, USA

Abstract

The existing data appears to provide hints of an underlying high scale theory. These arise from the gauge coupling unification, from the smallness of the neutrino masses, and via a non-vanishing muon anomaly. An overview of high scale models is given with a view to possible tests at the Large Hadron Collider. Specifically we discuss here some generic approaches to deciphering their signatures. We also consider an out of the box possibility of a four generation model where the fourth generation is a mirror generation rather than a sequential generation. Such a scenario can lead to some remarkably distinct signatures at the LHC.

[1]Based on lectures at the 46th Course - Erice -Sicily: 29 August -7 September, 2008

Introduction: The standard modelof electro-weak interactions is remarkably successful in explaining experimental data up to LEP energies, of ~ 100 GeV. However, this model cannot be extrapolated to the Planck scale,

$$M_{Pl} = (8\pi G_{Newton})^{-1/2} \sim 2.4 \times 10^{18}\ GeV. \qquad (1)$$

One of the purposes of high energy physics is to undertand the laws of physics from low energy up to M_{Pl}. Therefore much of the current effort in high energy physics is focussed on discovering what lies beyond the standard model. Although the standard model is remarkably successful it does suffer from some drawbacks. One obvious drawback is the appearance of many arbitrary parameters. On a more theoretical level the loop correction to the Higgs mass diverges quadratically, and there is a lack of unification with gravity. Other things the standard model does not explain is why the charges of quarks and leptons are quantized, and why the matter in the universe is made up mostly of protons and not of anti-protons, and what gives rise to non-baryonic dark matter and dark energy. Further, one might ask what explains the mass hierarchy. A variety of possibilities have been considered as one goes beyond the standard model. These include grand unification (GUT), compositeness, supersymmetry (SUSY) and supegravity (SUGRA) based models, strings and branes, extra (warped) dimensions, Ads/CFT, Stueckelberg and other U(1) extensions, unparticles, noncommutative geometry and many other possibilities. These extensions may be loosely classified as falling in two broad approaches: the top down and the bottom up. In the top down approach one starts with a presumed candidate for a unified theory and one works one's way downward to realize the standard model. In the bottom up approach, one starts at the electroweak scale and works upwards to high scales which could be either the grand unification scale or the string/ Planck Scale.

High scale theories: There are at least two important hints that one is dealing with high scale models as one goes beyond the standard model. The first of these is the unification of gauge coupling constants. Here one finds that the extrapolation of high precision LEP data produces a unification of the $SU(3)_C \times SU(2)_L \times U(1)$ gauge couplings g_3, g_2, g_1 at a high scale within SUGRA soft breaking models with the MSSM spectrum. (For reviews of high scale physics see[1, 2].). While unification of gauge couplings is also possible within the extra dimensions models due to a power law running, it is less predictive in that context. Another hint of a high scale arises from the smallness of neutrino masses. The astrophysical limits from the WMAP satellite experiment on the sum of neutrino masses is[3]

$$\sum_{i=1,2,3} |m_{\nu_i}| < .7 \text{ eV}. \tag{2}$$

A small neutrino mass O(eV) could arise from a See-Saw mechanism[4]

$$m_\nu \sim \frac{m^2}{M} \tag{3}$$

With $m \sim M_W \sim 10^2$ GeV and $M \sim 10^{16}$ GeV, one can generate neutrino masses in the sub eV region. Thus the smallness of neutrino masses points to a high scale within the SeeSaw mechanism. While small neutrino masses can also be generated by other mechanisms, the SeeSaw mechanism appears to be more natural one.

High scale physics points to SUSY: If one wishes to incorporate high scales along with the standard model in a common framework, then supersymmetry appears to be a logical possibility. The reason for this is because SUSY stabilizes scales. Incorporation of gravity in SUSY requires transition from global susy to local supersymmetry[5] and its modern form supergravity[6]. SUSY grand unification incorporates GUTs and SUSY but has no gravity. Supergravity grand unification has SUSY, GUTs and gravity. However, a viable scheme requires breaking of supersymmetry and in supergravity grand unification (SUGRA GUT)[7] supersymmetry is broken by gravity mediation[7, 8, 9]. An interesting point about supergravity is that it is the field point limit of string theory, and thus check on the validity of SUGRA GUT would be a pointer to the underlying unified theory of quantum gravity.

Implications of high scale models at the electroweak scale: In a supersymmetric model or string model, to make contact with low energy physics one must break the supersymmetry with soft terms, and typically the generation of soft terms involves three steps: (i) First one must generate SUSY breaking in a sector different from the visible sector; (ii) Then this breaking is communicated by messengers to the visible sector; and finally (iii) With (i) & (ii) one generates soft terms in the visible sector. The two main scenarios for supersymmetry breaking are the gravity mediated breaking[7, 8, 9, 10], and gauge mediated breaking[11]. The breaking of SUSY in the hidden sector could be arranged by a variety of methods. The simplest way it by use of a chiral superfield in the hidden sector. In this case the one finds

$$m_{soft} \sim F/M_{\rm Pl}, \tag{4}$$

where with $\sqrt{F} \sim 10^{10}$ GeV, one finds $m_{\rm soft} \sim 10^{2-3}$ GeV. Further, $m_{\rm soft} \sim 10^{2-3}$GeV can also arise from gaugino condensation[12] such that $m_{soft}^{string} \sim \langle\lambda\lambda\rangle/M_{\rm Pl}^2$ $\langle\lambda\lambda\rangle \sim (10^{13}{\rm GeV})^3$. In gauge mediated breaking one finds that soft SUSY breaking in the visible sector is generated radiatively[11]

$$m_{\rm soft} \sim (\alpha/4\pi)F/M_{\rm msg}, \tag{5}$$

where with $\sqrt{F} \sim M_{\rm msg} \sim 10^4$ GeV, one has $m_{\rm soft} \sim (10^2 - 10^3)$ GeV. A variety of other possibilities also exist for the breaking of supersymmetry such as anomaly breaking, hierarchical breaking of SUSY, and meta -stable supersymmetry breaking. We will discuss the hierarchical breaking of SUSY a bit later. Sugra and heterotic string models are high scale models. The model mSUGRA is defined by the soft terms $V_{SB} = m_0^2 \sum_i \phi_i^\dagger \phi_i + A_0 W^{(3)} + B_0 W^{(2)} + m_{1/2} \sum_{\alpha=3,2,1} \bar{\lambda}_\alpha \lambda_\alpha$, where $W^{(2)}$ is the quadratic part and $W^{(3)}$ is the cubic part of the superpotential. In MSSM $W^{(2)}$ takes the form $W^{(2)} = \mu\epsilon_{ij} H_1^i H_2^j$, where H_2 couples to the up quarks and H_1 couples to the down quarks and the leptons. Thus the parameter space of mSUGRA model is defined by $m_0, m_{1/2}, A_0, B_0$, and μ. An interesting phenomenon in SUGRA type models is the breaking of the electroweak symmetry by radiative corrections. Here one finds two conditions: one which determines $|\mu|^2$ in terms of the soft parameters and the other which determines B (which is B_0 at the electroweak scale) in terms of $\tan\beta$, where $\tan\beta$ is the ratio of the two VEVS, i.e., $\tan\beta =< H_2 > / < H_1 >$. Using these constraints one can choose the mSUGRA parameters to be $m_0, m_{1/2}, A_0, \tan\beta$, and sign$(\mu)$. Some of the early phenomenological implications of the SUGRA models can be found in [13] and some of the later works can be found in [14, 15, 16, 17]. Soft breaking in string models determines B. In a simple heterotic string model with compactification on three tori $T_2 \times T_2 \times T_2$ with moduli consisting of S, and $T_i = T$ (i=1,2,3) one has on using modular invariance the result[18]

$$B = m_0 b \frac{e^{D/2}}{(T+\bar{T})^3}, \tag{6}$$

where $D = -log(S + \bar{S} + ..)$, and $b = b(m_0, m_{1/2}, A_0)$ is a function of the other soft parameters $m_0, m_{1/2}, A_0$. Since $e^{-D} = 2/g_{string}^2$, one determines $\tan\beta$ in terms of α_{string} and other soft parameters. The loop corrections play an important role in the analysis[19, 20]. It should be noted that historically the first hint that the top quark may be much heavier than was then thought (i.e., in circa 1983) came from the analysis of radiative breaking in SUGRA models [21] and should be viewed as an important triumph of the high scale SUGRA models.

Hierarchical Breaking of SUSY: In string theory it often happens that some of the extra $U(1)$ factors are anomalous. It is then permissible to add FI- D terms one for each $U(1)$. To break SUSY one may add two scalars $\phi^{\pm}$ to MSSM[22], which are singlets under SM $SU(3)_C \times SU(2)_L \times U(1)_Y$, but with charges ± 1 under $U(1)_X$. Thus adding the term $W_{\pm} = m\phi^+\phi^-$ to the MSSM superpotential and minimizing the full potential $\mathcal{V} = m^2\left(|\phi^+|^2 + |\phi^-|^2\right) + \frac{g_X^2}{2}\left(\sum_i Q_X^i |\tilde{f}_i|^2 + |\phi^+|^2 - |\phi^-|^2 + \xi_X\right)^2$, one finds that $Q_X^i > 0$ drives the fields to $\langle\phi^+\rangle = 0$, $\langle\phi^-\rangle^2 = \xi_X - \frac{m^2}{g_X^2}$, $\langle F_{\phi^+}\rangle = m\sqrt{\xi_X} + \cdots$ which gives $< D >= (m^2/g_X)$ and the scalar masses m_i^2 and the gaugino masses m_λ are then given by

$$m_i^2 \simeq m^2 Q_X^i, m_\lambda \sim \frac{1}{M_{\rm Pl}^2}\langle F_{\phi^+}\phi^- + F_{\phi^-}\phi^+\rangle \sim m\frac{\xi_X}{M_{\rm Pl}^2}. \tag{7}$$

In heterotic string models one finds that the FI parameter ξ at one loop is[23, 24]

$$\xi_X \sim \frac{g_X^2 Trace(Q_X)}{192\pi^2} M_{\rm Pl}^2. \tag{8}$$

In type II string compactifications ξ_a can in principle be of any size[25].

We investigate now the sparticle mass hierarchy with many $U(1)'s$ each with an FI term which gives us the scalar potential

$$\mathcal{V}_D = \mathcal{V}_D^{\rm MSSM} + \sum_a \frac{g_a^2}{2}\Big(\sum_i Q_a^i |\tilde{f}_i|^2 + \alpha_a|\phi^+|^2 - \alpha_a|\phi^-|^2 + \xi_a\Big)^2. \tag{9}$$

In this case the VEVs of $\phi^{\pm}$ will absorb one FI term but the remaining set will make contributions to the scalar mass^2 proportional to ξ_a. Since $< D_a >\sim \xi_a$ one finds $m_i^2 \sim \sum_a g_a^2 Q_i^a \xi_a$.. Further, in heterotic string models since ξ_a can be order $O(M_{\rm Pl}^2)$ one can generate scalar masses of order $O(M_{\rm Pl})$. Thus with this mechanism one can generate a split SUSY scenario[26]. However, vacuum energy considerations put stringent upper limits on the scalar masses[27]. To consider this possibility we begin with the scalar potential in SUGRA and strings which is given by

$$\mathcal{V} = -\kappa^{-4}e^{-G}[G^{M\bar{N}}G_M G_{\bar{N}} + 3] + \mathcal{V}_D \; ; \; G = -\kappa^2\mathcal{K} - \ln(\kappa^6 W W^\dagger), \tag{10}$$

where $\mathcal{K}$ is the Kahler potential, and $\kappa = 1/M_{Pl}$. An important constraint here is the condition for the vanishing of the vacuum energy which is given by [27]

$$|\gamma_S|^2 + \sum_I |\gamma_I|^2 + |\gamma_+|^2 + |\gamma_-|^2 + \frac{1}{3m_{3/2}^2 M_{\rm Pl}^2} \sum_a \frac{g_a^2}{2} D_a^2 = 1 \ , \tag{11}$$

where $|\gamma_S|^2 = -\frac{1}{3}G^{S\bar{S}}G_S G_{\bar{S}}$, $|\gamma_I|^2 = -\frac{1}{3}G^{I\bar{I}}G_I G_{\bar{I}}$, and $|\gamma_\pm|^2$ are defined in a similar fashion. The above gives $\langle D_a \rangle \ < m_{3/2}M_{\rm Pl}$ which implies that the scalar mass^2 $\tilde{m}_i^2$ is bounded from above so that[27]

$$\tilde{m}_i^2 \leq m_{3/2}M_{Pl}. \tag{12}$$

Eq.(12) implies that with $m_{3/2} = O(\text{TeV})$, the sfermion mass cannot exceed the value (10^{10-13} GeV). Thus unless $m_{3/2}$ itself is of size the Planck scale, one cannot get the scalar mass to be of the Planck size. For an alternate scenario for generating a mass hierarchy in soft breaking see[28].

Ellipsoidal and Hyperbolic branch of radiative breaking of the electroweak symmetry: Sugra models resolve the problem of why the Higgs mass^2 at low scales is tachyonic. This is done via radiative electroweak symmetry breaking (REWSB) (For a review see [29]). However, there are two branches to REWSB which are as follows: (i) Ellipsoidal Branch: When the loop correction to the effective potential are small REWSB occurs so that one has the constraint

$$\frac{m_{1/2}'^2}{a^2} + \frac{m_0^2}{b^2} + \frac{A_0^2}{c^2} \simeq 1; \ \ m_{1/2}'^2 = m_{1/2} + cA_0. \tag{13}$$

Here one finds that the soft parameters lie on the surface of an ellipsoid for a fixed μ which fixes the radii a, b, c; (ii) Hyperbolic Branch: For large loop correction one of the terms on the right hand side of Eq.(13) can turn negative and one has a hyperbolic branch (HB) of REWSB so that [30]

$$\frac{m_{1/2}'^2}{\alpha^2(Q_0)} - \frac{m_0^2}{\beta^2(Q_0)} \simeq \pm 1. \tag{14}$$

On HB multi-TeV scalars can exist even with small fine tuning which is parametrized by μ[30]. Here one finds that m_0 can get rather large for fixed μ, which leads to multi TeV scalars. This region of multi TeV scalars is also sometimes labeled as the Focus Point region (FP). In the mSUGRA case one is dealing with the soft parameters which are universal at the high scale which we take to be the grand unification scale. However, the nature of physics at high scales is still largely unknown. So one must also consider sugra models with non-universalities in soft breaking (NUSUGRA)[17]. Such non-universalities can arise in the Higgs sector (NUH), in the gaugino sector (NUG), and in the third generation sector (NU3). Additionally one may also consider non-universalities in the first two generations.

Implications of Brookhaven experiment on $g_\mu - 2$: SUGRA models predict the existence of a sizable correction to the muon magnetic moment on $a_\mu = (g_\mu - 2)/2$. Defining Δa_μ to be $\Delta a_\mu = a_\mu^{exp} - a_\mu^{SM}$, one finds that the most recent analyses using the Brookhaven data gives for Δa_μ the result[31]

$$\Delta a_\mu = 27.5(8.4) \times 10^{-10}, \quad 3.3\sigma \text{ discrepancy} \tag{15}$$

Interestingly it was already predicted in the early eighties[32] within the framework of SUGRA models that the supersymmetric electroweak corrections to Δa_μ could be as large or larger than the SM electroweak corrections (For further analyses see [33]). Additionally it is known that an extension of the standard model such as the 5D models with one compact extra dimension on a half circle (S^1/Z_2) do not generate a significant contribution to Δa_μ[34]. However, some models based on extra dimensions, such as the universal extra dimension model (UED), do also allow for a sizable correction to $g_\mu - 2$[35].

Missing link are Sparticles: If the BNL experiment holds up, i.e., a 3.3σ discrepancy is present, then within SUSY/SUGRA it is predicted that some of the sparticles have an upper bound and must be seen at the LHC. There are 32 sparticle masses in MSSM. Including certain sum rule constraints one has in excess of 10^{25} mass hierarchies. Only one of these would be realized at the LHC if the mSUGRA or some variant of it is the correct model. We focus on the first four sparticle mass hierarchies aside from the light higgs. A mapping of the parameter space of mSUGRA under constraints from experiment reduces more than 10^4 4 particle hierarchies to very few [36, 37, 38, 39] minimal sugra patterns ($mSPs$). Specifically, one finds that only sixteen 4-particle patterns survive with $\mu > 0$ (these are labeled mSP1-mSP16) and only 6 additional 4-particle patterns survive for $\mu < 0$ (mSP17-mSP22). These patterns are displayed in Table 1. Comparing with the Snowmass[40] and the Postwamp3 benchmarks[41] one finds that these cover only 5 out of the 22 patterns listed above. Thus the analysis of [36] gives a more comprehensive mapping of the parameter space of the mSUGRA model than those given by the Snowmass benchmarks[40], Postwamp3 benchmarks[41] and the low mass (LM)/high mass (HM) benchmarks given by the CMS Collaboration. A similar mapping of NUSUGRA finds 15 additional NUSUGRA patterns which are labeled NUSP1-NUSP15 in [38]. These patterns are displayed in Table 2 and show some significant new features such as $\tilde{g}$ being the NLSP. Analyses similar to the above can be carried out using the soft breaking in strings and in D brane models[42] and partial results were reported in [37]. In these analyses we have not taken into account the effect of CP phases on the sparticle spectrum. Such phases can affect the spectrum strongly in some cases. However, one must also impose the constraints arising from the electric dipole moments of the electron and of the neutron when the CP phases are included (For a recent review see [43]).

From models to LHC signatures: At the LHC one collides two beams of protons with a center of mass energy of 14 TeV. No matter what the model the end result will be a bunch of leptons, jets, photons and missing energy. Out of these we have to extrapolate back to determine the underlying model. For SUGRA models with R parity, the signatures necessarily include a significant amount of missing energy. In the analysis we impose the constraints from $g_\mu - 2$ experiment, $b \to s + \gamma$[44], WMAP, and the experimental constraints from LEP and the Tevatron. We discuss some prominent signatures below. (Some recent analyses of signatures can be found in [45, 46, 47, 48, 49, 50, 51, 52, 53, 54, 55, 56, 57, 58]). The first of these is the excess of trileptonic events[59] at colliders. For example, in pp collisions one has $pp \to W^{\pm *} \to \chi^{\pm}\chi_2^0 \to (W^{\pm} + \chi_1^0) + \chi_2^0)$, $W^{\pm} \to l_1^{\pm}\nu, ..., \chi_2^0 \to l_2^+ l_2^- \chi_1^0, ...$ Thus the off shell production of W^* will lead to a trileptonic signature

$$p + p \to l_1^{\pm} l_2^{\pm} l_2^{\mp} + \text{jets} + \text{ missing energy} \tag{16}$$

Additionally there are many other sources of trileptonic signals with in SUGRA models. Like sign dileptons are produced in greater abundance than in SM.

$$\begin{aligned} \tilde{g} \to \bar{c} + \tilde{c}_L \to \bar{c} + e^+ + d, \ \tilde{g} \to c + \tilde{c}_L^* \to c + e^- + \bar{d}, \\ \tilde{g}\tilde{g} \to e^+e^+, e^-e^-, \cdots \end{aligned} \tag{17}$$

Thus one finds a significant excess of like sign dileptons relative to what one might expect in the standard model. There are actually a large number of signatures that one would generate from the LHC data. In Table 3 a list of the signatures most likely to lead to the discovery of new physics are listed.

Fuzzy signature vectors: Given a model one can define a signature vector $\xi = (\xi_1, .., \xi_{41})$, $\xi_i = n_i/N$, $N = \sum_i n_i$, where n_i is the number of events for i-th signature. For a pattern X one can define a fuzzy vector pattern vector

$$\Delta\xi^X = (\Delta\xi_1^X, .., \Delta\xi_{41}^X). \tag{18}$$

where $\Delta\xi_i^X$ is the range for signature i within the pattern X. Two patterns X and Y are distinguishable if at least one element $\Delta\xi_i^X$ does not overlap $\Delta\xi_i^Y$. Thus define

$$(\Delta\xi^X|\Delta\xi^Y) = 0(1) : \text{overlap (no overlap)}. \tag{19}$$

According to this simple criterion some patterns are distinguishable from others. The patterns are also constrained by $B_S \to \mu^+\mu^-$ data and by the dark matter cross sections[37]. We discuss the allowed parameter space consistent with dark matter constraints in further detail below.

Decoding the origin of dark matter using LHC data[60]: As is well known dark matter constitutes a significant part of our universe (for a review see[61]). For neutralino dark matter relic density constraints can be satisfied in a variety of ways. These include the coannihilation region[62, 63], the HB region[30], and the pole region[62, 64, 65]. The pole region could include the Z pole, the light Higgs pole h and the heavy CP even and the CP odd Higgs poles. One is interested in $\Omega_\chi \equiv \rho_\chi/\rho_c$ where ρ is the mass density of relic neutralinos in the universe and ρ_c is the critical mass density needed to close the universe, i.e. $\rho_c = \frac{3H_0^2}{8\pi G_N}$. Here H_0 is the Hubble parameter at current time and G_N is the newtonian constant and $\rho_c = 1.9h_0^2 \times 10^{-29}\text{gm/cm}^3$. In the analysis of $\Omega_\chi h_0^2$ we need to solve the Boltzman equation for n, the number density of neutralinos in the early universe, which is given by $\frac{dn}{dt} = -3Hn - \langle\sigma v\rangle(n^2 - n_0^2)$. In the above, n_0 is the value of n at thermal equilibrium, $\langle\sigma v\rangle$ is the thermal average of the neutralino annihilation cross section $\sigma(\tilde{\chi}_1^0\tilde{\chi}_1^0 \to X)$ and v is the relative $\tilde{\chi}_1^0$ velocity, and H is the Hubble parameter at time t.

The stau co-annihilation region is one of the important regions for the satisfaction of the relic density. It involves processes such as $\tilde{\tau}\tilde{\tau} \to \tau\tau$, $\tilde{\tau}\chi \to \tau Z, \tau h, \tau\gamma$, $\tilde{\tau}\tilde{\ell}_i(i \neq \tau) \to \tau\ell_i$, $\tilde{\tau}\tilde{\tau}^* \to f_i\bar{f}_i, W^+W^-, ZZ, \gamma Z, \gamma\gamma$. Here one must consider the total density $n = \sum_i n_i$ where i runs over all the sparticles that enter in the co-annihilations, where n now obeys the equation $\frac{dn}{dt} = -3Hn - \langle\sigma_{\text{eff}}v_{\text{rel}}\rangle(n^2 - n_{\text{eq}}^2)$,

$\sigma_{\text{eff}} = \sum_{ij} \sigma_{ij}\gamma_i\gamma_j$. Here σ_{ij} is the cross section for annihilation of particles i and j, and $\gamma_i = n^i_{\text{eq}}/n_{\text{eq}}$ where n^i_{eq} refers to the number density of sparticle i at thermal equilibrium. In the coannihilation region, the neutralino is mostly a bino. On the other hand in the HB region, the neutralino can have very significant higgsino content, while in other regions of the parameter space it could have varying portions of gaugino and higgsino content. It is interesting to ask if the LHC data will allow one to differentiate among the regions of the parameter space where the dark matter originates. In a recent work this issue was analyzed in some detail[60]. It was shown that the LHC data can indeed allow one to discriminate among dark matter models. Specifically, it was shown that using various signatures one can differentiate between dark matter originating on the stau coannihilation branch vs dark matter originating in the HB region.

An out of the box possibility: a 4th generation which is mirror: Essentially all of the model building within grand unification and in strings starts with the assumptions: (i) 3 generations of quarks and leptons; (ii) the gauge group $SU(3)\times SU(2)\times U(1)$ down to the electroweak scale. A relaxation of assumption (ii) with additonal $U(1)$ factors leads to interesting new predictions (see, e.g., [66, 67] and references therein). However, barring few exceptions the assumption of 3 generations is often taken without reservation, . The main reasons for the 3 generation assumption include: (i) the Z-width constraint, (ii) the CKM unitarity constraints on the CKM matrix element V_{ij} (i,j=1,2,3), (iii) the constraints on oblique parameters (S,T,U), and (iv) the gauge coupling unification constraint. The constraints on extra generations have been analyzed in a number of papers[68, 69, 70] and all of them can be overcome. Thus the Z width constraint is easily overcome by making the extra generation masses greater than $M_Z/2$. Regarding the CKM unitarity constraints, a careful analysis shows that there is a window for an extra generation consistent with the limits $|V_{14}| \leq .04|$, $|V_{41}| \leq .08$, $|V_{24}| \leq .17$ (see Kribs etal in [70]), and there are also windows for an extra generation in other CKM matrix elements. Actually, the most stringent constraints arise from the so called oblique parameters (S, T, U) and specifically, from the parameter S. An extra generation will contribute an amount $\Delta S = 0.21$ which is unacceptable. However, this problem too can be overcome in specific regions of the 4th generation mass parameters. Finally, the gauge coupling unification constraints for an extra generation can also be satisfied.

Let us suppose then that there is indeed a fourth generation and further that this generation rather than being sequential is a mirror generation (For early work on mirrors in model building see [71]). In this case one needs to examine the same restrictions as discussed above. Specifically the Z-width constraint, the CKM unitarity constraint, and the gauge coupling unification constraints are satisfied as for a sequential generation. To discuss the constraints on the oblique parameters, consider an $SU(2)_L$ multiplet with up and down fermion masses M_1, M_2. The constraint on the oblique parameter S is satisfied in much the same way as for the sequential 4th generation since the correction ΔS is invariant under the transformation[72]

$$\text{fermions } (\psi_1, \psi_2) \leftrightarrow \text{mirror fermions } (\psi_2^c, \psi_1^c);\ Y \leftrightarrow -Y, M_1 \leftrightarrow M_2 \tag{20}$$

Suppose then that there is a large GUT group, which unifies families and which breaks leaving a certain $U(1)_F$ subgroup unbroken under which the families and mirror families are charged and their charges do not pair up. Then one or more mirror families can remain mass less along with the sequential families down to the electroweak scale. An example of the above phenomenon is the analysis of [73] for $SO(18)$. In this analysis one finds that there are $V-A$ families with charges $Q_F=-1, Q_F=3$ and $V+A$ families with charges $Q_F=1, Q_F=-3, Q_F=5$. One finds that there are 3 families with charges $Q_F=3$, two mirror families with $Q_F=1$ and one mirror family with $Q_F=5$ which are light. All other families and mirror families become heavy. Thus down to the electroweak scale one finds light particles some of which are families and others mirror families. These light families and mirror families eventually gain masses at the electroweak scale.

We discuss now the implication for string model building. Allowing for a light mirror generation will modify very significantly string model building. For example, in $E_8 \times E_8$ heterotic string, one considers Calabi-Yau (CY) compactifications $M_{10} = M_4 \times K$, where K is a CY manifold. The resulting theory is a 4D $N=1$ theory with the gauge symmetry $E_8 \times E_6$. Typically one ends up with many families and mirror families, and one needs quotient manifolds K/G where G is a discrete symmetry of K which gives $n(27) - n(27^*) = \chi(K)/2N_G$, where $\chi(K)$ is Euler Characteristic, and $N(G)$ is the number of elements of G. With one light mirror generation, we should not impose the constraint $\chi(K)/2N_G = 3$ but rather $\chi(K)/2N_G = 2$. Many additional possibilities in string model buildling exist if one allows for a light mirror generation. For example, Kac-Moody level 2 heterotic string constructions are interesting in that they allow for adjoint Higgs representations to break the gauge symmetry. However, no known examples of 3 massless generations exist. For this reason not much model building has occurred for this class of models. However, this class of models could become viable if one allows for a light mirror generation and three light sequential generations since $n_f - n_{mf} = 2$

One can extend MSSM to accommodate a light mirror generation[74]. Thus in analogy to the ordinary lepton generation such as the 3rd generation of leptons

$$\psi_L \equiv \begin{pmatrix} \nu_L \\ \tau_L \end{pmatrix} \sim (1,2,-\frac{1}{2}), \tau_L^c \sim (1,1,1), \nu_L^c \sim (1,1,0), \tag{21}$$

where the quantum numbers correspond to the $SU(3)_C$, $SU(2)_L$, $U(1)_Y$ one has a mirror generation defined by

$$\chi^c \equiv \begin{pmatrix} E_{\tau L}^c \\ N_L^c \end{pmatrix} \sim (1,2,\frac{1}{2}), E_{\tau L} \sim (1,1,-1), N_L \sim (1,1,0). \tag{22}$$

Similarly corresponding to the ordinary quarks such as the third generation quarks

$$q \equiv \begin{pmatrix} t_L \\ b_L \end{pmatrix} \sim (3,2,\frac{1}{6}), t_L^c \sim (3^*,1,-\frac{2}{3}), b_L^c \sim (3^*,1,\frac{1}{3}),$$

one has mirror quarks defined by

$$Q^c \equiv \begin{pmatrix} B_L^c \\ T_L^c \end{pmatrix} \sim (3^*,2,-\frac{1}{6}), T_L \sim (3,1,\frac{2}{3}), B_L \sim (3^*,1,-\frac{1}{3}).$$

In order to discuss experimental implications of such an extension one needs to construct the couplings in mirMSSM involving standard model supermultiplets and mirror supermultiplets. A partial analysis of such couplings was given in [74]. Using these couplings one finds some very distinct signatures for mirMSSM. One important result is the posssibility of a very large contribution to the neutrino magnetic moment and the other consists of very distinct signatures at the LHC for mirror quarks and mirror leptons.

We discuss first the τ neutrino magnetic moment including the exchange of mirror leptons and their sparticle counterparts. In the standard model

$$\mu_{\nu_\tau} = \frac{3eg^2 m_\nu}{64\pi^2 M_W^2} \sim (m_{\nu_\tau}/1eV) \times 3 \times 10^{-19}\mu_B,$$

where m_ν is the neutrino mass and where μ_B is the Bohr Magneton. A similar size is expected in the supersymmetric extension from the exchange of the the charginos and sleptons. We will focus now on τ neutrino magnetic moment. The current limits on ν_τ is[75]

$$|\mu(\nu_\tau)| \leq 1.3 \times 10^{-7}\mu_B \tag{23}$$

Thus the neutrino magnetic moments predicted in the standard model lie far beyond the reach of experiment. However, the magnetic moments in mirMSSM lie within reach. For illustration we consider a simple model where there is a mixing only between the mirror generation and the 3rd generation. In this case including the mirror particle and sparticle exchange one gets for the τ neutrino magnetic moment the result[74]

$$\Delta\mu_{\nu_\tau} \sim \frac{g_2^2 m_e m_{\tau'}\mu_B}{48\pi^2} G_1(\frac{m_{\tau'}}{m_W}) + \cdots,$$
$$G_1(r) = \frac{4-r^2}{1-r^2} + \frac{3r^2}{(1-r^2)^2} ln(r^2), \tag{24}$$

where τ' is the mirror lepton. The modification produces a correction numerically so that a μ_{ν_τ} as large as $O(10^{-9})$ can arise and thus within the realm of experimental observation with improved experiment. At the same time one finds that the contribution of the mirrors to the τ magnetic moment is within the current experimental limits.

Next we discuss the LHC signatures for mirrors. In the analysis we will include a right handed singlet for each of the generations so that we will have Dirac neutrinos. By inclusion of a mirror generation we have added the following new set of fermionic particles: B, T, E, N, where all fields including N are Dirac. At the same time in the bosonic sector we have added the following set of scalars [74] $\tilde{B}_1, \tilde{B}_2, \tilde{T}_1, \tilde{T}_2, \tilde{E}_1, \tilde{E}_2, \tilde{\nu}_1, \tilde{\nu}_2, \tilde{\nu}_3$. The reason for the appearance of three extra sneutrino states is as follows: Within the third generation and the fourth generation, there are two Dirac neutrinos which contain four chiral states. These would lead to 4 chiral scalars. One of these in the usual sneutrino in the third generation in MSSM while the additional three sneutrinos are new and are listed as $\tilde{\nu}_1, \tilde{\nu}_2, \tilde{\nu}_3$ above. With inclusion of a mirror generation many new signatures are now possible. Thus, e.g., if $M_N > M_E + M_W$, one will have the decay signatures[74]

$$N \to E^- W^+, \; E^- \to \tau^- Z \to \tau^- e^+ e^-, \tau^- \mu^+ \mu^-, \tau^+ \tau^+ \tau^-. \tag{25}$$

The Drell-Yan process can generate interesting signatures. Thus, e.g., one has processes of the type[74]

$$pp \to Z^* \to E^+ E^- \to 2\tau 4l, 4\tau 2l, 6\tau, \tag{26}$$

where $l_1, l_2 = e, \mu$. In addition to the above one will have final states with taus, leptons and jets. Many other signatures are possible such as[74]

$$pp \rightarrow \tilde{\nu}_i \tilde{\nu}_i^* \rightarrow \tilde{E}_k^+ \tilde{E}_k^- W^+ W^-, \tilde{E}_k^+ E^- W^\mp \tilde{\chi}^\pm, \tag{27}$$

followed by the decays of the $\tilde{E}^+ \tilde{E}^-$ which give τs + leptons + jets + $E_T^{\rm miss}$ which may have as many as 8 leptons, where all the leptons could be τs.

The couplings of the heavy CP even and CP odd Higgs to mirrors are very different from those of an sequential 4th generation. Thus the couplings of the CP odd Higgs boson A^0 to a sequential fourth generation are given by

$$\mathcal{L}_{\rm 4th} = \frac{ig}{2M_W}(m_{4d}\bar{d}_4\gamma_5 d_4 \tan\beta + m_{4u}\bar{u}_4\gamma_5 u_4 \cot\beta + \cdot\cdot)A^0, \tag{28}$$

while for a mirror 4th generation one has[74]

$$\mathcal{L}_{\rm mir} = \frac{ig}{2M_W}(M_B\bar{B}\gamma_5 B \cot\beta + M_T\bar{T}\gamma_5 T \tan\beta + \cdot\cdot)A^0. \tag{29}$$

These give rise to decay branching ratios as follows[74]

$$\frac{\Gamma(A^0 \rightarrow u_4\bar{u}_4)}{\Gamma(A^0 \rightarrow d_4\bar{d}_4)} \simeq \frac{m_{d_4}^2}{m_{u_4}^2}\tan^4\beta, \quad \frac{\Gamma(A^0 \rightarrow T\bar{T})}{\Gamma(A^0 \rightarrow B\bar{B})} \simeq \frac{m_B^2}{m_T^2}\cot^4\beta. \tag{30}$$

The relative difference between the decay into a sequential fourth generation and into a mirror fourth generation in this case is a factor of $\tan^8\beta$ which is a remarkable signature that separates 4th sequential generation from a 4th mirror generation.

The branching ratios of the CP odd Higgs can also provide important signatures that differentiate a fourth generation from a mirror generation. Thus one may define the ratio of branching ratios $R_{d_4/u_4}^{H^0} = BR(H^0 \rightarrow d_4\bar{d}_4)/BR(H^0 \rightarrow u_4\bar{u}_4)$. Using the MSSM vertices one finds

$$R_{d_4/u_4}^{H^0} = \frac{m_{d_4}^2}{m_{u_4}^2}(\cot\alpha\tan\beta)^2 P_{d_4/u_4}^{H^0}, \tag{31}$$

where α is the Higgs mixing parameter and $P_{d_4/u_4}^{H^0}$ is a phase space factor given by $P_{d_4/u_4}^{H^0} = (1 - 4m_{d_4}^2/m_H^2)^{3/2}(1 - 4m_{u_4}^2/m_H^2)^{-3/2}$. In contrast, for the decay of the CP even heavy Higgs into the mirror quarks when $(m_{H^0} > 2m_Q, Q = B, T)$ one finds [74]

$$R_{B/T}^{H^0} = \frac{m_B^2}{m_T^2}(\tan\alpha\cot\beta)^2 P_{B/T}^{H^0}. \tag{32}$$

We note that in this case the dependence on α and β is much different relative to the case when H^0 decays into sequential fourth generation quarks.

Conclusions: The main message is that the study of the sparticle landscape and of patterns can be a useful tool in extrapolating data back to theory. The landscape with $O(10^4)$ patterns for the 4 lightest sparticles reduces down just to about 50, in SUGRA models under the WMAP3, LEP and Tevatron constraints. This is a significant progress. The analysis of lepton and jet events already allows one to separate many of these patterns. Additional discrimination arises from $B_s \to \mu\mu$ process, Higgs production cross sections, and from studies of dark matter limits. Thus SUGRA models predict a candidate for dark matter which is detectable in direct detection of dark matter experiments. A combined analysis of limits from this data along with data from LHC is very powerful in limiting theory models and may even lead us uniquely to the underlying model beyond the standard model. However, one needs to keep an open mind regarding what LHC may teach us. In this regard it is desirable that one consider out of the box possibilities. mirMSSM is one such possibility discussed in this lecture.

Acknowledgements: The author thanks Professor Gerard 't Hooft and Professor Antonino Zichichi for invitation to lecture at the 46th Course in Subnuclear Physics at Erice, and Professor Zichichi for the hospitality extended to him during the period of the course. The work reported here is based on collaboration with Richard Arnowitt, Ali H Chamseddine, Utpal Chattopadhyay, Daniel Feldman, Tarek Ibrahim and Zuowei Liu. This research is supported in part by NSF grant PHY-0757959.

mSP	Mass Pattern	$\mu > 0$	$\mu < 0$
mSP1	$\widetilde{\chi}_1^0 < \widetilde{\chi}_1^\pm < \widetilde{\chi}_2^0 < \widetilde{\chi}_3^0$	Y	Y
mSP2	$\widetilde{\chi}_1^0 < \widetilde{\chi}_1^\pm < \widetilde{\chi}_2^0 < A/H$	Y	Y
mSP3	$\widetilde{\chi}_1^0 < \widetilde{\chi}_1^\pm < \widetilde{\chi}_2^0 < \widetilde{\tau}_1$	Y	Y
mSP4	$\widetilde{\chi}_1^0 < \widetilde{\chi}_1^\pm < \widetilde{\chi}_2^0 < \widetilde{g}$	Y	Y
mSP5	$\widetilde{\chi}_1^0 < \widetilde{\tau}_1 < \widetilde{l}_R < \widetilde{\nu}_\tau$	Y	Y
mSP6	$\widetilde{\chi}_1^0 < \widetilde{\tau}_1 < \widetilde{\chi}_1^\pm < \widetilde{\chi}_2^0$	Y	Y
mSP7	$\widetilde{\chi}_1^0 < \widetilde{\tau}_1 < \widetilde{l}_R < \widetilde{\chi}_1^\pm$	Y	Y
mSP8	$\widetilde{\chi}_1^0 < \widetilde{\tau}_1 < A \sim H$	Y	Y
mSP9	$\widetilde{\chi}_1^0 < \widetilde{\tau}_1 < \widetilde{l}_R < A/H$	Y	Y
mSP10	$\widetilde{\chi}_1^0 < \widetilde{\tau}_1 < \widetilde{t}_1 < \widetilde{l}_R$	Y	
mSP11	$\widetilde{\chi}_1^0 < \widetilde{t}_1 < \widetilde{\chi}_1^\pm < \widetilde{\chi}_2^0$	Y	Y
mSP12	$\widetilde{\chi}_1^0 < \widetilde{t}_1 < \widetilde{\tau}_1 < \widetilde{\chi}_1^\pm$	Y	Y
mSP13	$\widetilde{\chi}_1^0 < \widetilde{t}_1 < \widetilde{\tau}_1 < \widetilde{l}_R$	Y	Y
mSP14	$\widetilde{\chi}_1^0 < A \sim H < H^\pm$	Y	
mSP15	$\widetilde{\chi}_1^0 < A \sim H < \widetilde{\chi}_1^\pm$	Y	
mSP16	$\widetilde{\chi}_1^0 < A \sim H < \widetilde{\tau}_1$	Y	
mSP17	$\widetilde{\chi}_1^0 < \widetilde{\tau}_1 < \widetilde{\chi}_2^0 < \widetilde{\chi}_1^\pm$		Y
mSP18	$\widetilde{\chi}_1^0 < \widetilde{\tau}_1 < \widetilde{l}_R < \widetilde{t}_1$		Y
mSP19	$\widetilde{\chi}_1^0 < \widetilde{\tau}_1 < \widetilde{t}_1 < \widetilde{\chi}_1^\pm$		Y
mSP20	$\widetilde{\chi}_1^0 < \widetilde{t}_1 < \widetilde{\chi}_2^0 < \widetilde{\chi}_1^\pm$		Y
mSP21	$\widetilde{\chi}_1^0 < \widetilde{t}_1 < \widetilde{\tau}_1 < \widetilde{\chi}_2^0$		Y
mSP22	$\widetilde{\chi}_1^0 < \widetilde{\chi}_2^0 < \widetilde{\chi}_1^\pm < g$		Y

Table 1: Hierarchical mass patterns in mSUGRA. Y stands for appearance of the pattern for the sub case. Taken from [38].

Signature	Description	Signature	Description
0L	0 Lepton	0T	0 τ
1L	1 Lepton	1T	1 τ
2L	2 Leptons	2T	2 τ
3L	3 Leptons	3T	3 τ
4L	4 Leptons and more	4T	4 τ and more
0L1b	0 Lepton + 1 b-jet	0T1b	0 τ + 1 b-jet
1L1b	1 Lepton + 1 b-jet	1T1b	1 τ + 1 b-jet
2L1b	2 Leptons + 1 b-jet	2T1b	2 τ + 1 b-jet
0L2b	0 Lepton + 2 b-jets	0T2b	0 τ + 2 b-jets
1L2b	1 Lepton + 2 b-jets	1T2b	1 τ + 2 b-jets
2L2b	2 Leptons + 2 b-jets	2T2b	2 τ + 2 b-jets
ep	e^+ in 1L	em	e^- in 1L
mp	μ^+ in 1L	mm	μ^- in 1L
tp	τ^+ in 1T	tm	τ^- in 1T
OS	Opposite Sign Di-Leptons	0b	0 b-jet
SS	Same Sign Di-Leptons	1b	1 b-jet
OSSF	Opposite Sign Same Flavor Di-Leptons	2b	2 b-jets
SSSF	Same Sign Same Flavor Di-Leptons	3b	3 b-jets
OST	Opposite Sign Di-τ	4b	4 b-jets and more
SST	Same Sign Di-τ	TL	1 τ plus 1 Lepton

Kinematical signatures
1. P_T^{miss}
2. Effective Mass = $P_T^{miss} + \sum_j P_T^j$
3. Invariant Mass of all jets
4. Invariant Mass of e^+e^- pair
5. Invariant Mass of $\mu^+\mu^-$ pair
6. Invariant Mass of $\tau^+\tau^-$ pair

Table 2: A list of 40 counting signatures along with the kinematical signatures analyzed in [38]. Each counting signature is accompanied in addition by at least two jets. Lepton $=e, \mu$. Taken from [38].

References

[1] K. R. Dienes, Phys. Rept. **287**, 447 (1997).

[2] A. Zichichi, S. Ferrara, R. M. Mössbauer, "Searching for the Superworld", World Scientific (2007), ISBN 9812700188, 9789812700186.

[3] D. N. Spergel *et al.* [WMAP Collaboration], Astrophys. J. Suppl. **148**, 175 (2003).

[4] R. N. Mohapatra *et al.*, "Theory of neutrinos: A white paper," Rept. Prog. Phys. **70**, 1757 (2007).

[5] P. Nath and R. Arnowitt, Phys. Lett. B **56**, 177 (1975); R. Arnowitt, P. Nath and B. Zumino, Phys. Lett. B **56**, 81 (1975); P. Nath and R. Arnowitt, Phys. Lett. B **65**, 73 (1976).

[6] D. Z. Freedman, P. van Nieuwenhuizen and S. Ferrara, Phys. Rev. D **13**, 3214 (1976); S. Deser and B. Zumino, Phys. Lett. B **62**, 335 (1976).

[7] A. H. Chamseddine, R. Arnowitt and P. Nath, Phys. Rev. Lett. **49** (1982) 970.

[8] R. Barbieri, S. Ferrara and C.A. Savoy, Phys. Lett. **B119**, 343 (1982).

[9] L. Hall, J. Lykken and S. Weinberg, Phys. Rev. **D27**, 2359 (1983); P. Nath, R. Arnowitt and A. H. Chamseddine, Nucl. Phys. B **227**, 121 (1983).

[10] For a review see, P. Nath, arXiv:hep-ph/0307123.

[11] G. F. Giudice and R. Rattazzi, Phys. Rept. **322**, 419 (1999).

[12] H. P. Nilles, Phys. Lett. B **115**, 193 (1982); S. Ferrara, L. Girardello and H. P. Nilles, Phys. Lett. B **125**, 457 (1983).

[13] A small sample of early works on the phenomenological implications of SUGRA models consists of S. Weinberg, Phys. Rev. Lett. **50**, 387 (1983); R. Arnowitt, A. H. Chamseddine and P. Nath, Phys. Rev. Lett. **50** (1983) 232; A. H. Chamseddine, P. Nath and R. Arnowitt, Phys. Lett. B **129** (1983) 445; P. Nath, R. Arnowitt and A. H. Chamseddine, Phys. Lett. B **121** (1983) 33; P. Nath, R. Arnowitt and A. H. Chamseddine, HUTP-83/A077; D. A. Dicus, S. Nandi, W. W. Repko and X. Tata, Phys. Rev. Lett. **51**, 1030 (1983);

Phys. Rev. D **29**, 67 (1984); Phys. Rev. D **29**, 1317 (1984); D. A. Dicus, S. Nandi and X. Tata, Phys. Lett. B **129**, 451 (1983); J. M. Frere and G. L. Kane, Nucl. Phys. B **223**, 331 (1983); J. R. Ellis, J. M. Frere, J. S. Hagelin, G. L. Kane and S. T. Petcov, Phys. Lett. B **132**, 436 (1983); J. R. Ellis, J. S. Hagelin, D. V. Nanopoulos and M. Srednicki, Phys. Lett. B **127**, 233 (1983); M. K. Gaillard, L. J. Hall, B. Zumino, F. del Aguila, J. Polchinski and G. G. Ross, Phys. Lett. B **122**, 355 (1983);

[14] U. Chattopadhyay, A. Corsetti and P. Nath, Phys. Rev. D **68**, 035005 (2003); J. R. Ellis, K. A. Olive, Y. Santoso and V. C. Spanos, Phys. Lett. B **565**, 176 (2003); H. Baer, A. Belyaev, T. Krupovnickas and J. O'Farrill, JCAP **0408**, 005 (2004); J. Edsjo, M. Schelke, P. Ullio and P. Gondolo, JCAP **0304**, 001 (2003); P. Gondolo, J. Edsjo, P. Ullio, L. Bergstrom, M. Schelke and E. A. Baltz, JCAP **0407**, 008 (2004); Y. Mambrini and E. Nezri, [hep-ph/0507263]; M. M. Nojiri, G. Polesello and D. R. Tovey, JHEP **0603**, 063 (2006); J. L. Feng, A. Rajaraman and B. T. Smith, Phys. Rev. D **74**, 015013 (2006); A. Belyaev, S. Dar, I. Gogoladze, A. Mustafayev and Q. Shafi, arXiv:0712.1049 [hep-ph].

[15] R. R. de Austri, R. Trotta and L. Roszkowski, JHEP **0605**, 002 (2006).

[16] B. C. Allanach, C. G. Lester and A. M. Weber, JHEP **0612**, 065 (2006); B. C. Allanach, K. Cranmer, C. G. Lester and A. M. Weber, JHEP **0708**, 023 (2007); B. C. Allanach and D. Hooper, JHEP **0810**, 071 (2008).

[17] P. Nath and R. Arnowitt, Phys. Rev. D **56**, 2820 (1997); A. Birkedal-Hansen and B. D. Nelson, Phys. Rev. D **67**, 095006 (2003); U. Chattopadhyay and P. Nath, Phys. Rev. D **65**, 075009 (2002); U. Chattopadhyay and D. P. Roy, Phys. Rev. D **68**, 033010 (2003); D. G. Cerdeno and C. Munoz, JHEP **0410**, 015 (2004); G. Belanger, F. Boudjema, A. Cottrant, A. Pukhov and A. Semenov, Nucl. Phys. B **706**, 411 (2005); H. Baer, A. Mustafayev, S. Profumo, A. Belyaev and X. Tata, JHEP **0507**, 065 (2005).

[18] P. Nath and T. R. Taylor, Phys. Lett. B **548**, 77 (2002).

[19] S. R. Coleman and E. J. Weinberg, Phys. Rev. D **7**, 1888 (1973).

[20] R. L. Arnowitt and P. Nath, Phys. Rev. D **46**, 3981 (1992).

[21] L. Alvarez-Gaume, J. Polchinski and M. B. Wise, Nucl. Phys. B **221**, 495 (1983).

[22] G. R. Dvali and A. Pomarol, Phys. Rev. Lett. **77**, 3728 (1996).

[23] M. Dine, N. Seiberg and E. Witten, Nucl. Phys. B **289**, 589 (1987).

[24] J. J. Atick, L. J. Dixon and A. Sen, Nucl. Phys. B **292**, 109 (1987).

[25] L. E. Ibanez, R. Rabadan and A. M. Uranga, Nucl. Phys. B **542**, 112 (1999).

[26] N. Arkani-Hamed, S. Dimopoulos, G. F. Giudice and A. Romanino, Nucl. Phys. B **709**, 3 (2005).

[27] B. Kors and P. Nath, Nucl. Phys. B **711**, 112 (2005).

[28] K. S. Babu, T. Enkhbat and B. Mukhopadhyaya, Nucl. Phys. B **720**, 47 (2005).

[29] L. E. Ibanez and G. G. Ross, Comptes Rendus Physique **8**, 1013 (2007).

[30] K. L. Chan, U. Chattopadhyay and P. Nath, Phys. Rev. D **58**, 096004 (1998); J. L. Feng, K. T. Matchev and T. Moroi, Phys. Rev. Lett. **84**, 2322 (2000); H. Baer, C. Balazs, A. Belyaev, T. Krupovnickas and X. Tata, JHEP **0306**, 054 (2003). For a review see, A. B. Lahanas, N. E. Mavromatos and D. V. Nanopoulos, Int. J. Mod. Phys. D **12**, 1529 (2003).

[31] M. Davier, Nucl. Phys. Proc. Suppl. 169, 288 (2007).

[32] T. C. Yuan, R. Arnowitt, A. H. Chamseddine and P. Nath, Z. Phys. C **26**, 407 (1984); D. A. Kosower, L. M. Krauss and N. Sakai, Phys. Lett. B **133**, 305 (1983); J. L. Lopez, D. V. Nanopoulos and X. Wang, Phys. Rev. D **49**, 366 (1994); U. Chattopadhyay and P. Nath, Phys. Rev. D **53**, 1648 (1996); T. Moroi, Phys. Rev. D **53**, 6565 (1996).

[33] T. Ibrahim and P. Nath, Phys. Rev. D **61**, 095008 (2000); Phys. Rev. D **62**, 015004 (2000); T. Ibrahim, U. Chattopadhyay and P. Nath, Phys. Rev. D **64**, 016010 (2001).

[34] P. Nath and M. Yamaguchi, Phys. Rev. D **60**, 116006 (1999). Phys. Rev. D **60**, 116004 (1999).

[35] T. Appelquist and B. A. Dobrescu, Phys. Lett. B **516**, 85 (2001).

[36] D. Feldman, Z. Liu and P. Nath, Phys. Rev. Lett. **99**, 251802 (2007).

[37] D. Feldman, Z. Liu and P. Nath, Phys. Lett. B **662**, 190 (2008).

[38] D. Feldman, Z. Liu and P. Nath, JHEP **0804**, 054 (2008).

[39] D. Feldman, Z. Liu and P. Nath, arXiv:0806.4683 [hep-ph].

[40] B. C. Allanach *et al.*, [arXiv:hep-ph/0202233].

[41] M. Battaglia, A. De Roeck, J. R. Ellis, F. Gianotti, K. A. Olive and L. Pape, Eur. Phys. J. C **33**, 273 (2004).

[42] B. Kors and P. Nath, Nucl. Phys. B **681** (2004) 77; D. Lust, S. Reffert and S. Stieberger, Nucl. Phys. B **706** (2005) 3; T. W. Grimm and J. Louis, Nucl. Phys. B **699** (2004) 387; G. L. Kane, P. Kumar, J. D. Lykken and T. T. Wang, Phys. Rev. D **71** (2005) 115017.

[43] T. Ibrahim and P. Nath, Rev. Mod. Phys. **80**, 577 (2008); arXiv:hep-ph/0210251; J. S. Lee, M. Carena, J. Ellis, A. Pilaftsis and C. E. M. Wagner, arXiv:0712.2360 [hep-ph]; J. R. Ellis, J. S. Lee and A. Pilaftsis, JHEP **0810**, 049 (2008).

[44] M. E. Gomez, T. Ibrahim, P. Nath and S. Skadhauge, Phys. Rev. D **74**, 015015 (2006); G. Degrassi, P. Gambino and P. Slavich, Comput. Phys. Commun. **179**, 759 (2008).

[45] A. Djouadi, M. Drees and J. L. Kneur, JHEP **0603**, 033 (2006). L. S. Stark, P. Hafliger, A. Biland and F. Pauss, JHEP **0508**, 059 (2005).

[46] J. R. Ellis, S. Heinemeyer, K. A. Olive and G. Weiglein, JHEP **0605**, 005 (2006).

[47] U. Chattopadhyay, D. Das, A. Datta and S. Poddar, Phys. Rev. D **76**, 055008 (2007); N. Bhattacharyya, A. Datta and S. Poddar, Phys. Rev. D **78**, 075030 (2008); N. Bhattacharyya, A. Datta and M. Maity, Phys. Lett. B **669**, 311 (2008); S. Bhattacharya, A. Datta and B. Mukhopadhyaya, arXiv:0809.2012 [hep-ph].

[48] N. Arkani-Hamed, G. L. Kane, J. Thaler and L. T. Wang, JHEP **0608**, 070 (2006); G. L. Kane, P. Kumar and J. Shao, J. Phys. G **34**, 1993 (2007).

[49] J. P. Conlon, C. H. Kom, K. Suruliz, B. C. Allanach and F. Quevedo, JHEP **0708**, 061 (2007).

[50] H. Baer, V. Barger, G. Shaughnessy, H. Summy and L. t. Wang, Phys. Rev. D **75**, 095010 (2007).

[51] P. G. Mercadante, J. K. Mizukoshi and X. Tata, Braz. J. Phys. **37** (2007) 549.

[52] R. Kitano and Y. Nomura, Phys. Rev. D **73**, 095004 (2006).

[53] W. Beenakker, M. Klasen, M. Kramer, T. Plehn, M. Spira and P. M. Zerwas, Phys. Rev. Lett. **83**, 3780 (1999).

[54] H. K. Dreiner, S. Grab and M. K. Trenkel, arXiv:0808.3079 [hep-ph].

[55] S. P. Martin, Phys. Rev. D **75**, 115005 (2007); Phys. Rev. D **76**, 095005 (2007); Phys. Rev. D **75**, 115005 (2007); Phys. Rev. D **78**, 055019 (2008).

[56] B. Altunkaynak, M. Holmes and B. D. Nelson, JHEP **0810**, 013 (2008).

[57] G. J. Gounaris, J. Layssac and F. M. Renard, Phys. Rev. D **77**, 093007 (2008).

[58] R. L. Arnowitt, B. Dutta, T. Kamon, N. Kolev and D. A. Toback, Phys. Lett. B **639**, 46 (2006); R. L. Arnowitt, B. Dutta, A. Gurrola, T. Kamon, A. Krislock and D. Toback, Phys. Rev. Lett. **100**, 231802 (2008).

[59] P. Nath and R. Arnowitt, Mod. Phys. Lett. A **2**, 331 (1987); H. Baer, C. h. Chen, F. Paige and X. Tata, Phys. Rev. D **50**, 4508 (1994); V. D. Barger, C. Kao and T. j. Li, Phys. Lett. B **433**, 328 (1998). For 4 lepton modes see, T. Ibrahim, Phys. Rev. D **77**, 065028 (2008).

[60] D. Feldman, Z. Liu and P. Nath, Phys. Rev. D **78**, 083523 (2008).

[61] G. Bertone, D. Hooper and J. Silk, Phys. Rept. **405**, 279 (2005).

[62] K. Griest and D. Seckel, Phys. Rev. D **43**, 3191 (1991).

[63] M. Drees and M. M. Nojiri, Phys. Rev. D **47**, 376 (1993). S. Mizuta and M. Yamaguchi, Phys. Lett. B **298**, 120 (1993); J. Edsjo and P. Gondolo, Phys. Rev. D **56**, 1879 (1997); J. R. Ellis, T. Falk and K. A. Olive, Phys. Lett. B **444** (1998) 367; J. R. Ellis, T. Falk, K. A. Olive and M. Srednicki, Astropart. Phys. **13**, 181 (2000); T. Nihei, L. Roszkowski and R. Ruiz de Austri, JHEP **0207**, 024 (2002).

[64] P. Gondolo and G. Gelmini, Nucl. Phys. B **360**, 145 (1991).

[65] P. Nath and R. L. Arnowitt, Phys. Rev. Lett. **70**, 3696 (1993); Phys. Lett. B **299**, 58 (1993) [Erratum-ibid. B **307**, 403 (1993)]; J. L. Lopez, D. V. Nanopoulos and K. j. Yuan, Phys. Rev. D **48**, 2766 (1993); H. Baer and M. Brhlik, Phys. Rev. D **53**, 597 (1996); V. D. Barger and C. Kao, Phys. Rev. D **57**, 3131 (1998).

[66] B. Kors and P. Nath, Phys. Lett. B **586**, 366 (2004); JHEP **0412**, 005 (2004); arXiv:hep-ph/0411406; JHEP **0507**, 069 (2005).

[67] D. Feldman, Z. Liu and P. Nath, Phys. Rev. Lett. **97**, 021801 (2006); JHEP **0611**, 007 (2006); K. Cheung and T. C. Yuan, JHEP **0703**, 120 (2007); arXiv:0710.2005 [hep-ph]; D. Feldman, Z. Liu and P. Nath, Phys. Rev. D **75**, 115001 (2007); AIP Conf. Proc. **939**, 50 (2007); D. Feldman, Z. Liu and P. Nath, arXiv:0806.4683 [hep-ph].

[68] P. H. Frampton, P. Q. Hung and M. Sher, Phys. Rept. **330**, 263 (2000); J. L. Hewett and T. G. Rizzo, Phys. Rev. D **35**, 2194 (1987).

[69] V. A. Novikov, L. B. Okun, A. N. Rozanov and M. I. Vysotsky, JETP Lett. **76**, 127 (2002) [Pisma Zh. Eksp. Teor. Fiz. **76**, 158 (2002)].

[70] G. D. Kribs, T. Plehn, M. Spannowsky and T. M. P. Tait, Phys. Rev. D **76**, 075016 (2007); R. Fok and G. D. Kribs, Phys. Rev. D **78**, 075023 (2008); B. Holdom, JHEP **0608**, 076 (2006); JHEP **0703**, 063 (2007); P. Q. Hung and M. Sher, Phys. Rev. D **77**, 037302 (2008); Z. Murdock, S. Nandi and Z. Tavartkiladze, Phys. Lett. B **668**, 303 (2008); O. Cakir, H. Duran Yildiz, R. Mehdiyev and I. Turk Cakir, Eur. Phys. J. C **56**, 537 (2008).

[71] J. Maalampi, J.T. Peltoniemi, and M. Roos, PLB 220, 441(1989); J. Maalampi and M. Roos, Phys. Rept. **186**, 53 (1990);

[72] H. J. He, N. Polonsky and S. f. Su, Phys. Rev. D **64**, 053004 (2001).

[73] G. Senjanovic, F. Wilczek and A. Zee, Phys. Lett. B **141**, 389 (1984).

[74] T. Ibrahim and P. Nath, Phys. Rev. D **78**, 075013 (2008).

[75] S. N. Gninenko, Phys. Lett. B, **452**, 414 (1999).

[76] P. Nath and P. Fileviez Perez, Phys. Rept. **441**, 191 (2007).

CHAIRMAN: P. NATH

Scientific Secretaries: A. Kaminzka, M. Goebel

DISCUSSION I

- Nicotri:

Apart from masses, what are the free parameters of MSSM and of mSUGRA?

- Nath:

The MSSM has a large parameter space consisting of an upward of 100 free parameters - masses, phases and mixing angles. The mSUGRA model decreases the number of free parameters to four, which are the masses m_0 and $m_{1/2}$, the tri-linear coupling, A_0, the bi-linear coupling, B_0, and the sign of the Higgs mixing parameter μ.

- Ferrara:

In supergravity mediated SUSY breaking the hidden sector depends on the compactification scenario and can be rather complicated (also possibly giving FCNC problem). What do you really need for minimal supergravity (SUSY breaking) model?

- Nath:

The hidden sector is constrained by phenomenology. One gets constraints on the Kähler potential and consequently on the sparticle masses. The interesting question is if you can derive these constraints from a more fundamental theory. This can happen in a variety of models at the fundamental level such as dilaton dominance in the soft breaking in heterotic strings or by an additional symmetry.

- Susskind:

It seems that we are seeing the very tiny tip of a huge iceberg. The hidden sector, the inflation, the cosmological constant or whatever cancels it, the U(1) problem: all of this suggests that there are many more degrees of freedom beyond the Standard Model?

- Nath:

Yes, I agree. What we will see is only a tip of the iceberg. First we build a minimal model consistent with phenomenological constraints. We add a little bit more to that model and test it and then see what happens next.

- Ni:

In the "bottom up" approach you mentioned an intermediated scale, "high scale", at 10^{16} GeV. How is this scale related to the neutrino mass?

- Nath:

What I wanted to show is that the seesaw mechanism connects the high energy scale with the low energy scale and generates the small value for neutrino masses from a large mass. There are other mechanisms resulting for instance from extra dimensions generating also small masses. However the seesaw mechanism is the most natural and the most elegant one in my opinion.

- Ni:

The neutrino mass from cosmology is just an upper bound, the actual mass can be even smaller than ~0.6 eV. Does this mean that the "high scale" can be even higher?

- *Nath:*

Yes, indeed. The high scale can be even larger than the 10^{16} GeV. I wanted only to point out the general idea of the seesaw mechanism. The neutrinoless double beta decay data can determine the size of the neutrino mass (if the neutrino is a Majorana particle).

- *Tannenbaum:*

(to P. Nath and. L. Susskind) What would you say if only the Higgs particle was found at the LHC consistent with the SM and nothing else? Would it be a victory or a defeat for the Standard Model?

- *Susskind:*

I do not know. Sidney [Coleman], can you help me?

- *Nath:*

I think that it is highly unlikely that the LHC experiment will not discover any new particles beyond the Higgs boson. It would be surprising that the Standard Model could be so fine tuned which is what is needed to stabilize the Higgs boson mass in the Standard Model if no new particles are present. Also it would be difficult to explain the BNL muon g-2 result if no new particles exist unless the experimental result on this does not hold up.

- *Tannenbaum:*

(regarding the answer that if no MSSM are found at LHC then BNL g-2 is wrong) I think that you can set your watch by the BNL g-2 result like you can with the hydrogen hyperfine structure. It is the theory that concerns me.

- *Nath:*

We really do not know what happens beyond the SM and the only criteria we have for a theory is naturalness, which means a low amount of fine tuning. This is what motivates us to go beyond the SM to something like SUSY.

- *Marshall:*

Why is there no Z+jets-background in the opposite-sign same-flavor dilepton invariant mass plot?

- *Nath:*

The background was provided by CMS. The tt-bar background seems to be the dominant background. I wanted to give a rough picture to show the difference between the chargino NLSP (next to lightest supersymmetric particle) model and the stau NLSP model. If the background becomes too large, there is the possibility that one can go to larger luminosities to reduce it.

- *Marshall*:

Why are we talking about 10^{16} GeV when the hidden sector is 10^{12}-10^{13} GeV?

- *Nath:*

I agree, and this goes back to L. Susskind's question. There is still a lot we have to understand about the hidden sector.

- *Nicotri:*

Is it possible that the neutrino mass is so small that the high scale is "degenerate" with the Planck scale? And if this is the case how the model you have described is modified?

- *Nath:*

There could be many scenarios. I just wanted to present the most convincing one. I personally believe there should be a high scale around 10^{16} GeV below the Planck scale.

- *Shakirov:*

You presented an analysis showing that if both the precise measurements of the anomalous magnetic moment and our picture of SUSY are correct then we will definitely see supersymmetric particles at the LHC. However, when they are not detected, does this mean that a serious modification of the model must be made?

- *Nath:*

Yes, if we do not observe supersymmetric particles at the LHC the model cannot hold in the present form. Then we have to modify the model in a very significant way.

- *Bonati:*

For the SM Higgs mass there is an upper bound imposed by triviality. Is there a similar bound also for other particles predicted by mSUGRA?

- *Nath:*

The lighter Higgs mass in mSUGRA must be lighter than approximately 135 GeV. The heavy Higgs boson is bounded by the soft Supersymmetry breaking. In most of the SUGRA models it turns out that the heavy Higgs boson must be lighter than around 1 TeV. However, it could also be very light just above the LSP mass as in the Higgs pattern (HP) scenario presented in my lecture. But these light Higgs scenarios models (HPs) are strongly constrained by the present dark matter searches.

- *Dubynskiy:*

One of the advantages of SUSY is the gauge unification of the three coupling constants at the GUT scale. Are experimental results and errors taken into account when calculating the running coupling constants?

- *Nath:*

Yes, experimental results and error are taken into account. However, the running of the couplings depends on the sparticle masses and on GUT threshold effects. So a precise knowledge of these will allow a more precise prediction on the meeting of the couplings.

- *Dubynskiy:*

You have shown the deviation (3.3σ) between the SM prediction and the measurement of g-2. How do low energy calculations affect the SM prediction

- *Nath:*

The g-2 SM prediction depends on the hadronic correction to the SM electroweak contribution. Over the past year an improvement in the contribution of the hadronic correction to the SM electroweak correction has been achieved. Hence, the uncertainty in the deviation between the experiment and the SM prediction has been reduced.

- *Weir:*

You mentioned how WMAP3 data constrains mSUGRA. Do you know how WMAP5 data improve these constraints.

- *Nath:*

I can give you just a rough guess. In the analysis I presented a 2σ error corridor is taken. Since the WMAP5 data presumably lie in this 2σ corridor, I believe that the results arising from taking the WMAP5 constraint can be accommodated in the analysis I presented.

- *Bernhard:*

Can the mSUGRA parameter space be constrained by the limits of $B_s \to \mu^+\mu^-$ measured at the Tevatron? So can you probe the indirect annihilation cross-section of the Neutralinos from that, as well?

- *Nath:*

The calculation for the indirect annihilation of Neutralinos has been computed by many authors. However, they are very model dependent. Concerning your first question, indeed the mSUGRA parameter space is constrained by the $B_s \to \mu^+\mu^-$ results at the Tevatron.

Predicted Signatures at the LHC from $U(1)$ Extensions of the Standard Model[1].

Pran Nath

Department of Physics, Northeastern University, Boston, MA 02115-5000, USA

Abstract

We discuss the $U(1)_X$ extensions of the standard model with focus on the Stueckelberg mechanism for mass growth for the extra $U(1)_X$ gauge boson. The assumption of an axionic connector field which carries dual $U(1)$ quantum numbers, i.e., quantum numbers for the hypercharge $U(1)_Y$ and for the hidden sector gauge group $U(1)_X$, allows a non-trivial mixing between the mass growth for the neutral gauge vector bosons in the $SU(2)_L \times U(1)_Y$ sector and the mass growth for the vector boson by the Stueckelberg mechanism in the $U(1)_X$ sector. This results in an extra Z' which can be very narrow, but still detectable at the Large Hadron Collider (LHC). The $U(1)_X$ extension of the minimal supersymmetric standard model is also considered and the role of the Fayet-Illiopoulos term in such an extension discussed. The $U(1)_X$ extensions of the SM and of the MSSM lead to new candidates for dark matter.

[1]Based on lectures at the 46th Course - Erice -Sicily: 29 August -7 September, 2008.

Introduction: In lecture I we have discussed high scale models based on supersymmetry, local supersymmetry and supergravity [1, 2]. Specifically we focused on supergravity grand unified models[3] and their experimental consequences, their extensions and tests at the large Hadron Collider (LHC)[4]. Here we discuss extensions of the standard model(SM) and of the minimal supersymmetric standard model(MSSM) to include extra $U(1)$ factors. Such extra $U(1)$ factors appear in a variety of unified models: in grand unified models, i.e., in string models and in D brane models. Thus in $SO(10)$ and E_6 grand unification one has

$$SO(10) \supset SU(5) \times U(1)_\chi; \ E_6 \supset SO(10) \times U(1)_\psi \tag{1}$$

Similarly D brane constructions naturally have many $U(1)$ factors since here on typically starts with a stack of n branes which has a $U(n)$ gauge symmetry. Since $U(n) \supset SU(n) \times U(1)$, one has $U(1)$ factors appearing. Thus, for example, to construct the standard model gauge group in D-brane models one starts with the gauge groups $U(3) \times U(2) \times U(1)^2$ which results in $SU(3) \times SU(2) \times U(1) \times U(1)^3$. Typically the breaking of the extra $U(1)$ symmetry will come about by a Higgs mechanism. Thus if a field has non-vanishing $U(1)$ quantum number, then spontaneous breaking which gives a vacuum expectation value (VEV) to that field will break the extra $U(1)$ symmetry and make the extra gauge boson massive. This is illustrated by the following example: Under $SU(5) \times U(1)_\chi$, the 45 plet of $SO(10)$ has the following decomposition, $45 = 1(0) + 10(4) + \overline{10}(-4) + 24(0)$. The fields which can gain vacuum expectation values without violating charge or color are $1(0)$ and $24(0)$, which however has vanishing $U(1)_\chi$ quantum numbers. Thus while the 45 plet breaks $SO(10)$ it does not reduce the rank and the $U(1)_\chi$ gauge boson remains massless. To reduce the rank one needs a 16 plet or a 126 plet of Higgs. Thus under $SU(5) \times U(1)_\chi$ the 16 plet of Higgs has the decomposition $16 = 1(-5) + \bar{5}(-3) + 10(-1)$, while the 126 plet has the decomposition $126 = 1(-10) + \bar{5}(-2) + 10(-6) + \overline{15}(6) + 45(2) + \overline{50}(-2)$. It is then clear that the VEV growth for 16 or 126 will lead to a rank reduction. Thus a combination of 45 and 16 or a combination of 45 and 126 can reduce $SO(10)$ to the standard model gauge group.

More recently it has been shown that one can accomplish the above with a single irreducible Higgs representation, specifically a 144 plet representation of $SO(10)$[5]. This is so because under $SU(5) \times U(1)_\chi$, the 144 plet decomposes as follows: $144 = \bar{5}(3) + 5(7) + 10(-1) + 15(-1) + 24(-5) + 40(-1) + \overline{45}(3)$. In this case the 24-plet of $SU(5)$ in the above decomposition carries a $U(1)$ quantum number, and thus the $SO(10)$ gauge group breaks directly to the SM gauge group after 144 plet develops a VEV. Further, it may happen that one or more of the extra $U(1)'s$ may remain unbroken at the GUT scale, and such breaking may only occur at the electroweak scale. In this case the mass of the extra gauge boson which we shall generically call a Z' will have a mass of the electroweak size. Further, if the extra $U(1)$ is coming from a grand unified group the size of the gauge coupling would typically be the size of the gauge couplings in the electroweak group, i.e., typically of size g_2. Thus one generally expects decay widths for the Z' gauge bosons in such scenarios to be size O(GeV).

The outline of the rest of the paper is as follows: We first discuss the Stueckelberg extension of the standard model and the electroweak constraints on it[6, 7] [8, 9]. Next we discuss the Stueckelberg extension of MSSM[10]. This is followed by a discussion of the LHC signatures of Stueckelberg extensions. Of specific interest here is the possibility of a narrow Z'. Finally we discuss the possibility of dark matter arising from the hidden sector of this theory[11, 12] consisitent with the WMAP[13] constraints. In this context we also discuss the Stueckelberg extension to include kinetic energy mixing. Conclusions are given at the end. There are many related works on $U(1)$ extensions (see, e.g., [14], and for a review see [15]) as well as works where hidden sectors play a central role[16, 17, 18, 19, 20, 21, 22, 23, 24]. Further recent works regarding Stueckelberg extensions in the context of the string and D brane models can be found in [25, 26, 27] and in [28, 29].

We begin by introducing the Stueckelberg mechanism[30, 31] for mass growth for $U(1)$ gauge fields. In its simplest form the Stueckelberg mechanism works as follows: consider the Lagrangian given by

$$L = -\frac{1}{4}F_{\mu\nu}F^{\mu\nu} - \frac{1}{2}(mA_\mu + \partial_\mu\sigma)^2 + gA_\mu J^\mu \tag{2}$$

where σ is an axionic like field and J^μ is a conserved current. The Lagrangian is invariant under the transformation

$$\delta A_\mu \rightarrow \partial_\mu\epsilon, \ \delta\sigma \rightarrow -m\epsilon. \tag{3}$$

The Stueckelberg mechanism is endemic in extra dimension models, in strings, and in D branes. Thus consider, for example, the compactification of a 5D theory on S^1/Z_2 (see, e.g., [32]). As an illustration consider the kinetic energy of a $U(1)$ gauge field $A_a(a = 0, 1, 2, 3, 5)$ in 5D,

$$L_5 = -\frac{1}{4}F_{ab}(z)F^{ab}(z), \ a = 0, 1, 2, 3, 5, \tag{4}$$

where $z^a = (x^\mu, y)$, $\mu = 0, 1, 2, 3$ and we may write $A_a = (A_\mu(z), \phi(z))$. After compactification on the fifth co-ordinate y on a half circle, one has the Lagrangian in 4D with one mass less mode and an infinite number of massive Kaluza-Klein modes

$$L_4 = -\frac{1}{4}\sum_{n=0}^{\infty} F_{\mu\nu}(x)^{(n)}F^{\mu\nu(n)}(x) - \frac{1}{2}\sum_{n} M_n^2(A_\mu^{(n)}(x) + \frac{1}{M_n}\partial_\mu\phi^{(n)}(x))^2 + \cdots. \quad (5)$$

In string models and in D brane models the Stueckelberg phenomenon arises naturally from the Green-Schwarz mechanism from a two form field $B_{\mu\nu}$. Thus consider the Lagrangian with a $U(1)$ gauge field A_μ, and a coupling with the two form field $B_{\mu\nu}$ so that [33, 34, 35]

$$L_0 = -\frac{1}{4}F_{\mu\nu}F^{\mu\nu} - \frac{1}{12}H^{\mu\nu\rho}H_{\mu\nu\rho} + \frac{m}{4}\epsilon^{\mu\nu\rho\sigma}F_{\mu\nu}B_{\rho\sigma}, \quad (6)$$

where $H_{\mu\nu\rho}$ is the field strength of the two form field $B_{\mu\nu}$ so that $H_{\mu\nu\rho} = \partial_\mu B_{\nu\rho} + \partial_\nu B_{\rho\mu} + \partial_\rho B_{\mu\nu}$. It is convenient to write the last term as [35] $-\frac{m}{6}\epsilon^{\mu\nu\rho\sigma}(H_{\mu\nu\rho}A_\sigma + \sigma\partial_\mu H_{\nu\rho\sigma})$. An intergration over σ and insertion back gives L_0. Next suppose one solves for $H^{\mu\nu\rho}$ which gives $H^{\mu\nu\rho} = -m\epsilon^{\mu\nu\rho\sigma}(A_\sigma + \partial_\sigma\sigma)$. An integration on $H^{\mu\nu\rho}$ gives L_0 in the form[35]

$$L_1 = -\frac{1}{4}F_{\mu\nu}F^{\mu\nu} - \frac{m^2}{2}(A_\sigma + \partial_\sigma\sigma)^2. \quad (7)$$

Thus we see that the presence of the Green-Schwarz term leads us to mass growth for the $U(1)$ gauge field by the Stueckelberg mechanism.

The Stueckelberg mechanism to give mass to the vector bosons works whether or not the extra $U(1)_X$ is anomalous. Thus in general, an anamolous $U(1)$ gives at one loop[36, 37]

$$\delta L_{1loop} = \lambda c Tr(G \wedge G), \ \delta A_\mu = \partial_\mu\lambda. \quad (8)$$

This term is cancelled by variation of the effective tree level term[36]

$$\begin{aligned} L_0 &= \cdot\cdot + mA^\mu\partial_\mu\sigma + \frac{c\sigma}{m}Tr(G \wedge G), \\ \delta L_0 &= -\lambda c Tr(G \wedge G), \quad \delta\sigma = -\lambda m, \end{aligned} \quad (9)$$

so that

$$\delta L_{1loop} + \delta L_0 = 0. \quad (10)$$

Thus one has the following two cases: (i) Anomalous $U(1)$ case, $c \neq 0$, $m \neq 0$: Here one cancels the anomaly and at the same time one has mass growth by the Stueckelberg mechanism for the $U(1)_X$ gauge vector boson; (ii) Non-anomalous case, $c = 0$, $m \neq 0$. In this case there is no need for anomaly cancellation, but there is still mass growth for the vector boson by the Stueckelberg mechanism. Anomalous $U(1)$ in the above context have been discussed by a number of authors (see, e.g.,[38, 26, 39, 40]).

Stueckelberg vs the Higgs mechanism: Next we explore the connection between the Stueckelberg mechanism and the Higgs mechanism[34]. Consider a $U(1)$ gauge theory with a Higgs potential so that

$$L_0 = -\frac{1}{4}F_{\mu\nu}F^{\mu\nu} - (D^\mu\phi)^\dagger D_\mu\phi - V(\phi) + L_{gf},$$
$$V(\phi) = \mu^2(\phi^\dagger\phi) + \lambda(\phi^\dagger\phi)^2. \tag{11}$$

With $\mu^2 < 0$, and $\lambda > 0$ a spontaneous breaking of the $U(1)$ gauge symmetry occurs, and one has

$$\phi = \frac{1}{\sqrt{2}}(\rho + v)e^{ia/v}, v = \sqrt{-\mu^2/\lambda}. \tag{12}$$

In the limit $(-\mu^2, \lambda) \to \infty$ with $M = ev$ fixed, ρ becomes infinitely heavy and decouples from the rest of the system, and the residual Lagrangian is given by[34]

$$L = -\frac{1}{4}F_{\mu\nu}F^{\mu\nu} - \frac{1}{2}M^2(A_\mu - \frac{1}{M}\partial_\mu a)^2 + L_{gf}. \tag{13}$$

Thus we see that in the limit considered above the Higgs mechanism leads to the Stueckelberg mechanism in a very direct way.

Stueckelberg extension of the Standard Model: Although the Stueckelberg mechanism has been around for a long time, a successful incorporation of this mechanism into particle physics was made only recently[6, 10, 36] and its phenomenological implications investigated[7, 8, 9, 41, 12, 42]. Specifically one can extend the standard model by considering the gauge group $SU(3)_C \times SU(2)_L \times U(1)_Y \times U(1)_X$. We will assume that the visible sector matter fields (quarks, leptons, Higgs) are neutral under the gauge group $U(1)_X$, while the hidden sector fields are neutral under the standard model gauge group. Let the gauge fields for $U(1)_Y$ and $U(1)_X$ be B_μ and C_μ. We assume that there exists a connector sector which carries dual quantum numbers, i.e., quantum numbers of the gauge group $U(1)_Y$ and of $U(1)_X$. Specifically we assume that the axionic field σ is indeed this connector field. Then we append to the SM Lagrangian the following Stueckelberg extension

$$L_{St} = -\frac{1}{4}C_{\mu\nu}C^{\mu\nu} + g_X C_\mu J^\mu_X - \frac{1}{2}(\partial_\mu\sigma + M_1 C_\mu + M_2 B_\mu)^2, \tag{14}$$

where J^μ_X is a conserved current arising from the hidden sector. The above Lagrangian is invariant under the $U(1)$ transformations

$$\delta_Y B_\mu = \partial_\mu\lambda_Y, \quad \delta_Y C_\mu = 0, \quad \delta_Y\sigma = -M_2\lambda_Y, \tag{15}$$

and under the $U(1)_X$ transformations

$$\delta_X C_\mu = \partial_\mu \lambda_X, \quad \delta_X B_\mu = 0, \quad \delta_X \sigma = -M_1 \lambda_X. \tag{16}$$

The total Lagrangian is of course $L_{SM} + L_{St}$. After spontaneous breaking of the electroweak symmetry, the above Lagrangian gives a mass^2 matrix of the form

$$M^2_{ab} = \begin{pmatrix} M_1^2 & M_1 M_2 & 0 \\ M_1 M_2 & M_2^2 + \frac{1}{4} g_Y^2 v^2 & -\frac{1}{4} g_Y g_2 v^2 \\ 0 & -\frac{1}{4} g_Y g_2 v^2 & \frac{1}{4} g_2^2 v^2 \end{pmatrix}, \tag{17}$$

where we use a basis $(V_\mu^{\rm T})_a = (C_\mu, B_\mu, A_\mu^3)_a$ and use the standard form $-\frac{1}{2} V_{a\mu} M^2_{ab} V_b^\mu$ for the mass term in the Lagrangian. In Eq.(17), g_2 and g_Y are the $SU(2)_L \times U(1)_Y$ gauge coupling constants and $v = 2M_{\rm W}/g_2 = (\sqrt{2}G_F)^{-\frac{1}{2}}$, where $M_{\rm W}$ is the mass of the W boson, and G_F is the Fermi constant.

The eigen modes of the mass matrix of the vector bosons give a mass less mode which is a photon γ and two massive modes which are Z and Z'. The definition of the electric charge is modified so that

$$\frac{1}{e^2} = \frac{1}{g_2^2} + \frac{1+\epsilon^2}{g_Y^2}, \tag{18}$$

where $\epsilon = M_2/M_1$. Fits to the precision electroweak data require that ϵ be very small, i.e., $\epsilon \leq .06$[8]. The smallness of ϵ leads to the existence of a very narrow resonance if there is no matter in the hidden sector, or if the Z' is forbidden kinematically to decay into the hidden sector matter. This can happen if, for example, $M_{Z'} < 2M_{hid}$ where M_{hid} is the mass of the hidden sector fermion. Further, it also follows that if there is matter in the hidden sector, such matter will couple to the photon with a milli charge strength, i.e., with strength $Q_\epsilon = \epsilon e$. Detailed fits to the LEP I data show that one can satisfy the LEP I constraints to essentially the same level of confidence as fits to the standard model.

LEP II constraints: These arise from the constraints on the contact interaction[43]

$$L_C \sim \frac{g^2 \eta_{sign}}{(1+\delta)\Lambda^2} \sum_{i,j=L,R} \bar{e}_i \gamma^\mu e_i \bar{f}_j \gamma_\mu f_j. \tag{19}$$

where $\delta = 0$ for $f \neq e$, and $\delta = 1$ for $f = e$, η_{ij} gives the relative contribution of the different chiralities, and η_{sign} tells us if the contribution is constructive or destructive relative to the SM contribution. The most stringent constraints arise from Λ_{VV} for which LEP II gives

$$\Lambda_{VV} > 21.7\text{GeV}, \tag{20}$$

while the Stueckelberg extended model gives[9]

$$\Lambda_{VV} = \frac{M_{Z'}}{M_Z}\left(\frac{4\pi}{\sqrt{2}G_F v_e'^2}\right)^{1/2}. \tag{21}$$

For the Stueckelberg extension the LEP II constraints are automatically satisfied once one satisfies the LEP I constraints at the 1% level.

The Stueckelberg extension differs from the previous models in that the photon field is now a linear combination of three gauge fields, A^3_μ, B_μ, C_μ so that

$$A^\gamma_\mu = -c_\theta s_\phi C_\mu + c_\theta c_\phi B_\mu + s_\theta A^3_\mu. \tag{22}$$

This is to be compared with the form of A^γ_μ in the standard model where one has $A^\gamma_\mu = c_\theta B_\mu + s_\theta A^3_\mu$, so it is a combination of only the fields B_μ and A^3_μ. It is precisely because A^γ_μ has a small dependence on the field C_μ that the photon is able to couple to the hidden sector matter with milli charge strength. The model predicts a very narrow Z' resonance in absence of its decay into the hidden sector matter which is in contrast to rather broad resonances arising from Kaluza-Klein excitation of a compact extra dimension[44, 45]. Electroweak tests of the Stueckelberg extension of the standard model can be found in [7, 8].

Stueckelberg extension of the minimal supersymmetric standard model: We consider now the Stueckelberg extension of the minimal supersymmetric standard model(MSSM), and to this end we introduce vector super fields for the $U(1)_Y$ and the $U(1)_X$ gauge groups which we label as B and C respectively. Additionally we introduce the chiral multiplets S and $\bar{S}$ which contain the axionic field and we choose for the Stueckelberg Lagrangian the form

$$\int d\theta^2 d\bar{\theta}^2 (M_1 C + M_2 B + S + \bar{S})^2. \tag{23}$$

The above Lagrangian is invariant under the following $U(1)_Y$ and $U(1)_X$ transformation

$$\begin{aligned} &\delta_Y B = \Lambda_Y + \bar{\Lambda}_Y,\ \delta_Y S = -M_2 \Lambda_Y, \\ &\delta_X C = \Lambda_X + \bar{\Lambda}_X,\ \delta_X S = -M_1 \Lambda_X. \end{aligned} \tag{24}$$

Additionally, of course, we have the gauge kinetic energy terms for gauge multiplet C while the kinetic energy terms for the B multiplet are contained in the MSSM Lagrangian. In the Wess-Zumino gauge the components of the gauge multiplet C are $(C_\mu, \lambda_C, \bar{\lambda}_C, D_C)$, and similarly for the gauge multiplet B, while the chiral multiplet S

has the components $S = (\rho + i\sigma, \chi, F_S)$ and similarly for $\bar{S}$. It is then possible to form two additional Majorana spinors beyond those in MSSM. Here one has

$$\psi_S = \begin{pmatrix} \chi_\alpha \\ \bar{\chi}^{\dot{\alpha}} \end{pmatrix}, \quad \lambda_X = \begin{pmatrix} \lambda_{C\alpha} \\ \bar{\lambda}_C^{\dot{\alpha}} \end{pmatrix}. \tag{25}$$

Including the above two, one has six Majorana basis states

$$\psi_S, \lambda_X, \lambda_Y, \lambda_3, \tilde{h}_1, \tilde{h}_2, \tag{26}$$

where $\lambda_Y, \lambda_3, \tilde{h}_1, \tilde{h}_2$ are the four familiar gaugino and higgsino states of MSSM. As in SUGRA models, there is mixing among the six neutral states which gives rise to the six neutralino states in the mass diagonal basis. We can label these states as

$$\xi_1^0, \xi_2^0, \chi_1^0, \chi_2^0, \chi_3^0, \chi_4^0, \tag{27}$$

where χ_i^0 $(i = 1 - 4)$ are the familiar four neutralino states in MSSM, and ξ_1^0 and ξ_2^0 are the two additional neutral states that arise from the Stueckelberg sector. An interesting situation arises when the LSP of the entire system is in the Stueckelberg sector, i.e., it is ξ_1 which we assume is the lighter of $\xi_i (i = 1, 2)$. We will discuss this possibility in the context of dark matter later.

Higgs sector in Stueckelberg extension of MSSM: The Higgs sector is affected in the Stueckelberg extension of MSSM. This is so because one has in the Stueckelberg extension of MSSM the field ρ which couples with the CP even MSSM higgs fields H^0, h^0 giving a 3×3 mass matrix. We display this mass matrix below[36]

$$(M^2_{H^0}) = \begin{pmatrix} (M_Z^2 c_\beta^2 + m_A^2 s_\beta^2) & -(M_Z^2 + m_A^2) s_\beta c_\beta & -s_\theta c_\beta M_Z M_2 \\ -(M_Z^2 + m_A^2) s_\beta c_\beta & (M_Z^2 s_\beta^2 + m_A^2 c_\beta^2) & s_\theta s_\beta M_Z M_2 \\ -s_\theta c_\beta M_Z M_2 & s_\theta s_\beta M_Z M_2 & m_\rho^2 \end{pmatrix}, \tag{28}$$

where $s_\beta (c_\beta) = \sin\beta (\cos\beta)$, $s_\theta = \sin\theta_W$ where θ_W is the weak angle. The above matrix in the neutral Higgs sector has the eigen states H_1^0, H_2^0, H_3^0. One may choose these so that $H_1^0, H_2^0, H_3^0 \to H^0, h^0, \rho$ for the case when the coupling to the Stueckelberg sector vanishes, i.e., $M_2 = 0$, where h^0 is the light neutral Higgs of MSSM and H^0 is the heavy neutral Higgs of MSSM.

The Stueckelberg mechanism and the Higgs mechanism in MSSM: In the non-supersymmetr case it was shown that the Higgs mechanism reduces to the Stueckelberg mechanism[30] in an appropriate limit with a constraint on the mass^2 and coupling constant parameters of the Higgs potential. In the supersymmetric case, it turns out that one needs

Fayet-Iliopoulos D-terms to accomplish this reduction[41]. Thus we begin by adding Fayet-Iliopoulos D terms to the scalar potential

$$L_{FI} = \xi_X D_C + \xi_Y D_Y \tag{29}$$

Including these the D-part of the scalar potential has the form

$$V_D = \frac{g_X^2}{2}(Q_X|\phi^+|^2 - Q_X|\phi^-|^2 + \xi_X)^2 + \frac{g_Y^2}{2}(Y_\phi|\phi^+|^2 - Y_\phi|\phi^-|^2 + \xi_Y)^2 \tag{30}$$

Minimization of the potential gives $\langle\phi^+\rangle = 0$, $\langle\phi^-\rangle \neq 0$ and

$$M_1 = \sqrt{2} g_X Q_X \langle\phi^-\rangle, M_2 = \sqrt{2} g_Y Y_\phi \langle\phi^-\rangle, \tag{31}$$

The limit $g_X Q_X \to 0, g_Y Y_\phi \to 0$ with M_1, M_2 fixed reduces the Higgs mechanism above to the Stueckelberg form[41].

Kinetic mixing: The kinetic energy of two $U(1)'s$ will mix if there are fields even with GUT size mass which carry dual quantum numbers[16]. Thus consider two $U(1)$ gauge fields A_1^μ, A_2^μ with the interaction

$$L_{km} = -\frac{1}{4}F_{1\mu\nu}F_1^{\mu\nu} - \frac{1}{4}F_{2\mu\nu}F_2^{\mu\nu} - \frac{\delta}{2}F_{1\mu\nu}F_2^{\mu\nu} + J'_\mu A_1^\mu + J_\mu A_2^\mu. \tag{32}$$

where J_μ is the source arising from the visible sector matter fields and J'_μ is the source containing fields in the hidden sector. We can go to the diagonal basis with the transformation

$$\begin{pmatrix} A_1^\mu \\ A_2^\mu \end{pmatrix} = K_0 \begin{pmatrix} A'^\mu \\ A^\mu \end{pmatrix}, \quad K_0 = \begin{pmatrix} \frac{1}{\sqrt{1-\delta^2}} & 0 \\ \frac{-\delta}{\sqrt{1-\delta^2}} & 1 \end{pmatrix}. \tag{33}$$

However, there is a degree of arbitrariness in the choice of the matrix that diagonalizes the kinetic energy matrix. Specifically one may take K instead of K_0 as the diagonalizing matrix where

$$K = K_0 R, \quad R = \begin{pmatrix} \cos\theta & -\sin\theta \\ \sin\theta & \cos\theta \end{pmatrix}. \tag{34}$$

A convenient choice of θ is

$$\theta = arctan[\delta/\sqrt{1-\delta^2}] \tag{35}$$

which gives the asymmetric solution

$$L_1^K = A^\mu \left[\frac{1}{\sqrt{1-\delta^2}}J_\mu - \frac{\delta}{\sqrt{1-\delta^2}}J'_\mu\right] + A^{\mu'} J'_\mu \tag{36}$$

We identify A^μ with the photon field which has interactions with the visible sector source J_μ with normal strength and also an interaction with the hidden sector source J'_μ with a strength which is proportional to δ and thus this interaction is milli charge size. The field A'_μ is another massless vector field which interacts with the hidden sector source J'_μ and has no interactions with the visible sector current J_μ.

Next let us consider the Stueckelberg mechanism with kinetic energy mixing. Here in addition to the kinetic mixing one also has mass mixings so that

$$L^m_{St} = -\frac{1}{2}M_1^2 A_{1\mu}A_1^\mu - \frac{1}{2}M_2^2 A_{2\mu}A_2^\mu - M_1 M_2 A_{1\mu}A_2^\mu. \tag{37}$$

Diagonalization of the mass matrix fixes θ so that

$$\theta = arctan\left(\frac{\epsilon\sqrt{1-\delta^2}}{1-\epsilon\delta}\right). \tag{38}$$

With the above one finds the following interactions in the basis where both the mass and the kinetic energy terms are diagonal

$$\begin{aligned} L^{int}_{St} &= \frac{1}{\sqrt{1-2\epsilon\delta+\epsilon^2}}\left(\frac{\epsilon-\delta}{\sqrt{1-\delta^2}}J_\mu + \frac{1-\epsilon\delta}{\sqrt{1-\delta^2}}J'_\mu\right)A^\mu_M \\ &+ \frac{1}{\sqrt{1-2\epsilon\delta+\epsilon^2}}\left(J_\mu - \epsilon J'_\mu\right)A^\mu_\gamma. \end{aligned} \tag{39}$$

Here A^μ_γ is the photonic field and A^μ_M is the massive vector boson field. We note that the coupling of the photon with the hidden sector is controlled by ϵ and vanishes when ϵ vanishes. This is in contrast to the case of the pure kinetic mixing where the coupling of the photon with the hidden sector is controlled by δ. One can carry out a similar extension of the Stueckelberg extension of the standard modelgauge group to $SU(3)_C \times SU(2)_L \times U(1)_Y \times U(1)_X$ with both kinetic mixing and mass mixing for the two $U(1)'s$. In this case one finds that the electric charge is modified so that

$$\frac{1}{e^2} = \frac{1}{g_2^2} + \frac{1-2\epsilon\delta+\epsilon^2}{g_Y^2}. \tag{40}$$

An interesting result arises in that in the absence of the hidden sector, the weak sector of the model depends only on the combination $\bar{\epsilon} = (\epsilon-\delta)/\sqrt{1-\delta^2}$. In this case the precision LEP data constrains $\bar{\epsilon}$[12]. For a discussion of other approaches to $U(1)$ probes of the hidden sector see [19, 20, 28].

Milli-weak and hidden sector dark matter in Stueckelberg extensions: The Stueckelberg extension gives rise to two new candidates for dark matter. One of these is milli weak or extra weak dark matter arising from the Stueckelberg extension of MSSM. The other is dark matter which arises from the hidden sector to which $U(1)_X$ couples. We discuss each of these cases in some detail below.

Milli weak dark matter: In the Stueckelberg extension of MSSM one finds 6 neutralino states four of which are the usual states from the MSSM sector while the remaining two arise from the Stueckelberg sector. Here there are two distinct possibilities[10]. The first one corresponds to the case when χ_1^0 is the LSP. In this case the analysis of dark matter and of the supersymmetric signatures would exactly be the same as for the case of the SUGRA models except for some minor corrections arising from the mixings of the Stueckelberg and the MSSM sector. The second possibility corresponds to the case when ξ_1^0 is the LSP. In this case the LSP is milli weak or extra weakly interacting. The LSP of the MSSM sector would decay into ξ_1^0 inside the detector, and thus one still has missing energy signals. Naively one might expect that because of the extra weak interactions of ξ_1^0 it might be difficult to achieve an efficient annihilation of the ξ_1^0's. However, this is not the case when one takes into account the coannihilations. It is then possible to satisfy the relic density constraints consistent with WMAP data[41, 46]. For other possible candidates arising from $U(1)$ extensions see[47, 48].

Hidden sector dark matter[11, 12]: Hidden sector fermions could be candidates for dark matter. Specifically let us assume that on has massive Dirac fermions with mass m_D in the hidden sector. These would be milli charged (for other recent works with milli charged particles see [49]

Detailed analyses then show that one can satisfy the WMAP relic density constraints[11 12]. There are two clear regimes for the relic density constraints to be satisfied. The first one is when $M_{Z'} > 2m_D$ in which case the Z' width will be normal size because Z' can decay with normal strength into the Dirac fermions of the hidden sector. However, the Z' coupling with the SM quarks and leptons is small because it is suppressed by a factor ϵ^2. Consequently in this scenario it is difficult to detect the Z' by Drell-Yan at the Large Hadron Collider. The second possibility corresponds to the case when $M_{Z'} < 2m_D$. Here it is still possible to satisfy the relic density constraints by a proper thermal averaging over the Z' pole. However, in this case the Z' cannot decay on shell into the Dirac fermions of the hidden sector but it can do so into the quarks and leptons of SM. Thus one can detect the Z' by the Drell -Yan at the Large Hadron Collider. The Dirac fermions in the hidden sector can explain the positron excess seen by the anti-matter satellite probe PAMELA[50]. An analysis shows that such an excess can be understood from the annihilation of Dirac fermions in the hidden sector[51]. An important issue emphasized in the analysis of [51] is the enhancement or boost of the ratio $\langle\sigma v\rangle_{Halo}/\langle\sigma v\rangle_{Freezeout}$ near a pole due to the fact that the temperature in the halo is much smaller compared to the temperature at the freezeout. The above phenomenon come about when twice the mass of the annihilating particle is close to the resonance mass in the channel in which they annihilate. For the annihilation of the Dirac particles considered in [51] the resonance is the Stueckelberg Z' pole. In this case the ratio is very sensitive to the temperature, and the velocity averaging in the halo gives larger results due to a much smaller temperature in the halo relative to the velocity averaging at the freezeout where the temperature is much larger $T \sim m_D/(20-30)$. Because of the above enhancement much smaller additional boost factors are needed to fit the positron data[51]. For other works on the Stueckelberg mechanism and applications see [52, 53, 21]. and for other works on dark matter from hidden sector see[49, 54].

Conclusion: Extra $U(1)$'s arise in a wide variety of GUT, string and D brane models, and the Stueckelberg extensions of SM provide a natural framework to incorporate them. The Stueckelberg mechanism is an alternative to the Higgs mechanism for the breaking of the $U(1)$ gauge symmetry. It has the advantage over the Higgs mechanism in that one does not need to construct a scalar potential or generate spontaneous breaking to give mass to the gauge vector boson. The Stueckelberg extension leads to new phenomena testable at colliders, specfically a sharp Z' resonance with a width in the MeV to sub GeV range in the absence of decay of the Z' into hidden sector matter. Two new types of dark matter emerge in the Stueckelberg extension. The first is a milli weak dark matter candidate whose interactions with matter are weaker than the weak interactions of the weakly interacting massive particles (WIMPS). The second new candidate for dark matter that emerges is the milli charged dark matter which arises solely from the hidden sector. It is found that both the milli weak and the milli charged dark matter candidates can generate the right amount of dark matter in the universe consistent with the WMAP relic density constraints. Since the Stueckelberg mechanism arises quite naturally in string and D brane models, a confirmation the Stueckelberg mechanism experimentally would lend support to the existence of a new regime of physics - perhaps string theory.

Acknowledgements: The author thanks Professor Gerard 't Hooft and Professor Antonino

Zichichi for invitation to lecture at the 46th Course in Subnuclear Physics at Erice, and Professor Zichichi for the hospitality extended him during the period of the course. The new work reported here is based on collaborative work with K.S. Babu, Ilia Gogoladze, Daniel Feldman, Boris Kors, Zuowei Liu, and Raza Syed. This research is supported in part by NSF grant PHY-0757959.

References

[1] P. Nath and R. Arnowitt, Phys. Lett. B **56**, 177 (1975); R. Arnowitt, P. Nath and B. Zumino, Phys. Lett. B **56**, 81 (1975); P. Nath and R. Arnowitt, Phys. Lett. B **65**, 73 (1976).

[2] D. Z. Freedman, P. van Nieuwenhuizen and S. Ferrara, Phys. Rev. D **13**, 3214 (1976); S. Deser and B. Zumino, Phys. Lett. B **62**, 335 (1976).

[3] A. H. Chamseddine, R. Arnowitt and P. Nath, Phys. Rev. Lett. **49** (1982) 970; L. J. Hall, J. D. Lykken and S. Weinberg, Phys. Rev. D **27**, 2359 (1983).

[4] D. Feldman, Z. Liu and P. Nath, Phys. Rev. Lett. **99**, 251802 (2007); Phys. Lett. B **662**, 190 (2008); JHEP **0804**, 054 (2008); arXiv:0806.4683 [hep-ph]; Phys. Rev. D **78**, 083523 (2008).

[5] K. S. Babu, I. Gogoladze, P. Nath and R. M. Syed, Phys. Rev. D **74**, 075004 (2006); Phys. Rev. D **72**, 095011 (2005); P. Nath and R. M. Syed, JHEP **0602**, 022 (2006).

[6] B. Kors and P. Nath, Phys. Lett. B **586**, 366 (2004).

[7] B. Kors and P. Nath, JHEP **0507**, 069 (2005).

[8] D. Feldman, Z. Liu and P. Nath, Phys. Rev. Lett. **97**, 021801 (2006).

[9] D. Feldman, Z. Liu and P. Nath, JHEP **0611**, 007 (2006).

[10] B. Kors and P. Nath, JHEP **0412**, 005 (2004).

[11] K. Cheung and T. C. Yuan, JHEP **0703**, 120 (2007).

[12] D. Feldman, Z. Liu and P. Nath, Phys. Rev. D **75**, 115001 (2007).

[13] D. N. Spergel *et al.* [WMAP Collaboration], Astrophys. J. Suppl. **170**, 377 (2007).

[14] M. S. Carena, A. Daleo, B. A. Dobrescu and T. M. P. Tait, Phys. Rev. D **70**, 093009 (2004).

[15] P. Langacker, arXiv:0801.1345 [hep-ph].

[16] B. Holdom, Phys. Lett. B **166**, 196 (1986), B **259**, 329 (1991).

[17] J. Kumar and J. D. Wells, Phys. Rev. D **74**, 115017 (2006);

[18] W. F. Chang, J. N. Ng and J. M. S. Wu, Phys. Rev. D **74**, 095005 (2006).

[19] T. Han, Z. Si, K. M. Zurek and M. J. Strassler, JHEP **0807**, 008 (2008).

[20] Y. Gershtein, F. Petriello, S. Quackenbush and K. M. Zurek, Phys. Rev. D **78**, 095002 (2008).

[21] M. Ahlers, H. Gies, J. Jaeckel, J. Redondo and A. Ringwald, Phys. Rev. D **77**, 095001 (2008); M. Ahlers, J. Jaeckel, J. Redondo and A. Ringwald, Phys. Rev. D **78**, 075005 (2008); S. A. Abel, J. Jaeckel, V. V. Khoze and A. Ringwald, Phys. Lett. B **666**, 66 (2008); S. A. Abel, M. D. Goodsell, J. Jaeckel, V. V. Khoze and A. Ringwald, JHEP **0807**, 124 (2008) [arXiv:0803.1449 [hep-ph]].

[22] A. Ibarra, A. Ringwald and C. Weniger, arXiv:0809.3196 [hep-ph].

[23] H. Georgi, Phys. Rev. Lett. **98**, 221601 (2007); K. Cheung, W. Y. Keung and T. C. Yuan, Phys. Rev. Lett. **99**, 051803 (2007).

[24] H. Goldberg and P. Nath, Phys. Rev. Lett. **100** (2008) 031803; arXiv:0706.3898 [hep-ph].

[25] C. Coriano', N. Irges and E. Kiritsis, Nucl. Phys. B **746**, 77 (2006).

[26] P. Anastasopoulos, Fortsch. Phys. **55**, 633 (2007); P. Anastasopoulos, F. Fucito, A. Lionetto, G. Pradisi, A. Racioppi and Y. S. Stanev, Phys. Rev. D **78**, 085014 (2008).

[27] R. Armillis, C. Coriano and M. Guzzi, JHEP **0805**, 015 (2008).

[28] J. Kumar, A. Rajaraman and J. D. Wells, Phys. Rev. D **77**, 066011 (2008).

[29] C. P. Burgess, J. P. Conlon, L. Y. Hung, C. H. Kom, A. Maharana and F. Quevedo, JHEP **0807**, 073 (2008).

[30] E. C. G. Stueckelberg, Helv. Phys. Acta **11**, 225 (1938).

[31] For a history of the Stueckelberg mechanism see, F. Cianfrani and O. M. Lecian, Nuovo Cim. **122B**, 123 (2007) [arXiv:physics/0701340].

[32] P. Nath and M. Yamaguchi, Phys. Rev. D **60**, 116004 (1999).

[33] M. Kalb and P. Ramond, Phys. Rev. D **9**, 2273 (1974).

[34] T. J. Allen, M. J. Bowick and A. Lahiri, Mod. Phys. Lett. A **6**, 559 (1991).

[35] D. M. Ghilencea, L. E. Ibanez, N. Irges and F. Quevedo, JHEP **0208**, 016 (2002) [arXiv:hep-ph/0205083].

[36] B. Kors and P. Nath, arXiv:hep-ph/0411406.

[37] R. Blumenhagen, B. Kors, D. Lust and S. Stieberger, Phys. Rept. **445**, 1 (2007).

[38] C. Coriano', M. Guzzi, A. Mariano and S. Morelli, arXiv:0811.3675 [hep-ph]; C. Coriano, M. Guzzi, N. Irges and A. Mariano, arXiv:0811.0117 [hep-ph]; R. Armillis, C. Coriano', M. Guzzi and S. Morelli, arXiv:0809.3772 [hep-ph]; R. Armillis, C. Coriano and M. Guzzi, JHEP **0805**, 015 (2008); C. Coriano, N. Irges and S. Morelli, Nucl. Phys. B **789**, 133 (2008);

[39] M. Zagermann, arXiv:0801.1666 [hep-th].

[40] J. De Rydt, T. T. Schmidt, M. Trigiante, A. Van Proeyen and M. Zagermann, arXiv:0808.2130 [hep-th].

[41] D. Feldman, B. Kors and P. Nath, Phys. Rev. D **75**, 023503 (2007).

[42] D. Feldman, Z. Liu and P. Nath, AIP Conf. Proc. **939**, 50 (2007) [arXiv:0705.2924 [hep-ph]].

[43] [LEP Collaboration and ALEPH Collaboration and DELPHI Collaboration and], arXiv:hep-ex/0312023.

[44] I. Antoniadis, K. Benakli and M. Quiros, Phys. Lett. B **460**, 176 (1999).

[45] P. Nath, Y. Yamada and M. Yamaguchi, Phys. Lett. B **466**, 100 (1999).

[46] F. Fucito, A. Lionetto, A. Mammarella and A. Racioppi, arXiv:0811.1953 [hep-ph].

[47] V. Barger, C. Kao, P. Langacker and H. S. Lee, Phys. Lett. B **600**, 104 (2004).

[48] T. Hur, H. S. Lee and S. Nasri, Phys. Rev. D **77**, 015008 (2008).

[49] J. H. Huh, J. E. Kim, J. C. Park and S. C. Park, Phys. Rev. D **77**, 123503 (2008) [arXiv:0711.3528 [astro-ph]].

[50] O. Adriani *et al.*, arXiv:0810.4995 [astro-ph] (Submitted to Nature).

[51] D. Feldman, Z. Liu and P. Nath, arXiv:0810.5762 [hep-ph].

[52] Y. Zhang, S. Z. Wang and Q. Wang, JHEP **0803**, 047 (2008).

[53] C. P. Burgess, J. P. Conlon, L. Y. Hung, C. H. Kom, A. Maharana and F. Quevedo, JHEP **0807**, 073 (2008).

[54] E. J. Chun and J. C. Park, arXiv:0812.0308 [hep-ph].

CHAIRMAN: P. NATH

Scientific Secretaries: T. Budd, G. Stavenga

DISCUSSION II

- *Marshall*:

You suggested that a Z' decaying to the visible rather than the hidden sector, would be detectable at the LHC. Is there a natural argument for that to be the case?

- *Nath*:

Well what I discussed were two possibilities, and which one occurs depends on the mass spectrum of matter in the hidden sector. At the moment we have no idea exactly what the matter content of the hidden sector is. So what I did was discuss both cases. The first one is the case where the Z' could decay into the hidden sector particles. In this case it will have a large width and therefore the branching ratio into the visible sector particles will be suppressed and the Z' will not be easily detectable. On the other hand for the second case the Z' is below the quark-anti-quark threshold of the hidden sector. In this case the Z' cannot decay into the hidden sector particles and will decay only into the visible particles. And that would give you a Z' decay signal that is detectable.

- *Marshall*:

Is there one that is favored in the spectrum of possibilities? Or are these equally likely?

- *Nath*:

At this point I would say that they are equally likely unless one has a specific theory of the hidden sector which computes the mass spectrum, which one doesn't have at this point.

- *Goebel*:

I have two questions regarding the fit of the electroweak precision data. Why don't you use the W mass and the Z mass or the top mass for the fit?

Nath:

Our inputs are exactly as in the LEP analysis (aside from the additional inputs of the Stueckelberg sector) and we fit many observables. So we follow exactly what the LEP group did, the only difference being that we do the analysis including the new sector and so what I presented here was parallel to the LEP analysis. Just so that we can compare how close we come to that analysis and what we find is that our chi square fits are very close the chi square fits of the LEP group. One can use other inputs and redo the analysis but that would not change the excellent quality of the fits. Basically the quality of the fits including the new sector is as good as the standard model fits.

- *Goebel*:

So do you use an external fitting package like ZFitter?

- *Nath*:

We did the old style actual hand calculations.

- *di Mauro*:

Could you give us more details about the Stuckelberg mechanism in anomalous U(1) theories?

- *Nath:*

We haven't looked at that in detail, so I cannot give you any more detail on that. However there have been many followup works, specifically there have been a whole series of papers in string theory D brane models where they have gotten the type of models that I have been talking about here. The string compactifications typically have an anomalous U(1). In such situations in addition to the massive Z prime there are other signatures. For more details you can review those papers which are subsequent to our work and discuss the anomalous case.

- *Carrasco*:

There are a lot of very impressive numerical explorations of the consequences of various models looking for identifying signals. These numerical results are going to be dependent to varying degrees on the chosen input parameters. How do you control to get useful robust signal?

- *Nath*:

First of all once you open a new sector it will bring in new parameters. And that is something you have to live with. Given that, then the question is what you get in return and so what I was trying to show was that in return you get signals that are unlike any others. If such signals are present they will stick out like a sore thumb. The main issue is if our searches will be sensitive enough to pick those signals. Because if you remember the J/psi was missed in the initial search because the search was too coarse. So what we have here is a similar analogue but with a very sharp resonance in the several hundred GeV region, and it is possible that if you are not careful you could miss it. Returning to your question, while there are free parameters in the model, the signals predicted by the model are detectable at the LHC. For the Z' of the Stuckelberg extension the mass range that can be explored is about 3.5 TeV.

- *Arnowitt*:

You suggested an alternate dark matter candidate, the XWIMP, which coannihilate with ordinary WIMPS to satisfy the WMAP amount of dark matter. Are the two WIMPs near degenerate as in ordinary coannihilation? Can this near degeneracy be used as a signal?

- *Nath*:

The answer to the first part is yes. Just like in ordinary coannihilation the mass difference between the XWIMP and the WIMP have to be small, typically one-twentieth of the average mass of the two.

- *Arnowitt*:

Which one can decay into which? Which is the lightest?

- *Nath*:

In principle any order can occur, but for the case in which the XWIMP is the lightest one, that will become the LSP of the whole system and therefore will be a dark matter candidate. What we have noticed is that the WIMP will decay into the XWIMP in a short period, namely it will decay within the detector itself, and it is not a long lived object. And so the dark matter will be XWIMPs and therefore very difficult to detect. We haven't actually computed the cross sections, but if you do a quick analysis on the back of the envelope you find that the cross sections will be suppressed by a couple of orders of magnitude. Therefore it will make the job of the dark matter searches even harder. With respect to the second question, which is actually an interesting question, as to what possible signal colliders might see for this kind of scenarios, that phenomenology has not been pursued yet.

- *Arnowitt*:

Is the near degeneracy natural like the coannihilation of the stau-neutralino?

- *Nath*:

In the mSUGRA case the renormalisation group analysis brings the stau and the lightest neutralino close to each other in mass in certain regions of the mSUGRA parameter space. This is not the case for the XWIMPs. Here you have to actually choose the parameters so their masses lie close to each other.

- *Lockyer*:

Please comment on models in which the Z' decays with flavor violation.

- *Nath:*

The Z' decays with flavor violation are suppressed in the type of models I have discussed. However, it is possible to modify these models with additional interactions which could allow processes of the type you mention.

- *Susskind*:

What exactly is gravity mediation between the hidden and the visible sector? Is it literally graviton exchange or is it multiple graviton exchange or something else in the gravitational multiplet?

- *Nath*:

The way it works is that you look at the scalar potential and that is where you compute all the soft terms. The basic assumption of gravity mediation is that the superpotential is separated in two sectors: the visible sector and the hidden sector. That is one writes the superpotential W into the sum W1+W2, where W1 contains the visible part that is all the matter fields that we observe, and W2 contains all the hidden sector fields. Now you break supersymmetry through whatever means in the hidden sector. Originally we did the breaking using a chiral field, but you can do it any other way. For our analysis it leads to a vacuum expectation value for W2. The VEV of W2 gives you already the gravitino mass. You can write a simple formula for that. So then what you do, having gotten a VEV for W2, you go back to the supergravity scalar potential and expand it, which gives you the soft terms.

- *Dubynskiy*:

Could you please comment on the form of the Z' peak? What is responsible for the deep drop at the left side of the peak?

- *Nath:*

This is a very well-known effect, and it is the interference of the background with the pole. So if you have an amplitude A plus an amplitude B, then just below the pole there will be a negative interference, at the pole there will be an enhancement, and above the pole there will be a positive interference. This is a well-known effect already seen in gamma-Z interference experiments. So what we see here is the same phenomenon for the Z' case.

Ellis:

Progress on Ultraviolet Finiteness of Supergravity

Z. Bern

Department of Physics and Astronomy
UCLA, Los Angeles, CA
90095-1547, USA

Abstract

In this lecture we summarize recent calculations pointing to the possible ultraviolet finiteness of $\mathcal{N} = 8$ supergravity in four dimensions. We outline the modern unitarity method, which makes possible multi-loop calculations in this theory and which allows us to exploit a remarkable relation between tree-level gravity and gauge-theory amplitudes. We also discuss a link between observed cancellations at loop level and improved behavior of tree-level amplitudes under large complex deformations of momenta.

1 Introduction

For over 25 years the prevailing wisdom has been that it is impossible to construct a perturbatively ultraviolet finite point-like quantum field theory of gravity in four dimensions (see *e.g.* refs. [1]). In this lecture we will describe recent concrete calculations that call into question this belief.

Of all unitary quantum gravity field theories, maximally supersymmetric $\mathcal{N} = 8$ supergravity [2] is the most promising one to investigate for possible ultraviolet finiteness. Its high degree of supersymmetry suggests that it has the best ultraviolet properties of any gravity field theory with two derivative couplings. Moreover, with the modern unitarity method [3, 4, 5, 6, 7, 8, 9], the high degree of supersymmetry can be exploited to greatly simplify calculations. In fact, the striking simplicity of the theory recently led Arkani-Hamed, Cachazo and Kaplan to suggest that $\mathcal{N} = 8$ supergravity may in a sense be the simplest quantum field theory [10]. The unitarity method allows us to exploit a remarkable relation between gravity and gauge theory tree amplitudes [11, 12, 13], allowing us to map gravity calculations into algebraically simpler gauge-theory calculations.

In a classic paper 't Hooft and Veltman showed that gravity coupled to matter generically diverges at one loop in four dimensions [14, 15]. Due to the dimensionful nature of the coupling, the divergences cannot be absorbed by a redefinition of the original parameters of the Lagrangian, rendering the theory non-renormalizable. Pure Einstein gravity does not possess a viable counterterm at one loop, delaying the divergence to two loops [14, 16, 17]. The two-loop divergence of pure Einstein gravity was established by Goroff and Sagnotti and by van de Ven, through direct computation [18].

Over the years supersymmetry has been studied extensively as a mechanism for delaying the onset of divergences in gravity theories. No supergravity theory can diverge until at least three loops [19, 20, 1]. With additional assumptions this can be delayed further, perhaps even to nine loops [21, 22, 23, 24]. But protection against divergences

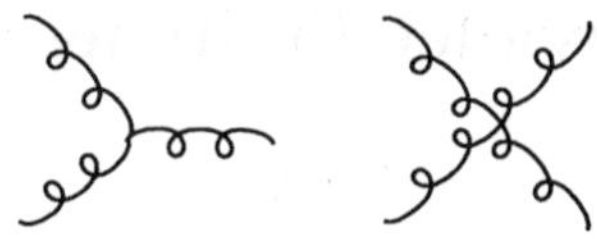

Figure 1: Gauge theories have three- and four-point vertices in a Feynman diagrammatic description.

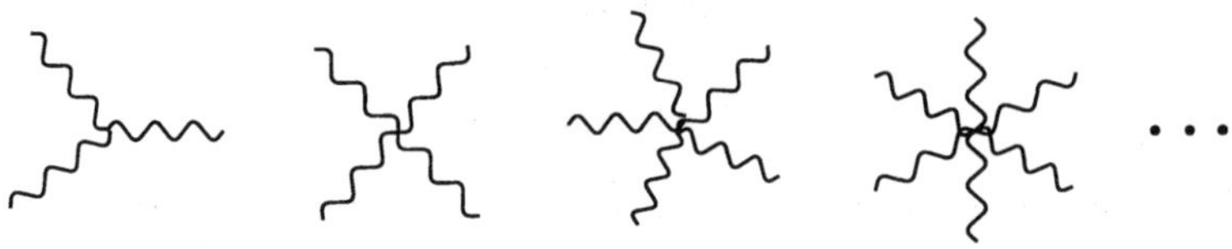

Figure 2: Gravity theories have an infinite number of higher-point contact interactions in a Feynman diagrammatic description.

from supersymmetry alone should fail at some loop order, because the dimensionful nature of the gravity coupling implies that each order of perturbation theory should be more divergent than the previous order. An additional mechanism is required to prevent divergences is needed.

Here we review direct evidence at three loops [25, 26] that such a mechanism does indeed exist in gravity theories. It is striking that to date *all* explicit complete calculations of $\mathcal{N} = 8$ supergravity scattering amplitudes [6, 27, 28, 29, 25, 30, 31, 32, 10, 26] find a power counting identical to that of $\mathcal{N} = 4$ super-Yang-Mills theory, which is known to be finite in four dimensions [33, 34]. Interestingly, M theory has also been used to argue for the finiteness of $\mathcal{N} = 8$ supergravity [35], though issues with decoupling towers of massive states [36] may alter these conclusions.

Here we focus only on the possible order-by-order finiteness of the perturbative series of $\mathcal{N} = 8$ supergravity, leaving aside non-perturbative issues or how one might go about constructing a realistic ultraviolet finite theory. The present goal is to demonstrate a set of non-trivial ultraviolet cancellations and to identify their origin. We are interested in the ultraviolet finiteness of gravity because the existence of cancellations sufficient to render the theory finite would imply the existence of a new symmetry or dynamical mechanism to enforce these. A proper understanding of the mechanism behind such cancellations should have a profound impact on our understanding of gravity.

2 Quantum gravity

Before turning to the on-shell methods used in modern calculations, it is helpful to first survey gravity Feynman diagrams. We compare Feynman diagrams of gravity to those of gauge theory. Consider the Einstein-Hilbert and Yang-Mills Lagrangians,

$$\mathcal{L}_{\rm YM} = -\frac{1}{4} F^a_{\mu\nu} F^{a\,\mu\nu}\,, \qquad\qquad \mathcal{L}_{\rm EH} = \frac{2}{\kappa^2}\sqrt{-g}R\,. \tag{1}$$

Following standard Feynman diagrammatic methods we gauge fix and then expand the Lagrangians in a set of vertices. As illustrated in figs. 1 and 2, with standard gauge

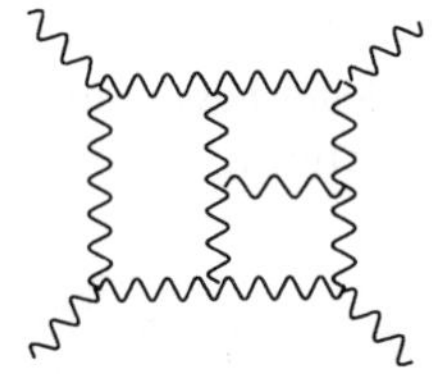

Figure 3: A sample diagram at three loops. A three-graviton vertex is quadratic in the momenta while a gauge theory vertex in linear in the momenta, implying that gravity theories should in general be worse behaved in the ultraviolet compared to gauge theories.

choices for gauge theory there are three- and four-point interactions, while for gravity there are an infinite number of contact interactions. (As we shall discuss below, all interactions beyond three points in fact are unnecessary when using on-shell methods.) Perhaps more impressive than the infinite number of interactions is the complexity of these interactions. Consider, for example, the three-graviton interaction. In standard de Donder gauge the three-graviton vertex is of the form,

$$\begin{aligned} &G_{3\mu\alpha,\nu\beta,\sigma\gamma}(k_1,k_2,k_3) \\ &\quad = i\frac{\kappa}{2}\mathrm{Sym}\Big[-\frac{1}{2}P_3(k_1\cdot k_2\eta_{\mu\alpha}\eta_{\nu\beta}\eta_{\sigma\gamma}) - \frac{1}{2}P_6(k_{1\nu}k_{1\beta}\eta_{\mu\alpha}\eta_{\sigma\gamma}) + \frac{1}{2}P_3(k_1\cdot k_2\eta_{\mu\nu}\eta_{\alpha\beta}\eta_{\sigma\gamma}) \\ &\qquad + P_6(k_1\cdot k_2\eta_{\mu\alpha}\eta_{\nu\sigma}\eta_{\beta\gamma}) + 2P_3(k_{1\nu}k_{1\gamma}\eta_{\mu\alpha}\eta_{\beta\sigma}) - P_3(k_{1\beta}k_{2\mu}\eta_{\alpha\nu}\eta_{\sigma\gamma}) \\ &\qquad + P_3(k_{1\sigma}k_{2\gamma}\eta_{\mu\nu}\eta_{\alpha\beta}) + P_6(k_{1\sigma}k_{1\gamma}\eta_{\mu\nu}\eta_{\alpha\beta}) + 2P_6(k_{1\nu}k_{2\gamma}\eta_{\beta\mu}\eta_{\alpha\sigma}) \\ &\qquad + 2P_3(k_{1\nu}k_{2\mu}\eta_{\beta\sigma}\eta_{\gamma\alpha}) - 2P_3(k_1\cdot k_2\eta_{\alpha\nu}\eta_{\beta\sigma}\eta_{\gamma\mu})\Big]\,, \end{aligned} \tag{2}$$

where "Sym" signifies a symmetrization in $\mu \leftrightarrow \alpha$, $\nu \leftrightarrow \beta$ and $\sigma \leftrightarrow \gamma$, and P_3 and P_6 signify a symmetrization over the three external legs, generating three or six terms respectively. Here the coupling κ is related to Newton's constant by $\kappa^2 = 32\pi^2 G_N$. In total the vertex has on the order of 100 terms. The k_i are the momenta of the three gravitons and $\eta_{\mu\nu}$ the flat metric. The precise form of the vertex depends on the gauge, but in general the three vertex is rather complicated. We may contrast this to the relatively simple three-gluon vertex in Feynman gauge,

$$V^{abc}_{3\mu,\nu,\sigma}(k_1,k_2,k_3) = gf^{abc}\Big[(k_1-k_2)_\sigma\eta_{\mu\nu} + \text{cyclic}\Big]\,. \tag{3}$$

Obviously, gravity seems much more complicated than gauge theory.

One can do better with special gauge choices and appropriate field redefinitions [18, 12], considerably simplifying the Feynman rules, but multiloop calculations in gravity based on Feynman diagrams are extremely difficult, if not impossible on any computer. Because of this difficulty, few explicit determinations of counterterms have been carried out using Feynman diagrams, with some notable exceptions [14, 15, 18].

2.1 Power counting in gravity theories

In four dimensions Newton's constant is dimensionful, implying that gravity theories in four dimensions are non-renormalizable by power counting. The standard argument that gravity theories are badly behaved in the ultraviolet follows easily from loop-level Feynman diagrams. Consider a loop level Feynman diagram such as the three-loop one in fig. 3. If the diagram is for gluon scattering of Yang-Mills theory, we obtain a Feynman integral of the form,

$$\int \prod_{j=1}^{L} \frac{d^D p_j}{(2\pi)^D} \frac{\cdots (g f^{abc} p_i^\mu) \cdots}{\prod_m (l_m^2 + i\epsilon)}, \tag{4}$$

where the numerator factor of the form $(g f^{abc} p_i^\mu)$ signifies a vertex factor, given by a coupling g, color factor f^{abc} and a momentum p_j^μ. The denominators are the Feynman propagators of the diagram, carrying momenta depending on the independent loop and external momenta p_i and k_i. This may be contrasted with the corresponding gravity diagram, whose form is similar except the vertices have two powers of momenta in the numerator for each vertex,

$$\int \prod_{j=1}^{L} \frac{d^D p_j}{(2\pi)^D} \frac{\cdots (\kappa p_i^\mu p_k^\nu) \cdots}{\prod_m (l_m^2 + i\epsilon)}. \tag{5}$$

Generically, the momenta appearing in the vertices will be loop momenta. Because of the larger number of loop momenta in the numerators we may expect each gravity Feynman integral to be worse behaved in the ultraviolet, compared to the corresponding gauge theory diagram, unless there are non-trivial cancellations. Moreover, based on these simple power-counting considerations, as the number of loops increases we may expect an ever worsening behavior compared to gauge theory.

We can easily determine the structure of a counterterm given a divergence at a given loop order from power counting. Since every loop gains an extra power of $G_N \sim 1/M_{\rm planck}^2$ every additional loop must gain two powers of mass dimension to compensate. For graviton amplitudes this corresponds to an additional power of the Riemann tensor, $R_{\mu\nu\sigma\rho}$, or to a covariant derivative $\mathcal{D}^2$. At one-loop pure-gravity counterterms in the Lagrangian involve two powers of the Riemann tensor, but by an accident of four dimensions, the Gauss-Bonnet theorem can be used to eliminate the potential on-shell counterterm. However, if matter is added, then generically a one-loop divergence appears [14, 15]. At two loops the potential counterterm is of the form $R^{\lambda\rho}{}_{\mu\nu} R^{\mu\nu}{}_{\sigma\tau} R^{\sigma\tau}{}_{\lambda\rho}$. In pure gravity, thanks to calculations by Goroff and Sagnotti and by van de Ven [18], we know for a fact that the theory diverges and the coefficient of this counterterm is nonvanishing.

The first divergence in any supergravity theory can be no earlier than three loops, since the potential one- and two-loop counterterms cannot be made consistent with supersymmetry [19]. In this case, the potential counterterm consistent with supersymmetry is an R^4 term [1] with indices appropriately contracted, corresponding to the square of the Bel-Robinson tensor. With additional assumptions we can raise the loop order where a divergence is first expected in $\mathcal{N} = 8$ supergravity. For example, if an off-shell superspace with $\mathcal{N} = 6$ supersymmetries manifest were to exist, potential divergences in four dimensions would be delayed to at least five loops [21]. Similarly, the existence of a superspace with $\mathcal{N} = 7$ supersymmetries manifest would delay the first potential divergence to at least six loops [21]. If one were to assume the existence of a fully covariant off-shell superspace with $\mathcal{N} = 8$ supersymmetries manifest, then the first potential divergence would be pushed to the seven loop order [37]. It is important to note that since no off-shell superspace beyond $\mathcal{N} = 4$ has been constructed, these bounds are speculative. A full superspace invariant, which could act as potential a counterterm, has been constructed at eight loops, suggesting that a divergence might appear at this loop order, if it does not appear earlier [38]. The appearance of the first potential divergence can even be pushed to nine loops, with an additional speculative assumption that all fields respect ten-dimensional general coordinate invariance [24]. This bound coincides with the one argued [23] from the type II string theory non-renormalization theorem of Berkovits [22]. Beyond this order, no purely supersymmetric mechanism has been suggested for preventing the onset of divergences. Indeed, for a supergravity theory to be ultraviolet finite to all orders of the perturbative expansion, novel cancellations beyond the known supersymmetric ones *must* exist.

It is important to note that no power counting arguments can prove the existence of a divergence, only only that divergences cannot appear prior to a certain loop order. If the theory were to possess a hidden symmetry, not accounted for in the power counting bound, the bound may suggest the appearance of a divergence when there is none. The possibility of hidden symmetries is real: the three-loop divergence thought to appear in the three-loop four-point amplitude in the studies from 1980's [1] is now known to not be present [25, 26]. As already mentioned in the introduction there are a number of good reasons to reexamine the ultraviolet behavior of $\mathcal{N} = 8$ supergravity using modern methods.

3 On-shell methods

The computations we summarize in the next section were obtained using on-shell methods. A key feature of these methods is that the elementary building blocks for obtaining new amplitudes are previously obtained on-shell amplitudes. These methods fall into two basic categories: on-shell recursion [39] and the modern unitarity method [3]. On-shell recursion is suitable for obtaining complete tree level amplitudes while the unitarity method is suitable at loop level. At one-loop a hybrid bootstrap approach making use of both unitarity and on-shell recursion has also been developed [40]. The use of these methods have become much more widespread in recent years because of their computational efficiency, especially at loop level. Here we only briefly review these, focusing on the salient features for gravity, and refer the reader to various reviews [41] for more extensive expositions.

3.1 On-shell vertices

As an explicit example of the inherent simplification of on-shell methods, consider an on-shell version of the three-graviton vertex in eq. (2), where we dot the three legs with physical polarizations tensors, satisfying the on-shell conditions, $\varepsilon_i^{\mu\nu} k_{i\mu} = \varepsilon_i^{\mu\nu} k_{i\nu} = \varepsilon^{\mu}{}_{\mu} = 0$. This gives the simplified vertex,

$$G_3(k_1, k_2, k_3) = -i\kappa\varepsilon_1^{\mu\alpha}\varepsilon_2^{\nu\beta}\varepsilon_3^{\sigma\gamma}\Big[(k_1)_\sigma\eta_{\mu\nu} + \text{cyclic}\Big]\Big[(k_1)_\gamma\eta_{\alpha\beta} + \text{cyclic}\Big], \tag{6}$$

which is not much more complicated than the corresponding on-shell Yang-Mills vertex,

$$V_3^{abc}(k_1,k_2,k_3) = \varepsilon_1^\mu \varepsilon_2^\nu \varepsilon_3^\sigma 2g f^{abc}\Big[(k_1)_\sigma \eta_{\mu\nu} + \text{cyclic}\Big]\,, \tag{7}$$

where the polarization vector satisfied $\varepsilon_i^\mu k_{i\mu} = 0$. As we shall discuss below, the structure of the on-shell graviton vertex as a product of the kinematic part of gauge-theory vertices is not accidental, but reflects a profound and important property of gravity.

If we proceed naively with an on-shell formalism, we encounter a difficulty: the on-shell three vertex for gravitons or gluons actually vanishes, because the process is kinematically forbidden. The solution is to use complex momenta [18, 42, 7], which allows us to satisfy the on-shell conditions and momentum conservation, yet define a nonvanishing vertex.

In general, in four dimensions it is best to express gauge and gravity amplitudes in terms of spinors. We define the two helicity configurations gluons as

$$\varepsilon_\pm^\mu(k_i,q_i) = \pm\frac{\langle q_i^-|\,\gamma^\mu\,|k_i^-\rangle}{\sqrt{2}\langle q_i^\mp|k_i^\pm\rangle}\,, \tag{8}$$

where $|k_i^\pm\rangle \equiv u_\pm(k_l)$ are Weyl spinors. The graviton polarization tensors are simply products of these,

$$\varepsilon_\pm^{\mu\alpha}(k_i,q_i) = \varepsilon_\pm^\mu(k_i,q_i)\,\varepsilon_\pm^\alpha(k_i,q_i)\,. \tag{9}$$

For later reference we define spinor inner products, as

$$\langle j\, l\rangle = \langle k_j^-|k_l^+\rangle\,, \qquad\qquad [j\, l] = \langle k_j^+|k_l^-\rangle\,. \tag{10}$$

These products are antisymmetric, $\langle j\, l\rangle = -\langle l\, j\rangle$, $[j\, l] = -[l\, j]$. For massless momenta, spinor inner products are complex square roots of Lorentz inner products, satisfying $\langle i\, j\rangle\, [j\, i] = 2k_i\cdot k_j$. For further details, see refs. [43].

3.2 Behavior of amplitudes under large complex shifts

On-shell recursion [39, 44, 45] gives us a simple means for systematically constructing new tree-level amplitudes in gravity and in gauge theory using previously constructed lower-point amplitudes. The proof of the recursion relations employs a complex shift of two momenta,

$$\begin{aligned} k_j^\mu &\to k_j^\mu(z) = k_j^\mu - \frac{z}{2}\langle j^-|\,\gamma^\mu\,|l^-\rangle, \\ k_l^\mu &\to k_l^\mu(z) = k_l^\mu + \frac{z}{2}\langle j^-|\,\gamma^\mu\,|l^-\rangle\,, \end{aligned} \tag{11}$$

so that they remain massless, $k_j^2(z) = k_l^2(z) = 0$, and overall momentum conservation is maintained. An on-shell amplitude containing the momenta k_j and k_l then depends on the parameter z,

$$A(z) = A(k_1,\ldots,k_j(z),k_{j+1},\ldots,k_l(z),k_{l+1},\ldots,k_n)\,, \tag{12}$$

where the physical amplitude is recovered by setting $z = 0$.

A crucial step in the proof of on-shell recursion relations is the requirement that $A(z)$ vanishes as $z\to\infty$. This property has been demonstrated for a wide class of shifts in gauge [39, 46] and gravity theories [44, 45]. This property is especially surprising for gravity, where the larger number of powers of loop momentum in the vertices would tend to make the diagrams ill-behaved as $z \to \infty$. In general, individual Feynman diagrams are not well behaved as $z \to \infty$; only the complete amplitudes vanish in this limit because of a strong cancellation between diagrams. However, with a space-like gauge similar to light-cone gauge [47] it is possible to arrange the diagrams in both gauge theory [48] and gravity [45] such that each diagram is well behaved at large z. The improved large z behavior appears connected to an enhanced Lorentz symmetry appearing for large z [45].

An important consequence of the on-shell recursion relations is that we can recursively construct all tree-level scattering amplitudes starting from on-shell three-point vertices. This can even be carried out in D dimensions [45]. Using the unitarity method we can then construct all loop amplitudes. The four- and higher-point vertices in both gauge or gravity theories are irrelevant as far as scattering amplitudes are concerned. Thus, all the vertices displayed in figs. 1 and 2, beyond the three-point ones, are unnecessary.

Note the behavior of the amplitudes as $z \to \infty$ is connected to the high-energy behavior of the theory, albeit in a complex direction. As we shall discuss below, the vanishing of $A(z)$ as $z \to \infty$ is connected to loop-level ultraviolet cancellations.

3.3 Kawai-Lewellen-Tye tree-level relations

From the perspective of Lagrangians or off-shell Feynman rules it is very difficult to discern any simple relations between gravity and gauge theory amplitudes. Nevertheless, tree-level gravity amplitudes can be rewritten in terms of gauge-theory amplitudes, a fact first uncovered in string theory by Kawai, Lewellen and Tye (KLT) [11, 12, 13] in string theory. These relations also hold in field theory, as the low-energy limit of string theory. In this limit, the KLT relations for four-, and five-point amplitudes are

$$
\begin{aligned}
M_4^{\rm tree}(1,2,3,4) &= -is_{12}A_4^{\rm tree}(1,2,3,4)\,\widetilde{A}_4^{\rm tree}(1,2,4,3)\,, \qquad (13)\\
M_5^{\rm tree}(1,2,3,4,5) &= is_{12}s_{34}A_5^{\rm tree}(1,2,3,4,5)\,\widetilde{A}_5^{\rm tree}(2,1,4,3,5)\\
&\quad + is_{13}s_{24}A_5^{\rm tree}(1,3,2,4,5)\,\widetilde{A}_5^{\rm tree}(3,1,4,2,5)\,. \qquad (14)
\end{aligned}
$$

Here the M_n's are amplitudes in a gravity theory stripped of couplings, the A_n's and $\widetilde{A}_n$'s are the color-ordered amplitudes in two, possibly distinct, gauge theories, and the Mandelstam invariants are $s_{ij} \equiv (k_i + k_j)^2$ with k_i being the outgoing momentum of leg i. The color-ordered [43] gauge-theory amplitudes correspond to the coefficient of color traces with a given ordering of matrices. The gravity states are direct products of gauge-theory states for each external leg. Explicit formulæ for n-point amplitudes may be found in refs. [27, 49].

As a simple example, consider the four-graviton amplitude $M_4^{\rm tree}(1^-,2^-,3^+,4^+)$, where the $\pm$ labels refer to the helicity of the gravitons in an all outgoing convention. Using the gauge theory expressions [43],

$$
\begin{aligned}
A_4^{\rm tree}(1^-,2^-,3^+,4^+) &= i\frac{\langle 1\,2\rangle^3}{\langle 2\,3\rangle\,\langle 3\,4\rangle\,\langle 4\,1\rangle}\,,\\
A_4^{\rm tree}(1^-,2^-,4^+,3^+) &= i\frac{\langle 1\,2\rangle^3}{\langle 2\,4\rangle\,\langle 4\,3\rangle\,\langle 3\,1\rangle}\,, \qquad (15)
\end{aligned}
$$

from the KLT relation (13), we obtain immediately,

$$\mathcal{M}_4^{\text{tree}}(1^-,2^-,3^+,4^+) = i\left(\frac{\kappa}{2}\right)^2 s_{12}\, \frac{\langle 1\,2\rangle^3}{\langle 2\,3\rangle\,\langle 3\,4\rangle\,\langle 4\,1\rangle}\, \frac{\langle 1\,2\rangle^3}{\langle 2\,4\rangle\,\langle 4\,3\rangle\,\langle 3\,1\rangle}\,, \tag{16}$$

where we have restored the gravity coupling constant. We invite the reader to evaluate this four-graviton amplitude using Feynman rules starting from the Einstein action; the result will match eq. (16). These relations hold not only for any theory corresponding to the low energy limit of a string theory, but appear to hold for a much broader range of gravity theories [13].

The KLT relations have recently been clarified, giving a more transparent form of the relation [49]. Consider the diagrammatic expansion of color-dressed gauge-theory amplitudes,

$$\begin{aligned} \frac{1}{g^{n-2}}\mathcal{A}_n^{\text{tree}}(1,2,3,\ldots,n) &= \sum_i \frac{n_i c_i}{(\prod_j p_j^2)_i}\,,\\ \frac{1}{g^{n-2}}\widetilde{\mathcal{A}}_n^{\text{tree}}(1,2,3,\ldots,n) &= \sum_i \frac{\tilde{n}_i c_i}{(\prod_j p_j^2)_i}\,, \end{aligned} \tag{17}$$

where the $\mathcal{A}_n^{\text{tree}}$ and $\widetilde{\mathcal{A}}_n^{\text{tree}}$ are two distinct gauge theory amplitudes. The sum runs over all diagrams with only cubic vertices, so each diagram has exactly $n-2$ propagators. As discussed in ref. [49] the n_i and $\tilde{n}_i$ numerators can be constrained to satisfy a kinematic numerator identity similar to the Jacobi identity satisfied by the c_i color factors. If the numerators satisfy this constraint, the gravity amplitudes are then given by,

$$-iM_n^{\text{tree}}(1,2,3,\ldots,n) = \sum_i \frac{n_i \tilde{n}_i}{(\prod_j p_j^2)_i}\,, \tag{18}$$

where the sum runs over the same set of diagrams as in eq. (17). Although unproven, these equations have been checked explicitly through eight points in ref. [49].

Although we use the KLT relations merely as a technical trick for efficiently evaluating multiloop gravity amplitudes, there is little doubt that the relations point towards a unification of gravity and gauge theory. It is interesting to note that string theory automatically encodes this unification.

3.4 Modern unitarity method for obtaining loop amplitudes

The modern unitarity method [3] gives us a systematic means for constructing loop amplitudes using on-shell tree amplitudes as input. This method has been applied to a wide variety of problems, including QCD (see refs [50] for examples of recent applications), AdS/CFT studies of $\mathcal{N}=4$ super-Yang-Mills amplitudes [51, 52], and to quantum gravity as discussed here.

To obtain an amplitude one first constructs an initial ansatz in terms of integral functions that reproduces one cut. Then, subsequent cuts of the amplitude are compared against the corresponding cuts of the ansatz. If any discrepancy is found in a later cut, additional terms that vanish when all the earlier cut conditions are imposed are added

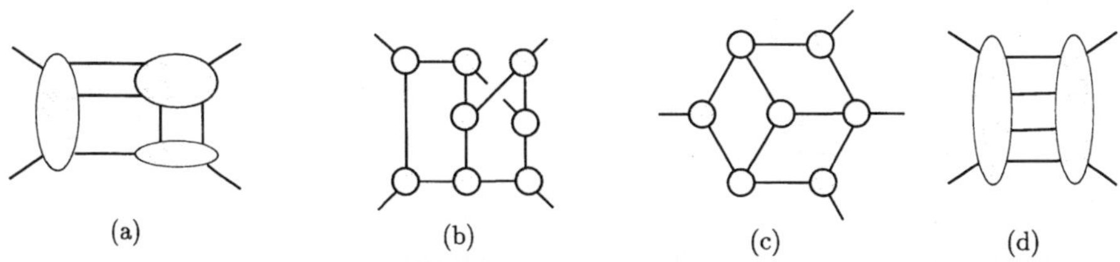

Figure 4: Examples of generalized cuts for determining a three-loop four-point amplitude. Each blob represents an on-shell tree amplitude, and the displayed intermediate lines are all on shell.

to the ansatz. Once a complete set of cuts have been checked, this procedure gives an integral representation of the loop amplitude with the correct cuts in all channels. The result is the complete amplitude, equivalent to what would have been obtained with Feynman diagrams.

At one loop, massless amplitudes in supersymmetric theories are determined completely by their four-dimensional cuts [3]. Unfortunately, this has not been demonstrated at higher loops. Even if the theory is ultraviolet finite, infrared singularities are present in four dimensions requiring use of a version of dimensional regularization compatible with supersymmetry [53]. To guarantee that no terms are dropped in the construction, the unitarity cuts must be evaluated in D dimensions. Unfortunately, evaluating the cuts in D dimensions [4] makes the calculation significantly more difficult, because powerful four-dimensional spinor methods [43] can no longer be used. Some of this additional complexity may be avoided by performing internal-state sums in terms of the gauge supermultiplet of $D = 10$, $\mathcal{N} = 1$ super-Yang-Mills theory instead of the $D = 4$, $\mathcal{N} = 4$ multiplet; this counts the same states but simplifies the bookkeeping.

In practical calculations it is useful to first construct an ansatz for the amplitude based on using four dimensional momenta in the generalized cuts, since we can use powerful spinor methods. Recently, there has been considerable improvement in bookkeeping the superpartners crossing unitarity cuts in four dimensions [54, 10], based on extensions of Nair's on-shell superspace [55] to general amplitudes. After an ansatz for the complete amplitudes is constructed based on four-dimensional cuts, to guarantee that no terms are dropped, it must be check against the D-dimensional cuts. It turns out that for four point amplitudes there is considerable evidence that all terms are detected in the four-dimensional cuts [56, 6, 57, 8], but for higher points there are additional terms [52].

We note that the KLT relations give us a rather efficient means of evaluating gravity generalized cuts. For gravity cuts reducing the loop amplitude to products of tree amplitudes summed over intermediate states, the KLT relations allow us to re-express these cuts as sums of products of gauge theory cuts which are generally simpler to evaluate. An important feature of this construction is that once the superpartner sums are performed for $\mathcal{N} = 4$ super-Yang-Mills generalized cuts, the corresponding super-partner sum in $\mathcal{N} = 8$ supergravity follows easily from the KLT relations. This holds in D dimensions as well. More generally, any simplifications performed on the gauge theory generalized cuts can be immediately carried over to gravity cuts.

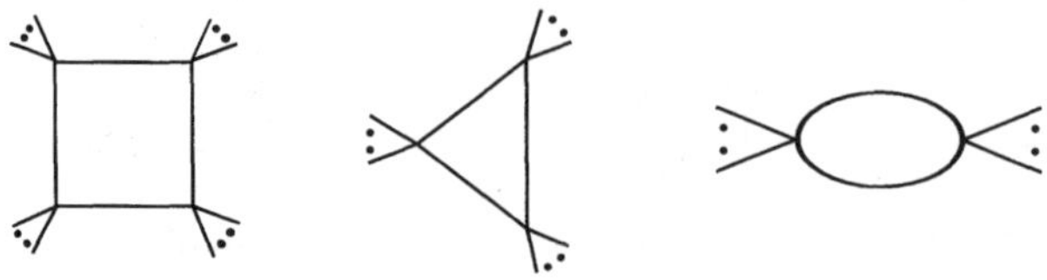

Figure 5: Any one-loop n point amplitude can be expressed as a linear combination of scalar box, triangle and bubble integrals. In $\mathcal{N}=4$ super-Yang-Mills and $\mathcal{N}=8$ supergravity after reducing all integrals to this basis, the coefficients of all triangle and bubble integrals vanish.

4 Ultraviolet properties of $\mathcal{N}=8$ supergravity

4.1 Early higher-loop computations

In ref. [56, 6] the two-loop four-point amplitudes of $\mathcal{N}=4$ and $\mathcal{N}=8$ supergravity were determined in terms of scalar integrals using the unitarity method. A class of unitarity cuts—the "iterated two-particle cuts"—were evaluated to all loop orders, yielding a partial reconstruction of the four-point amplitudes to all loop orders. The iterated cuts are fairly simple to evaluate because the same sewing algebra appears at each loop order [56]. For the case of $\mathcal{N}=4$ super-Yang-Mills theory, examining the powers of loop momenta in the numerator of the generic iterated two-particle cut contributions suggests the finiteness bound [6],

$$D < \frac{6}{L} + 4 \qquad (L > 1)\,, \tag{19}$$

where D is the dimension of space-time and L the loop order. (The case of one loop, $L=1$, is special, with the amplitudes finite for $D<8$, not $D<10$.) The bound (19) differs somewhat from earlier superspace power counting [34], although all bounds confirm the ultraviolet finiteness of $\mathcal{N}=4$ super-Yang-Mills theory in $D=4$. This all-loop-order bound (19) has since been confirmed [21] using $\mathcal{N}=3$ harmonic superspace [58]. Explicit computations demonstrate this bound is saturated through at least four loops [56, 6, 59].

In ref. [6], the iterated two-particle cuts of $\mathcal{N}=8$ supergravity amplitudes were also analyzed, leading to the proposal that the four-point $\mathcal{N}=8$ supergravity amplitude should be ultraviolet finite for

$$D < \frac{10}{L} + 2 \qquad (L > 1)\,. \tag{20}$$

(Again the one-loop case is special with the $\mathcal{N}=8$ supergravity amplitudes being finite for $D<8$ [32].) This bound implies that in $D=4$ the first potential divergence in $\mathcal{N}=8$ supergravity may appear at five loops. The formula is also consistent with bounds obtained by Howe and Stelle [21], assuming the existence of an $\mathcal{N}=6$ harmonic superspace [58]. Since this proposed bound is based on a partial calculation, if there are additional cancellations with uncalculated terms then the true finiteness bound would be improved compared to eq. (20).

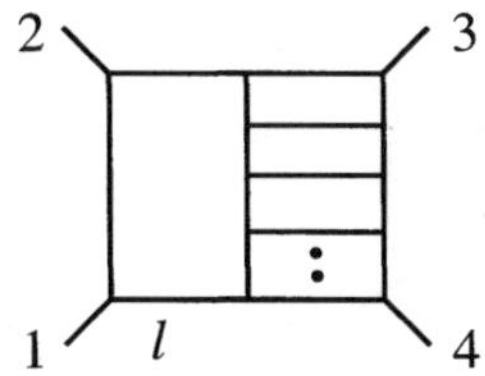

Figure 6: An L-loop contribution as proposed in ref. [6] based on a partial analysis of unitarity cuts. In $\mathcal{N}=4$ super-Yang-Mills theory the numerator factor is $[(l+k_4)^2]^{(L-2)}$, after dividing out factors which are independent of loop momenta. In the $\mathcal{N}=8$ supergravity case the proposed corresponding factor was $[(l+k_4)^2]^{2(L-2)}$. This proposal violates the no-triangle property of the one-loop amplitude isolated by the cut in fig. 7(a), implying additional cancellations with other contributions.

4.2 Novel cancellations at one loop

A key clue for the existence of novel cancellations comes from one loop. A one-loop theorem [60] states that near four dimensions any n-point massless amplitude can be expressed as

$$A_n = \sum_i d_i I_4^i + \sum_i c_i I_3^i + \sum_i b_i I_2^i \,, \tag{21}$$

where the scalar integrals $I_{2,3,4}$, are respectively bubbles, triangles, and boxes and d_i, c_i, b_i are (possibly dimension dependent) rational coefficients. These basis scalar integrals are displayed in fig. 5 and are the same ones that would appear in φ^3 theory. Some time ago for the special case of maximally helicity violating one-loop amplitudes of $\mathcal{N}=8$ supergravity, a curious cancellation was observed, which set the coefficients of all triangle and bubble integrals to zero [27]. More recently, it became clear that the vanishing of the triangle and bubble integral coefficient was a general property of all one-loop $\mathcal{N}=8$ supergravity amplitudes [28, 29, 31], with proofs given in refs. [32]. This has been called the "no-triangle property", although it also implies lack of bubble integrals. For $\mathcal{N}=8$ supergravity this one-loop behavior is rather surprising and suggests the existence of hidden cancellations not apparent in naive power counting. We can use this to limit the number of powers of loop momenta which can appear in the numerator of one-loop amplitudes in $\mathcal{N}=8$ supergravity. Under the Brown-Feynman or Passarino-Veltman reduction [60] each power of $l\cdot k_i$—where l is a loop momentum and k_i an external momentum—will give a linear combination of integrals with either no or one canceled propagator. For example, an amplitude with a hexagon integral with a numerator of the form $(l\cdot k_i)^3$, after integral reductions will contain box and triangle integrals. With an additional power of $l\cdot k_i$, bubble integrals will also appear.

Interestingly, the observed novel cancellations appear to be generic in gravity theories, as suggested by the one-loop study of ref. [30]. In fact, these type of cancellations may be generic to any theory where color factors do not prevent cancellations between contributions with a differing ordering of legs [61]. The novel one-loop cancellations were also shown to follow from the remarkably good high-energy behavior of gravity tree amplitudes under the complex deformations used to prove on-shell recursion relations in gravity [39, 44, 10].

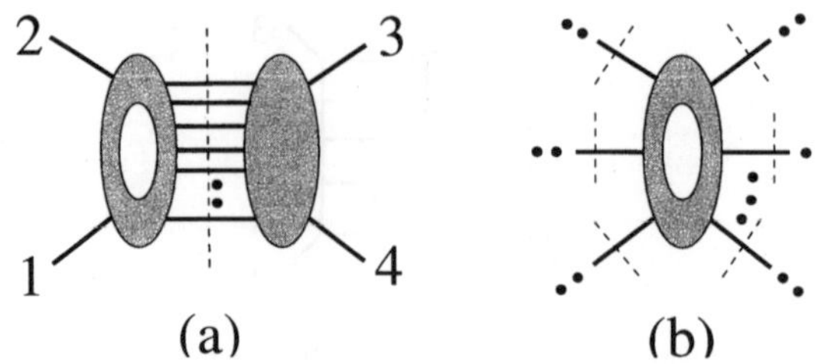

Figure 7: The no-triangle property ensures that all one-loop subamplitudes appearing in a multi-loop $\mathcal{N} = 8$ supergravity amplitude have the same degree of divergence as that of $\mathcal{N} = 4$ super-Yang-Mills theory. The cut (a) is an L-particle cut of an L-loop amplitude. Cut (b) makes use of generalized unitarity to isolate a one-loop subamplitude. These cuts may be used to demonstrate the existence of novel cancellations to all loop orders.

By itself, the cancellation of triangle and bubble integrals at n-points does not directly say anything about the ultraviolet properties of the theory. Indeed it was already known for a long time that the four-point amplitude is composed of only a box integral [62], with the divergence first appearing in $D = 8$; one should not be surprised that the higher-point amplitudes have the same degree of divergence. However, the one-loop cancellations to give non-trivial consequences are higher higher.

4.3 Novel cancellations to all loop orders

As pointed out in ref. [59], at least in a certain class of terms, we can use the unitarity method we can non-trivially constrain the form of multiloop amplitudes to *all* loop orders [59], starting from the one-loop no-triangle cancellations. In particular, consider the diagrammatic contributions to the L-loop four-point amplitude of the form in fig. 6. Based on evaluating the iterated two-particle cuts, ref. [6] suggested that for $\mathcal{N} = 4$ super-Yang-Mills theory the numerator of the integral corresponding to this diagram be dressed by a factor of

$$[(l + k_4)^2]^{(L-2)}\,, \tag{22}$$

after removing factors which are independent of the loop momenta. For $\mathcal{N} = 8$ supergravity, the proposed corresponding factor was,

$$[(l + k_4)^2]^{2(L-2)}\,. \tag{23}$$

which is much worse behaved as L increase than the super-Yang-Mills numerator. Power counting the integral in fig. 7 with this numerator gives the $\mathcal{N} = 8$ finiteness bound in eq. (20).

However, additional cancellations must exist because the proposed numerator (23) the high power of loop momenta would lead to a violation of the no triangle integral property in the one-loop subamplitude isolated by the unitarity cut in fig. 7(a). This violation would start at three loops. The cancellations would need to be with other integrals, effectively reducing the number of powers of loop momenta in the numerators. In fact, after the cancellations take place, at three loops only two powers of loop momenta should remain, matching the number appearing in the $\mathcal{N} = 4$ super-Yang-Mills case (22).

Besides constraints from the cut in fig. 7(a), non-trivial additional constraints follow from the set of all generalized cuts isolating a one-loop subamplitude schematically depicted in fig. 7(b).

It is interesting to note that there does not appear to be a purely supersymmetric explanation of these cancellations. If it is not entirely supersymmetry then what might be behind the cancellations? The unitarity method implies that cancellations must have their origin in tree-level cancellations. An obvious guess is that the loop-level cancellations should be related to the tree-level ones under large complex deformations of the momenta discussed in section 3.2. Indeed this is the case [29, 30]. From an analysis of one-loop cancellations using an integration formalism due to Forde [63], the following picture emerges: Most of the one-loop cancellations observed in $\mathcal{N}=8$ supergravity leading to the no-triangle property are already present in non-supersymmetric pure gravity and are directly tied to the tree level cancellations [30]. Schematically, in $\mathcal{N}=8$ supergravity we find that a one-loop integral with the $2n$ maximum number of powers of loop undergoes a cancellations of $n+4$ powers, leaving $n-4$. Of the canceled powers of loop momentum 8 are due to supersymmetry while $n-4$ are generic and cancel also in pure Einstein gravity. The generic cancellations have recently been attributed to the unordered nature of the graviton amplitudes [32, 61].

Although the above arguments demonstrate the existence of novel cancellations to all loop orders, they do not demonstrate that the cancellations hold for all terms. In particular, contributions not detectable by isolating a one loop subamplitude will not be constrained. We therefore need to inspect such terms. A logical next step therefore is to explicitly confirm that all terms in the three-loop four point amplitude—which is the first nontrivial case—have cancellations of the sort found by inspecting the cuts in fig. 7.

4.4 Superfiniteness at three loops

In order to check whether the cancellations are complete, in refs. [25, 26] the full three-loop four-point amplitude of $\mathcal{N}=8$ supergravity was obtained using the unitarity method. The result of this computation proves that the cancellations observed in terms sensitive to cuts isolating a one loop subamplitude, occur in all terms composing the three-loop amplitudes.

In the original calculation of the three-loop four-point amplitude [25], the result was given in a form where individual diagrams have a worse power counting than the complete answer. An improved form, was subsequently found using the method of maximal cuts [8].[1] In the latter form the individual contributions are no worse behaved than the contributions of the corresponding super-Yang-Mills amplitude [26].

The three-loop amplitudes can be described in terms of the "parent diagrams" displayed in fig. 8 dressed by appropriate numerator factors. The integrals composing the amplitudes are of the form,

$$I^{(x)} = (-i)^3 \int \left[\prod_{i=1}^{3} \frac{d^D q_i}{(2\pi)^D}\right] \frac{N^{(x)}(q_j, k_j)}{\prod_{m=5}^{14} l_m^2}, \tag{24}$$

[1]A related method is the leading singularity method, which has also been used in various $\mathcal{N}=4$ multiloop calculations [9].

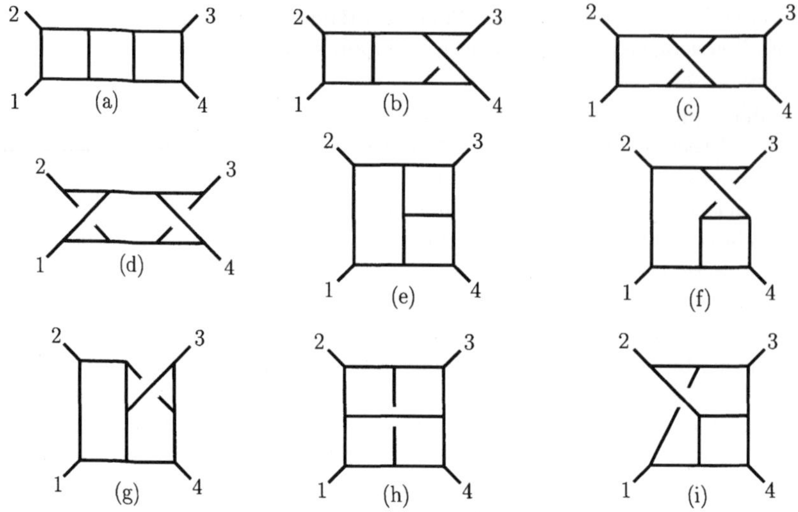

Figure 8: The different parent integrals in terms of which four-point three-loop amplitudes may be expressed.

where the q_i's are three independent loop momenta, the l_m's are the momenta carried by the propagators of the diagrams, and the $N^{(x)}(q_j, k_j)$ are numerator factors.

In the improved form given in ref. [26], no numerator factors of $\mathcal{N} = 8$ supergravity have a worse power count than the corresponding factors of $\mathcal{N} = 4$ super-Yang-Mills theory. That is, numerators in either theory satisfy,

$$\partial_{q_i} \partial_{q_j} \partial_{q_k} N^{(x)}(q_m, k_m) = 0 \,, \tag{25}$$

for the independent loop momenta with i, j, k taking on the values $1, 2, 3$. Thus, no more than two powers of loop momenta appear and the power counting of the three-loop four-point $\mathcal{N} = 8$ supergravity amplitude is identical to that of the corresponding $\mathcal{N} = 4$ super-Yang-Mills amplitude. Not only does this computation prove that the three-loop divergence is not present, the amplitude exhibits "superfiniteness" or cancellations beyond those required for finiteness. Instead of the finiteness condition (20), at $L = 3$ the correct power count is the one matching super-Yang-Mills theory (19). Explicit calculations show that the finiteness bound (19) is saturated in both theories, so at three loops both theories diverge in six dimensions.

5 Conclusions

Maximally supersymmetric $\mathcal{N} = 8$ supergravity theory is the most promising theory for finding a unitary perturbatively ultraviolet-finite point-like quantum field theory of gravity. The modern unitarity gives us a powerful means to probe the multiloop ultraviolet behavior of this theory, avoiding the rather rapid increase in complexity of Feynman diagrams at higher-loop orders. We summarized the results of refs. [25, 26], where a loop-integral representation of the three-loop four-point amplitude of $\mathcal{N} = 8$ supergravity was presented, exhibiting cancellations beyond those needed for finiteness. The three-loop power counting matched the one for $\mathcal{N} = 4$ super-Yang-Mills theory, which is known to be ultraviolet finite.

While these computations reveal nontrivial cancellations, their origin remains to be fully unraveled. For a subset of contributions, all-loop cancellations [59] follow from the one-loop "no-triangle property" [27, 28, 29, 64, 31, 32, 10]. In refs. [29, 30, 10] one-loop cancellations in generic theories of gravity were linked to unexpectedly soft behavior of tree-level gravity amplitudes under large complex shifts of their momenta [44, 45]. This mechanism was also proposed as a source of all-loop cancellations, which may be sufficiently strong to render $\mathcal{N} = 8$ supergravity ultraviolet finite in conjunction with supersymmetric cancellations. Improved ultraviolet properties in $\mathcal{N} = 8$ supergravity have also been linked to M theory dualities [35, 36] and to string theory non-renormalization theorems [22, 23].

An important question is whether any of the known symmetries of $\mathcal{N} = 8$ supergravity can be used to constrain its multiloop ultraviolet behavior. For example, $\mathcal{N} = 8$ supergravity contains a non-compact $E_{7(7)}$ duality symmetry [2, 65]. Its explicit action on fields has been presented recently [66, 10].

A next step in unraveling the UV properties of $\mathcal{N} = 8$ supergravity, will be the calculation of the four-loop four-point amplitude. We have every reason to believe it will be finite, but crucial guiding information will come from identifying the smallest dimension in which the divergence appears. It will allow us to test whether the cancellations at three-loops might have an explanation in terms of the possible existence of extended off-shell superspaces [21]. This calculation is in progress, with the key step of obtaining the corresponding $\mathcal{N} = 4$ super-Yang-Mills amplitude–including non-planar contributions–already carried out [67].

A key open problem is to develop a physical interpretation of the observed ultraviolet cancellations. With such a description it should be much more straightforward to construct a proof of finiteness for $\mathcal{N} = 8$ supergravity. More generally, it would be important to explore the surprising and profound relations between gravity and gauge theories. In particular, the KLT relations hint at a unification between gravity and gauge theory of the sort implied by string theory.

Acknowledgments

The author thanks John Joseph Carrasco, Lance Dixon, Darren Forde, Harald Ita, Henrik Johansson, David Kosower and Radu Roiban for many helpful discussions and collaborations on the results summarized here. This work was supported by the US Department of Energy under contract DE–FG03–91ER40662.

References

[1] P. S. Howe and K. S. Stelle, Int. J. Mod. Phys. A **4**, 1871 (1989); N. Marcus and A. Sagnotti, Nucl. Phys. B **256**, 77 (1985).

[2] E. Cremmer, B. Julia and J. Scherk, Phys. Lett. B **76**, 409 (1978); E. Cremmer and B. Julia, Phys. Lett. B **80**, 48 (1978); Nucl. Phys. B **159**, 141 (1979).

[3] Z. Bern, L. J. Dixon, D. C. Dunbar and D. A. Kosower, Nucl. Phys. B **425**, 217 (1994) [hep-ph/9403226]; Z. Bern, L. J. Dixon, D. C. Dunbar and D. A. Kosower, Nucl. Phys. B **435**, 59 (1995) [hep-ph/9409265]. Z. Bern, L. J. Dixon and D. A. Kosower, Nucl. Phys. B **513**, 3 (1998) [hep-ph/9708239].

[4] Z. Bern and A. G. Morgan, Nucl. Phys. B **467**, 479 (1996) [hep-ph/9511336]; Z. Bern, L. J. Dixon, D. C. Dunbar and D. A. Kosower, Phys. Lett. B **394**, 105 (1997) [hep-th/9611127]; Z. Bern, L. J. Dixon and D. A. Kosower, JHEP **0001**, 027 (2000) [hep-ph/0001001]. Z. Bern, L. J. Dixon and D. A. Kosower, JHEP **0408**, 012 (2004) [hep-ph/0404293].

[5] Z. Bern, L. J. Dixon and D. A. Kosower, Nucl. Phys. B **513**, 3 (1998) [hep-ph/9708239]; Z. Bern, V. Del Duca, L. J. Dixon and D. A. Kosower, Phys. Rev. D **71**, 045006 (2005) [hep-th/0410224].

[6] Z. Bern, L. J. Dixon, D. C. Dunbar, M. Perelstein and J. S. Rozowsky, Nucl. Phys. B **530**, 401 (1998) [hep-th/9802162].

[7] R. Britto, F. Cachazo and B. Feng, Nucl. Phys. B **725**, 275 (2005) [hep-th/0412103];

[8] Z. Bern, J. J. M. Carrasco, H. Johansson and D. A. Kosower, Phys. Rev. D **76**, 125020 (2007) [0705.1864 [hep-th]].

[9] E. I. Buchbinder and F. Cachazo, JHEP **0511**, 036 (2005) [hep-th/0506126]; F. Cachazo and D. Skinner, 0801.4574 [hep-th]; F. Cachazo, 0803.1988 [hep-th]; F. Cachazo, M. Spradlin and A. Volovich, 0805.4832 [hep-th]; M. Spradlin, A. Volovich and C. Wen, 0808.1054 [hep-th].

[10] N. Arkani-Hamed, F. Cachazo and J. Kaplan, 0808.1446 [hep-th].

[11] H. Kawai, D. C. Lewellen and S. H. H. Tye, Nucl. Phys. B **269**, 1 (1986); N. E. J. Bjerrum-Bohr, Phys. Lett. B **560**, 98 (2003) [hep-th/0302131]; Nucl. Phys. B **673**, 41 (2003) [hep-th/0305062]; N. E. J. Bjerrum-Bohr and K. Risager, Phys. Rev. D **70**, 086011 (2004) [hep-th/0407085]; S. Ananth and S. Theisen, Phys. Lett. B **652**, 128 (2007) [0706.1778 [hep-th]]; H. Elvang and D. Z. Freedman, JHEP **0805**, 096 (2008) [0710.1270 [hep-th]].

[12] Z. Bern and A. K. Grant, Phys. Lett. B **457**, 23 (1999) [hep-th/9904026];

[13] Z. Bern, A. De Freitas and H. L. Wong, Phys. Rev. Lett. **84**, 3531 (2000) [hep-th/9912033];

[14] G. 't Hooft and M. J. G. Veltman, Annales Poincare Phys. Theor. A **20**, 69 (1974).

[15] S. Deser and P. van Nieuwenhuizen, Phys. Rev. D **10**, 401 (1974); S. Deser, H. S. Tsao and P. van Nieuwenhuizen, Phys. Rev. D **10**, 3337 (1974); P. van Nieuwenhuizen and J. A. M. Vermaseren, Phys. Rev. D **16**, 298 (1977); D. C. Dunbar and P. S. Norridge, Class. Quant. Grav. **14**, 351 (1997) [hep-th/9512084]; D. C. Dunbar and P. S. Norridge, Nucl. Phys. B **433**, 181 (1995) [hep-th/9408014].

[16] R. E. Kallosh, Nucl. Phys. B **78**, 293 (1974).

[17] P. van Nieuwenhuizen and C. C. Wu, J. Math. Phys. **18**, 182 (1977).

[18] M. H. Goroff and A. Sagnotti, Phys. Lett. B **160**, 81 (1985); Nucl. Phys. B **266**, 709 (1986); A. E. M. van de Ven, Nucl. Phys. B **378**, 309 (1992).

[19] M. T. Grisaru, Phys. Lett. B **66**, 75 (1977); E. Tomboulis, Phys. Lett. B **67**, 417 (1977);

[20] S. Deser, J. H. Kay and K. S. Stelle, Phys. Rev. Lett. **38**, 527 (1977);

[21] P. S. Howe and K. S. Stelle, Phys. Lett. B **554**, 190 (2003) [hep-th/0211279].

[22] N. Berkovits, Phys. Rev. Lett. **98**, 211601 (2007) [hep-th/0609006].

[23] M. B. Green, J. G. Russo and P. Vanhove, Phys. Rev. Lett. **98**, 131602 (2007) [hep-th/0611273].

[24] K. S. Stelle, talk presented at the *UCLA workshop: Is* $\mathcal{N} = 8$ *Supergravity Finite?*, http://www.physics.ucla.edu/tep/workshops/supergravity/agenda/Talks/talk2.pdf

[25] Z. Bern, J. J. Carrasco, L. J. Dixon, H. Johansson, D. A. Kosower and R. Roiban, Phys. Rev. Lett. **98**, 161303 (2007) [hep-th/0702112].

[26] Z. Bern, J. J. M. Carrasco, L. J. Dixon, H. Johansson and R. Roiban, 0808.4112 [hep-th].

[27] Z. Bern, L. J. Dixon, M. Perelstein and J. S. Rozowsky, Nucl. Phys. B **546**, 423 (1999) [hep-th/9811140].

[28] Z. Bern, N. E. J. Bjerrum-Bohr and D. C. Dunbar, JHEP **0505**, 056 (2005) [hep-th/0501137]; N. E. J. Bjerrum-Bohr, D. C. Dunbar and H. Ita, Phys. Lett. B **621**, 183 (2005) [hep-th/0503102];

[29] N. E. J. Bjerrum-Bohr, D. C. Dunbar, H. Ita, W. B. Perkins and K. Risager, JHEP **0612**, 072 (2006) [hep-th/0610043].

[30] Z. Bern, J. J. Carrasco, D. Forde, H. Ita and H. Johansson, Phys. Rev. D **77**, 025010 (2008) [0707.1035 [hep-th]].

[31] N. E. J. Bjerrum-Bohr and P. Vanhove, JHEP **0804**, 065 (2008) [0802.0868 [hep-th]].

[32] N. E. J. Bjerrum-Bohr and P. Vanhove, 0805.3682 [hep-th]; 0806.1726 [hep-th].

[33] S. Mandelstam, Nucl. Phys. B **213**, 149 (1983).

[34] P. S. Howe and K. S. Stelle, Phys. Lett. B **137**, 175 (1984).

[35] G. Chalmers, hep-th/0008162; M. B. Green, J. G. Russo and P. Vanhove, JHEP **0702**, 099 (2007) [hep-th/0610299].

[36] M. B. Green, H. Ooguri and J. H. Schwarz, Phys. Rev. Lett. **99**, 041601 (2007) [0704.0777 [hep-th]].

[37] M. T. Grisaru and W. Siegel, Nucl. Phys. B **201**, 292 (1982) [Erratum-ibid. B **206**, 496 (1982)].

[38] P. S. Howe and U. Lindstrom, Nucl. Phys. B **181**, 487 (1981); R. E. Kallosh, Phys. Lett. B **99**, 122 (1981).

[39] R. Britto, F. Cachazo and B. Feng, Nucl. Phys. B **715**, 499 (2005) [hep-th/0412308]; R. Britto, F. Cachazo, B. Feng and E. Witten, Phys. Rev. Lett. **94**, 181602 (2005) [hep-th/0501052].

[40] Z. Bern, L. J. Dixon and D. A. Kosower, Phys. Rev. D **73**, 065013 (2006) [hep-ph/0507005]; C. F. Berger, Z. Bern, L. J. Dixon, D. Forde and D. A. Kosower, Phys. Rev. D **74**, 036009 (2006) [hep-ph/0604195].

[41] Z. Bern, L. J. Dixon and D. A. Kosower, Ann. Rev. Nucl. Part. Sci. **46**, 109 (1996) [hep-ph/9602280]; Z. Bern, Living Rev. Rel. **5**, 5 (2002) [gr-qc/0206071]; F. Cachazo and P. Svrček, PoS **RTN2005**, 004 (2005) [hep-th/0504194]; L. J. Dixon, PoS **HEP2005**, 405 (2006) [hep-ph/0512111]; Z. Bern, L. J. Dixon and D. A. Kosower, Annals Phys. **322**, 1587 (2007) [0704.2798 [hep-ph]]; L. F. Alday and R. Roiban, Phys. Rept. **468**, 153 (2008) [0807.1889 [hep-th]].

[42] E. Witten, Commun. Math. Phys. **252**, 189 (2004) [hep-th/0312171];

[43] M. L. Mangano and S. J. Parke, Phys. Rept. **200**, 301 (1991); L. J. Dixon, in *QCD & Beyond: Proceedings of TASI '95*, ed. D. E. Soper (World Scientific, 1996) [hep-ph/9601359].

[44] J. Bedford, A. Brandhuber, B. J. Spence and G. Travaglini, Nucl. Phys. B **721**, 98 (2005) [hep-th/0502146]; F. Cachazo and P. Svrček, hep-th/0502160; N. E. J. Bjerrum-Bohr, D. C. Dunbar, H. Ita, W. B. Perkins and K. Risager, JHEP **0601**, 009 (2006) [hep-th/0509016]; A. Brandhuber, S. McNamara, B. Spence and G. Travaglini, JHEP **0703**, 029 (2007) [hep-th/0701187]; P. Benincasa, C. Boucher-Veronneau and F. Cachazo, JHEP **0711**, 057 (2007) [hep-th/0702032]; P. Benincasa and F. Cachazo, 0705.4305 [hep-th]; A. Hall, Phys. Rev. D **77**, 124004 (2008) [0803.0215 [hep-th]]; C. Cheung, 0808.0504 [hep-th]; H. Elvang, D. Z. Freedman and M. Kiermaier, 0808.1720 [hep-th].

[45] N. Arkani-Hamed and J. Kaplan, JHEP **0804**, 076 (2008) [0801.2385 [hep-th]];

[46] S. D. Badger, E. W. N. Glover and V. V. Khoze, JHEP **0601**, 066 (2006) [hep-th/0507161]; S. D. Badger, E. W. N. Glover, V. V. Khoze and P. Svrcek, JHEP **0507**, 025 (2005) [hep-th/0504159]; C. Schwinn and S. Weinzierl, JHEP **0704**, 072 (2007) [hep-ph/0703021]; A. Hall, Phys. Rev. D **77**, 025011 (2008) [0710.1300 [hep-ph]].

[47] G. Chalmers and W. Siegel, Phys. Rev. D **59**, 045013 (1999) [hep-ph/9801220];

[48] D. Vaman and Y. P. Yao, JHEP **0604**, 030 (2006) [hep-th/0512031]; 0805.2645 [hep-th].

[49] Z. Bern, J. J. M. Carrasco and H. Johansson, Phys. Rev. D **78**, 085011 (2008) [0805.3993 [hep-ph]].

[50] C. F. Berger *et al.*, Phys. Rev. D **78**, 036003 (2008) [0803.4180 [hep-ph]]; 0808.0941 [hep-ph]; R. Britto, B. Feng and G. Yang, JHEP **0809**, 089 (2008) [0803.3147 [hep-ph]]. T. Binoth, G. Ossola, C. G. Papadopoulos and R. Pittau, JHEP **0806**, 082 (2008) [0804.0350 [hep-ph]]; R. K. Ellis, W. T. Giele, Z. Kunszt, K. Melnikov and G. Zanderighi, 0810.2762 [hep-ph].

[51] C. Anastasiou, Z. Bern, L. J. Dixon and D. A. Kosower, Phys. Rev. Lett. **91**, 251602 (2003) [hep-th/0309040]; Z. Bern, L. J. Dixon and V. A. Smirnov, Phys. Rev. D **72**, 085001 (2005) [hep-th/0505205]. F. Cachazo, M. Spradlin and A. Volovich, Phys. Rev. D **74**, 045020 (2006) [hep-th/0602228]; Z. Bern, M. Czakon, D. A. Kosower, R. Roiban and V. A. Smirnov, Phys. Rev. Lett. **97**, 181601 (2006) [hep-th/0604074]; L. F. Alday and J. Maldacena, JHEP **0706**, 064 (2007) [0705.0303 [hep-th]]; JHEP **0711**, 068 (2007) [0710.1060 [hep-th]]; J. M. Drummond, J. Henn, G. P. Korchemsky and E. Sokatchev, 0712.1223 [hep-th].

[52] Z. Bern, L. J. Dixon, D. A. Kosower, R. Roiban, M. Spradlin, C. Vergu and A. Volovich, Phys. Rev. D **78**, 045007 (2008) [0803.1465 [hep-th]].

[53] W. Siegel, Phys. Lett. B **84**, 193 (1979); Z. Bern and D. A. Kosower, Nucl. Phys. B **379**, 451 (1992); Z. Bern, A. De Freitas, L. J. Dixon and H. L. Wong, Phys. Rev. D **66**, 085002 (2002) [hep-ph/0202271].

[54] J. M. Drummond, J. Henn, G. P. Korchemsky and E. Sokatchev, 0807.1095 [hep-th]; A. Brandhuber, P. Heslop and G. Travaglini, 0807.4097 [hep-th]; J. M. Drummond, J. Henn, G. P. Korchemsky and E. Sokatchev, 0808.0491 [hep-th]; J. M. Drummond and J. M. Henn, 0808.2475 [hep-th]. M. Bianchi, H. Elvang and D. Z. Freedman, JHEP **0809**, 063 (2008) [0805.0757 [hep-th]]; H. Elvang, D. Z. Freedman and M. Kiermaier, 0808.1720 [hep-th]. M. Kiermaier, H. Elvang and D. Z. Freedman, 0811.3624 [hep-th].

[55] V. P. Nair, Phys. Lett. B **214**, 215 (1988).

[56] Z. Bern, J. S. Rozowsky and B. Yan, Phys. Lett. B **401**, 273 (1997) [hep-ph/9702424].

[57] Z. Bern, M. Czakon, L. J. Dixon, D. A. Kosower and V. A. Smirnov, Phys. Rev. D **75**, 085010 (2007) [hep-th/0610248].

[58] A. Galperin, E. Ivanov, S. Kalitsyn, V. Ogievetsky and E. Sokatchev, Class. Quant. Grav. **1**, 469 (1984); Class. Quant. Grav. **2**, 155 (1985).

[59] Z. Bern, L. J. Dixon and R. Roiban, Phys. Lett. B **644**, 265 (2007) [hep-th/0611086].

[60] L. M. Brown and R. P. Feynman, Phys. Rev. **85** (1952) 231; L.M. Brown, Nuovo Cim. **21**, 3878 (1961); G. Passarino and M. J. G. Veltman, Nucl. Phys. B **160**, 151 (1979).

[61] S. Badger, N. E. J. Bjerrum-Bohr and P. Vanhove, 0811.3405 [hep-th].

[62] M. B. Green, J. H. Schwarz and L. Brink, Nucl. Phys. B **198**, 474 (1982).

[63] D. Forde, Phys. Rev. D **75**, 125019 (2007) [0704.1835 [hep-ph]].

[64] R. Kallosh, 0711.2108 [hep-th].

[65] M. K. Gaillard and B. Zumino, Nucl. Phys. B **193**, 221 (1981); B. de Wit and H. Nicolai, Nucl. Phys. B **208**, 323 (1982).

[66] L. Brink, S. S. Kim and P. Ramond, JHEP **0806**, 034 (2008) [0801.2993 [hep-th]]; R. Kallosh and M. Soroush, Nucl. Phys. B **801**, 25 (2008) [0802.4106 [hep-th]]; R. Kallosh, C. H. Lee and T. Rube, 0811.3417 [hep-th].

[67] Z. Bern, J. J. M. Carrasco, L. J. Dixon, H. Johansson and R. Roiban, to appear.

CHAIRMAN: Z. BERN

Scientific Secretaries: J.J. Carrasco, H. Johansson

DISCUSSION

- Nicotri:

Is the unitarity method a redefinition of quantum field theory (QFT) or a new way of evaluating diagrams? Are you redefining QFT or is it just a rearrangement? There was a point where you stated that the only relevant vertex interaction term in gravity and YM theory is the three vertex. How can the effect of the four vertex rule be reproduced with only a three vertex?

- Bern:

This is just a reorganization of the QFT calculation. If we ever got an answer that was not in agreement with a correct Feynman diagram calculation of the same quantity, it would be wrong. The claim is that we get the same answer as Feynman diagrams, but you carry around only the physical quantities during intermediate steps of the calculation.

Regarding the four vertex, let me turn the question around and ask "Why does a four-vertex contribution show up as a Feynman rule in the off-shell formulation?" The answer is because we require off-shell gauge-invariance, and in the off-shell formalism if you just had the three vertex with no four vertex, your calculations would be off by a little piece.

The extra higher-point vertices in gravity, their role in life is to enforce general coordinate invariance off shell when you've introduced unnecessary degrees of freedom. When you construct the correct on-shell formalism with no unphysical degrees of freedom, they are simply not necessary.

- Nicotri:

You claimed that the eventual finiteness of N=8 supergravity would indicate the presence of another symmetry different from supersymmetry. Is it possible that such a symmetry could "save" also General Relativity, producing a finite Quantum General Relativity?

- Bern:

We know for a fact, from the calculation of Goroff and Sagnotti, that pure gravity diverges at two loops. That being said, we believe there exists a shared symmetry between pure gravity and N=8 supergravity that is responsible for most of the cancellations we see in N=8. It would be the symmetries of pure gravity plus the additional symmetry due to supersymmetry which would be where N=8's finiteness would come from.

- Hoen:

You mentioned the conjecture that N=8 supergravity is now being conjectured to be the simplest QFT. In what sense the simplest or simpler than others?

- Bern:

One way that Arkani-Hamed and collaborators mean simplest is: N=8 supergravity may be completely determined directly by the behaviour of three-point vertices even at loop level. This would make the S matrix very easy to determine.You might have grown up believing that phi^3 theory is the simplest theory. But in fact this is not true. If you look at its S matrix, it basically has no structure: it is just a collection of poles with no special features. If you look at N=8 supergravity, there are many special features. For example: Tree amplitudes have delta-function support on non-trivial curves in twistor space.Twistor space is obtained by taking a particular Fourier transform of spinors.They also have arguments for simplicity of N=8 supergravity based on holography and the simple structure of its moduli space.

- *Hoen:*

From your calculations you drew the conclusion that the key cancellations are generic to gravity. In what sense is this true? As far as I understood you drew this conclusion from supergravity, which is only one way of treating gravity.

- *Bern:*

The analysis we did was actually pure gravity, later Arkani-Hamed and collaborators did the same sort of analysis in the supersymmetry case. What you discover at one loop, with n-legs, is that the power counting that you thought would grow as $l^{(2n)}$ - where l is a loop momentum - really behaves as $l^{(n+4)}$. Sitting on top, you can have supersymmetry which helps by some fixed amount to reduce the power count.

- *Shakirov:*

Let us imagine for a moment that you have proved the UV finiteness of N=8 supergravity. Is there any way to relate it to ordinary non-supersymmetric physical gravity, by means of some supersymmetry breaking?

- *Bern:*

I believe the answer is yes, but obviously we need to deal with one problem at a time. Assuming finiteness, we'll need to understand in detail what is responsible for finiteness in N=8 before we could answer anything like that. Based upon the cancellations we see in pure gravity, we suspect that if N=8 supergravity is finite, N=5, and N=6 could also be finite. (N=7 is trivially N=8). These are technically much more challenging to explore, so it makes sense to address N=8 first.

- *Schwarz:*

In a paper last year with Green and Ooguri (ArXiv:0704.0777 [hep-th]), we studied a limit of string theory that seems to give N=8 supergravity. We discovered that non-perturbative states become massless and build back a dual torus. We concluded that the non-perturbative completion of N=8 supergravity is string theory. So it does not give a new quantum theory of gravity, rather leads back to string theory non-perturbatively.

- *Bern:*

In principle, that might be possible. At the moment, our analysis is only perturbative.

- *Nath:*

If N=8 supergravity turns out to be a finite theory, it would seem to provide an example of a quantum theory of gravity which is not a string theory.

- *Schwarz:*

We claim that N=8 supergravity non-perturbatively leads you back to string theory.

- *Pacholek:*

Is the unitary method with the on shell formalism restricted only to supergravity or can it be used in any other QFT?

- *Bern:*

It can be used in any QFT. In fact, in QCD at one loop this is the way state of the art one-loop calculations are being performed today with the goal to provide the necessary background calculations for LHC observations. Ordinary Feynman diagrams run into technical difficulties with the needed computer power and with numerical stability. The on-shell methods have been demonstrated to avoid these difficulties.

- Arnowitt:

This is a question regarding theories which are good in the ultra-violet. For theories like this you might have problems in the infrared due to massless propagators in the denominator. What is the infrared status of the theory?

- Bern:

You are of course right that the S matrix for gravity contains infrared singularities, as was discussed by Weinberg a long time ago. We understand infrared singularities quite well from our experience in QCD. Any scattering amplitude will have infrared singularities, but we understand what their sources are and how to cancel them, just as we do in QCD and QED by combining real emission and virtual contributions. It's more complicated, of course, but the same basic ideas hold.

- Arnowitt:

Does that relation giving the gravity amplitude as a sum over products of two Yang-Mills amplitudes hold to arbitrary loop order or just to the fourth loop?

- Bern:

The Kawai-Lewellen-Tye relationship between gravity and Yang-Mills amplitudes holds only at tree level. There is no relationship like that even at one loop. However the key trick is that the unitarity method for obtaining loop results relies only on tree-level information as input. So we can use this tree-level relationship to construct all loop level amplitudes to any order of perturbation theory.

- Antoniadis:

This is more of a comment. It seems to me that you have no real understanding of any UV finiteness of N=8 supergravity but just based on technical observations.

- Bern:

Your point is well taken: this is all driven by technical observations. We do not yet have an explanation that we would consider satisfactory. On the other hand, the scaling behaviour of tree level amplitudes in certain complex directions is undoubtedly related to why the theory would be UV finite. So I think we've got a handle to grab, which is to first understand these tree-level properties in a deeper way.

- Antoniadis:

The comparison with N=4 is not very instructive because N=4 Yang-Mills is already a renormalizable QFT and finiteness is only related to the vanishing of the beta-function.

- Bern:

Yes, supergravity is much more complicated in terms of its ultraviolet behaviour. At each loop level you're fighting ever increasing UV trouble. This is why for 25 years essentially no one believed that it could be finite. If it is finite, you need a symmetry that gets ever more powerful as the number of loops increases, completely unlike N=4 Yang-Mills theory.

- Slobodenick:

In your theory we can calculate diagrams with different topologies. Can we calculate these topologies analytically or must we use the computer?

- *Bern:*

Even when we use a computer we are doing things analytically. The modern way is to use *Mathematica* like a pen and paper. Certainly it's quicker to check our ansatze numerically against unitarity cuts, but all the construction is done essentially analytically. When you're faced with 50 diagrams having complex topologies, certainly using computers is crucial to reduce the headaches associated with bookkeeping.

- *Dubynskiy:*

You said that the properties of N=8 supergravity could be analyzed by means of N=4 super-Yan-Mills theory. Also you said that supersymmetry cannot protect supergravity from divergences in all orders and the mechanism of cancellation in supergravity is yet to be understood. Is the role of supersymmetry absolutely crucial for the finiteness of N=4 SYM and is there any hope that the finiteness in N=4 super-Yang-Mills and N=8 supergravity have similar origin.

- *Bern:*

If you are trying to build a finite theory of anything then by far the easiest place to look is at supersymmetry, because supersymmetry automatically brings the power counting down. Adding supersymmetry is the best place to start. Supersymmetry is not the whole story but it helps you to get within range of finiteness. The other reason why we started with N=8 supergravity has to do with technical simplicity. The more symmetry you have the simpler the theory. You noticed in John Joseph Carrasco's talk he could fit a four-loop super-Yang-Mills amplitude on a few transparencies. If you try this in a theory with less supersymmetry then it is hopeless, you have to bring in a billion of pages or so. In the future we can think about relaxing the amount of supersymmetry, but it is kind of hard to think of the details of that until we have a complete and firm understanding of finiteness in N=8 supergravity. Regarding the question of whether supersymmetry is necessary for finiteness of super-Yang-Mills theory, there are plenty of finite models with less supersymmetry than the maximal N=4 theory, but I don't remember off-hand if there are well known examples of finite gauge theories with no supersymmetry.

- *'t Hooft:*

What your work seems to indicate is that at every order, the effects of very high energy intermediate states are strongly suppressed, so that no infinities occur. But of course, at high energies, perturbation expansion still diverges badly: we still do not know what happens at the Planck scale. Did you make attempts to understand this better? What is the physical picture at the Planck length, and what is the physical reason as to why the high wave numbers are suppressed?

- *Bern:*

Unfortunately, the basic problem is that in order to answer that question you need to have all orders in perturbation theory under control. We can go to high loop orders, but not infinite loop orders. Without a resummation it's hard to make any kind of definite statement. There might be one helpful thing. I could talk about the way in which we understand how the cancellations work at one loop, but this is very technical. But I shouldn't be standing here talking about integrals and residues. I should be talking about some physical symmetry, but I'm not because we don't know the symmetry. There might be a hint in this recent paper by Arkani-Hamed and Kaplan. They point out that there is an enhanced Lorentz symmetry when you scale the momenta of these tree amplitudes in large complex directions. But no one has gone from that to something we could really put our hands on physically.

- *'t Hooft:*

There's something else. In a sense, having an integral over an anticommuting variable is like a differentiation. So say you have an N-dimension supersymmetric theory. Then N goes down if you have anticommuting variables to integrate over. So if N=8 and you have N generators, you might combine them into rotational singlets, if you map into rotational isospin variables. And then you could show N=8 is equivalent to a 0-spin theory which of course is trivial to integrate. Something of this sort that could be the symmetry you're looking for, but of course this is specific to supersymmetry.

- *Bern:*

Yes, that is very interesting, but at the moment we don't believe it is supersymmetry that is responsible for the key cancellations. I hope in a few years we will have a better answer regarding the physical origin. It is clearly the number one question we are after. At the moment, we have a collection of technical observations.

- *Budd:*

Although we were not supposed to ask about non-perturbative finiteness, I will ask something that may be related. Do these calculations depend on the background geometry? Are there any ways to generalize the calculations to more general backgrounds? Or is this irrelevant to the UV finiteness?

- *Bern:*

It should be irrelevant to the calculation of ultraviolet divergences. Without question, from the technical point of view the best place to do the calculation is in flat-space. A reason is that only in flat space do we know the Kawai-Lewellen-Tye relations. In fact, I consider it to be another important problem, how we can generalize those Kawai-Lewellen-Tye relations away from flat space. But the first problem in trying to do this, speaking of things we don't understand, is that no one properly understands the Kawai-Lewellen-Tye relations even in flat space where we know they work. Getting back to your question, in the ultraviolet limit it shouldn't really matter what kind of background you have, so I don't think that that has any effect on the ultraviolet properties.

- *Budd:*

So these cancellations hold in the UV regime only?

- *Bern:*

Yes. It has to do with the power counting as the loop momenta becomes very large, i.e. at very very short distances, so any kind of geometry at longer distances is not going to matter.

- *Stavenga:*

I am still a little bit mystified by the unitarity method. The physical reasoning is that you use only physical quantities. What I know from ordinary Feynman perturbation theory is that you have two ways. You have the time ordered formalism where you use physical quantities but where energy conservation is not conserved at every vertex, or you do it in the relativistic formalism, where mass shell condition is not enforced. But you do both.

- *Bern:*

Those are excellent points! The points you make reflect the Heisenberg uncertainty principle which has precisely to do with energy violation in loop calculations. Obviously we need to be compatible with the uncertainty principle. You have to violate energy conservation or for a short amount of time in order, or equivalently violate the relationship between energy and momentum, to get the correct answer. That is part of the explanation why you have to create those unphysical particles in the intermediate states. The more precise statement of the unitarity method is that we can fully reconstruct the analytic function that is generated by the Feynman diagram calculation, including the off-shellness you were talking about, using only on-shell quantities in all intermediate steps of the calculation. Only at the very last line of the calculation we put back integrals with their off-shell Feynman propagators. So we do in fact violate on-shellness in the final answer, but to arrive at the final answer we do not need to violate it in the intermediate steps.

- *Stavenga:*

Another question related to this. The on-shellness is a property of Minkowski space, but normally you want to do a Wick rotation, which takes you to Euclidian space. And in Euclidian space there is no such on-shell property. All fluctuations are basically off-shell, so is there a way to do Wick rotation here in a consistent way that preserves the on-shell properties?

- *Bern:*

Yes. What happens when you put the on-shell conditions on the intermediate states is that you are getting rid of the Feynman propagator. And when you assemble the final answer at the end of the calculation you put back the propagator, and then only at that point you do your Wick rotations. The procedure for constructing the amplitude does not interfere with the Wick rotation. It is only necessary when you are actually integrating, when you are explicitly evaluating the loop integrals in terms of polylogarithms. At that point is when you do the Wick rotation, but beforehand you do not need it, so there is no interference whatsoever. The basic claim is that we reconstruct exactly the integrals, exactly as you would have out of the Feynman diagram calculation. The key difference is that there are no intense cancellations between terms, having to do with non-gauge invariance of Feynman diagrams.

Status of Supersymmetry : Foundations and Applications

Sergio Ferrara[◇♣] **and Alessio Marrani**[♡♣]

◇ *Physics Department, Theory Unit, CERN, CH 1211, Geneva 23, Switzerland*
sergio.ferrara@cern.ch

♣ *INFN - Laboratori Nazionali di Frascati, Via Enrico Fermi 40,00044 Frascati, Italy*
marrani@lnf.infn.it

♡ *Stanford Institute for Theoretical Physics, 382 Via Pueblo Mall, Varian Lab, Stanford University, Stanford CA, 94305-4060, USA*

Abstract

We review some aspects of the foundations of supersymmetry as a conjectural invariance of the Laws of Nature.

This symmetry bypasses the Coleman-Mandula Theorem by enlarging the fundamental space-time symmetries to the Superworld.

Application of supersymmetry to particle physics requires its spontaneous breaking, as it happens for the electroweak symmetry of the Standard Model.

1 Introduction

The 46th Course of the prestigious *Erice International School of Subnuclear Physics* (*ISSP*) could have been an historical one, because the *Superworld* could have been at the eve of its discovery.

Indeed, on September 10 2008 the first proton beams were successfully circulating in the *CERN Large Hadron Collider* (*LHC*) ring, and the *LHC* could have become fully working in a few months, possibly providing scientific community with the first hints for *New Physics*.

Unfortunately and unexpectedly, an accident happened on September 19, due to a faulty electrical connection in a region between two of the accelerator magnets, which resulted in mechanical damage and release of helium from the magnet cold mass into the *LHC* tunnel itself. This incident will result in a postponement of at least one year of the first physics results from the *LHC* experiments.

Meanwhile, a great theoretical physicist passed away, Professor Sidney Coleman from Harvard University, and the 46th Course of *ISSP* has been dedicated to his memory.

Interestingly enough, Coleman's name and ideas are closely related to the *Superworld*, in view of the fact that *Supersymmetry* is a symmetry of relativistic field theories and of the S-matrix which evades the famous *Coleman-Mandula Theorem* [1]. Such a Theorem refers to the impossibility (under some mild assumptions) of combining non-trivially internal and space-time symmetries. After the advent of *Supersymmetry*, Haag, Lopuszansky and Sohnius generalized *Coleman-Mandula Theorem* to the case of *super-Lie algebras* (named pseudo Lie algebras by the Authors themselves). Such a result is usually named *Haag-Lopuszansky-Sohnius Theorem* [2].

Nowadays, we know that both such Theorems can be circumvented by theories which also include extended objects (rather than only pointlike particles). Indeed, superPoncaré algebras can naturally contain *brane* charges of different kinds. *Branes* democratically appear together with point particles in candidate theories of quantum gravity, such as *superstring theories* or *M-theory* (for recent developments in such theories, the reader is addressed *e.g.* to Prof. J. Schwarz's Contribution to the present Proceedings volume [3], and Refs. therein).

Remarkably, this year is also marked by the 30th Anniversary of the first lectures at 16th Course of Erice *ISSP* (1978) [4].For further references, see also [5, 6].

This contribution is organized as follows.

In Sect. 2 some links between possible *New Physics* and the *Superworld* are reviewed, treating experimental (Subsect. 2.1) as well theoretical (Subsect. 2.2) evidences. Then, in Sects. 3 and 4 the discovery of *Supersymmetry* in particle physics and some results on *Supermultiplets* and *Superfields* in *Superspace* are treated, respectively. Sect. 5 deals with the discovery of the supersymmetric version of *General Relativity*, called *Supergravity*. The supersymmetric version of the *Higgs mechanism*, named *super-Higgs effect* is considered in Sect. 6. Then, Sect. 7 is devoted to a brief report of some applications of *Supersymmetry* to the real world, namely to the *(Minimally) Supersymmetric Standard Model* (*MSSM*) and to supesymmetry breaking effects. Sect. 8 contains an outlook, with some final comments and remarks.

2 Will New Physics be the *SuperWorld*? Theoretical and Experimental Motivations for Supersymmetry

Since *superpartners* of the fundamental constituents of matter (*photinos* for photons, *gauginos* - or *gluinos* - for gluons, *W-inos* and *Z-ino* for vector bosons, *squarks* for quarks, *sleptons* for leptons,...,and *Higgsinos* for Higgses) have *yet* to be discovered, all experimental signals are *indirect* evidences.

One must fairly say that such indirect evidences can have, in all cases, an alternative explanation.

However, let us take an optimistic, yet biased, point of view, and let us list some of them below.

2.1 Experimental Evidence

1. **Dark Matter Problem**. Almost 90% of the missing mass of the Universe is *not* conventional matter. The *Superworld* offers a natural candidate for *dark matter*: a stable (the lightest) superparticle, called *neutralino* (for some reviews, see *e.g.* [7], as well as Prof. Kolb's Lectures at ISSP07 [8]).

2. **Gauge Coupling Unification**. The precision measurements of the $SU(3)$, $SU(2)$ and $U(1)$ gauge couplings at LEP and other Laboratories all around the world indicate a better convergence towards an *Unification Energy* of order $\mathcal{O}(10^{16}$ Gev), if in their running the *superpartners* of ordinary particles (*e.g.* in the framework of *Minimal Supersymmetric Standard Model*) are included, with masses of order $\mathcal{O}(10^2 - 10^3$ Gev). In this respect, it is important the critical analysis performed by A. Zichichi *et al.* in [9].

3. **Fundamental (Light) Scalar Particle (Higgs)**. [10] In a renormalizable quantum field theory description of the *Standard Model*, a fundamental scalar is really justified only in the framework of *supersymmetry*. In fact, normally the classical description of *spontaneous symmetry breaking* (and subsequent *Higgs mechanism*) of a gauge theory receives severe quantum corrections, which make unnatural to understand the *hierarchy* of scales given by the ratio of Fermi energy E_F and Planck mass M_P ($\frac{E_F}{M_P} \sim \mathcal{O}(10^{-16})$). Instead, this *hierarchy* is explained in the *Superworld*, if the effective supersymmetry breaking scale M_{susy} satisfies

$$M_{susy} \lesssim \mathcal{O}(1 \text{ TeV}). \tag{2.1}$$

In other words, *supersymmetry* is the only symmetry which predicts the existence of fundamental scalars in Nature, and which further fixes the *"quartic"* coupling of scalar fields (charged under a gauge symmetry) in terms of the electroweak (*ew*) gauge coupling [11, 12]:

$$\lambda\phi^4 \overset{susy}{\longrightarrow} \lambda \sim g_{ew}^2. \tag{2.2}$$

Some other indirect possible experimental evidences are:

4. **Large Top Quark Mass**, needed for radiative electroweak-supersymmetry breaking [13].

5. **Suppression of the Proton Decay Channel** [14, 15]

$$p \longrightarrow e^{+}\pi^{0}. \tag{2.3}$$

6. **Natural Candidate for *"Scalar"* Inflaton Field** (see *e.g.* [16], and Refs. therein).

2.2 Theoretical Evidence

A crucial ingredient of the *Superworld* is the enlargement of the space-time (*bosonic, commuting*) dimensions with *"fermionic"* (*anticommuting*) dimensions.

In a d-dimensional space-time (with coordinates x^{μ} and momentum operator P_{μ}, $\mu = 1, ..., d$), some spinor *supercharge operators* Q_{α} are introduces, with the spinor index α ranging $1, ..., 2^{d/2}$ (or $2^{d/2-1}$ in the case of chiral Majorana) for d even and $1, ..., 2^{(d-1)/2}$ for d odd. As P_{μ} is associated to bosonic (commuting) coordinates x^{μ}, so the *supercharges* Q_{α} are associated to *Grassmann* (anticommuting) coordinates ϑ^{α}. The resulting space, parameterized by the set of coordinates (x^{μ}, ϑ^{α}) is named *Superspace* [12, 17], and it is the *"ambient space"* of the *Superworld.*

Spinor algebras in d dimensions have been studied in [18]. As mentioned above, for d even the cardinality of the range of the spinor index α (or equivalently the dimension of a spinor *supercharge*) is $2^{d/2}$ and $2^{d/2-1}$ for Majorana and chiral Majorana spinors, respectively. In particular, for $d = 2$ (*modulo* 8) the spinor is chiral Majorana, for $d = 6$ (*modulo* 8) the spinor is chiral pseudo-Majorana, and for $d = 4$ (*modulo* 8) the spinor can be regarded either as Majorana (non-chiral) or as complex chiral (such that its complex conjugation yields the antichiral spinor). On the other hand, for d odd the dimension of a spinor is $2^{(d-1)/2}$. In particular, for $d = 1, 3$ (*modulo* 8) the spinor is Majorana, whereas for $d = 5, 7$ (*modulo* 8) the spinor is pseudo-Majorana[1].

Concerning the *automorphism symmetry* of the supersymmetry algebra, called $\mathcal{R}$-*symmetry*, here we limit ourselves to mention that it is also d-dependent. For $d = 2, 3, 9, 10, 11$ the $\mathcal{R}$-*symmetry* group is $SO(\mathcal{N})$, for $d = 5, 6, 7$ it is $USp(2\mathcal{N})$, and for $d = 4, 8$ it is $U(\mathcal{N})$ [19]. In general the maximum number of dimensions in which a supersymmetry algebra admits an interacting field theory realization is $d = 11$ [20]. The $d = 11$ theory is $\mathcal{N} = 1$ supergravity [21], which in modern terms is supposed to be the low-energy limit of M-*theory* (see *e.g.* [22], and Refs. therein). In particular, the maximum number of independent supersymmetry spinor charge components is 32 [20, 23], which, in light of previous statements on the dimension of *supercharges*, in $d = 4$ corresponds to $\mathcal{N} = 8$ independent *supercharges.*

[1]Pseudo-Majorana and complex spinors have twice the (real) dimension of a real spinor.

The enlargement of space-time to *Superspace* is the crucial step in order to build the supersymmetric theories of fundamental interactions: *General Relativity* in *Superspace* originates *Supergravity* [24]; *Yang-Mills* gauge theory in *Superspace* originates the *supersymmetric Yang-Mills* gauge theory [11, 12], on which the *Minimally Supersymmetric Standard Model* is built; finally, *String Theory* in *"ambient Superspace"* originates *Superstring Theory* [25, 26]. Actually, superstrings are twice supersymmetric, because they also have a super-worldsheet (*i.e.* a super-Riemann surface) [25, 26].

Let us now ask the following question: *does the Superworld yield finite theories of gravity? Or, in other words, do quantum gravity solutions exist in the Superworld*

The answer is yes, and there are various examples, some of them listed below:

1. By Maldacena's *AdS/CFT correspondence*, *finite* $\mathcal{N} = 4$, $d = 4$ Yang-Mills gauge theory is *dual* to a quantum theory of gravity, namely to Type IIB on $AdS_5 \times S^5$ [27, 28, 29].

2. $\mathcal{N} = 8$, $d = 4$ supergravity shares many properties of $\mathcal{N} = 4$, $d = 4$ Yang-Mills gauge theory, and it may be *ultraviolet-finite* (see e.g. [30] and [31], as well as the Prof. Z. Bern's Contribution to the present Proceedings volume).

3. The ultraviolet completion of supergravity theories is provided by *superstrings* in $d = 10$ [25] and *M-theory* [21, 32] in $d = 11$.

Actually, approaches to quantum gravity alternative to *supergravity* and *superstring/M-theory* do exist, see *e.g. Loop Quantum Gravity* [33].

However, the main virtue of *superstring* theory is that it provides a framework in which computations can be performed, and stringy corrections to the low-energy effective *supergravity* theory can be calculated. In general, the number of massless states predicted by *supersymmetry* in any compactification of $d = 10$ *superstring* theory or $d = 11$ M-theory down to $d = 4$ space-time dimensions depends on the topological properties of the *internal* manifold $\mathcal{M}_6$ or $\mathcal{M}_7$.

3 The Discovery of Space-Time Rigid Supersymmetry

We here give an introductory review to the results of [34, 35, 36], to whom the reader is addressed for further elucidation.

Let us start by recalling that the basic algebraic structure underlying the notion of *supersymmetry* is a *graded* Lie algebra (GLA). GLAs are extensions of ordinary Lie algebras (LAs), in which a *grading* is introduced, namely a distinction among *even* and *odd* elements. Even elements belong to an ordinary LA, and they obey commutation relations; odd elements, responsible for the grading, obey anti-commutation relations among themselves and commutations relations with the even elements. These latter relations indeed specify that the odd elements are a representation of the LA, namely the *grading representation.*

By denoting the set of elements of a D-dimensional LA by A_m ($m = 1, ..., D$) and the set of elements of an its d-dim. grading representation by Q_α ($\alpha = 1, ..., d$), the following set of commutation rules define the resulting $(D + d)$-dim. GLA:

$$[A_m, A_n] = f_{mn}^{\ \ l} A_l, \quad LA; \tag{3.1}$$

$$[A_m, Q_\alpha] = S_{m\alpha}^{\ \ \beta} Q_\beta, \quad grading\ \ repr; \tag{3.2}$$

$$\{Q_\alpha, Q_\beta\} = F_{\alpha\beta}^{\ \ m} A_m, \quad odd\ \ sector, \tag{3.3}$$

where $f_{mn}^{\ \ l}$ are the structure constants of the LA, and $[\cdot,\cdot]$ and $\{\cdot,\cdot\}$ denote the usual commutators and anticommutators throughout. Notice that $f_{mn}^{\ \ l} = f_{[mn]}^{\ \ \ l}$, $S_{m\alpha}^{\ \ \beta} = S_{[m\alpha]}^{\ \ \ \beta}$ but $F_{\alpha\beta}^{\ \ m} = F_{(\alpha\beta)}^{\ \ \ m}$, where round and square brackets denote symmetrization and skew-symmetrization with respect to enclosed indices. Usually, Q_α are named *supercharges.*

The set of commutation rules (3.1)-(3.3) can be written compactly by introducing a *graded commutator* as follows:

$$[X_A, X_B] = X_A X_B - (-1)^{AB} X_B X_A = f_{AB}^{\ \ C} X_C, \tag{3.4}$$

where $A, B = 0, 1$ for even and odd elements respectively ($X_0 \equiv \{A_m\}$, $X_1 \equiv \{Q_\alpha\}$), and

$$f_{AB}^{\ \ C} \equiv \{f_{mn}^{\ \ l}, S_{m\alpha}^{\ \ \beta}, F_{\alpha\beta}^{\ \ m}\} \tag{3.5}$$

is the corresponding set of *graded structure constants,* which must satisfy certain relations as a consequence of the Jacobi identities. The symmetry in lower indices of $F_{\alpha\beta}^{\ \ m}$ appearing in the left-hand side of Eq. (3.3) yields the additional information (absent in ordinary LAs) that (no sum on repeated indices here)

$$Q_\alpha = \frac{1}{2} F_{\alpha\alpha}^{\ \ m} A_m. \tag{3.6}$$

Eq. (3.6) plays a crucial role in supersymmetric theories, in connection with the positivity of the energy and with spontaneous symmetry breaking.

The super-Poincaré (or *graded* Poincaré) algebra is a graded extension of the Poincaré LA (whose generators are the space-time displacements P_μ and the Lorentz transformations $M_{\mu\nu}$). The *grading* representation and the *odd* sector of $\mathcal{N} = 1$ super-Poincaré GLA[2] read as follows:

$$[P_\mu, Q_\alpha] = 0, \quad [M^{\mu\nu}, Q_\alpha] = -i\sigma^{\mu\nu}_{\alpha\beta} Q_\beta, \quad grading\ \ repr; \tag{3.7}$$

$$\{Q_\alpha, Q_\beta\} = -2\left(\gamma^\mu C^{-1}\right)_{\alpha\beta} P_\mu, \quad odd\ \ sector, \tag{3.8}$$

[2]Here and throughout below, $\mathcal{N}$ denotes the number of spinorial *supercharges* Q_α.

in which the gamma matrices γ^μ were introduced, $\sigma^{\mu\nu} \equiv \frac{1}{4}[\gamma^\mu, \gamma^\nu]$, and the Lorentzian metric $g_{\mu\nu} = (-,+,+,+)$ is used to raise and lower space-time indices (as well as to define the Lorentzian scalar product $\cdot$ introduced further below). $\mathcal{C}$ denotes the charge conjugation matrix, which in the Majorana representation is nothing but γ^0.

The GLA (3.7)-(3.8) was considered in [34] and [35]. In [36] Wess and Zumino studied a larger GLA, which actually is the one arising from string theory, being nothing but a graded extension of the conformal algebra of space-time.

It is here worth recalling that the symmetries generated by a traceless symmetric divergenceless energy-momentum tensor $\theta_{\mu\nu}(\chi)$ in d-dim. space-time (with coordinates x^μ) correspond to the following conserved vector current (see *e.g.* [37]):

$$J_\mu^\xi \equiv \theta_{\mu\nu}\xi^\nu, \tag{3.9}$$

where ξ^μ is the parameter of the infinitesimal transformation. J_μ^ξ is conserved provided the traceless part of $\partial_\mu \xi_\nu$ does vanish, implying that

$$2\partial_{(\mu}\xi_{\nu)} - \frac{2}{d} g_{\mu\nu}\partial \cdot \xi = 0. \tag{3.10}$$

The solution to the differential Eq. (3.10) defines the parameter $\xi_\mu(x)$ of a conformal transformation in a d-dim. Lorentzian space-time. For $d \neq 2$, $\xi_\mu(x)$ has a finite number of parameters, being quadratic in x^μ:

$$\xi_\mu(x) = c_\mu + \omega_{\mu\nu}x^\nu + \varepsilon x_\mu + a_\mu x^2 - 2(a \cdot x)x_\mu, \qquad \omega_{\mu\nu} = \omega_{[\mu\nu]}. \tag{3.11}$$

On the other hand, for $d = 2$ Eq. (3.10) reduces to the Klein-Gordon equation, and it gives an infinite-dimensional LA.

As mentioned above, world-sheet symmetries of string theory correspond to a grading of the infinite-dimensional conformal algebra in $d = 2$. Instead, the *graded* extension of the conformal algebra of $d = 4$ Lorentzian space-time introduces an $\mathcal{N} = 1$ *grading* with generators Q_α and S_α, and it reads as follows [36] :

$$\begin{cases} [Q_\alpha, P_\mu] = 0, \quad [Q_\alpha, M_{\mu\nu}] = i(\sigma_{\mu\nu}Q)_\alpha, \\ [Q_\alpha, K_\mu] = -i(\gamma_\mu S)_\alpha, \quad [Q_\alpha, D] = \frac{i}{2}Q_\alpha, \\ [S_\alpha, P_\mu] = i(\gamma_\mu Q)_\alpha, \quad [S_\alpha, M_{\mu\nu}] = i(\sigma_{\mu\nu}S)_\alpha, \qquad \textit{grading repr;} \\ [S_\alpha, K_\mu] = 0, \quad [S_\alpha, D] = -\frac{i}{2}S_\alpha, \\ [Q_\alpha, \Pi] = -\frac{3}{4}i(\gamma_5 Q)_\alpha, \quad [S_\alpha, \Pi] = \frac{3}{4}i(\gamma_5 S)_\alpha, \end{cases} \tag{3.12}$$

$$\begin{cases} \{Q_\alpha, Q_\beta\} = 2(\gamma^\mu\gamma^0)_{\alpha\beta}P_\mu, \quad \{S_\alpha, S_\beta\} = 2(\gamma^\mu\gamma^0)_{\alpha\beta}K_\mu, \\ \{Q_\alpha, S_\beta\} = -2(\sigma^{\mu\nu}\gamma^0 M_{\mu\nu} - \gamma^0 D + 2\gamma_5\gamma^0\Pi)_{\alpha\beta}, \end{cases} \qquad \textit{odd sector,} \tag{3.13}$$

where D and K_μ are the generators of dilatations and conformal boosts, respectively, and Π is the generator of a $U(1)$ chiral transformation.

By introducing anticommuting spinor constant parameters ε_α and η_α ($\{\varepsilon_\alpha, \varepsilon_\beta\} = 0$, $\{\eta_\alpha, \eta_\beta\} = 0$), the anticommutators (3.13) can be rewritten as commutators for the elements $\overline{\varepsilon}_\alpha Q_\alpha$ and $\overline{\eta}_\alpha S_\alpha$ (no sum on repeated indices).

The two constant spinors ε and η can be embedded in linearly x-dependent spinor

$$\zeta(x) \equiv \varepsilon + \not{x}\eta \tag{3.14}$$

which is the general solution to the differential Eq. (in $d = 4$):

$$\left(2\gamma_{(\mu}\partial_{\nu)} - \frac{1}{2}g_{\mu\nu}\gamma \cdot \partial\right)\zeta(x) = 0. \tag{3.15}$$

By ranging *e.g.* $\alpha = 1, 2$, this ensures that the bilinear object

$$\xi_\mu(x) \equiv 2i\overline{\zeta}_1(x)\gamma_\mu\zeta_2(x) \tag{3.16}$$

is a solution to Eq. (3.10).

The $\mathcal{N} = 1$, $d = 4$ superconformal algebra (3.12)-(3.13) implies that all conformally invariant field theories which are invariant under the GLA (3.7)-(3.8) are also invariant under the larger GLA (3.12)-(3.13).

A remarkable property of the $\mathcal{N} = 1$, $d = 4$ superconformal algebra (3.12)-(3.13) is the presence of the generator Π, related to the existence of a conserved $U(1)$ charge (named $\mathcal{R}$-symmetry).

At the level of *supercurrents*, while for super-Poincaré it holds

$$\partial^\mu J_{\mu\alpha} = 0, \tag{3.17}$$

after [38] in the superconformal case it can be shown that (recall Eq. (3.14))

$$\partial^\mu J_\mu^\zeta = 0, \quad J_\mu^\zeta \equiv \overline{\zeta}^\alpha(x) J_{\mu\alpha}. \tag{3.18}$$

In quantum field theory, the superconformal invariance (eventually present at the classical level) is spoiled by quantum anomalies. Exceptions are provided *e.g.* by $\mathcal{N} = 4$ and some $\mathcal{N} = 2$, $d = 4$ Yang-Mills theories, as well as $d = 2$ conformal field theories underlying string theory. In all such cases, the absence of conformal (and superconformal) anomalies is deeply related to the finiteness of the theories themselves.

4 Supermultiplets and Superfields

A major advance in supersymmetric theories, which allows one to construct manifestly invariant interactions and study the multiplet structure of such theories, is due to the work by Salam and Strathdee on *superfields* and unitary representations of supersymmetry on one-particle states [39, 40].

Further advances obtained in subsequent papers led to a simpler and deeper understanding of the classification of GLAs (see *e.g.* [41, 42, 18], and Refs. therein). The classification of supersymmetry Poincaré algebras yields, as first observed by Gell-Mann [43], that only a finite number of *extended* supersymmetry algebras (*i.e.* superalgebras with more than one *supercharge*) can be used to describe the fundamental interactions of massless point-like quanta.

Superfields are nothing but fields in an enlarged space, called *superspace*. To each space-time point x^μ one associates a Grassmann anti-commuting coordinate θ_α. As a consequence, a supersymmetry transformation is nothing but a part of the following group motion in *superspace*:

$$x'^\mu = x^\mu + i\bar{\varepsilon}\gamma^\mu\theta, \tag{4.1}$$

$$\theta'_\alpha = \theta_\alpha + \varepsilon_\alpha. \tag{4.2}$$

As $d = 4$ space-time can be regarded as the coset space

$$M = \frac{ISO\,(3,1)}{SO\,(3,1)}, \tag{4.3}$$

where $ISO\,(3,1)$ and $SO\,(3,1)$ respectively denote the inhomogeneous and homogeneous special Lorentz group in $d = 4$, so the $\mathcal{N} = 1$, $d = 4$ *superspace* is the supercoset space

$$\mathcal{M} = \frac{SuperISO\,(3,1)}{SO\,(3,1)}, \tag{4.4}$$

with $SuperISO\,(3,1)$ denoting the $\mathcal{N} = 1$ *super-extension* of $ISO\,(3,1)$. *Supefields* are therefore representations of $SuperISO\,(3,1)$ (also named super-Poincaré group) on *superspace*[3]. On this space, Q_α acts as a differential operator:

$$Q_\alpha = -\frac{\partial}{\partial\overline{\theta}^a} + i\,(\gamma^\mu\theta)_\alpha\,\partial_\mu, \tag{4.5}$$

and analogously the operators P_μ and $M_{\mu\nu}$ are realized as follows:

$$P_\mu = -i\partial_\mu; \tag{4.6}$$

$$M_{\mu\nu} = -i\left(2x_{[\mu}\partial_{\nu]} + \overline{\theta}\sigma_{\mu\nu}\frac{\partial}{\partial\theta}\right). \tag{4.7}$$

Also superconformal fields enjoy a full realization in *superspace*, and similar expression for the operators S_α, D, K_μ and Π cam be obtained. Although the superconformal algebra has not been used often in $d = 4$ space-time dimensions, it has been intensively studied in $d = 2$, due to its fundamental role in superstring theory.

---

[3]More general *superspaces* with $\mathcal{N} > 1$ can be considered. They were called *harmonic superspaces* by their inventors (see *e.g.* [44], and Refs. therein), and they are better suited to describe field theories with extended supersymmetries.

Another relevant issue is the construction of massless and massive representations of the supersymmetry algebra on one-particle states, carrying irreducible representations of the Poincaré group *à la* Wigner.

The multiplet structure of supersymmetric theories can be studied by exploiting Wigner's method of *induced representations* [40], in order to analyze the irreducible representations of the Poincaré group.

Considering one-particle states with a given fixed value p_μ of the momentum P_μ, the GLA (3.7)-(3.8) reduces to a Clifford algebra, because the momentum operator is replaced by its eigenvalue:

$$P_\mu |p\rangle = p_\mu |p\rangle . \tag{4.8}$$

The rest frame GLA (*stability* - or *"little"* - *superalgebra*) in this case reads:

$$\begin{aligned} \{Q_\alpha, Q_\beta\} &= 2\left(\not{p}\gamma^0\right)_{\alpha\beta}; & (4.9)\\ \{Q_\alpha, W_\mu\} &= \frac{i}{4}\left(\left[\gamma_\mu, \not{p}\right]\gamma_5\gamma^0 Q\right)_\alpha , & (4.10) \end{aligned}$$

where W_μ is the *Pauli-Lubanski-Bargmann vector operator*

$$W_\mu \equiv \frac{1}{4!}\epsilon_{\mu\nu\rho\sigma}P^\nu M^{\rho\sigma}. \tag{4.11}$$

In the rest frame for massive particles $p^\mu = (m, \underline{0})$, and the *"little"* superalgebra (4.9)-(4.10) reduces to a Clifford algebra of two creation and two annihilation operators carrying spin $S = \frac{1}{2}$.

By using Weyl spinors Q_α and $\overline{Q}_{\dot\alpha}$, the *induced representation* is constructed as follows:

$$|\Omega_p\rangle , \quad \overline{Q}_{\dot\alpha}|\Omega_p\rangle , \quad \overline{Q}^{\dot\alpha}\overline{Q}_{\dot\alpha}|\Omega_p\rangle , \tag{4.12}$$

where $|\Omega_p\rangle$ is defined as

$$P_\mu |\Omega_p\rangle = p_\mu |\Omega_p\rangle , \quad Q_\alpha |\Omega_p\rangle = 0. \tag{4.13}$$

If $W_i^2|\Omega_p\rangle = 0$, one gets a massive Wess-Zumino multiplet with spin content[4] $S = \left(0^+, 0^-, \frac{1}{2}\right)$ [36]. On the other hand, if $W_i^2|\Omega_p\rangle = m^2 s(s+1)|\Omega_p\rangle$, one obtain all higher s-spin representations, with dimension $4s(s+1)$ and spin content $S = \left(s^+, s^-, s+\frac{1}{2}, s-\frac{1}{2}\right)$, with the upperscripts having the same meaning as in the case $s = 0$ mentioned above.

For massless particles $W_\mu = \lambda P_\mu$, where λ is the *helicity* of the particle, and (in $d = 4$) $P_\mu = (E, 0, 0, E)$, E denoting the energy of the particle. In this case, the *"little"* superalgebra (4.9)-(4.10) reduces to

$$\{Q, Q^\dagger\} = 2E, \quad \{Q, Q\} = \{Q^\dagger, Q^\dagger\} = 0, \tag{4.14}$$

i.e. a Clifford algebra of one creation operator and one annihilation operator with helicity $\lambda = -\frac{1}{2}$. The particle states are $|\Omega_\lambda\rangle$ and $Q^\dagger|\Omega_\lambda\rangle$, as well as those of the PCT-conjugated

[4]The upperscripts for 0 denote the real and imaginary part of a complex scalar, whereas $\frac{1}{2}$ stands for the helicity of a Weyl spinor.

multiplet, *i.e.* $|\Omega_{-\lambda}\rangle$ and $Q^\dagger|\Omega_{-\lambda}\rangle$. Such four states describe two particles with helicity λ and $\lambda-\frac{1}{2}$, and their antiparticles with opposite helicity. It is worth noticing that in the case of massless representations the Clifford algebra has half of the elements of the massive case, because of the projection operator $\not{P}$ (with $P^2=0$) annihilates the remaining operators, which are indeed represented by null operators in the Hilbert space of one-particle states.

The construction outlined above can be performed for extended superalgebras, as well. In such a case, there are $\mathcal{N}$ spinorial supercharges $Q_\alpha^{\ i}$, whose anticommutators read

$$\{Q_\alpha^{\ i}, Q_\beta^{\ j},\} = 2\left(\gamma^\mu\gamma^0\right)_{\alpha\beta} P_\mu \delta^{ij}, \tag{4.15}$$

with the possible inclusion of *central charges* Z^{ij} on the right-hand side [2].

For brevity's sake, we will confine our discussion to massless representations of extended Poincaré superalgebras; for a thorough treatment of massive $\mathcal{N}$-extended representations, the reader is addressed *e.g.* to [45]. In the massless case the central charges Z^{ij} must necessarily vanish (see Sect. 8), and the $\mathcal{N}$-extension of the *"little"* superalgebra (4.14) reads as follows ($i,j=1,...,\mathcal{N}$):

$$\left\{Q^{\ i}, Q^\dagger_{j},\right\} = 2\delta^i_j E, \quad \{Q^{\ i}, Q^{j},\} = \left\{Q^\dagger_i, Q^\dagger_j,\right\} = 0, \tag{4.16}$$

i.e. a Clifford algebra of $\mathcal{N}$ creation operators and $\mathcal{N}$ annihilation operators. Its smallest non-trivial representation is $2^{\mathcal{N}}$-dimensional, and it is given by the $\mathcal{N}$-extension of the $\mathcal{N}=1$ case reported above: the $2^{\mathcal{N}}$ states are

$$|\Omega_\lambda\rangle, Q^\dagger_{i_1}|\Omega_\lambda\rangle, ..., Q^\dagger_{i_1}...Q^\dagger_{i_k}|\Omega_\lambda\rangle, ..., Q^\dagger_{i_1}...Q^\dagger_{i_{\mathcal{N}}}|\Omega_\lambda\rangle. \tag{4.17}$$

For a given $1\leqslant k\leqslant\mathcal{N}$, there are $\binom{\mathcal{N}}{k}$ states (sitting in the k-fold antisymmetric representation of $SU(\mathcal{N})$) with helicity $\lambda-\frac{k}{2}$, thus the last state on the right of Eq. (4.17) has helicity $\lambda-\frac{\mathcal{N}}{2}$. The PCT-conjugate states are obtained by *doubling* the representation (4.17), starting with a PCT-conjugate *Clifford vacuum* with helicity $\frac{\mathcal{N}}{2}-\lambda$.

In this way, one obtains a set of $2^{\mathcal{N}+1}$ particle and antiparticle massless states. The representation with the lowest helicity states contain particles whose helicity range from $\lambda=0$ to $\lambda=\frac{\mathcal{N}}{4}$ ($\frac{\mathcal{N}+1}{4}$ for $\mathcal{N}$ odd).

It is here worth remarking that for a Clifford vacuum with helicity $\lambda=\frac{\mathcal{N}}{4}$ (and $\mathcal{N}$ multiple of 4) the PCT *doubling* is not needed, and the resulting massless representation is PCT-*self-conjugate*. This phenomenon occurs in $\mathcal{N}=4$ and $\mathcal{N}=8$ theories, for Clifford vacua with helicity $\lambda=1$ and $\lambda=2$, respectively.

The very fact that the maximal helicity state in a given supermultiplet cannot be lower than $\frac{\mathcal{N}}{4}$ determines a bound for the possible supersymmetric theories describing massless quanta up to spin $S=2$. The maximal superalgebra containing at most a spin-2 massless state (to be identified with the graviton) is the $\mathcal{N}=8$ extended superalgebra. Furthermore, the maximal superalgebra with representations containing at most spin-1 states is the $\mathcal{N}=4$ superalgebra. Finally, the maximal superalgebra containing multiplets only with spin-0 and spin-$\frac{1}{2}$ states is the $\mathcal{N}=2$ superalgebra.

Thus, when gravitational interactions are absent, interacting supersymmetric theories exist up to $\mathcal{N}=4$. In such theories Yang-Mills gauge symmetries can be introduced, provided they commute with the spinor charges $Q_\alpha^{\ i}$. This fact implies that particles within the same supermultiplets transform in the same representation of the gauge internal symmetry.

The most important consequence of this assignment of internal quantum numbers is that chiral fermions can occur only in supermultiplets of the $\mathcal{N}=1$ supersymmetry algebra. Within such a framework, the chiral representation is provided by the Wess-Zumino multiplet (with spin content $S=\left(0^+,0^-,\frac{1}{2}\right)\right)$ [36] mentioned above, with the particles sitting in the complex representation of the internal symmetry group.

Also in the $\mathcal{N}=2$ case the Clifford vacuum can be in a complex representation of the internal symmetry group. However, the application of two supercharges to such a vacuum transforms it into a *mirror* particle (rather than in the antiparticle), with opposite helicity but in the *same* complex representation (rather than in the complex conjugate one) of the internal symmetry.

The treatment of the massive representation is rather different, and, as mentioned above, we will not dwell on it. Here we just limit ourselves to pointing out that, in absence of *central charges*, the smallest reprsentation of the algebra given by Eq. (4.15) contains $2^{2\mathcal{N}}$ particles with spin $S=0,\frac{1}{2},1,\frac{3}{2},...,\frac{\mathcal{N}}{2}$, whose degeneracies (indexed by k) lie in the k-fold antisymmetric (skew-traceless) representation of $USp\left(2\mathcal{N}\right)$ [45]. However, for $\mathcal{N}>1$ a *multiplet shortening* can occur in presence of *central charges* (see *e.g.* [45]). Indeed, massive multiplets with non-vanishing central charges can have particles with spins ranging from $S=0$ up to $S=\frac{\mathcal{N}}{4}$ in $\mathcal{N}=2$ and $\mathcal{N}=4$ supersymmetric theories. In general, such *shortened* massive multiplets with central extensions of the supersymmetry algebra (4.15) naturally occur in those extended supersymmetric Yang-Mills theories which exhibit a spontaneous gauge symmetry breaking [46].

5 The Discovery of Supergravity

This Section deals with the origin of *supergravity*, and it is mainly based on [24, 47, 48, 6], to whom the reader is addressed for further elucidation.

Supergravity theories are the gauge theories of the supersymmetry algebra. Extensive investigation of higher-spin fields led to the general belief that consistent interactions for massless particles with a finite number of states exist only up to spin 2 (for some reviews see *e.g.* [49, 50], and Refs. therein). Moreover, particles with spin larger than $\frac{1}{2}$ must be related to gauge symmetries in order for the associated *ghosts* to be decoupled from the S-matrix. For spin-1 and spin-2 bosons the related gauge symmetries re Yang-Mills and general coordinate invariance, respectively. For spin-$\frac{3}{2}$ fermions (also named *Rarita-Schwinger fields*) the related gauge symmetry is *local* supersymmetry. The corresponding massless spin-$\frac{3}{2}$ quantum has been named *"gravitino"* by Prof. Sidney Coleman.

Let us start by considering the free spin-$\frac{3}{2}$ massless wave equation

$$\epsilon^{\mu\nu\rho\sigma}\gamma_5\gamma_\nu\partial_\rho\psi_\sigma\left(x\right)=0. \tag{5.1}$$

As originally noticed by Rarita and Schwinger [51], it is gauge-invariant under the following infinitesimal transformation:

$$\delta\psi_\mu(x) = \partial_\mu\varepsilon(x), \tag{5.2}$$

where $\varepsilon(x)$ is an arbitrary spinor function of space-time. In presence of interactions, Eq. (5.1) gets modified as follows:

$$\epsilon^{\mu\nu\rho\sigma}\gamma_5\gamma_\nu\partial_\rho\psi_\sigma(x) = J^\mu(x), \tag{5.3}$$

with the constraint

$$\partial_\mu J^\mu(x) = 0. \tag{5.4}$$

The spin-$\frac{3}{2}$ field, as any other gauge field, must be coupled to a conserved spinor current. Thus, one reaches the conclusion that a massless spin-$\frac{3}{2}$ particle can be consistently coupled only in a supersymmetric theory.

Conversely, the local supersymmetry transformation (5.2) and the consistent coupling of a massless spin-$\frac{3}{2}$ field to a conserved spinor current can be derived by requiring local supersymmetry invariance of a supersymmetric Lagrangian density $\mathcal{L}$. Indeed, the variation of $\mathcal{L}$ under a local supersymmetry transformation reads

$$\delta_{\varepsilon(x)}\mathcal{L} = i\partial_\mu\overline{\varepsilon}(x)J^\mu(x) + \textit{total derivative}. \tag{5.5}$$

In Eq. (5.5) $J^\mu(x)$ is the (on-shell) Noether current of the matter system. This term can be compensated for by introducing a new coupling, namely

$$\mathcal{L}' \equiv -ik\overline{\psi}_\mu(x)J^\mu(x), \tag{5.6}$$

with $\psi_\mu(x)$ transforming as in (5.2). From dimensional arguments, it follows that the new interaction term (5.6) must have a coupling constant k of dimension $[m]^{-1}$ (see also Sect. 6).

In order to relate $\mathcal{L}'$ and k to gravity, it suffices to use the supersymmetric properties of the *supercurrent* J^μ. From current algebra considerations, supersymmetry algebra does require the supercurrent to transform into the energy momentum tensor $\theta_{\mu\nu}$:

$$\delta_{\varepsilon(x)}J_\mu = \theta_{\mu\nu}\gamma^\nu\varepsilon + ..., \tag{5.7}$$

where "..." involves other fields which do not contribute to the $\{Q,Q\}$ anticommutator. By recalling Eq. (5.6), Eq. (5.7) yields that

$$\delta_{\varepsilon(x)}\mathcal{L}' = ik\overline{\varepsilon}\gamma_\mu\psi_\nu\theta^{\mu\nu}. \tag{5.8}$$

Such a term can be canceled only by the gravitational coupling

$$\mathcal{L}_{grav} = -g_{\mu\nu}\theta^{\mu\nu}, \tag{5.9}$$

involving the metric tensor $g_{\mu\nu}$, with local supersymmetry infinitesimal variation given by

$$\delta_{\varepsilon(x)}g_{\mu\nu} = ik\overline{\varepsilon}\gamma_{(\mu}\psi_{\nu)}. \tag{5.10}$$

These simple considerations were the basic ingredients used in [24] in order to construct the first supergravity theory as a gauge theory of local supersymmetry. Eqs. (5.2) and (5.10) dictate the starting supersymmetry transformations of the (spin-2) graviton and its supersymmetric (spin-$\frac{3}{2}$) partner, the *gravitino* ψ:

$$\delta_{\varepsilon(x)} V_\mu^a = \frac{i}{2} k \bar{\varepsilon} \gamma^a \psi_\mu, \quad \delta_{\varepsilon(x)} \psi_\mu = \frac{1}{k} D_\mu \varepsilon, \tag{5.11}$$

where V_μ^a is the *Vielbein* field and D_μ denotes the covariant gravitational differential operator (containing the spin connection).

In addition, general gauge theory considerations require a *quartic* self-coupling term of the gravitino (the so-called *seagull* term), together with a correction (*quadratic* in the gravitino field) of the $\delta_{\varepsilon(x)}\psi_\mu$ transformation law itself.

Such two terms can also be obtained in a *first-order formulation* of the theory, in which the *spin connection* is considered as an independent field variable. In this approach, the gravitino field gives a torsion contribution to the space-time geometry, and the supergravity theory can be elegantly reformulated as an *Einstein-Cartan theory* of gravity coupled to spin-$\frac{3}{2}$ massless Rarita-Schwinger field(s) [47].

However, it should be remarked that the *first order approach* to supergravity is less convenient in presence of coupling to matter fields. In particular, when matter is present, local supersymmetry does require new couplings which are not predict by the *Einstein-Cartan* formulation underlying the *first order approach.*

6 Spontaneously Broken Supergravity: the Super-Higgs Effect

Mainly relying on the results of [52, 53, 54, 55, 56], this Section deals with the spontaneous breaking of local supersymmetry.

In supergravity theories, as in any other gauge theory, the spontaneous breakdown of supersymmetry is accompanied by the mass growth of the spin-$\frac{3}{2}$ Rarita-Schwinger *gravitino* field, which becomes massive, absorbing the $\pm\frac{1}{2}$ helicity states of the would-be massless *goldstino* fermion field. This is usually called *super-Higgs effect.*

The *super-Higgs effect* can be described on general grounds by generalizing the usual treatment performed in gauge field theory to the case of local supersymmetry. This approach has the advantage of being valid for any $\mathcal{N}$-extended supergravity theory with arbitrary coupling to matter.

Here we briefly consider the case of $\mathcal{N} = 1$, $d = 4$ supergravity coupled to an arbitrary matter system with scalar fields ϕ_a and spin-$\frac{1}{2}$ fields χ_i. Following [53], in order to extract relevant information concerning the *super-Higgs effect* it suffices to demand the invariance of the action *up to quadratic terms* in the fermion fields. In other words, one starts from the Lagrangian density

$$\frac{1}{\sqrt{-g}}\mathcal{L}^{(2)} = -i \left[\begin{array}{c} \frac{1}{2}\overline{\chi}_i \mathcal{Z}_{ij} \not{\partial} \chi_j - \frac{1}{2\sqrt{-g}}\epsilon^{\mu\nu\rho\sigma}\overline{\psi}_\mu \gamma_5 \gamma_\nu D_\rho \psi_\sigma + \\ + \mathcal{A}_i \overline{\chi}_i \gamma \cdot \psi + \mathcal{M} \overline{\psi}_\mu \sigma^{\mu\nu} \psi_\nu + \frac{1}{2}\overline{\chi}_i \mathcal{N}_{ij} \chi_j \end{array} \right] - \mathcal{V}(\phi), \tag{6.1}$$

where $\mathcal{Z}_{ij}$, $\mathcal{A}_i$, $\mathcal{M}$ and $\mathcal{N}_{ij}$, as well as the scalar potential $\mathcal{V}$, are scalar fields-dependent quantities, and $g < 0$ denotes the determinant of the $d = 4$ metric tensor. Then, the cancellation of terms linear in the fermion fields is demanded in the expression

$$\int d^4x \delta_{\varepsilon(x)} \mathcal{L}^{(2)} = 0, \tag{6.2}$$

where the following local supersymmetry (infinitesimal) variations are used:

$$\delta_{\varepsilon(x)}\chi_i = C_i(\phi)\,\varepsilon; \quad \delta_{\varepsilon(x)}\psi_\mu = -D_\mu\varepsilon + \mathcal{M}\gamma_\mu\varepsilon; \quad \delta_{\varepsilon(x)}\phi_a = i\overline{\chi}_i L_{ia}(\phi)\,\varepsilon. \tag{6.3}$$

By using Eqs. (6.1)-(6.3), the cancellation of the terms proportional to $\chi\delta\varepsilon$, $\psi\delta\varepsilon$, $\chi\varepsilon$ and $\psi\varepsilon$ in Eq. (6.2) yields the following Eqs.:

$$\mathcal{A}_i = \mathcal{Z}_{ij} C_j; \tag{6.4}$$

$$C_i \mathcal{A}_i - 3\mathcal{M}B = \frac{1}{2}\mathcal{V}; \tag{6.5}$$

$$\mathcal{N}_{ij} C_j + 4\mathcal{A}_i B + \frac{\partial \mathcal{V}}{\partial \phi_a} L_{ia} = 0, \tag{6.6}$$

where $B = \mathcal{M}/2$. Such Eqs. further yield the two basic Eqs.:

$$2C_i(\phi)\,\mathcal{Z}_{ij}(\phi)\,C_j(\phi) - 3\mathcal{M}^2(\phi) = \mathcal{V}(\phi); \tag{6.7}$$

$$\mathcal{N}_{ij}(\phi)\,C_j(\phi) + 2\mathcal{Z}_{ij}(\phi)\,C_j(\phi)\,\mathcal{M}(\phi) = -\frac{\partial \mathcal{V}(\phi)}{\partial \phi_a} L_{ia}(\phi). \tag{6.8}$$

The dependence on the scalars ϕ_a has been explicited, in order to stress that Eqs. (6.7)-(6.8) hold for arbitrary scalar field configurations.

Eq. (6.7) shows that the scalar potential is no longer positive-definite in local supersymmetry, owing to the gravitino contribution. Indeed, the origin of such a term can be traced back to the auxiliary fields S and P of the gravitational multiplet.

Unbroken supersymmetry corresponds to

$$\left\langle \delta_{\varepsilon(x)}\chi_i \right\rangle = 0 \Leftrightarrow C_i(\phi_{cr}) = 0, \tag{6.9}$$

where, through Eq. (6.8) and due to the invertibility of L_{ia}, the critical scalar configuration $\phi_{a,cr}$ is defined by

$$\left(\frac{\partial \mathcal{V}}{\partial \phi_a} L_{ia} \right)_{\phi=\phi_{cr}} = 0; \tag{6.10}$$

$$\Updownarrow$$

$$\frac{\partial \mathcal{V}}{\partial \phi_a}(\phi_{cr}) = 0, \ \forall a. \tag{6.11}$$

If $\mathcal{M}(\phi_{cr}) = 0$, a Minkowski vacuum (vanishing cosmological constant) is originated. If $\mathcal{M}(\phi_{cr}) \neq 0$ an anti-de Sitter vacuum is determined (with strictly negative cosmological constant). Indeed, these two situations correspond to the two possible global $\mathcal{N} = 1$, $d = 4$ supersymmetry algebras.

On the other hand, *broken supersymmetry* corresponds to

$$\langle \delta_{\varepsilon(x)} \chi_i \rangle \neq 0 \tag{6.12}$$

at the critical scalar configurations defined by Eq. (6.11). Since the scalar potential $\mathcal{V}$ is not positive-definite, in general it can take any value and any sign at such critical scalar values.

For physical applications, the most interesting case of broken supersymmetry is provided by the case with vanishing cosmological constant (Minkowski background), namely when Eq. (6.12) is fulfilled, along with

$$\phi_a = \phi_a^0 : \begin{cases} \mathcal{V}(\phi^0) = 0; \\ \frac{\partial \mathcal{V}}{\partial \phi_a}(\phi^0) = 0, \ \forall a; \end{cases} \overset{Eq.\ (6.7)}{\Longleftrightarrow} 3\mathcal{M}^2(\phi^0) = 2C_i(\phi^0)\, \mathcal{Z}_{ij}(\phi^0)\, C_j(\phi^0). \tag{6.13}$$

By using dimensionful notation, Eq. (6.13) becomes ($\mathcal{M}$ real positive)

$$\mathcal{M}(\phi^0) = \frac{1}{M_P}\sqrt{\frac{8\pi}{3}}\sqrt{2C_i(\phi^0)\, \mathcal{Z}_{ij}(\phi^0)\, C_j(\phi^0)}, \tag{6.14}$$

where M_P denotes the Planck mass, and $C_i(\phi^0)$ are constants of dimension $[m]^2$. Eq. (6.14) relates the *supersymmetry breaking scale* E to the *gravitino* mass $m_{3/2} \equiv \mathcal{M}(\phi^0)$ and to the Planck mass as follows:

$$m_{3/2} = \sqrt{\frac{8\pi}{3}} \frac{E^2}{M_P}, \tag{6.15}$$

where

$$E^4 \equiv 2C_i(\phi^0)\, \mathcal{Z}_{ij}(\phi^0)\, C_j(\phi^0). \tag{6.16}$$

It should be here recalled that, as in any gauge theory exhibiting a spontaneous symmetry breaking and thus an *Higgs effect*, the gauge field mass is given by

$$m_{gauge} \sim g \langle \varphi \rangle, \tag{6.17}$$

where g is the gauge coupling and $\langle \varphi \rangle$ is the vacuum expectation value of the would-be Goldstone boson. By comparing Eqs. (6.15) and (6.17), it is easy to conclude that the *gravitino* ψ is the gauge field of local supersymmetry and, in presence of spontaneous breaking of such a gauge symmetry, it acquires a mass through the *super-Higgs effect*, where $g = k = \frac{\sqrt{8\pi}}{M_P}$ and $\langle \varphi \rangle$ denoting the vacuum expectation value of some auxiliary field satisfying Eq. (6.12).

Eqs. (6.4)-(6.8) can be generalized to the case of $\mathcal{N}$-extended supersymmetry, thus providing a description of *super-Higgs effect* in $\mathcal{N}$-extended supergravities (see *e.g.* [54, 55, 56]).

In particular, the scalar potential $\mathcal{V}$ turns out to satisfy the following matrix relation, generalizing Eq. (6.7) (A, $B = 1, ..., \mathcal{N}$):

$$2C_{iA}(\phi)\,\mathcal{Z}_{ij}(\phi)\,C_{jB}(\phi) - 3\mathcal{M}_{AC}(\phi)\,\mathcal{M}_{CB}(\phi) = \mathcal{V}(\phi)\,\delta_{AB}, \tag{6.18}$$

implying

$$\begin{aligned} 2C_{i1}(\phi)\,\mathcal{Z}_{ij}(\phi)\,C_{j1}(\phi) - 3\mathcal{M}_{1C}(\phi)\,\mathcal{M}_{C1}(\phi) &= \\ 2C_{i2}(\phi)\,\mathcal{Z}_{ij}(\phi)\,C_{j2}(\phi) - 3\mathcal{M}_{2C}(\phi)\,\mathcal{M}_{C2}(\phi) &= \ldots = \\ 2C_{i\mathcal{N}}(\phi)\,\mathcal{Z}_{ij}(\phi)\,C_{j\mathcal{N}}(\phi) - 3\mathcal{M}_{\mathcal{N}C}(\phi)\,\mathcal{M}_{C\mathcal{N}}(\phi) &= \mathcal{V}(\phi), \end{aligned} \tag{6.19}$$

where the mass matrix of $\mathcal{N}$ gravitinos is symmetric ($\mathcal{M}_{AB} = \mathcal{M}_{(AB)}$). Thus, for instance $C_{i1}(\phi) = 0$ does *not* imply $C_{iA}(\phi) = 0$, and, in light of the reasoning performed above, one can conclude that *partial* spontaneous breaking of local supersymmetry is possible in an $\mathcal{N}$-extended framework.

Furthermore, by diagonalizing the *gravitino* mass matrix

$$\mathcal{M}_{AB} = m_{3/2,A}\delta_{AB}, \tag{6.20}$$

one obtains a general formula for the masses of *gravitinos* generated by the *super-Higgs effect* in $\mathcal{N}$-extended supergravities with vanishing cosmological constant, generalizing Eqs. (6.14) and (6.15) to the case $\mathcal{N} > 1$:

$$\begin{aligned} m_{3/2,A}\left(\phi^0\right) &= \frac{1}{M_P}\sqrt{\frac{8\pi}{3}}\sqrt{2C_{iA}\left(\phi^0\right)\mathcal{Z}_{ij}\left(\phi^0\right)C_{jA}\left(\phi^0\right)} = \\ &= \sqrt{\frac{8\pi}{3}}\frac{E_A^2}{M_P}, \end{aligned} \tag{6.21}$$

where the *supersymmetry breaking scale* E_A (related to the A-th *supercharge* spinor generator) has been defined as follows (generalizing Eq. (6.16)):

$$E_A^4 \equiv 2C_{iA}\left(\phi^0\right)\mathcal{Z}_{ij}\left(\phi^0\right)C_{jA}\left(\phi^0\right). \tag{6.22}$$

It is worth noticing that the E_A's can all be different, and some of them may vanish, in which case some supersymmetry is left unbroken.

As pointed out above, since $\mathcal{V}$ is not positive-definite, in general $\mathcal{V}(\phi_{cr}) \gtreqless 0$. For $\mathcal{V}(\phi_{cr}) > 0$ (de Sitter background, positive cosmological constant) *all* supersymmetries are broken, otherwise Eq. (6.19) would not hold. Instead, for $\mathcal{V}(\phi_{cr}) < 0$ (anti-de Sitter background, negative cosmological constant) a *partial breaking* of supersymmetry may occur, depending on whether $C_{iA}(\phi) = 0$ for some A.

Finally, it is worth remarking that in $\mathcal{N} = 1$, $d = 4$ supergravity theories one can give more explicit formulæ, because the *"standard supergravity Lagrangian"* which describes the locally supersymmetric coupling of matter and gauge fields is parameterized in terms of two functions, namely (see *e.g.* [57, 58, 59])

$$f_{\alpha\beta}(z) \tag{6.23}$$

and

$$G(z,\overline{z}) \equiv K(z,\overline{z}) + ln\left(|f(z)|^2\right), \tag{6.24}$$

where $f_{\alpha\beta}(z)$ is the kinetic spin-1 matrix ($\alpha, \beta = 1, ..., n_V$, with n_V denoting the number of matter *vector* supermultiplets), and K is the Kähler potential of the scalar manifold parameterized by the complex scalars z_i and $\overline{z}^i$. As a consequence, in this framework the spin-$\frac{1}{2}$ and spin-1 kinetic matrices can be arranged in the following "vector":

$$\left(\mathcal{Z}^i_{\overline{j}} = \frac{\partial^2 G(z,\overline{z})}{\partial z_i \partial \overline{z}^j},\ Re\left(f_{\alpha\beta}(z)\right)\right). \tag{6.25}$$

Moreover, the *gravitino* mass term reads

$$m_{3/2}(z,\overline{z}) \equiv \mathcal{M}(z,\overline{z}) = e^{G/2} = |f(z)|\, e^{K/2}. \tag{6.26}$$

From the reasoning performed above, the *spontaneous breakdown of supersymmetry* with vanishing vacuum energy (*i.e.* in a Minkowski background) can occur *iff*

$$|f(z_0)| \neq 0 \tag{6.27}$$

at

$$z^i = z^i_0 : \begin{cases} \mathcal{V}(z_0, \overline{z}_0) = 0; \\ \frac{\partial \mathcal{V}}{\partial z_i}(z_0, \overline{z}_0) = 0, \ \forall i. \end{cases} \tag{6.28}$$

Furthermore, at the scalar configurations satisfying Eqs. (6.27) and (6.28) it also holds that

$$e^G G^i \left(G^{-1}\right)^j_i G_j + \frac{1}{2} Re\left(\left(f^{-1}\right)^{\alpha\beta}\right) D_\alpha D_\beta = 3e^G, \tag{6.29}$$

where D_α's denote the *D-terms*, and $G^i \equiv \frac{\partial G}{\partial z_i}$, $G_i \equiv \frac{\partial G}{\partial \overline{z}^i} = \overline{G^i}$ and $\left(G^{-1}\right)^j_i \equiv \frac{\partial^2 G(z,\overline{z})}{\partial z_j \partial \overline{z}^i}$, which is nothing but Eq. (6.13) specialized to the considered framework.

Thus, in the cases in which $D_\alpha = 0$ when Eq. (6.28) holds, the condition for vanishing vacuum energy in $\mathcal{N} = 1$, $d = 4$ spontaneously broken supergravity (or in other words, for the *super-Higgs effect* to occur in $\mathcal{N} = 1$ local supersymmetry with vanishing cosmological constant) reads

$$G^i \left(G^{-1}\right)^j_i G_j = 3. \tag{6.30}$$

For a generic model, Eq. (6.30) requires some *fine-tuning* of parameters of the superpotential in order to be satisfied. However, there is a large class of $\mathcal{N} = 1$, $d = 4$ supergravity theories, exhibiting semi-definite positive scalar potential $\mathcal{V}$, for which no *fine-tuning* is necessary. A subclass of such theories is given by the so-called *no-scale supergravities* (with undetermined *gravitino* mass; see *e.g.* [60]), which naturally appear in the point-field limit of certain superstring compactifications (see *e.g.* [61, 62]).

7 Supersymmetry Breaking and the Minimally Supersymmetric Standard Model

Local Supersymmetry, or equivalently *Supergravity*, is the only known extension of Einstein's *General Relativity*, based on a new space-time symmetry, which extends the *Principle of General Covariance* to diffeomorphisms in *Superspace*, named *superdiffeomorphisms*.

Surprisingly enough, unlike General Covariance, Supercovariance can be broken, and its spontaneous breaking predicts its gauge field, the spin-$\frac{3}{2}$ gravitino, to become massive, through the so-called *super-Higgs mechanism* (see Sect. 6).

The *minimal* supersymmetric extension of the *Standard Model* of fundamental interactions (apart from gravity), named *Minimally Supersymmetric Standard Model* (MSSM) is viable only if *soft-breaking* terms are introduced [63, 64]. These terms can be naturally obtained by coupling MSSM to supergravity, and performing the limit with the *gravitino* mass $m_{3/2}$ fixed and the Planck mass M_P large [65, 66, 67]. A particularly remarkable class of supergravity theories are the so-called *no-scale supergravities*, with a tree-level *"sliding"* mass of the *gravitino* [60, 68] (see the end of Sect. 6).

Spontaneous symmetry breaking requires *"messenger fields"* to exist. Such *"messenger quanta"* break supersymmetry (either for gravity effects or for other hidden - renormalizable - sectors) and communicate the *"effective supersymmetry breaking"* to the observable sector (MSSM) trough either supergravity interactions (*gravity mediation*, see *e.g.* [65, 66, 67, 60, 68]) or through effective interactions originating from (supersymmetry-violating) loop graphs (*gauge mediation*, see *e.g.* [69, 70, 71, 72], see below).

The *"cosmological constant problem"* plaguing the Standard Model (see *e.g.* [73], and Refs. therein) also remains in the *MSSM*, even if at tree-level it can be explained either if supersymmetry is broken or in *no-scale supergravity* models mentioned above. Indeed, depending on the various models considered, the *cosmological constant* Λ reads:

$$\Lambda^4 \sim M_P^4, \text{ or} \tag{7.1}$$

$$\Lambda^4 \sim m_{3/2}^2 M_P^2, \text{ or} \tag{7.2}$$

$$\Lambda_\alpha^4 \sim \left(\frac{m_{3/2}}{M_P}\right)^\alpha m_{3/2}^4 \ (\alpha \text{ large}) . \tag{7.3}$$

As shortly mentioned above, primordial supersymmetry breaking E is due to the super-Higgs effect, which in Minkowski space yields

$$m_{3/2} = \sqrt{\frac{8\pi}{3}}\frac{E^2}{M_P} = \sqrt{\frac{8\pi}{3}}\frac{|F_u|}{M_P}, \tag{7.4}$$

where the *supersymmetry breaking scale* E is nothing but the absolute value of the (vacuum expectation value of the) complex field $\mathcal{F}_u \sim \mathcal{S} + i\mathcal{P}$, whose real and imaginary parts are the supergravity *"auxiliary"* fields (completing the *off-shell* boson-fermion counting). It is immediate to note that at $|F_u|$ fixed

$$lim_{M_P \to \infty} m_{3/2} = 0, \tag{7.5}$$

thus for models in which F_u comes from a non-gravitational sector (and it is therefore small with respect to M_P) $m_{3/2}$ is small compared to the electroweak scale.

Concerning the *goldstino* λ_g, the *Goldstone fermion* of spontaneously broken supersymmetry, it enters the supercurrent as [74, 75]

$$J_{\mu\alpha} = E^2 \gamma_\mu \lambda_g + \dots . \tag{7.6}$$

From the (*Yukawa*) coupling of λ_g to ordinary supermatter (the subscript "M" stands for "*matter*")

$$g_M \overline{\lambda}_g \psi_M \Phi_M, \tag{7.7}$$

where $(\psi_M, \Phi_M \equiv A_M + iB_M)$ is the generic Wess-Zumino matter supermultiplet, the conservation of the supercurrent $(\partial^\mu J_{\mu\alpha} = 0)$ implies (the subscript "*sm*" stands for *Standard Model*) [74, 75, 76, 5]

$$E_{sm}^2 = g_M E^2 = m_{A_M}^2 - m_{\psi_M}^2 \equiv \Delta m^2, \tag{7.8}$$

so that (recalling Eq. (7.4))

$$E_{sm}^2 = g_M m_{3/2} M_P. \tag{7.9}$$

Since $g_M \sim \mathcal{O}(1)$ in *gauge mediated supersymmetry breaking*, whereas $g_M \sim \mathcal{O}(10^{-15})$ in *gravity mediated supersymmetry breaking*[5], it follows that *goldstino* almost decouples in presence of *gravity mediation* (*softly broken* theory).

It is here worth recalling that the difficulty to use *tree-level supersymmetry breaking* in the *MSSM* relies in the following mass formula, which is true for any spontaneously broken renormalizable gauge theory[6] [78]:

$$SuperTrace\left(M^2\left(\Phi_M\right)\right) \equiv \sum_S (-1)^{2S} (2S+1)\, m_S^2\left(\Phi_M\right) = 0, \tag{7.10}$$

where $M^2(\Phi_M)$ denotes the scalar-dependent square mass matrix of all states in the considered theory. This formula is evaded when either by coupling the gauge theory to supergravity (*gravity mediation*, see *e.g.* [65, 66, 67, 60, 68]) or by considering large loop effects (*gauge mediation*, see *e.g.* [69, 70, 71, 72]).

In the contest of $\mathcal{N} = 1$, $d = 4$ supergravity coupled to n_C chiral and n_V vector supermultiplets, Eq. (7.10) gets generalized as follows [79, 80] $(\alpha, \beta = 1, ..., n_V,\ i, j = 1, ..., n_C)$:

$$SuperTrace\left(M^2\right) = 2\left(n_C - 1\right) m_{3/2}^2 - F^i \mathcal{Q}_{i\bar{j}} \overline{F}^{\bar{j}}, \tag{7.11}$$

[5] Actually, other mechanisms of *mediation of supersymmetry breaking* have been proposed in literature, such as the *anomaly mediation* or the *"mirage" mediation*. Their treatment falls outside the scope of this brief report; the interested reader is *e.g.* addressed to [77] for an introductory report and a list of references.

[6] Without $U(1)$ *D-term* contributions.

where F^i are the *F-terms*. Furthermore, $\mathcal{Q}_{i\bar{j}} \equiv R_{i\bar{j}} + V_{i\bar{j}}$, where

$$R_{i\bar{j}} = \partial_i \overline{\partial}_{\bar{j}} ln \left(det \left(g_{m\bar{n}} \right) \right) \tag{7.12}$$

is the Ricci tensor of the non-linear scalar sigma model (whose metric is $g_{i\bar{j}}$, appearing in $g_{i\bar{j}} \left(\partial_\mu z^i \right) \partial^\mu \overline{z^{\bar{j}}}$), and

$$V_{i\bar{j}} \equiv -\partial_i \overline{\partial}_{\bar{j}} ln \left(det \left(Ref_{\alpha\beta} \right) \right), \tag{7.13}$$

where $Ref_{\alpha\beta}$ is the kinetic matrix of the $\mathcal{F}^2$ vector term (*i.e.* appearing in $Ref_{\alpha\beta}\mathcal{F}^{\alpha}_{\mu\nu}\mathcal{F}^{\mu\nu|\beta}$). Note that the r.h.s. of Eq. (7.11) is due to gravitational effects, in the sense that $m_{3/2}$ vanishes in absence of gravity couplings and $\mathcal{Q}_{i\bar{j}}$ vanishes for *renormalizable* gauge interactions (because $f_{\alpha\beta}$ and $g_{i\bar{j}}$ are constant in this case). In the case of *gravity mediation* (see *e.g.* [65, 66, 67, 60, 68]), the non-vanishing of $SuperTrace\left(M^2\right)$ is due to soft-breaking terms in the limit in which $M_P \to \infty$ and $m_{3/2}$ is kept fixed.

Let us finally dwell shortly on the above mentioned *gauge mediated supersymmetry breaking* (see *e.g.* [69, 70, 71, 72]). In such a framework, *supersymmetry breaking* is communicated to the supermultiplets of the MSSM via heavy chiral supermultiplets Ξ that are *charged* (sitting in *real* representations) under the Standard Model gauge symmetries. If the masses of such chiral multiplets are not equal (due to a *supersymmetry breaking* through an *O'Raifertaigh mechanism*), one has a mass term with structure

$$X\Xi^2, \tag{7.14}$$

where $X = M + \vartheta^2 F$ is a gauge-singlet *"spurion"* superfield (because $F \neq 0$), and Ξ is a chiral superfield. X and Ξ are not directly coupled to the chiral fields of the Standard Model, thus the breaking of supersymmetry is transferred through loop graphs, determining

$$m_{gaugino} \sim \frac{g^2}{16\pi^2}\frac{F}{M} \tag{7.15}$$

at 1-loop, and

$$m_{scalar} \sim \left(\frac{g^2}{16\pi^2}\right)^2 \left(\frac{F}{M}\right)^2 \tag{7.16}$$

at 2-loops, where M is the mass of the messenger superfield Ξ, with $\frac{F}{M} \sim \mathcal{O}\left(8 - 10\ TeV\right)$.

If one requires that gauge mediation gives a contribution to the MSSM mass splitting larger than the non-renormalizable higher-dimensional operators, suppressed by inverse powers of M_P, then upper limits on $\langle M \rangle$ and $\langle F \rangle$ can be computed.

Needless to say, while *gauge mediation* is mostly justified by a *"down-top"* approach to higher energies, the *gravity mediated scenario* is determined by a *"top-down"* approach, because the hidden sector has only gravitational interactions, and probably comes from a more fundamental theory in higher dimensions, where the several scalars involved in the supersymmetry breaking sector are *moduli* fields.

8 Conclusions

The evasion of the *Coleman-Mandula Theorem* [1] is merely due to the fact that continous symmetries considered by Coleman and Mandula are based on Lie algebras rather than on Lie superalgebras (graded Lie algebras). Other than this assumption, the important fact is that the relativistic symmetry contains the Poincaré symmetry and that there is a finite number of massive particles. The latter are described, through the *Wigner's method of induced representations*, by two Casimir operators, namely the square mass $P_\mu P^\mu$ and the relativistic spin $W_\mu W^\mu$ (with W_μ defined in Eq. (4.11)).

In the supersymmetric generalization of the *Coleman-Mandula Theorem*, given by the *Haag-Lopuszansky-Sohnius Theorem* [2], the only additional allowed symmetry is a *central extension* (implemented through a skew-symmetric *central charge matrix* $Z^{AB} = -Z^{BA}$, without considering *brane charges*) of the original $\mathcal{N}$-extended ($d = 4$) Wess-Zumino Poincaré superalgebra. It is given by

$$\left\{\begin{array}{ll} \left[Q_\alpha^A, P_\mu\right] = 0 = \left[Q_\alpha^A, Z^{BC}\right]; & \\ \left[Z^{AB}, Z^{CD}\right] = 0 = \left[Z^{AB}, P^\mu\right]; & \textit{grading repr;} \end{array}\right.$$
$$\left\{\begin{array}{ll} \left\{Q_\alpha^A, Q_{\dot{\alpha}B}\right\} = 2\left(\sigma_\mu\right)_{\alpha\dot{\alpha}} P^\mu \delta_B^A; & \\ \left\{Q_\alpha^A, Q_\beta^B\right\} = \epsilon_{\alpha\beta} Z^{AB}. & \textit{odd sector.} \end{array}\right. \tag{8.1}$$

Positivity requirements imply the following BPS-bound to be satisfied [81, 45]:

$$M \geqslant Z_1 \geqslant ... \geqslant Z_r, \tag{8.2}$$

where Z_i are the real positive skew-eigenvalues of Z^{AB}, and they are $\mathcal{N}/2$ for $\mathcal{N}$ even and $(\mathcal{N}-1)/2$ for $\mathcal{N}$ odd (*i.e.* $i = 1, ..., r = [\mathcal{N}]/2$, the square brackets denoting the integer part of the enclosed number) [45].

In the massless case (*i.e.* for $M = 0$), the bound (8.2) implies $Z^{AB} = 0$, and in this case another extension of the Poincaré superalgebra is possible, namely the $\mathcal{N}$-extended superconformal algebra (already considered by Wess and Zumino [36] for $\mathcal{N} = 1$). It is here worth remarking that the $\mathcal{N}$-extended superconformal algebra is a particular case of real form of the unitary sequence of simple superalgebras $\mathfrak{su}(\mathcal{M} \mid \mathcal{N})$ [41], whose even part is $\mathfrak{su}(\mathcal{M}) \oplus \mathfrak{su}(\mathcal{N}) \oplus \mathfrak{u}(1)$ (with no $\mathfrak{u}(1)$ for $\mathcal{M} = \mathcal{N}$).

Central charges are important in spontaneously broken $\mathcal{N} = 2$ and $\mathcal{N} = 4$ super Yang-Mills gauge theories [46], in the description of the massive spectrum, both at perturbative (*massive vectors*) and non-perturbative (*monopoles, dyons*) level. They are relevant for the *Seiberg-Witten solution* of Yang-Mills effective actions in the Coulomb phase [82].

On the gravity side, *moduli-dependent* central charges play a key role in the explanation of the *Bekenstein-Hawking entropy* [83, 84] of *extremal* black holes, as well as for the determination of the *ADM mass* [85] of such black holes. The *Attractor Mechanism* [86, 88, 89, 87, 90] (for recent reviews, see *e.g.* [91] and [92]) relates these two quantities, which in principle describe data at the event horizon (entropy) and at the spatial infinity (*ADM mass*).

Finally, it is here worth mentioning that super-Poincaré algebra allows for a further evasion of both the *Coleman-Mandula* [1] and *Haag-Lopuszanski-Sohnius* [2] *Theorems*, with the inclusion of additional *"central" charges* $Z_{\mu_1 \ldots \mu_p}$ (which are central only with respect to super-translations), related to extended objects (*"branes"*). In the simplest example of $\mathcal{N}=1$ super-Poincaré algebra, the charge $\mathcal{M}_{\mu\nu}$ is associated to BPS *domain walls* in supersymmetric gauge theories (see *e.g.* [93], and Refs. therein). *"Central" charges* $Z_{\mu_1 \ldots \mu_p}$ are associated to (*electric*) p-branes, which couple to $(p+1)$-rank gauge fields. On the other hand, by indicating with d the number of space-time dimensions, *"central" charges* $Z_{\mu_1 \ldots \mu_{d-p-4}}$ are associated to (*magnetic*) $(d-p-4)$-branes, coupling to $(d-p-3)$-rank gauge fields. In this framework, the condition of *self-duality* (*dyonic nature*) of a p-brane is $p=(d-4)/2$. Furthermore, it can be shown [94] that the condition of *asymptotical flatness* of the brane solutions yields the constraint $p \leqslant d-4$. A celebrated example is provided by the 2-brane (*membrane*) and its *dual* 5-brane in $d=11$ *M-theory.* Their simultaneous presence requires a quantization condition *à la Dirac-Schwinger-Zwanziger*, analogous to the one of electric charges and magnetic monopoles in electromagnetism.

Acknowledgments

A. M. would like to thank the Department of Physics, Theory Unit Group at CERN, Geneva CH, where part of this work was done, for kind hospitality and stimulating environment.

The work of S. F. has been supported in part by D.O.E. grant DE-FG03-91ER40662, Task C.

The work of A. M. has been supported by an INFN visiting Theoretical Fellowship at SITP, Stanford University, Stanford, CA, USA.

References

[1] S. R. Coleman and J. Mandula, *All Possible Symmetries Of The S Matrix*, Phys.Rev. **159**, 1251 (1967).

[2] R. Haag, J. T. Lopuszanski and M. Sohnius, *All Possible Generators of Supersymmetries of the S Matrix*, Nucl. Phys. **B88**, 257 (1975).

[3] J. H. Schwarz, *Status of Superstring and M-Theory*, Lectures presented at the Erice 2008 International School of Subnuclear Physics, `arXiv:0812.1372`.

[4] *"The New Aspects Of Subnuclear Physics"*, Proceedings of the 16th International School Of Subnuclear Physics, Erice, Italy, July 31 - August 11, 1978, A. Zichichi Ed., Plenum (New York), 1980.

[5] P. Fayet and S. Ferrara, *Supersymmetry*, Phys. Rept. **32**, 249 (1977).

[6] *"Supersymmetry"*, S. Ferrara Ed., 2 Vols., North-Holland Publishing (Amsterdam) and World Scientific (Singapore), 1987.

[7] G. Bertone, D. Hooper and J. Silk, *Particle dark matter: Evidence, candidates and constraints*, Phys. Rept. **405**, 279 (2005), hep-ph/0404175.

[8] E. W. Kolb, *Cosmology and the Unexpected*, Lectures given at International School of Subnuclear Physics: 45th Course: Searching for the "Totally Unexpected" in the LHC Era, Erice, Sicily, Italy, 29 Aug. - 7 Sep. 2007, arXiv:0709.3102.

[9] *"Searching for the Superworld", A Volume in Honor of Antonino Zichichi on the Occasion of the Sixth Centenary Celebrations of the University of Turin, Italy*, S. Ferrara and R. M. Mössbauer Eds., World Scientific (Singapore) 2007.

[10] S. Weinberg, *"The quantum theory of fields. Vol. 3: Supersymmetry"*, Cambridge University Press (Cambridge), 2000.

[11] S. Ferrara and B. Zumino, *Supergauge Invariant Yang-Mills Theories*, Nucl. Phys. **B79**, 413 (1974).

[12] A. Salam and J. A. Strathdee, *Supersymmetry and Nonabelian Gauges*, Phys. Lett. **B51**, 353 (1974).

[13] L. Alvarez-Gaumé, J. Polchinski and M. B. Wise, *Minimal Low-Energy Supergravity*, Nucl. Phys. **B221**, 495 (1983).

[14] J. R. Ellis, M. K. Gaillard, D. V. Nanopoulos and S. Rudaz, *Uncertainties in the Proton Lifetime*, Nucl. Phys. **B176**, 61 (1980).

[15] S. Dimopoulos, S. Raby and F. Wilczek, *Proton Decay in Supersymmetric Models*, Phys. Lett. **B112**, 133 (1982).

[16] P. Binetruy and M. K. Gaillard, *Candidates for the Inflaton Field in Superstring Models*, Phys. Rev. **D34**, 3069 (1986).

[17] S. Ferrara, J. Wess and B. Zumino, *Supergauge Multiplets and Superfields*, Phys. Lett. **B51**, 239 (1974).

[18] R. D'Auria, S. Ferrara, M. A. Lledó and V. S. Varadarajan, *Spinor algebras*, J. Geom. Phys. **40**, 101 (2001), hep-th/0010124.

[19] J. A. Strathdee, *Extended Poincaré Supersymmetry*, Int. J. Mod. Phys. **A2**, 273 (1987).

[20] W. Nahm, *Supersymmetries and their Representations*, Nucl. Phys. **B135**, 149 (1978).

[21] E. Cremmer, B. Julia and J. Scherk, *Supergravity Theory in Eleven-Dimensions*, Phys. Lett. **B76**, 409 (1978).

[22] K. Becker, M. Becker and J. H. Schwarz, *"String theory and M-theory: A modern introduction"*, Cambridge University Press (Cambridge), 2007.

[23] M. Gell-Mann, P. Ramond and R. Slansky, *Complex Spinors And Unified Theories*, in : *"Supergravity"*, P. van Nieuwenhuizen and D. Z. Freedman Eds., North-Holland Publishing (Amsterdam), 1979.

[24] D. Z. Freedman, P. van Nieuwenhuizen and S. Ferrara, *Progress Toward a Theory of Supergravity*, Phys. Rev. **D13**, 3214 (1976).

[25] M. B. Green, J. H. Schwarz and E. Witten, *"Superstring Theory"*, 2 Vols., Cambridge Monographs On Mathematical Physics, Cambridge University Press (Cambridge), 1987.

[26] J. Polchinski, *"String theory"*, 2 Vols., Cambridge University Press (Cambridge), 1998.

[27] J. M. Maldacena, *The Large $\mathcal{N}$ limit of superconformal field theories and supergravity*, Adv. Theor. Math. Phys. **2**, 231 (1998), and Int. J. Theor. Phys. **38**, 1113 (1999), hep-th/9711200.

[28] S. S. Gubser, I. R. Klebanov and A. M. Polyakov, *Gauge theory correlators from non-critical string theory*, Phys. Lett. **B428**, 105 (1998), hep-th/9802109.

[29] E. Witten, *Anti-de Sitter space and holography*, Adv. Theor. Math. Phys. **2**, 253 (1998), hep-th/9802150.

[30] Z. Bern, J. J. Carrasco, L. J. Dixon, H. Johansson, D. A. Kosower and R. Roiban, *Three-Loop Superfiniteness of $\mathcal{N}$= 8 Supergravity*, Phys. Rev. Lett. **98**, 161303 (2007), hep-th/0702112.

[31] R. Kallosh, *On a possibility of a UV finite $\mathcal{N}$= 8 supergravity*, arXiv:0808.2310.

[32] E. Witten, *String theory dynamics in various dimensions*, Nucl. Phys. **B443**, 85 (1995), hep-th/9503124.

[33] C. Rovelli, *"Quantum Gravity"*, Cambridge Monographs on Mathematical Physics, Cambridge University Press (Cambridge), 2004.

[34] Y. A. Gol'fand and E. P. Likhtman, *Extension of the Algebra of Poincaré Group and Violation of P Invariance*, JETP Lett. **13**, 323 (1971).

[35] D. V. Volkov and V. P. Akulov, *Is the Neutrino a Goldstone Particle?*, Phys. Lett. **B46**, 109 (1973).

[36] J. Wess and B. Zumino, *Supergauge Transformations in Four Dimensions*, Nucl. Phys. **B70**, 39 (1974).

[37] S. Coleman, *"Aspects of Symmetry: Selected Erice Lectures"*, Cambridge University Press (Cambridge), 1988.

[38] S. Ferrara and B. Zumino, *Transformation Properties of the Supercurrent*, Nucl. Phys. **B87**, 207 (1975).

[39] A. Salam and J. Strathdee, *Super-Gauge Transformations*, Nucl. Phys. **B76**, 477 (1974).

[40] A. Salam nd J. Strathdee, *Unitary Representations of Supergauge Symmetries*, Nucl. Phys. **B80**, 499 (1974).

[41] L. Frappat, P. Sorba and A. Sciarrino, *Dictionary on Lie Superalgebras*, `hep-th/9607161`.

[42] A. Van Proeyen, *Tools for Supersymmetry*, Lectures given at Spring School on Quantum Field Theory: Supersymmetry and Superstrings, Calimanesti, Romania, 24-30 Apr. 1998, `hep-th/9910030`.

[43] M. Gell-Mann, unpublished note, 1976.

[44] A. S. Galperin, E. A. Ivanov, V. I. Ogievetsky and E. S. Sokatchev, *"Harmonic Superspace"*, Cambridge University Press (Cambridge), 2001.

[45] S. Ferrara, C. A. Savoy and B. Zumino, *General Massive Multiplets In Extended Supersymmetry*, Phys. Lett. **B100**, 393 (1981).

[46] P. Fayet, *Spontaneous Generation of Massive Multiplets and Central Charges in Extended Supersymmetric Theories*, Nucl. Phys. **B149**, 137 (1979).

[47] S. Deser and B. Zumino, *Consistent Supergravity*, Phys. Lett. **B62**, 335 (1976).

[48] P. Van Nieuwenhuizen, *Supergravity*, Phys. Rept. **68**, 189 (1981).

[49] D. Francia and A. Sagnotti, *On the geometry of higher spin gauge fields*, Class. Quant. Grav. **20**, S473 (2003), `hep-th/0212185`.

[50] X. Bekaert, S. Cnockaert, C. Iazeolla and M. A. Vasiliev, *Nonlinear higher spin theories in various dimensions*, Lectures given at Workshop on Higher Spin Gauge Theories, Brussels, Belgium, 12-14 May 2004, `hep-th/0503128`.

[51] W. Rarita and J. S. Schwinger, *On a theory of particles with half integral spin*, Phys. Rev. **60**, 61 (1941).

[52] D. V. Volkov and V. A. Soroka, *Higgs Effect for Goldstone Particles with Spin 1/2*, JETP Lett. **18**, 312 (1973), and Pisma Zh. Eksp. Teor. Fiz. **18**, 529 (1973).

[53] S. Deser and B. Zumino, *Broken Supersymmetry and Supergravity*, Phys. Rev. Lett. **38**, 1433 (1977).

[54] S. Ferrara and L. Maiani, *An Introduction To Supersymmetry Breaking In Extended Supergravity*, Based on lectures given at SILARG V, 5th Latin American Symp. on Relativity and Gravitation, Bariloche, Argentina, Jan. 1985.

[55] S. Cecotti, L. Girardello and M. Porrati, *Constraints On Partial Superhiggs*, Nucl. Phys. **B268**, 295 (1986).

[56] R. D'Auria and S. Ferrara, *On fermion masses, gradient flows and potential in supersymmetric theories*, JHEP **0105**, 034 (2001), hep-th/0103153.

[57] J. Bagger and E. Witten, *The Gauge Invariant Supersymmetric Nonlinear Sigma Model*, Phys. Lett. **B118**, 103 (1982).

[58] J. A. Bagger, *Coupling the Gauge Invariant Supersymmetric Nonlinear Sigma Model to Supergravity*, Nucl. Phys. **B211**, 302 (1983).

[59] E. Cremmer, S. Ferrara, L. Girardello and A. Van Proeyen, *Yang-Mills Theories with Local Supersymmetry: Lagrangian, Transformation Laws and SuperHiggs Effect*, Nucl. Phys. **B212**, 413 (1983).

[60] E. Cremmer, S. Ferrara, C. Kounnas and D. V. Nanopoulos, *Naturally Vanishing Cosmological Constant in $\mathcal{N}=1$ Supergravity*, Phys. Lett. **B133**, 61 (1983).

[61] S. B. Giddings, S. Kachru and J. Polchinski, *Hierarchies from fluxes in string compactifications*, Phys. Rev. **D66**, 106006 (2002), hep-th/0105097.

[62] A. R. Frey and J. Polchinski, *$\mathcal{N}=3$ warped compactifications*, Phys. Rev. **D65**, 126009 (2002), hep-th/0201029.

[63] S. Dimopoulos and H. Georgi, *Softly Broken Supersymmetry and $SU(5)$*, Nucl. Phys. **B193**, 150 (1981).

[64] L. Girardello and M. T. Grisaru, *Soft Breaking of Supersymmetry*, Nucl. Phys. **B194**, 65 (1982).

[65] A. H. Chamseddine, R. L. Arnowitt and P. Nath, *Locally Supersymmetric Grand Unification*, Phys. Rev. Lett. **49**, 970 (1982).

[66] R. Barbieri, S. Ferrara and C. A. Savoy, *Gauge Models with Spontaneously Broken Local Supersymmetry*, Phys. Lett. **B119**, 343 (1982).

[67] L. J. Hall, J. D. Lykken and S. Weinberg, *Supergravity as the Messenger of Supersymmetry Breaking*, Phys. Rev. **D27**, 2359 (1983).

[68] J. R. Ellis, A. B. Lahanas, D. V. Nanopoulos and K. Tamvakis, *No-Scale Supersymmetric Standard Model*, Phys. Lett. **B134**, 429 (1984).

[69] M. Dine, Y. Nir and Y. Shirman, *Variations on minimal gauge mediated supersymmetry breaking*, Phys. Rev. **D55**, 1501 (1997), hep-ph/9607397.

[70] S. Dimopoulos, G. F. Giudice and A. Pomarol, *Dark matter in theories of gauge mediated supersymmetry breaking*, Phys. Lett. **B389**, 37 (1996), hep-ph/9607225.

[71] G. F. Giudice and R. Rattazzi, *Theories with gauge mediated supersymmetry breaking*, Phys. Rept. **322**, 419 (1999), hep-ph/9801271.

[72] C. Beasley, J. J. Heckman and C. Vafa, *GUTs and Exceptional Branes in F-theory - I*, arXiv:0802.3391.

[73] S. Weinberg, *The Cosmological Constant Problem*, Morris Loeb Lectures in Physics, Harvard University, May 2, 3, 5, and 10, 1988, Rev. Mod. Phys. **61**, 1 (1989).

[74] P. Fayet, *Mixing Between Gravitational and Weak Interactions Through the Massive Gravitino*, Phys. Lett. **B70**, 461 (1977).

[75] P. Fayet, *Relations Between the Masses of the Superpartners of Leptons and Quarks, the Goldstino Couplings and the Neutral Currents*, Phys. Lett. **B84**, 416 (1979).

[76] P. Fayet, *Spontaneously Broken Supersymmetric Theories of Weak, Electromagnetic and Strong Interactions*, Phys. Lett. **B69**, 489 (1977).

[77] M. A. Luty, *2004 TASI lectures on supersymmetry breaking*, hep-th/0509029.

[78] S. Ferrara, L. Girardello and F. Palumbo, *A General Mass Formula in Broken Supersymmetry*, Phys. Rev. **D20**, 403 (1979).

[79] M. T. Grisaru, M. Rocek and A. Karlhede, *The Superhiggs Effect In Superspace*, Phys. Lett. **B120**, 110 (1983).

[80] S. Ferrara, C. Kounnas and F. Zwirner, *Mass formulae and natural hierarchy in string effective supergravities*, Nucl. Phys. **B429**, 589 (1994); Erratum-ibid. **B433**, 255 (1995), hep-th/9405188.

[81] G. W. Gibbons and C. M. Hull, *A Bogomol'ny Bound for General Relativity and Solitons in $\mathcal{N}= 2$ Supergravity*, Phys. Lett. **B109**, 190 (1982).

[82] N. Seiberg and E. Witten, *Electric - magnetic duality, monopole condensation, and confinement in $\mathcal{N}= 2$ supersymmetric Yang-Mills theory*, Nucl. Phys. **B426**, 19 (1994), Erratum-ibid. **B430**, 485 (1994), hep-th/9407087.

[83] S. W. Hawking, *Gravitational Radiation from Colliding Black Holes*, Phys. Rev. Lett. **26**, 1344 (1971).

[84] J. D. Bekenstein, *Black Holes and Entropy*, Phys. Rev. **D7**, 2333 (1973).

[85] R. Arnowitt, S. Deser and C. W. Misner, *Canonical Variables for General Relativity*, Phys. Rev. **117**, 1595 (1960).

[86] S. Ferrara, R. Kallosh and A. Strominger, *$\mathcal{N}= 2$ extremal black holes*, Phys. Rev. **D52**, 5412 (1995), hep-th/9508072.

[87] A. Strominger, *Macroscopic entropy of $\mathcal{N}= 2$ extremal black holes*, Phys. Lett. **B383**, 39 (1996), `hep-th/9602111`.

[88] S. Ferrara and R. Kallosh, *Supersymmetry and attractors*, Phys. Rev. **D54**, 1514 (1996), `hep-th/9602136`.

[89] S. Ferrara and R. Kallosh, *Universality of supersymmetric attractors*, Phys. Rev. **D54**, 1525 (1996), `hep-th/9603090`.

[90] S. Ferrara, G. W. Gibbons and R. Kallosh, *Black Holes and Critical Points in Moduli Space*, Nucl. Phys. **B500**, 75 (1997), `hep-th/9702103`.

[91] S. Bellucci, S. Ferrara, R. Kallosh and A. Marrani, *Extremal Black Hole and Flux Vacua Attractors*, contribution to the Proceedings of Winter School on Attractor Mechanism (SAM 2006), Frascati, Italy, 20-24 March 2006, Lect. Notes Phys. **755**, 115 (2008), `arXiv:0711.4547`.

[92] S. Ferrara, K. Hayakawa and A. Marrani, *Erice Lectures on Black Holes and Attractors*, contribution to the Proceedings of International School of Subnuclear Physics: 45th Course: Searching for the "Totally Unexpected" in the LHC Era, Erice, Sicily, Italy, 29 Aug. - 7 Sep. 2007, `arXiv:0805.2498`.

[93] A. Kovner, M. A. Shifman and A. V. Smilga, *Domain walls in supersymmetric Yang-Mills theories*, Phys. Rev. **D56**, 7978 (1997), `hep-th/9706089`.

[94] G. W. Gibbons and P. K. Townsend, *Vacuum interpolation in supergravity via super p-branes*, Phys. Rev. Lett. **71**, 3754 (1993), `hep-th/9307049`.

CHAIRMAN: S. FERRARA

Scientific Secretaries: E. Musaev, A. Rebesh

DISCUSSION I

- Nicotri:

You indicated three links between gauge theories and gravity in one of your last slides of this morning. Can you comment more on them? Thank you.

- Ferrara:

I talked about instances in which gravity is UV finite. Of course, a first framework is provided by superstring theory, which yields finite field theory of gravity through the famous AdS/CFT correspondence introduced by Maldacena. Another realization is probably provided by N=8 supergravity (SUGRA), which is a strong candidate for a finite theory of gravity. It is believed so because this theory is the "square" of N=4 super Yang-Mills (SYM) theory, at least concerning the kinematic sector. In other words, the SUGRA multiplet is made up by two N=4 SYM multiplets. Furthermore, such a relation also holds at the level of tree amplitudes. Works by Bern, Dickson, Kosower and collaborators showed that the tree-level relation between squared SYM amplitudes and SUGRA amplitudes is kept at higher-loop orders, too. Hopefully, these relations and the finiteness of N = 4 SYM theory should yield the finiteness of N=8 SUGRA. This would be a framework evading the usual expectation that the only theories free of UV divergences are theories whose spin content goes at most up to one. If rigorously proved to be UV finite, N=8 SUGRA would be the first example of a supersymmetric theory which includes spin-2 field(s) and it is free of UV divergences.

- Susskind:

Let me give you another example of finite theory of gravity: M(atrix)-theory.

- Samsonov:

My question might be considered as the continuation of the previous one. Prof. Ferrara, you mentioned at the end of your lecture that M-theory might be renormalizable in some sense. What did you exactly mean?

- Ferrara:

Well, I said that nobody knows how to really make sense to the 11 space-time dimensions of M-theory, even if there is a consistent formulation of M-theory as a matrix theory, as mentioned by L. Susskind. Through a compactification on a circle, M-theory should give all superstring theories. Thus, for instance, M-theory is believed to be related to E_8 x E_8 heterotic or Type IIA superstrings compactified on a circle. Thus, at least in some particular background, this theory should correspond to a renormalizable theory. It is here worth remarking that M-theory realizes exactly the maximal dimension of space-time, according to the representation theory of super-Poincaré algebra. There is an easy way to prove that M-theory is the theory with the maximal number of space-dimensions (at least if you think about conventional higher-dimensional physical theories). Indeed, M-theory superalgebra only contains simple charges, namely the ones pertaining to 3-branes and 5-branes. In particular, no Kaluza-Klein(KK)-like charges arise. As a consequence, M-theory cannot be obtained by a dimensional reduction of some higher-dimensional theory. If this were instead the case, some central charges, determined by KK modes, would appear in the corresponding superalgebra. Analogously, one can state that there are no SUGRAs in space-time dimensions higher than 11.

- *Di Mauro:*

You told us that supersymmetric theories can be seen as field theories defined on a superspace, i.e. on a space-time with (some) non-commutative coordinates. So, I guess they might be regarded as particular cases of non-commutative field theories. If this were the case, they might exhibit peculiar phenomena, such as vacuum Cerenkov radiation (typical of non-commutative QED) or UV/IR mixing (generally related to non-commutative theories). Do these phenomena actually occur in supersymmetric theories?

- *Ferrara:*

Well, in the considered framework a very particular kind of non-commutativity arises, namely anti-commutativity. Indeed, the superspace underlying space-time supersymmetry is described by coordinates, which split in usual (commuting) numbers, plus Grassmannian numbers which do anti-commute. In such a context, anti-commutative variables are essentially a tool to describe a finite number of commutative fields, along with fermionic anti-commutative fields. Thus, strictly speaking, a supersymmetric theory is not anti-commutative, but rather it has a commutator form with a variable sign in its right-hand side (minus in the bosonic sector, plus in the fermionic one). In other words, a standard supersymmetric set of coordinates is composed by commutative and anti-commutative numbers. Consequently, I do not think your arguments apply.

- *Di Mauro:*

It is known that an ordinary Poincaré algebra can be deformed into a twisted Poincaré algebra, or in a k-Poincaré algebra, or something similar. Is it possible to deform a Poincaré superalgebra in an analogous fashion?

- *Ferrara:*

Yes, it is indeed possible to construct a non-commutative superalgebra. This procedure is mostly used when studying superstring backgrounds, as for instance in the works by Seiberg and Witten. There is a consistent mathematical machinery to introduce non-commutative field theories, also with non-anti-commutativity in the fermionic sector, implemented through Grasmannian variables. This gives rise to some nice new classes of theories.

- *Hoehn:*

At the beginning of your lecture you pointed out that supersymmetry is needed in a renormalizable quantum field theory, if one wants fundamental scalar fields (such as the Higgs one) to naturally arise in the spectrum. Could you please comment more on this? What is the crucial role of supersymmetry?

- *Ferrara:*

Well, actually the story goes slightly different. I said that, in the currently widely accepted interpretation of elementary particle physics, the only instance where a fundamental scalar field arises out is in the description of the spontaneously broken sector of a gauge theory. Actually, such a particle is still unobserved. Generally, supersymmetry provides a consistent framework in which scalar fields are elementary fields of the spectrum. Indeed, in the Minimal Supersymmetric Standard Model (MSSM) itself supersymmetry actually unifies the gauge sector to the Higgs sector. As I showed to you this morning, the quartic coupling of the Higgs field is not an arbitrary parameter (as in the usual Standard Model), but rather it is directly related to the square of the combination of the gauge couplings of SU(2) x U(1). Furthermore, in the supersymmetric limit the Higgs particle becomes partner of a massive gauge boson. In other words, in the current implementation of supersymmetry in the MSSM, the Higgs sector is crucial for a consistent description of the gauge sector.

- *Nath:*

I would here add that supersymmetry plays a key role also in the issues of the dark matter and of the heaviness of top quark, subject of intense research activity by Arnowitt, myself and you, among various other authors. The fact that the t-quark was predicted to be heavy is particularly remarkable. Furthermore, in a very interesting paper H. Goldberg proposed that the neutralino is a possible candidate for dark matter.

- *Ni:*

You mentioned about proton decay in supersymmetric theories. Which constraints can be put on supersymmetry, based on current experimental limits on the proton decay? Which are the decay channels predicting a proton lifetime longer than the current limit?

- *Ferrara:*

I am sorry, I am not expert on this issue, so I am not be able to answer to your question in a satisfactory way. All I can say is that there are various supersymmetric models, so the answer, whatever it may be, is necessarily model-dependent. I can add that when the Grand Unified Theories were firstly elaborated by Georgi and Glashow, the channel they worked about was actually suppressed. However, other dangerous terms, possibly yielding very fast proton decay, were later discovered in the model. They are there, unless certain relations between Yukawa couplings are imposed.

- *Goebel:*

You lectured about supergravity and its relevance for the gravity mediated supersymmetry breaking mechanism. Could you also shortly comment on the anomaly mediated supersymmetry breaking? Which is the most promising model?

- *Ferrara:*

Anomaly mediation of supersymmetry breaking takes into account the superconformal anomaly. So, the resulting effect is really tiny in the gravity sector. On the other hand, anomaly mediation shares some features with the gravity mediation, and also with the gauge mediation, mechanisms. Thus, I would say that it is an "intermediate" mechanism of supersymmetry breaking.

- *Bruno:*

What is the role of Majorana spinors in supersymmetry?

- *Ferrara:*

Majorana spinors are crucial for phenomenologically viable supersymmetric field theories. Indeed, if Dirac, rather than Majorana-Weyl, spinors were used to formulate supersymmetry, one would not discover N=1 superalgebra, which plays a key role in the phenomenology of supersymmetry. Indeed, N=1 supersymmetry is the only one consistent with chiral fermions, and it deals with a single Majorana supercharge. On the other hand, Dirac fermions naturally lead to N=2 supersymmetry, naturally arising in superstring theory from Calabi-Yau compactifications. However, N=2 supersymmetry cannot properly describe our world, because at low energies it implies mirror fermions, which are not observed. Actually, it is very difficult to make a phenomenologically consistent theory based on N=2 supersymmetry.

Furthermore, the classification of various SUGRAs based on the number of spinorial supercharge generators is strictly connected with the reality properties of spinors in higher space-time dimensions, and with their eventual Majorana nature.

- *Kopnin:*

I have a question about gravity mediated supersymmetry breaking. Is there a simple explanation why it induces only particular soft breaking terms?

- *Ferrara:*

Well, I did not lecture about any kind of soft breaking terms, because I did not deal with the phenomenology of supersymmetric models. The phenomenology of gravity mediated supersymmetry breaking has been covered in the lectures by Professor P. Nath. I only meant to provide the audience with some facts, discriminating between gravity mediated and gauge mediated supersymmetry breaking mechanisms. Gauge mediated supersymmetry breaking is one of the issues treated in the lectures by Prof. J. Schwarz. In this approach, formalized in the so-called F-Theory, the gravity sector is completely decoupled. In the corresponding renormalizable Lagrangian, the gravity does not communicate with the Standard Model sector at the tree level. However, there are evidences that in such a sector supersymmetry is spontaneously broken, and there are some fields which fill the gauge interactions of the Standard Model. In this framework, supersymmetry breaking is connected to loop-generated interactions. A detailed treatment of gravity mediated supersymmetry breaking models falls beyond the aim of my lectures.

CHAIRMAN: S. FERRARA

Scientific Secretaries: A. Gnecchi, A. Marrani

DISCUSSION II

- Shakirov:

As we saw today, the properties of supersymmetry crucially depend on the dimension of spacetime. In particular, if it is even or odd, divisible by 4, etc. .For this reason, it is natural to ask: are there any problems with dimensional regularization in supersymmetric theories?

- Ferrara:

This is a good question, I think also Gerard has been involved on work about this. Well, as far as I know, people use dimensional reduction, which is a method which manifestly violates supersymmetry. But this does not mean that what you compute through dimensional reduction is not consistent with supersymmetry. For instance, in gauge theory you can take a Lorentz-violating gauge, but this does not mean that the theory violates Lorentz invariance at all. As far as I know, there is currently no supersymmetric regularization procedure, at least for supersymmetric chiral gauge theory. However, people can still make computations using non-supersymmetric regularization. For example, dimensional regularization violates supersymmetry, and for non-chiral gauge theories you can use Pauli-Villars supersymmetric regularization scheme, as well. Dimensional reduction can also be used instead of dimensional regularization. In dimensional reduction, one introduces some fake degrees of freedom, depending on the epsilon parameter, where epsilon is the "distance" from four dimension. For instance, we know that string theory is a finite theory of gravity, and the natural parameter therein is the string cutoff. I think that Zvi Bern could make a comment about supersymmetric regularization.

- Bern:

There is definitely a problem about supersymmetric regularization procedure in chiral supersymmetric theories, and one can actually spend an infinite amount of time worrying about that. Such a problem manifestly shows up through peculiar pathologies. For example, divergences that you encounter in doing calculations are not real divergences. By performing explicit computations, you see that there is always such a problem. This problem seems to be peculiar to chiral frameworks. Indeed, I and my collaborators have gone rather far in higher-loop computations, namely at least until five loops, and we did not find any problem in non-chiral supersymmetric theories.

- Samsonov:

Today you lectured about models with 32 supersymmetries, and you said that they have a universal form in any spacetime dimension. You also mentioned about the importance of central charges. Is it possible to build a model with 32 supersymmetries with central charges in 4 spacetime dimensions? If yes, what is this model?

- Ferrara:

This is N=8, d=4 supergravity, also called maximal supergravity. Such an answer was already given in my seminar, when I mentioned that central charges correspond to configurations of the underlying theory, which has massive particles which may saturate the so called BPS bound. The usual stringy realization of maximal is the toroidal compactification of 10-dimensional Type II string. In N=8 supergravity a consistent black hole solution exists and, as a particular case, it corresponds to massive black holes which are BPS-saturated states.

- *Zichichi:*

How much is the mass of such a massive black hole? What do you mean by massive black hole?

- *Ferrara:*

In the considered theory everything is measured in Planck length unit, so the mass of the black hole is of the order of a Planck scale. Ultimately, it depends on the order of magnitude of the black hole charges. The background of a BPS-saturated black hole preserves only a part of the 32 supersymmetries pertaining to the asymptotical background. The black holes whose entropy is classically given by the celebrated Bekenstein-Hawking entropy-area formula, i.e. the so-called "large" black holes, are 1/8-BPS black holes in this framework. The black holes which preserve more than 4 supersymmetries, namely ¼- and ½-BPS black holes, have vanishing classical entropy.

- *Shcherbakov:*

A supersymmetry algebra with a central charge might be equivalent to another superalgebra without central charge in a different number of spacetime dimensions. Does this fact ever occur in black hole physics?

- *Ferrara:*

Yes, it may happen. For instance, one can actually obtain black holes in four dimensions starting from an higher-dimensional theory which does not contain any black hole solution. In such a case, four dimensional black holes are originated by the so-called Kaluza-Klein monopoles. The electromagnetic field appearing in such a black hole is the so-called graviphoton, which is nothing but the Kaluza-Klein vector. This is what happens when you interpret four-dimensional central charge as the fifth component of the five-dimensional momentum.

- *Bonati:*

In d-dimensional field theory with finite temperature, it is customary to use effective field theory in d-1 spacetime dimensions, through a dimensional reduction procedure. Since the features of supersymmetric theories strongly depend on the dimension d, is it also possible to use dimensional reduction, or are there problems with that?

- *Ferrara:*

Well, maybe it can be done, but I don't really know the answer.., good question anyway. I believe that non-vanishing temperature breaks supersymmetry, and probably what one achieves at high temperature is rather similar to what one may obtain in a non-supersymmetric framework; but this is my personal guess. That is because temperature splits bosons and fermions, similarly to what happens when you have a spontaneous symmetry breaking.

- *Korthals-Altes:*

You're right. Supersymmetry is broken by non-vanishing temperature. However, this is only a very soft breaking, in the sense that supersymmetry breaking is only a boundary effect.

- *Ferrara:*

I can also add that a temperature-dependent potential explicitly breaks supersymmetry.

- *Korthals-Altes:*

That is true, but I still point out that it is a very soft supersymmetry breaking effect.

- *Zichichi:*

That is an interesting topic, but I would say exactly the opposite. Of course, only by guess, because I am not working in the field.

- *Ferrara:*

Well, to make things clearer, you can imagine to add a mass term, which explicitly breaks supersymmetry. By sending the mass to infinity, the corresponding breaking effect clearly dominates on all the other ones, and I believe that at very high temperature physics is very similar to the one you would obtain in a non-supersymmetric theory. The more you increase the temperature, the more you break supersymmetry. In other words, introducing soft-breaking terms in a supersymmetric Lagrangian, and sending the soft-breaking masses to infinity, the Lagrangian clearly is no longer supersymmetric, because the fermions would decouple from the bosons.

- *Zichichi:*

What if supersymmetry is broken at, or very near to, the Planck scale? For example, at the GUT scale?

- *Ferrara:*

It could be that, even if supersymmetry is broken at the Planck scale, there is some mechanism which would keep the supersymmetry all the way down. There are ten-dimensional string models which are tachyon-free and classically consistent, ad they are non-supersymmetric. By analyzing the spectrum of such models, one finds massive spin-3/2 states, named gravitinos. In some sense, these models (related to heterotic strings) can be seen as models which break supersymmetry at the Planck scale. The problem is that they are not consistent against quantum corrections, because they exhibit an unstable vacuum. When adding quantum corrections, such theories generally yield divergences. Thus, the finiteness of superstring theory is crucially related to the fact that supersymmetry is unbroken at the Planck scale. So, one should believe that supersymmetry is unbroken at the Planck scale, and that when the Universe was born it was precisely and completely supersymmetric. Such a belief is also natural from the point of view which we usually adopt, when we think that supersymmetry is restored at very high energies. In other words, the higher energy is, the simpler things become. If the above holds, there is no reason why supersymmetry should be broken at three orders of magnitude down from the Planck scale, and this is connected to the other scale present in Nature, namely the Fermi scale.

- *Zichichi:*

I think that this is due to the fact that there is a gap of two orders of magnitude between alpha-GUT and alpha-string. Could you explain why?

- *Ferrara:*

Well, the explanation depends on the model under consideration. Such a gap is actually there if you use heterotic Grand Unification models, but in other models (such as the ones John Schwarz treated in his lectures) I think that such a gap is somehow different.

- *Zichichi:*

This is also related to the fact that alpha-GUT is not a theoretical prediction, but rather a "postdiction".

- *Nath:*

I think there is nothing particularly mysterious about supersymmetry. One can actually think about it as described by an ordinary Lagrangian with certain relationships among the couplings. Thus, there is no particular reason for which you cannot have such a symmetry present at the Planck scale, or at any other scale, unless you break it in some way. In other words, one starts by assuming the existence of a certain symmetry, and then it is crucial to investigate whether the potential appearing in the Lagrangian maintains or breaks such a symmetry. So the key issue is the existence of the symmetry itself, which however we usually assume dealing with supersymmetry.

- *Ferrara:*

I remember some lectures delivered by Steven Weinberg on supersymmetry breaking. He used the following argument based on the Standard Model: what would Standard Model be without gauge symmetry breaking? The clear answer is that it would be nothing but one of the most unphysical theory ever existed. Indeed, it would correspond to the original Yang-Mills theory, with four massless vectors, carrying long-range interactions (other than the electromagnetic one) never observed in Nature. Essentially, all the Physics and all the testable predictions of the Standard Model, as well as the fact that at low energies it reproduces the Fermi energy phenomenology, rely entirely on the phenomenon of gauge symmetry breaking. So, no surprise if there is a theory like supersymmetry, in which the supersymmetric phase yields a completely unphysical theory (as the Standard Model without the Higgs sector also is), and that all the Physics, which in this case is what we hope to discover at LHC, comes from the supersymmetry breaking sector. This is not really artificial, otherwise also the Standard Model might be thought to be completely artificial.

- *Zichichi:*

The trouble is that you are not able to theoretically predict supersymmetry breaking. In the gauge breaking there is at least a hint to justify what could be called theoretical predictions. All in all, this symmetry breaking is *ad hoc*, as Sidney Coleman used to emphasize.

- *Ferrara:*

It is here worth remarking that the symmetry breaking of the Standard Model is also *ad hoc,* in the sense that there is *no* reason why the parameters and the coupling constant of the phi^4 interaction have any knowledge of the gauge sector. This is ultimately related to the so-called hierarchy problem. In a spontaneous symmetry breaking the situation is similar, but one can still consider models such as the ones Nath lectured about, which are completely consistent from a mathematical point of view. When coupled to gravity, they determine a consistent Super-Higgs mechanism, and predictions about the mass spectrum, as well as on other observables, can be worked out. What really is missing in such a "top-down" approach, is the reason why you choose a particular breaking, and this might be fully justified only if one had a precise knowledge of the theory at the Planck scale. I think that the nice property of supersymmetry is that at relatively low energies, such as the ones explored by LHC, it can actually be disproved, or discovered. Unfortunately, string theory is not at that level. As far as I know, there are no stringy signal that LHC might disprove, or discover. Assuming that the completion of a supersymmetric field theory at the Planck scale is string theory or M-theory might provide a consistent scenario, in which the only thing we do not know is how to relate the low energy parameters with the Planck scale Physics. I think, however, that for the moment it would be sufficient to discover supersymmetry, which would be the most revolutionary thing in Physics after the observation in 1919 of the deflection of light due to the Sun, one of the most important experimental successes of Einstein's General Relativity.

- *Gazis:*

I would like to add just a comment to this discussion, namely the point of view of an experimentalist. I am not happy at all that most of the experiments at ATLAS and CMS have adopted the supergravity breaking mechanism. Having in mind that R-parity is conserved, they take into account the mass effect of the multi-jets leptons' energy missing, which results only in the possibility to detect some more events than what foreseen. From my personal point of view, this is a definitely unclear signal, because it might be an evidence for supersymmetry, but it also might be determined by the noise of the detector itself. Thus, I would really look for a new theory, maybe string theory, that could give us some hints of isolated multi-jets, or something that can be really disproved by the experimental evidence.

- *Ferrara:*

Well, string theory can only give you a particular model at low energy, and this is related to the question which I asked to Dimitri Nanopoulos during his lectures, namely whether in the eventual discovery of supersymmetry one can disentangle, and discriminate among different scenarios of supersymmetry breaking, like supergravity-mediated and gauge-mediated (this latter related to the local F-theory models, discussed by John Schwarz) ones. In other words: if supersymmetry is observed, is there a way to disprove one or the other mechanism of supersymmetry breaking? On top of that, I would also add you cannot obtain additional information from string theory, other than the ones given by a consistent low-energy theory.

- *Nath:*

In my opinion, it is here worth pointing out that it was quite usual in the history of Physics that a theoretical missing link lead to the discovery of new phenomena. At least two examples of that are provided by the Standard Model. The first one is the discovery of neutral currents, which lead to the discovery of the charm. The second one is the issue of the anomaly cancellation; in the Standard Model, it lead physicists to predict the existence of the top quark when the bottom quark was discovered. In the context of the theoretical puzzles related to the Higgs boson, supersymmetry might be one of the answers. So, in the history of theoretical Physics a missing link has often lead physicists to ask new, deep questions, to be later verified through experiments.

- *Zichichi:*

Well, let me now defend experimental Physics. I agree with what you say, Prof. Nath, but we should not forget that the Adler-Bell-Jackiw (ABJ) anomaly originated by the pi0-gamma-gamma decay, and this was a previously existing experimental fact. Trying to understand such a decay, the ABJ anomaly was discovered, and then theoretically extended to the thorough formulation of the anomaly-free theories. The anomaly-free condition definitely is the reason leading to the prediction of the top quark mass.

- *Musaev:*

Prof. Ferrara, you lectured about Majorana and Dirac spinors, and also about the fact that the real or complex nature of a spinor depends on the signature of spacetime. Are there any geometrical reasons for such a difference? Maybe related to geometrical invariants?

- *Ferrara:*

Well, let me start answering by saying that there are some people attempting to construct a spacetime with more than one timelike dimension, but the resulting theories are not so elegant in my opinion. Indeed, in such contexts one has to get rid of the extra timelike dimension by advocating extra symmetries, which also allow for the decoupling of ghosts, and so on. Furthermore, let me mention that, starting from the fact 32 is the maximum number of supersymmetries one can have without considering high spin field theories, there is still room for a Majorana chiral spinor in twelve dimension, when the spacetime signature is (10,2). However, if one tries to write down the supersymmetry algebra in twelve dimension, the anticommutator of two supercharges turns out not to contain the corresponding momentum. This happens simply because the gamma matrices have not the right properties to fit in the right-hand side of the anticommutator. Consequently, such a superalgebra cannot be interpreted as Poincaré superalgebra, but nevertheless it is a mathematically consistent superalgebra. All in all, theories in twelve dimensions are unphysical. It is still true that if you start from twelve dimensions with signature (10,2) and you perform a dimensional reduction along a timelike direction, you obtain the eleven-dimensional M-theory superalgebra. However, such an algebra cannot be consistently uplifted to twelve dimensions. In other words, in M-theory there are not states which can be interpreted as Kaluza-Klein states. So, there is no possibility to perform an uplifting to a twelve-dimensional theory in the ordinary sense. The only possible way one can construct a consistent twelve-dimensional theory is unrelated to the reasoning considered above, and it is given by the F-theory construction. In such a framework, described by John Schwarz in his lectures, one starts from the ten-dimensional spacetime where superstring theory lives, and at each point of the internal manifold one attaches a two-dimensional torus. Thus, F-theory is a somewhat twelve-dimensional theory, but not in the sense of what we are describing here.

R. LOLL

"Quantum Gravity from Dynamical Triangulation"

The Slides of the Lectures can be found at:

http://www.ccsem.infn.it/issp2008/

CHAIRMAN: R. LOLL

Scientific Secretaries: A. Nikiforov, D. Weir

DISCUSSION I

- Lockyer:

Why do the proton and the antiproton annihilate each other? Before the antiproton was discovered, it was not obvious that if you made something like the antiproton it would actually annihilate [the proton], because it's such a complicated object. Do you have any thoughts on that?

- Zichichi:

The Dirac equation refers to electrons -- not to protons -- and the annihilation is supposed to be universal. Anything should annihilate its own antiparticle; at the particle/antiparticle level this is well established.

- Lockyer:

But the proton is not a particle.

- Zichichi:

The proton contains a universe inside -- it is a miracle that you get spin-1/2 to begin with. The same miracle is repeated in the antiproton, because you assume that the forces which act inside the subnuclear world obey the CPT theorem. It is CPT that guarantees elementary particles and forces have the corresponding antiparticles and antigluons, and that is why the annihilation is granted. Nevertheless it is a problem to predict the anomalous magnetic moment of the proton, because even when you go to $q^2=0$ you do have a problem -- why is it what it is? I will give you my paper on antiparticles, which contains discussions with people including the father of all this, Paul Dirac. In some papers you may read that antihydrogen is antimatter; it is not antimatter, because the concept of antimatter is to be given to the proper objects -- this means, particles and glues.

- Stavenga:

You described a "sum-over-histories" where you write down all the ways to connect simplexes, divided by a symmetry factor. I know that algorithms to check the symmetry factor of configurations of graphs are exponential, so for 10000 simplexes it would be very difficult. Is this a problem?

- Loll:

This is not a problem because the factor relates to accidental symmetries of these very irregular structures, which are very rare.

- Stavenga:

Do you just ignore them?

- Loll:

You can completely ignore them. For even a handful of triangles, the factors are almost always one.

- Stavenga:

The series looks like a Feynman series, so it would seem to me that you are doing a perturbative expansion for a field theory.

- *Loll:*

There is no immediate way that you can think of it as a perturbative Feynman series, just because you are summing over a certain class of objects. You can think of it as a statistical mechanical model.

- *Stavenga:*

With Feynman graphs, you can define a quantum field theory as the sum of combinatorical objects with symmetry factors.

- *Loll:*

That's right, and our approach has some similarities. In 2D our approach is entirely combinatorial and you can almost do the exact path integral, so you have some sort of handle. Usually we evaluate these quantities for a finite volume, and then what is the small parameter? If it is a perturbation series there should be a small parameter, and that is something that does not appear at this stage.

- *Condeescu*:

What is the definition of equivalent sewings of simplexes and why is it a good idea to replace diffeomorphism equivalence?

- *Loll:*

Two such spaces are different if I do not have a one-to-one correspondence between them: if they are not identical as curved, piecewise flat spaces, and if I don't have a one-to-one map that preserves all the gluings. This is a good idea, because I really want to ergodically cover the space of all geometries. If you go to such a construction then you capture all the local curvature degrees of freedom. You are not, therefore, leaving out any local geometric degrees of freedom in any obvious way. Another point is that, strictly speaking, there are no theorems that prove that this is the correct way of doing it; you have to look at the end result and decide whether this makes sense. In the example for 2D, gluing triangles and taking the sum gives you a result that is equivalent to a continuum calculation. In higher dimensions you cannot do that because you do not have an analogous continuum calculation. You see what your regularised theory gives you in some continuum limit and whether that makes sense physically.

- *Susskind:*

Am I correct in saying that you are doing a euclidean study?

- *Loll:*

Today, in order to set the stage I focused on the Euclidean approach but we will see next time that it doesn't actually work for the Euclidean signature, you have to bring in causality.

- *Susskind:*

So in the end, it is a Minkowski space construction?

- *Loll:*

Yes, you start by writing down a path integral over Lorentzian piecewise flat spaces.

- *Susskind:*

Is it possible to maintain Lorentz signature?

- *Loll:*

No, and of course in field theories it is exactly the same. You have a Wick rotation -- a slight abuse of the phrase -- which maps all of these configurations to unique Euclidean spaces, and you actually evaluate the sum in the Euclidean sector. Then, you look at the end result and see whether you can translate that back to Minkowskian signature.

- *Susskind:*

So it involves an analytic continuation at some point that you cannot do numerically?

- *Loll:*

You do the sum in the Euclidean domain: there is no other way of extracting that nonperturbative information. Then, you stare at the end result. Some results immediately translate back to the Lorentzian sector but for general observables it is a highly nontrivial thing to do. That's also true of time-dependent observables in lattice QCD.

- *Nicotri:*

Can you please comment more on the diffeomorphism invariance and the problems with the functional measure.

- *Loll:*

One of the nasty things when you write down this continuum path integral as a function of the $g_{\mu\nu}$'s is that you have to quotient out by the symmetry, so there is no other way if you start with a coordinate system, writing down field variables, they could be $g_{\mu\nu}$'s or they could be vierbeins. You don't really know how to do the quotient: you could do it perturbatively, but that's not enough. You are stuck in the continuum theory, at that point. Doing it perturbatively you have to evaluate the Fadeev-Popov determinants. This is already impossible in two dimensions and you do it by bootstrap. In higher dimensions, you can't solve the functional determinants at all.

It is really a democracy among distinct curved spacetimes of a certain kind: all the possible spacetimes you can build from some elementary building blocks. There are not many natural classes of objects which you could write down. You take some gluing rules then go for all possible objects of this kind. People started by looking at Euclidean triangulations -- which are the most natural thing -- and you see what you get. Unfortunately, it doesn't make any sense but we will see the way out in the next lecture.

- *Di Mauro:*

The CDT program is based on a Lorentzian functional integral. Given that it is not possible to make a Wick rotation on a general spacetime, is there a relationship between causal dynamical triangulations and Hawking's Euclidean quantum gravity program?

- *Loll:*

The impossibility of defining a Wick rotation refers to the continuum theory. If you start with general $g_{\mu\nu}$'s no-one has come up with a general map. Of course, the map has to be compatible with the diffeomorphisms, for example. Of course you can do easy Wick rotations for stationary spacetimes -- that's straightforward but for a general time-dependent $g_{\mu\nu}$ there is no such thing. My statement is that, for the class of piecewise flat spacetimes where you no longer have this gauge redundancy, there is a nice map from piecewise linear Lorentzian objects to Euclidean ones. Then, you can ask: after you have done the Wick rotation are you not back to the Euclidean formulation, the Hawking-style configuration space? The answer is no, because these spaces do know about causality constraints that were implicit in the Lorentzian construction and the arrow of time. If you look at them purely combinatorially as objects made from building blocks, then my Wick-rotated set is inside the space of all Euclidean objects you could have written down in the first place, but it is a set of measure zero on which you perform the same path integral, and this gives a completely different result.

- *Reeb:*

You mentioned that dimensionality is dynamical, but how can spaces of lower dimension crumple up to form some higher dimensional space? Doesn't that require an embedding? How can objects move in this higher-dimensional space if they are confined to some lower-dimensional manifold?

- *Loll:*

Of course, the analogy with the crumpled paper had its weak points. One of them was, as you pointed out, that I was using an embedding picture. I wanted to illustrate that you had some microscopic building blocks: suppose that they are four-dimensional. Then, you can just put them together in such a way that at some larger scale they pretend to be a space of a different dimension. It can of course go both ways. If you glue 4-dimensional building blocks in the Euclidean approach, you find there are two phases. One of them, the ``branched-polymer" phase, has couplings such that the preferred arrangements -- an entropic notion, if you like -- are locally four-dimensional, but if you go to a slightly larger scale and compare length scales to volumes they behave like a one-dimensional space. By contrast a ``crumpled space", which has some analogy to the crumpled paper example, has a distinguished vertex around which every other building block crowds. Every building block wants to share that vertex. Triangulations of these crumpled spaces contain thousands of building blocks that all crowd together into a little space whose diameter never extends beyond the cutoff scale. In this way you get an object that is very high-dimensional.

- *Musaev:*

In quantum gravity one integrates over all manifolds, and it is an integral over histories. Am I right in saying that to make this calculation you introduce simplexes and you move the simplexes in some ambient space?

- *Loll:*

No, there is no ambient space whatsoever.

- *Musaev:*

And how do you sum over all triangulations?

- *Loll:*

We evaluate these sums by Monte Carlo simulations, a Markov process in the space of all configurations. How is such a triangulation given to me in completely intrinsic terms? I take a bunch of N triangles, I give them labels and write down a big $N \times N$ matrix. Whenever two triangles share an edge, I put a `1' in the matrix. I have to check that the resulting object is locally a manifold -- there are screwy ways to glue them which don't satisfy that -- but you can formulate this as conditions on the so-called connectivity matrix. If I give you that matrix, you can uniquely reconstruct the 2D triangulation, knowing nothing about the embedding space. This works in any dimension.

- *Musaev:*

So it is a different calculation compared to that of a point particle?

- *Loll:*

There you have a given spacetime and you look at your path. That is not the case here, nevertheless one still has notions of time that are intrinsically given. We can define a proper time: the time for freely falling observers. This does not need an embedding space to be well defined.

- *Musaev:*

Can we not always introduce an ambient space in which our triangulations live?

- *Loll:*

For curved, smooth manifolds there are embedding theorems which state that you can always embed these things if you go to a large enough vector space. But Einstein gravity is about intrinsic geometry -- that's how we understand it. Even 4D classical Einstein gravity requires that everything is intrinsic. If you introduced an embedding space you might able to do that, but things might depend on it in the end. That brings in additional degrees of freedom for which we have no use.

- *Bonati:*

Could the fact that the dimension became dynamical be related to the theory not being perturbatively renormalisable?

- *Loll:*

There might be many reasons. It took a while to understand that these are not nice spacetimes; they are given to you in a screwy way and you have to deduce the right questions in order to understand then that you are unfortunately just staring at a branched polymer. The answer to your question will be given to some extent by my next lecture. My collaborators and I have formulated a conjecture as to what might be the reason for that. We tried to verify it, and it seems to work. Of course, a priori there are many reasons why it might not work.

- *Bonati:*

More technically, in lattice QCD you can use the Euclidean formulation because of the Osterwalder-Schrader reconstruction theorem. Can you use that here?

- *Loll:*

If you want to evaluate these partition functions and look for their convergence properties, I don't see any way around having to Euclideanise them, to make these expressions real. The closest to what you are saying -- an Osterwalder-Schrader axiom -- is an analysis by Ashtekhar and collaborators made in the continuum. They asked themselves: what is the analogue of the Osterwalder-Schrader axioms if I am not working in fixed Minkowski space? It was found that such an object exists, but you do need a distinguished notion of time. Then, you can do all the usual things.

- *Shakirov:*

You mentioned a quantum Einstein equation. What are the variables in this equation -- is it similar to the Schroedinger equation?

- *Loll:*

No, I didn't mean this literally. I just used it to be synonymous with the given dynamical principle. I am looking for the nonperturbative ground state of the ensemble of geometries that are summed over.

- *Shakirov:*

When one has a functional integral one can calculate correlators, are there any sensible correlators to calculate in quantum gravity?

- Loll:

Yes, there are, but naive *n*-point functions are ill-defined because positions are not well-defined. There are more basic observables that are easier to measure which characterise the space to some extent.

- Shakirov:

Something like curvature?

- Loll:

Not just curvature, but simpler things like Hausdorff dimension, spectral dimension and fractal dimension. These things are relatively easy to measure. In principle you can define higher-order *n*-point functions where you integrate over all locations of the points. These are very nonlocal objects and although you would think that in principle you could construct all interesting physical information from them, in practice it is very difficult. To measure them on small lattices is also very difficult.

- Dent:

How do these triangulations behave during a rapid phase of expansion such as inflation? Would they be stretched or self-replicating? We know that inflation happened at high energy scales so perhaps this would have some implications.

- Loll:

We simulate the entire spacetime in a Euclidean regime. Nonperturbatively, the ground state generated is that of a de Sitter space. When you rotate back to Lorentzian signature, it is an exponentially expanding space.

- Slobodeniuk:

Are there some constraints on the manifold, perhaps you require a smooth metric? Can you use manifolds with conical singularities?

- Loll:

These piecewise flat manifolds are full of conical singularities. When I showed the six triangles, they are individually flat but the 2D intrinsic Gaussian curvature is located at a vertex in a delta function-like way. This generalises to higher dimensions so if you look at a triangulation made of *d*-dimensional simplexes then the curvature is distributed again in a delta function-like manner. It sits at the *D-2*-dimensional subsimplexes. So, in 4D the curvature is peaked on 2D triangles.

- Slobodeniuk:

Must we know the classification of 4-dimensional manifolds to carry out these calculations?

- Loll:

It is often said that the path integral is very nice because we can include a sum-over-topologies. In practice, though, this is impossible because the number of states for a given volume grows super-exponentially: you get a wildly diverging path integral that cannot be renormalised by normal methods. Although de facto you can write this down formally, it simply doesn't work when you try to do it. We stick with one given spacetime topology.

- Nagy:

The Einstein-Hilbert action is unbounded from below. But I recall that the principle of least action in classical field theory says that the action is minimal for the classical solution.

- *Loll:*

Classical solutions need only extremise the action, so there is no problem classically. Looking at the pure action, it is unbounded both above and below. It's normal for actions to be unbounded above from the dot x^2 term, which grows without bound. A model with an action unbounded below can still be well-defined because you get a nontrivial contribution from the measure which you don't see perturbatively. That is the key point: if you do it nonperturbatively, the number of times a given weight is realised contributes to the overall path integral. The effective dynamics is therefore a combination of a naive exponential weight plus the measure.

- *Ferrara:*

The Planck scale is only three orders of magnitude higher than the grand unification scale of other interactions. A large part of the community believes that you cannot separate the problem of quantum gravity from the problem of the other forces in the high energy domain. In some sense your approach is related to that of Ashtekar whereby quantum gravity can be solved by itself without unification. Don't you think that if the measure chose this approach, it would be very boring? Also, how would you compute the black hole entropy or any of the calculable quantum corrections that you can do in string theory?

- *Loll:*

You have posed two questions with many sub-questions. Firstly, I never said that I was solely interested in gravity. But, as you say, looking only at gravity you already find very nontrivial results close to the Planck scale. It may just be two and a half orders away from the GUT scale, but it is just a hypothesis that we have everything under control at the GUT scale.

- *Ferrara:*

That is why we must put everything together.

- *Loll:*

What we have already found is that there is a natural way of how we can reintroduce physical units. It turns out that the universes we generate are about 20 Planck lengths across. We look at certain fluctuations and find that, even for these small universes, a semiclassical description applies well. However, when we go to even smaller scales -- of the order of a few Planck lengths -- then things become very different. For example, the spectral dimension of these spacetimes smoothly changes to 2. Even within one order of magnitude there is still room for things to happen, for things to decouple -- the jury is still out on that. We could also couple matter and see which scenario is realised on the Planck scale. How do these fluctuations actually interact, does gravity drown out everything else? These are questions that can quantitatively be investigated in this framework. Loop quantum gravity is a long way from answering these questions. As for difficult questions about observables in continuum formulations, there are thousands of difficult questions that you can ask me. It is maximally difficult coming from a purely nonperturbative approach to answer these -- even to say how these match to perturbative computations that we can do, such as the black hole entropy. It is not impossible in principle, but we don't know whether all these semiclassically inspired questions still make sense at the Planck scale; they may cease to be the important questions when one is in the nonperturbative regime.

- *Reeb:*

Your method of dynamical triangulations looks like a way of regularising the path integral in terms of smooth metrics that otherwise we have no control over. Do you want to emphasise that your formalism adds some fundamentally new features?

- *Loll:*

There is nothing smooth in here, at the regularised level this is full of curvature singularities. It is like the piecewise straight paths of a nonrelativistic particle. If I take the limit $a \rightarrow 0$, then these spaces would be full of singularities. Individually, they are extremely nonclassical objects. This is the expectation you would have even in simple quantum field theories. It is not the smooth things that contribute most to the path integral, it is the rough objects.

- *Nath:*

You are talking about the Planck scale, but one typically arrives at the Planck scale by looking at the perturbative limit of interactions with matter and you make a connection with Newton's constant. Are you using two different definitions of the Planck mass? Then, what is the connection of your Planck length and mass to those connected with Newton's constant?

- *Loll:*

When I speak here of the Planck scale I put $\hbar$, c and G_{Newton} together to form a scale at which we extrapolate things should happen. Another question one might ask is, how do the coupling constants change with scale? That is a sort of renormalisation group question; behind this approach is a nontrivial vacuum. It is not the naive vacuum where the coupling goes to zero and we have a Gaussian theory. This is a different vacuum in coupling-constant space and it is the one I see when I perform this nonperturbative superposition.

- *Nath:*

You will have to make a connection with computing the forces between particles, then you can determine that connection.

- *Loll:*

Certainly that is one thing what one wants to extract from such a formalism: the effective G_{Newton} on all scales.

CHAIRMAN: R. LOLL

Scientific Secretaries: E. Ciuffoli, A. Gnecchi

DISCUSSION II

- *Shakirov:*

I have two questions.

1) The first one: you have mentioned this very interesting change of dimension in your lecture. Is it possible to put it in a cosmological context, saying that in the first moments after the Big Bang our universe was two-dimensional?

2) The second one: how do you specify the Minkowskian structure in every particular simplex?

- *Loll*

1) It is definitely possible, it's a statement of what happens on very short scales, and of course this we could only probe in very extreme circumstances. Right now we don't have any experiments that can probe this scale, so our only chance to "probe" it is to say "this is how it must have been at the beginning of the universe" and then look for ramifications, what observable consequences it might have now.

2) I just declare it to be cut out of Minkowski space. It just means that I give you all the edge lengths, and, depending on the way I set it up, some of the edges would be space-like, some would be time-like, and I declare it to be flat in the interior. That uniquely fixes the metric structure, I mean, the coordinate-invariant part of the metric structure, including the light cones.

- *Dubynskiy:*

When the 4D space arises after the transition from the 2D space does it arise with arbitrarily small size or does it have some critical size after it expands to macroscopical size? Is there oscillation of the dimension?

- *Loll:*

The curve I showed you was a continuous curve and this is what we measured from the random walk of the diffusion process. The approach is asymptotic, so there is no distinguished scale involved; it cannot jump from one to the other.

- *Dubynskiy:*

So if it starts to expand from two, it expands towards four dimensions, it never shrinks, so there are no critical sizes?

- *Loll:*

There is no distinguished size that I can see, apart from the particular asymptotic values that get me to some number – two or four – numerically.

- *Reeb:*

I am worried that your set-up with choosing time slices is maybe too restricted and maybe can capture the simple structure of de Sitter space but not more complex things, so can you for example build in matter into your model and then look at gravitational collapse to see if your CDT proposal can handle more difficult spacetimes? Have you looked at singularities in your proposal?

- *Loll:*

There are two questions. The first was about the time slicing: no, we don't see anything that constrains us. It's this intrinsic proper time that is there. There seems to be no way of not having it and obtaining the same nice results, so presumably that tells us something important. For the case of de Sitter, you could have said "well, clearly you are completely breaking any notion of covariance by introducing a preferred time", but of course what is really nice about the Euclidean de Sitter solution is that it is a round four-sphere.

- *Reeb:*

Maybe de Sitter comes out because you put in the time slicing?

- *Loll:*

No, S^4 is the maximally symmetric solution in the Euclidean domain. It is maximally symmetric with respect to interchange of time and space. I am exactly breaking this symmetry in my regularization, so it's very nice to see that it is restored in the vacuum state. We found this solution without putting it in, so we are sure it is not such a bad approximation. Part two of your question concerned matter coupling. It's straightforward to couple matter to this; it has been done in the Euclidean models with all kinds of matter: scalar matter, gauge matter. It's not difficult in principle, of course there are numerical limitations. What is difficult is again to say what are the diffeomorphism-invariant observables you are going to study in this setting, and what physical information you are going to expect from these. This is not straightforward. Another setting we're thinking about implementing is to put isolated point masses, one or several, into this. One may think it would be a nice way of verifying Newton's law. Again, one has to be careful because we are in a diffeomorphism-invariant setting, so again it's not as easy as saying "I have a mass at point x, a mass at point y and let me verify Newton's law", because there is no such thing as a point x and a point y.

- *Reeb:*

Did in any one of these scenarios ever come anything other than de Sitter space, like some gravitational collapse or something else?

- *Loll:*

Look, nothing that makes any sense ever came out before we derived de Sitter space, so it's a starting point. It's pretty good, and of course if you put matter in, other spaces might come out. For the time being I'm really looking at the gravitational vacuum, and this is what it seems to be. There is another very interesting point I would like to make in this context. The topology we're putting in is S^3 for the spatial slices cross a time interval, which we have identified cyclically. So what we have is S^3xS^1. Now what is very interesting is that it dynamically emerges that the system wants to be an S^4! You may remember the picture: the way it achieves this is by pinching off dynamically above and below. Although I start with a particular topology, you see dynamically that the system wants to be something different and we can actually map it onto S^4, the Euclidean de Sitter space. So again it's dynamics that is at work, and we really see some dynamical outcomes of the quantum fluctuations at small scales.

- *Nicotri:*

Can you please explain how one can put field theory on a spacetime with evolving dimension and in general with a non-integer dimension? What happens to field theory when the dimension of the space evolves? And, after you have seen the de Sitter space emerges, can you find some intuitive motivation which could let you guess from the beginning that this kind of space would have emerged?

- *Loll:*

Regarding the first question, as I said, it's very straightforward to couple matter to the system. Individually the geometries are kind of regular lattices, so you can just put whatever on the subsimplices, like the vertices and edges. For example, gauge fields you would put on the edges, just as you do in standard lattice field theory. Now, you ask what influence would the dynamical reduction of dimension have on the matter. My first guess would be, because it's a diffusion and therefore probes properties of the Laplacian of the space, that this "two going to four" would – at least to first approximation – be the dimension felt by a scalar field put on it, as long as it's not too strong itself. Of course, there could be a phase where geometry and matter interact strongly and where maybe this result will change, but certainly for a kind of test matter or a test particle I would imagine that some of these dynamical dimensions will reflect the behaviour of observables that have to do with matter.

Concerning the second question, could I have guessed that it was de Sitter space? It's a very nice outcome I would say. I wouldn't have dared to hope that it was de Sitter, but there you go.

- *Bonati:*

Computer simulations use dimensionless numbers; in lattice QCD the dimensional scale is determined for example by calculating string tension and identifying it with the experimental value; what do you use to set the scale?

- *Loll:*

I said that you determine from a scaling analysis that half a lattice spacing corresponds to one Planck length. This we obtained through a correlator, namely, between fluctuations of the three-volume. When you compute it explicitly it depends on Newton's constant. When we make this calculation on the lattice, we are also imagining that the outcome has a multiplicative renormalized Newton constant in front of it, and this is the way we make the identification between dimensionless lattice quantities and physical ones.

- *Bonati:*

Is it possible to obtain something like nonperturbative beta function for gravity, and eventually confront it with the nonperturbative renormalization group calculations you cited before?

- *Loll:*

In our attempts to compare this treatment with the renormalization group flow calculations that people have been doing I mentioned this morning, which get similar results, we have started to think about how to map certain flows of couplings. The set-up we have here is a little different from what you do in the continuum, where you speak of a single momentum scale k, that appears everywhere in your effective action and everything depends on k. Here we have the various constants floating around, the lattice spacing, and the total volume, but one can relate these things. It hasn't been completely worked out, but we think we will be able to phrase our results in this language eventually.

- *Johansson:*

Do you have an intuitive physical understanding for why space should be 2D at small scales?

- *Loll:*

No, we don't. You have to create this space in your computer and you then start to test its properties. First you think "do I know of any other quantities that behave in this way which I could use as a starting point for my imagination of what this thing might be?" I don't know of such a thing. We see some evidence for the presence of a fractal structure, and a non-integer spectral dimension is of course something very common in fractals. The type of fractals which may be relevant in our case are of a random nature. For fractals we have a good computational control over things like the Sierpinski gasket in any dimension. These are very regular fractals, which you get from subdividing certain structures in certain ways. So we haven't really found a very good way to think about our short-scale structure, but you're invited to do that and come up with a good suggestion.

- *Johansson:*

Why did you use 4D microscopic building blocks as opposed to 3D or 5D? You say the properties of the building blocks should not affect the macroscopic properties. Have you studied this?

- *Loll:*

Well, I said that the *details* of our set-up shouldn't affect the macroscopic properties. Of course, the dimension of the building blocks doesn't fall under that because certainly we *have* studied the two- and three-dimensional models and they give nothing like four-dimensional gravity. So, 4D is the next-simplest thing to try. Of course, at least at the discretized level, you can make an argument that a reason for doing it with 4D building blocks is the fact that you roughly speaking catch the local degrees of freedom that you know are in gravity. Now, in principle you could go to higher-dimensional building blocks and you would have to discuss what action it has, how do you weight everything. What may of course happen is that the geometry collapses dynamically down to four dimensions. Maybe that is a scenario a string theorist would embrace or like, I don't know, but it's of course a logical possibility. It is genuinely challenging to go one dimension up, just because the linear size of your universe becomes quite small and it becomes even harder to extract reliable physical results from it. Coming back to your previous question about two-dimensionality, we may be able to find a kind of two-dimensional effective theory, maybe it is something that has very little to do with gravity that determines the two-dimensional dynamics. That would be of course very interesting, and something you can test if you have a concrete proposal.

- *Slobodeniuk:*

Can we have in this theory objects like black holes or other objects with singularities?

- *Loll:*

Since I'm summing over all configurations and most of them are highly singular, certainly configurations with small black holes in some places are contained in that huge sum somewhere. So that is part of the answer. Another thing you might ask is if there is any way to get a macroscopic limit in which you obtain a macroscopic black hole, about which you might ask all kinds of interesting questions like how much entropy it might have. In order for this to come out you would have to tell the boundaries something about the existence of a black hole, in the sense that you would have to encode this as a boundary data in an appropriate, geometrical way. Everything that's encoded here is geometrical, there are no coordinates anywhere, which makes it a little harder. So yes, we have already thought a little bit about it, it's very nontrivial, I can tell you more about this in private if you like.

- *Weir:*

We heard from Prof. Kenway how lattice QCD had made great progress thanks to new hybrid Monte Carlo algorithms. Obviously a hybrid algorithm is not appropriate for gravity but how much scope is there for algorithms that allow you to simulate larger universes?

- *Loll:*

The methods we are using have been around for some time because the case of *Riemannian* random geometries has been studied for the last 20 years in various dimensions. What certainly distinguishes them from standard lattice computations is that they are not parallelizable, and that has to do again with the dynamical aspects of geometry. You cannot divide the lattice into regions that only talk to each other a little bit, because the lattice is of course constantly “on the move”. What we have looked at are lattices of sizes of about 350.000 simplices. That is the kind of maximal size we can look at currently, and one could push that somewhat. It seems that in principle there is room for improvement, to make them a bit larger and reduce the error bars of whatever observable you want to look at.

- *Arnowitt:*

Can you calculate the ratio of bare cosmological constant to the clothed one that emerges in the large-scale de Sitter results? And, another question: you do not sum over all topologies (there are too many of them), but suppose you summed over just two topologies. Does the partition function just factor or is it more complicated and can quantum transitions occur between topologies?

- *Loll:*

We have a bare regularized set-up where I have a bare cosmological constant and I need it to be positive, because in the Euclidean action it appears as an exp(-Lambda times volume) term. Why do I need a term that suppresses large volumes? Because I have an entropy term – coming from counting microstates – that diverges exponentially with the volume, so I need a sufficiently large bare cosmological constant Lambda for the partition function to be convergent. In taking the continuum limit, I will exactly tune my bare Lambda to the critical value that sits exactly on the radius of convergence of the partition function as N goes to infinity. The way I approach this critical boundary determines what my renormalized cosmological constant is, and that is a priori a free parameter. It dictates the rate of approach to the critical surface in my coupling constant space. It appears in observables, so if I evaluate a simple observable like the total space-time volume of the universe (in my case the de Sitter four-sphere in Euclidean space), of course Lambda enters; in this way I can then relate the renormalized coupling to a specific observable, to something I might observe in the real world.

- *Arnowitt:*

So you can do it in the AdS case or not?

- *Loll:*

No, it does not converge, I really do need a positive cosmological constant. As for your question about topology, the problem is: how do you give meaning to that in a dynamical way? How do you encode that it should have one out of two topologies, how do you tell the local dynamics about it. It seems there's a problem here; if you allow topology change as a function of time then you get something like a third quantization formalism, but then it's just locally allowed that things split up into two. But once you allow this as a process, it overwhelms everything else. So there is no very good way of phrasing what you want to do, you want to have a mild, domesticated version of topology change, maybe just a finite number of topologies. You can try, but it doesn't make too much sense at a fundamental level it seems to me.

- *Maier:*

In your first lecture you stated that the Lagrangian had no lower bound. What happens if you derive a Hamiltonian?

- *Loll:*

You don't see that because the nice thing about the Hamiltonian formalism, about performing a canonical analysis, is that you see explicitly that the conformal factor, by virtue of the presence of constraints, can be solved for. So, it is expressed as a function of the other variables, thus it doesn't seem to be there when you do things canonically. However, in the path integral it has obviously been there, and has been a reason for concern, but as I showed, this problem does find a nonperturbative resolution, so things are just fine.

CALT-68-2714

STATUS OF SUPERSTRING AND M-THEORY

John H. Schwarz[1]
California Institute of Technology
Pasadena, CA 91125, USA

Abstract

The first lecture gives a colloquium-level overview of string theory and M-theory. The second lecture surveys various attempts to construct a viable model of particle physics. A recently proposed approach, based on F-theory, is emphasized.

Lectures presented at the Erice 2008 International School of Subnuclear Physics

[1]jhs@theory.caltech.edu

1 Introduction

The plan for my two lectures is as follows: the first lecture (in section 2) will give a colloquium level overview of string theory. The presentation will mention some of the history of the subject at the same time that basic concepts (such as compactification of extra dimensions, dualities, and branes) are introduced. There is a lot of ground to cover, and it will be necessary to be somewhat sketchy. Since the theme of this school concerns discoveries that might be made at the LHC, it seems appropriate to emphasize string theory approaches to particle physics model building. Therefore this is the subject of the second lecture (in section 3). It will survey several different approaches that have been developed, describing some of the successes and problems of each of them. One approach that has seen a lot of recent progress, and which looks especially promising, will be emphasized.

For reasons that will be explained, the emphasis is on supersymmetrical string theories (called *superstring theories*), which naturally are associated with ten-dimensional spacetime. In certain cases, the strong coupling limit gives an eleven-dimensional theory, called *M-theory*, which does not contain strings. For the student who wants to learn more, there are many textbooks on string theory. Some of them are [1, 2, 3, 4]. The one by Zwiebach [3] is addressed to advanced undergraduates, but it is suitable for all physicists who are not trying to become experts. The other three attempt to bring the reader up to the state of the art at the time when they were written. The subject is fast-moving and much was learned in the intervals between their appearances, so a lot of new material is included in each subsequent book, and a lot of the older material is not repeated.

This school pays homage to Sidney Coleman, a great scholar and teacher, whose contributions to theoretical physics in general, and the Erice schools in particular, are legendary. I always enjoyed interacting with him. Although I did not know him very well, we did cross paths in Aspen and elsewhere on a number of occasions. I recall him once saying that there are three things that he does not like, all of which are becoming popular: supersymmetry, strings, and extra dimensions. Obviously, my views are quite different, but this did not lessen my regard for him, nor did it harm our personal relationship. In fact, I respected his honesty, especially as he did not try to impose his prejudices on the community.

2 Lecture 1: Overview of Superstring and M-Theory

String theory arose in the late 1960s as a radical alternative to conventional quantum field theory in which the fundamental objects are one-dimensional strings rather than zero-dimensional points. This proposal arose in the context of S-matrix theory, a subject that has been much maligned, but whose imprint is indelible. From a modern perspective, it is clear that theories of point particles and theories of strings are related by dualities. Therefore the two classes of theories, which used to seem completely at odds, actually are deeply intertwined and do not have a sharp separation. Thus, I would claim that the subject area that is currently called string theory should be viewed as the logical completion of quantum field theory. That being the case, string theory has a certain inevitability, and there is nothing radical about it at all.

2.1 Basic Concepts of String Dynamics

Let us start by sketching what it means to construct a theory based on strings. To do this it is convenient to review first point-particle theory (*i.e.*, quantum field theory) from a first-quantization viewpoint. This is not the way that it is usually taught. The reason for doing this is that the analogous formulation of string theory is much better understood, and much easier to describe, than the string theory analog of second-quantized quantum field theory.

For a *point particle* the classical motion makes the invariant length of the world-line extremal. The corresponding action, proportional to this length, is given by

$$S = -m \int ds.$$

Here ds is the invariant line element, given in a general curved spacetime by $ds^2 = g_{\mu\nu} dx^\mu dx^\nu$, and m is the mass of the particle in question. (We always set $\hbar = c = 1$.) The motion of a particle is given by a *world line* $x^\mu(\tau)$, where τ is an arbitrary parameter for the trajectory. The action is independent of the choice of this parametrization.

To pass to the quantum theory one computes an amplitude for propagation from an initial spacetime point $x_{\rm i}^\mu = x^\mu(\tau_1)$ to a final spacetime point $x_{\rm f}^\mu = x^\mu(\tau_2)$. As Feynman taught us, this is given by the path integral (or sum over histories)

$$A_{\rm if} = \int_1^2 Dx^\mu(\tau)\, e^{iS}.$$

This is slightly oversimplified in one respect. The τ reparametrization invariance is a type of gauge invariance that need to be accounted for. One possibility, which is not always the most convenient one, is to choose a gauge in which τ is time, *i.e.*, $x^0(\tau) = \tau$.

Interactions can be incorporated in this formalism by allowing world lines to join or bifurcate and associating a coupling constant g whenever this occurs. In this way, one can reproduce the perturbation expansion of a second-quantized quantum field theory. One shortcoming of this first-quantized approach is that it is not very convenient for studying nonperturbative phenomena.

We can now "invent" string theory by doing the same thing for one-dimensional extended objects. If the string's topology is that of a circle it is called a *closed string*. If the topology is that of a line segment, it is called an *open string*. A string sweeps out a two-dimensional surface in spacetime, called the *world sheet* of the string. In the case of a closed string, the topology of the world sheet is that of a cylinder, and in the case of an open string it is that of a strip. For a *string* the motion makes the invariant area of the world-sheet extremal

$$S = -T \int dA.$$

The invariant area element dA is given by a simple formula analogous to that of ds in the point-particle case. The coefficient T is identified as the *tension* (or energy per unit length) of the string. The world-sheet is described by embedding functions $x^\mu(\sigma, \tau)$. Here σ is interpreted as a coordinate along the string (periodic in the case of a closed string) and τ is a timelike parameter. The choice of coordinates is arbitrary in that there is now a two-dimensional reparametrization invariance.

The basic idea of the extension to the quantum theory is the same as before. Namely, one defines amplitudes for string propagation by a path integral, properly taking account of the reparametrization invariance. However, there are some very interesting differences between the two cases: First of all, after covariant gauge fixing, the world-sheet theory has a conformal symmetry. (The conformal group in two dimensions is infinite dimensional.) Consistency of the quantum theory requires cancellation of a conformal anomaly. The most straightforward way to achieve this is to take the spacetime dimension to be 26. This defines *critical string theory*. Alternative approaches give *noncritical string theory*.

The second important difference from the point-particle case concerns interactions. In the case of strings, these are uniquely determined by the free string theory. They arise for purely a topological reason. For example, the process in which a single closed string turns into two closed strings is given by the *pants diagram*. This is a smooth world sheet without any singularity associated with the string junction. Such junctions need to be incorporated in string path integrals. The smoothness of the world sheet has a remarkable consequence: loop amplitudes have no ultraviolet divergences! In the case of point-particle theories these can be traced to the world line junctions, which are short-distance singularities in the one-dimensional network of world lines. Moreover, at least in the case of oriented closed strings, there is just one Feynman diagram at each order of the perturbation expansion!

The actions S described above can be generalized to objects with p dimensions, called p-branes. However, the quantum analysis breaks down for $p > 1$, because their world-volume theories are nonrenormalizable. Later we will argue that various p-branes do arise in string theories and M-theory as nonperturbative excitations. The significance of this nonrenormalizability does not concern the existence of p-branes. Rather, it means that they cannot be treated as the fundamental objects on which to base a perturbation expansion. The fact that this is possible for strings is the feature that distinguishes them from higher-dimensional p-branes. However, at strong coupling, when a perturbation expansion is not helpful, this distinction evaporates and all branes are more or less equal.

2.2 A Brief History of String Theory

2.2.1 Hadronic String Theory

String theory arose in the late 1960s in an attempt to understand the strong nuclear force. This is the force that holds quark and gluons together inside the strongly interacting particles (hadrons). The story has a long and complex history that has not been assembled yet in a single scholarly study. However, many of the participants in this history presented their own personal recollections in a 2007 meeting at the Galileo Institute in Florence. My contribution to the proceedings is available [5].

Veneziano's discovery of a simple formula for a 2-particle to 2-particle scattering amplitude with many attractive features (such as Regge behavior) quickly led to the realization that a theory based on strings, rather than point-like particles, could account for various observed features of the strong nuclear force [6]. The original string theory, sketched in the previous subsection, contains only bosons, and (as discussed above) it is consistent for 26 spacetime dimensions. 25 of the dimensions are spatial and 1 is time.

Another string theory containing fermions (as well as bosons) was constructed in 1971 by Pierre Ramond, André Neveu, and me [7, 8]. It requires 10 dimensions. Its development led to the discovery of *supersymmetry*, a symmetry that relates bosons and fermions. Strings in theories with this symmetry are called *superstrings*.

The unrealistic dimension of spacetime was obviously a problem, but not the only one. Another one was that the spectrum of string excitations in both of the string theories includes *massless particles*, whereas all hadrons have positive mass. In the early 1970s a better theory of the strong nuclear force, called *quantum chromodynamics* (or QCD), was developed. As a result, string theory fell out of favor. What had been an active community with several hundred participants rapidly shrunk to a backwater of theoretical physics kept alive by only a few diehards.

2.2.2 Unification

One of the massless particles in the closed-string spectrum of the 26-dimensional bosonic string theory, as well as in the closed-string spectrum of ten-dimensional superstring theories, has precisely the right properties to be the *graviton* – the particle responsible for transmitting the gravitational force. In other words, it has spin two, which means that its polarization tensor belongs to a symmetric and traceless representation of the little group. Moreover, the requisite local symmetries are incorporated, namely general coordinate invariance and local Lorentz invariance. These properties ensure precise agreement with general relativity at energies low compared to the string scale.

To be completely honest, it should also be mentioned that there is also a massless scalar (called the dilaton). This field is very important, because its vacuum expectation value determines the string coupling constant. Somehow, the dynamics should give mass to this field to avoid contradictions with experiments that show that the long-range gravitational force is pure spin two to high accuracy. Presumably, this is achieved at the same time as the extra dimensions are compactified. I will discuss this more later.

The open-string spectrum contains massless spin one gauge fields, like those in the standard model. Again, the usual gauge symmetries are incorporated, which ensures agreement with Yang–Mills gauge theory at energies low compared to the string scale.

For several years, string theorists (including myself) tried to modify string theory so as to give mass to these fields, since we knew that there are no massless hadrons. However, these attempts always led to inconsistencies. Eventually it became clear that these were essential features of the theory, and that we should let the theory guide us, rather than trying to impose our will on it. Accordingly, in 1974 Joël Scherk and I proposed to use string theory as a unified theory of all forces (including gravity), rather than just the strong nuclear force [9]. Thus, we stumbled upon an approach to achieving *Einstein's dream* – a unified theory of all fundamental forces. The connection to general relativity at low energies was described independently by Yoneya [10, 11], though he did not propose using string theory as a unified theory.

The interpretation of string theories as unified theories of gravity and other forces has several advantages, which were immediately apparent. All prior attempts to describe quantum corrections to Einstein's theory of gravity assumed point particles. They gave nonrenormalizable ultraviolet divergences, which makes the theory unacceptable, at least within the context of perturbation theory. String theory, on the other hand, was known to be UV finite to all orders in perturbation theory. The intuitive reason for this, is that the string world sheet is smooth, without any short-distance singularities, no matter how complicated the topology.

A second advantage of using string theory as a theory of gravity is that the extra dimensions can be an advantage rather than a disadvantage. Extra spatial dimensions can be compact in a gravitational theory, where the spacetime geometry is determined by the dynamics. So the geometry can appear four-dimensional at low energies. It then becomes a dynamical question whether the theory has acceptable quantum vacua of this type.

2.2.3 The Size of Strings

When strings were supposed to describe hadrons their typical size needed to be roughly that of a typical hadron, namely 10^{-13}cm. However, in a relativistic quantum theory of gravity there is a characteristic length, known as the *Planck length*

$$L_{\rm P} \sim \sqrt{\hbar G/c^3} \sim 10^{-33}\text{cm}.$$

This corresponds to a mass scale of about 10^{19} GeV. This is the natural first guess for the fundamental string length scale in a string theory containing gravity. This shrinkage by 20 orders of magnitude was a very large conceptual change. Fortunately, the mathematics remained pretty much the same. More recently, it has been appreciated that this first guess might be modified for various reasons. There are possible corrections involving factors of the coupling constant, the volume of compactified extra dimensions, or even warpage in the geometry of extra dimensions. Therefore, it makes sense for experimentalists to look for signs of the string scale at the LHC, though I would be astonished if it did show up.

2.2.4 The First Superstring Revolution

String theory became a hot subject in the mid-1980s following some important discoveries: anomaly cancellation [14], heterotic strings [15], and Calabi–Yau compactification [16]. By the time the dust settled, it appeared that there are exactly five consistent superstring theories, each of which requires *ten spacetime dimensions and supersymmetry.* Three of them, called *Type I, Type IIA, Type IIB*, were introduced by Green and me [13] building on earlier work of Gliozzi, Scherk, and Olive [12]. The other two, called $SO(32)$ *heterotic and* $E_8 \times E_8$ *heterotic*, were formulated in [15]. They incorporate the gauge groups identified in [14]. The anomaly-free type I superstring also has $SO(32)$ gauge symmetry.

Each of these five theories has no adjustable dimensionless parameters. All dimensionless parameters arise either

- *dynamically* as the expectation values of scalar fields

  or

- as *integers that count something* such as topological invariants, physical objects (branes), or quantized fluxes.

It appeared that the quest for a unique theory was almost at hand. The only mystery was why there should be five theories when only one is required.

One scheme looked particularly promising. Specifically, the $E_8 \times E_8$ heterotic theory has consistent vacuum solutions in which six spatial dimensions form a compact *Calabi–Yau manifold.* Calabi–Yau manifolds are a class of six-dimensional manifolds that can be described as having three complex dimensions. The precise mathematical statement is that they are Kähler manifolds with $SU(3)$ *holonomy.*[2]

The other four dimensions form Minkowski spacetime, $M_{3,1}$. Thus, the ten-dimensional spacetime is a product space

$$\mathcal{M}_{10} = \mathrm{CY}_6 \times M_{3,1}.$$

[2]Kähler manifolds are complex manifolds that have no torsion. The restriction to the $SU(3)$ subgroup of the generic $U(3)$ holonomy group corresponds to the vanishing of the first Chern class. It also implies the existence of a covariantly constant spinor, which in turn implies that some supersymmetry is preserved.

Intuitively, one can imagine a Calabi–Yau space attached to every point in four-dimensional spacetime. If the CY space is small, then its structure cannot be probed at low energy. Experimental detection requires energies greater than the inverse size of the compact CY space. I will say more about these constructions in my second lecture.

2.2.5 The Second Superstring Revolution

In the mid-1990s there was a remarkable burst of progress in addressing some of the issues that we have raised. A few of the key initial contributions were [17, 18, 19]. The main lessons were the following:

- There is just one theory! What had been viewed as five theories are actually five different corners of a space of solutions to a unique underlying theory. The five superstring theories are related by various surprising equivalences, called dualities:

  T-duality ($R \rightarrow 1/R$) relates the two type II superstring theories, and also the two heterotic string theories. Thus, for example, if the type IIA superstring theory is considered in a spacetime that is nine-dimensional Minkowski spacetime times a circle of radius R ($M_{8,1} \times S^1$), this is equivalent to the type IIB superstring theory in a spacetime that is nine-dimensional Minkowski spacetime times a circle of radius $1/R$. (Here we are using units in which the fundamental string length scale L_s has been set equal to one.) Thus, the two type II theories in ten-dimensional Minkowski spacetime are two limiting cases of a continuum of consistent possibilities that are connected by letting the radius R range from zero to infinity. This radius is determined by the vacuum expectation value of a modulus field (one of the components of the ten-dimensional metric).

  S-duality ($g_s \rightarrow 1/g_s$) relates the type I superstring theory and the $SO(32)$ heterotic string theory. The string coupling constant g_s is given by the vacuum expectation value of a modulus field called the *dilaton*. Prior to this discovery one only knew how to analyze string theory at weak coupling using perturbation theory. S-duality implies that this pair of theories is continuously connected by varying g_s from zero to infinity. It also implies that the strong coupling expansion of one theory is given by the weak coupling expansion of the other theory. The type IIB superstring theory is related to itself in a similar manner, which determines its strong coupling expansion as well.

- Knowing the strong coupling expansion of three of the five superstring theories, it was natural to explore the strong coupling behavior of the other two superstring theories. The conclusion is quite striking. At strong coupling the type IIA and the $E_8 \times E_8$ theories become 11-dimensional. More precisely, the eleventh dimension has size proportional to $g_s L_s$. In the IIA case it is circle whereas in the $E_8 \times E_8$ case it is a line interval with two endpoints. In the latter case the 11-dimensional spacetime has two ten-dimensional faces. It turns out that one E_8 gauge group is localized on one face and the other E_8 on the other face, so that they are spatially separated.

In either the type IIA or $E_8 \times E_8$ theory the $g_s \to \infty$ limit leads to the same 11-dimensional Minkowski spacetime theory. This 11-dimensional theory, which is still not yet very well understood, is called *M-theory.* M-theory is approximated at low-energies by 11-dimensional supergravity, a theory that was formulated many years earlier [20].

- New stable objects, called *p-branes*, arise nonperturbatively. p is the number of spatial dimensions they occupy. Thus, in this notation, a point-particle is a 0-brane, a string is a 1-brane, and so forth. p-branes can carry generalized conserved charges that are sources for antisymmetric tensor gauge fields with $p+1$ indices, which generalize the Maxwell field in the $p = 0$ case.[3] The conserved charges satisfy generalized Dirac quantization conditions. In D spacetime dimensions, the magnetic dual of a p-brane is a $(D-p-4)$-brane.

  The main categories of p-branes are called D-branes, M-branes, and NS-branes. D-branes are characterized by Dirichlet boundary conditions for open strings. In the type IIA theory stable D-branes exist for even values of p and in the type IIB theory they exist for odd values of p. M-theory has a three-form gauge field, which can couple electrically to an M2-brane and magnetically to an M5-brane. The NS5-brane is the magnetic dual of the fundamental string in the heterotic and type II superstring theories.[4]

2.3 AdS/CFT: Holographic Duality

In 1997 Maldacena proposed a new class of dualities relating certain string theory and M-theory solutions in anti de Sitter geometries[5] (AdS_{d+1} times a compact space) to conformally invariant quantum field theories (CFT_d) [21]. The first piece of evidence for such a relationship is the fact that the conformal symmetry group in d dimensions, $SO(d,2)$, is the same as the isometry group of anti de Sitter space in $d+1$ dimensions. The precise relationship between amplitudes/correlation functions in the two pictures was spelled out [22, 23]. Such dualities are called *holographic*, because string theory in AdS_{d+1}, which has $d+1$ dimensions, is equivalent (dual) to a conformally invariant quantum field theory in d dimensions: the

[3]These gauge fields can also be regarded as differential forms $A = \frac{1}{(p+1)!} A_{\mu_1 \mu_2 \cdots \mu_{p+1}} dx^{\mu_1} \wedge dx^{\mu_2} \wedge \ldots \wedge dx^{\mu_{p+1}}$. In this notation the coupling of the gauge field to the brane world volume is given by the simple expression $\mu \int A$. The parameter μ is the brane charge.

[4]The type I superstring does not carry a conserved charge, and therefore it can break. It does not have a magnetic dual.

[5]Anti de Sitter space is a maximally symmetric spacetime with constant negative curvature.

dual CFT$_d$. This is reminiscent of recording a three-dimensional image on a two-dimensional emulsion. The reason that two constructions that look so radically different can nonetheless be equivalent is that when one of them is weakly coupled, the other is strongly coupled. If it were easy to understand both of them at the same time, there would be a paradox.

This is an enormous subject, and I will just say a little bit about it here. Three maximally symmetric examples are given in Maldacena's original paper: Type IIB superstring theory in an $AdS_5 \times S^5$ background geometry with N units of five-form flux threading the sphere is dual to $\mathcal{N} = 4$ super Yang–Mills theory with an $SU(N)$ gauge group. This by far the most studied example. The leading term in the 't Hooft large-N expansion of the gauge theory (the planar approximation) is supposed to be dual to the string theory in the tree approximation. This much of the duality is quite well understood, and I think it is likely that it will be completely proved someday. The higher-order terms in the $1/N$ expansion correspond to string loop corrections. This duality can be generalized to examples with less supersymmetry in which the five-sphere is replaced by another five-dimensional *Sasaki–Einstein space*. It is also possible to replace the AdS space by a space that is only asymptotically anti de Sitter. In this case the conformal field theory is perturbed by relevant operators.

The other two maximally supersymmetric examples relate M-theory in an $AdS_4 \times S^7$ background geometry with flux to a three-dimensional conformal field theory and M-theory in an $AdS_7 \times S^4$ background geometry with flux to a six-dimensional conformal field theory. Within the past year there has been a great deal of progress in understanding the first case (as well as an orbifold generalization) [24]. On the other hand, there has been very little progress in understanding the second M-theory duality.

The general AdS/CFT framework is being used to construct brane configurations that capture many of the essential features of QCD, even though it isn't precisely the same theory [25]. From the modern viewpoint, such constructions can be regarded as spinoffs of string theory rather than its central goal. Even so, this brings string theory back to its historical origins. It also makes the reasons for the early failure to achieve the original goal of describing hadrons much clearer. For one thing, it is now apparent that a string theory dual of QCD must involve at least one extra spatial dimension [26]. There are also spinoffs to other fields: 1) There have been striking applications of AdS$_5$/CFT$_4$ duality to studies of the quark-gluon plasma (RHIC, LHC), especially the computation of the viscosity to entropy density ratio η/s [27, 28]. Taking the field theory at finite temperature corresponds to including a black hole in the dual AdS geometry! 2) Similar constructions could have condensed matter applications. Systems that show promise are ones in which there is a phase transition at which the conformal field theory is strongly coupled. Then the dual black-hole description could be helpful. A possible application of this type is to high T_c superconductors. For a recent overview see [29]. 3) Nonrelativistic versions of AdS/CFT may be relevant to Bose-condensed cold-atom systems under special conditions [30, 31].

2.4 Some Other Topics in String Theory

2.4.1 Brane Worlds

As we already mentioned, the defining property of D-branes is that strings can end on them. Away from D-branes, a string must have the topology of a circle, but the string can break provided that its ends are attached to D-branes. The lowest excitation mode of the open string is a massless gauge field. This has the crucial consequence that Yang–Mills gauge theories (like the standard model) can live on stacks of D-branes. For example, in the case of N coincident type II D-branes, the world-volume theory of the branes is a supersymmetric $U(N)$ gauge theory.

This picture raises new possibilities for the physics of the extra dimensions, which is rather different from the usual Kaluza–Klein picture. Rather than having a wave function that is uniformly spread over the compact extra dimensions, a low-energy field can have its wave function concentrated on (or close to) a D-brane, which is a defect in the extra dimensions that fills the noncompact spacetime dimensions. The simplest such possibility is that the observable Universe is actually a stack of D3-branes, which are points embedded in the six extra spatial dimensions. A generalization of this approach uses intersecting stacks of higher-dimensional D-branes. For example, one could consider stacks of D6-branes wrapping 3-cycles in the six extra dimensions.

2.4.2 Flux Compactifications

As has already been mentioned, the type II superstring theories contain various massless antisymmetric-tensor (or differential-form) gauge fields

$$A_n = \frac{1}{n!} A_{\mu_1\mu_2\ldots\mu_n} dx^{\mu_1} \wedge dx^{\mu_2} \wedge \ldots \wedge dx^{\mu_n}.$$

p-branes with $p = n - 1$ are sources for these fields. These differential-form gauge fields have gauge-invariant field strengths of the form $F_{n+1} = dA_n$, which are invariant under gauge transformations of the form $\delta A_n = d\Lambda_{n-1}$. This is a straightforward generalization of Maxwell theory (the $n = 1$ case).

If the compact dimensions contain nontrivial $(n + 1)$-cycles, C_{n+1}, it is possible to have (quantized) flux threading the cycle:

$$\int_{C_{n+1}} F_{n+1} = 2\pi N, \qquad N \in \mathbb{Z}.$$

Such flux can be present without a p-brane source. The possibility of constructing superstring vacua containing such fluxes greatly increases the number of possible vacua, even though there are constraints that must be satisfied. Mike Douglas analyzed one particular CY compactification of the type IIB theory and estimated the number of distinct flux vacua to be about 10^{500} [32]. For a review of this subject, see [33].

The *moduli problem* — the occurrence of massless scalar fields ϕ_i with continuously adjustable vacuum expectation values — can be solved in the context of flux compactification: The fluxes induce a nontrivial potential energy function for the moduli $V(\phi_i)$. This *landscape* function has isolated minima, which are what Douglas counted. The moduli are massive at an isolated minimum.

2.4.3 Warped Compactification

One of the important properties of flux compactifications is that they give rise to warped geometries. This means that the ten-dimensional geometry is no longer a direct product of the form

$$\mathcal{M}_{10} = M_{3,1} \times K_6.$$

Instead, the four-dimensional Minkowski part of the ten-dimensional metric is multiplied by a *warp factor* $h(y)$ that depends on the position y in the internal manifold

$$ds_{10}^2 = h(y)dx \cdot dx + ds_6^2.$$

In 1999 Randall and Sundrum proposed that an exponential y dependence could give a large ratio between the warp factors at the positions of two 3-branes, which they called the *standard model brane* and the *Planck brane* [34]. They proposed that this could solve the hierarchy problem. This kind of exponential warping is exactly what one has in the case of an AdS spacetime. The new idea was to only consider a slice of AdS spacetime between two branes. This is only an approximation to what can be achieved in honest string theory solutions. Nonetheless, the RS scenario can be made rather precise in the context of flux compactifications with branes added [35]. If this is how Nature works, quantum gravity effects might be accessible at the LHC. The experimentalists intend to look for them.

2.5 String Cosmology

An important question, which was not discussed much in the 20th century, is "What does string theory have to say about cosmology?" Nowadays, string cosmology is a very active research area. There are even entire conferences devoted to it. Clearly, progress in this area requires understanding *time-dependent* solutions of string theory, which is technically challenging.

There is a lot of evidence that the very early Universe underwent a period of inflation during which the scale factor grew exponentially by a factor of at least e^{60} ("60 e-foldings"). In simple field theory models this is described in terms of a "slowly rolling" scalar field called an *inflaton*. Proposals for the string-based origins of inflation, or possible alternatives, are being explored extensively. One scenario is based on CY compactification with flux and warped throats in the geometry [36, 37]. In this scenario inflation takes place as a D3-brane moves down a throat, attracted to an anti-D3-brane at the bottom until they *collide and annihilate.* A scalar mode of an open string connecting the branes is the inflaton. The annihilation releases the brane tension energy. It heats up the Universe to start the *hot big bang epoch.* All sorts of strings are produced, and some might survive to be observable as *cosmic superstrings* [38].

3 Lecture 2: Superstring Phenomenology

Since the focus of this school is the physics of the LHC, it seems appropriate to discuss attempts to relate string theory to the particle physics phenomenology in greater detail than other topics. That is the purpose of this lecture. However, before getting into the details of such efforts, let me emphasize that the unique feature of string theory is that it unifies gravitation and particle physics. Moreover, its most successful prediction (or postdiction, if you wish) so far is that *Gravity Exists!*

It is not yet clear how predictive string theory is for particle physics. The discovery of the *string theory landscape* has led to some pessimism in this regard. However, as I will argue, there are good reasons to believe that if one makes certain experimentally motivated assumptions and inputs some experimental details, string theory could be quite predictive. I don't think it is reasonable to expect to predict all experimental facts from first principles. It would already be a great success, if one could predict a large number of facts starting from a small number of experimentally motivated assumptions. How far one can go in this direction remains to be seen.

In perturbative string theory there are two basic ways to obtain non-abelian gauge symmetry in four dimensions: 1) It is already present in 10 dimensions (type I and heterotic strings) and broken partially by compactification. 2) It arises as the symmetry of compact extra dimensions (Kaluza–Klein). The Kaluza–Klein approach does not give anything realistic. Nonperturbatively, there are other possibilities that we will discuss later.

3.1 Perturbative Heterotic String

The gauge groups of the standard model and grand unified models embed nicely in E_8 but not in $SO(32)$. In the period 1985–95 only one scheme looked promising for phenomenology: Calabi–Yau compactification of the $E_8 \times E_8$ heterotic string. This entails taking the ten-dimensional geometry to be a product of four-dimensional Minkowski spacetime and a six-dimensional Calabi–Yau space, $\mathcal{M}_{10} = \mathrm{CY}_6 \times M_{3,1}$. There must also be nontrivial gauge fields, satisfying certain consistency conditions (related to anomaly cancellation), in the internal space. This gives a GUT-like effective four-dimensional theory at low energy with $\mathcal{N} = 1$ supersymmetry. These effective theories have a number of attractive features:

- They have the structure of *supersymmetric grand unified theories.* The advantages of low-energy supersymmetry and grand unification are therefore naturally incorporated.

- Each solution has a definite number of families of quarks and leptons determined by the topology of the CY space.

- The standard model gauge symmetry can be embedded in one E_8 factor. Then there is a *hidden sector*, associated to the second E_8 factor. Supersymmetry can break dynamically in the hidden sector. This breaking is communicated by singlet fields, which have gravitational strength interactions, to the visible sector. This is usually called *gravity mediation*, though more than gravity is involved.

- There are several good *dark-matter* candidates: The lightest supersymmetric particle (or LSP) is absolutely stable in schemes with unbroken R-parity. In gravity mediation the leading dark-matter candidates are the lightest neutralino (a mixture of the partners of the photon, Z boson, and Higgs bosons) and hidden-sector particles.[6]

These developments created a lot of excitement in the mid-80s. In my view a large part of the theory community made a phase transition from being too skeptical of string theory to being too optimistic about the short-term prospects for constructing a realistic theory. The successes were qualitative, and there were many problems and puzzling questions:

- Hundreds of Calabi–Yau manifolds were known (now many thousands). Which one of them, if any, is the *right one*? Is there a theoretical principle for deciding?

- Why are there four other consistent superstring theories? Can they give interesting solutions, too?

- The CY compactification scenario described above is based on perturbation theory. What new *nonperturbative features* appear at strong coupling?

- These CY solutions typically give a large number of massless scalar fields (called *moduli*).[7] The moduli fields have gravitational strength interactions, but long-range scalar forces of that strength are ruled out by standard tests of general relativity. Thus, all these CY solutions are unsatisfactory unless there is some way of giving mass to the moduli. (Recall that in the first lecture I reported that this can be achieved in flux compactifications of type II superstring theories.)

[6]In other mediation schemes, such as gauge mediation, the gravitino (the supersymmetry partner of the graviton) and saxions (supersymmetry partners of axions) are candidates.

[7]Moduli fields have a flat potential. This means that the vacuum energy does not depend on their values.

- In a gravitational theory the value of the minimum of the potential also matters, because it determines the *vacuum energy density*, which is the effective cosmological constant. For many years it was assumed (without a good theoretical argument) that this should be zero. However, about a decade ago it was discovered in studies of distant supernovas (and confirmed by other means) that the vacuum energy density (or dark energy) is very small and positive. The value is roughly 10^{-120} in Planck units, which is the smallest measured number in all of science. Typically, one would expect to compute a result of order unity, so it is a great challenge to try to account for this result.[8] Supersymmetry can help, but once it is broken at the TeV scale, one still expects to obtain a value of order 10^{-60} in Planck units, which is 60 orders of magnitude too large.
- The ratio of the Planck scale to the GUT scale comes out an order of magnitude too small.
- Extra family-antifamily pairs typically occur.

3.2 Other Approaches

Following the second superstring revolution in the mid-1990s several new approaches to particle phenomenology opened up. These exploit discoveries such as

- 11d M-theory
- Singular geometries (e.g. orbifolds and orientifolds)
- D-branes and other branes
- Flux compactifications and warped geometries
- Nonperturbative type IIB superstring solutions (F-theory)

3.2.1 Heterotic M-Theory

Hořava and Witten discovered that the $E_8 \times E_8$ heterotic string at strong coupling is 11-dimensional [39]. In particular they showed that anomaly cancellation requires that E_8 gauge fields live on each boundary component [40]. In heterotic M-theory the geometry is a solid slab with two ten-dimensional boundaries. One E_8 lives on each boundary.

Calling the coordinate that runs across the slab y, the shape and size of the CY_6 manifold can be y-dependent. In this way one incorporates nonperturbative improvements of the perturbative heterotic constructions. Taking this possibility into account improves the range

[8]If the true value were zero, and there were no dark energy, this would still be a big problem.

of possible ratios of the Planck scale to the GUT scale [41]. Fairly realistic examples have been constructed by the Penn group. (See [42, 43] and references therein). However, despite their many realistic features, these examples still have massless moduli.

3.2.2 M-Theory on a G_2 Manifold

The strong coupling limit of type IIA superstring theory is eleven-dimensional M-theory, which is approximated at low energies by 11-dimensional supergravity. To obtain $\mathcal{N} = 1$ supersymmetry in four dimensions at low energies requires a compactification of the form

$$\mathcal{M}_{11} = K_7 \times M_{3,1},$$

where K_7 is a compact 7-manifold of G_2 holonomy. (G_2 is the smallest of the five exceptional Lie groups. It has 14 generators.) If the G_2 manifold is smooth, then the effective four-dimensional theory at low energy has no non-abelian gauge symmetry or parity violation. To overcome this problem it is necessary to consider G_2 manifolds with certain types of singularities [44]. While the necessary types of singularities are well understood, there has been little progress on constructing compact 7-manifolds of G_2 holonomy that contain such singularities. Such manifolds, having an odd dimension, are obviously not complex manifolds. The math is much tougher when techniques of complex analysis cannot be used. Thus no quasi-realistic models of this type have been constructed, though they presumably exist.

3.2.3 Intersecting D-branes in Type II

Consider Type IIA superstring theory compactified on a six-torus. Suppose that a stack of M D6-branes wraps a 3-cycle of the torus and fills the four-dimensional spacetime. This gives a supersymmetric $U(M)$ gauge theory. Suppose now that a second stack of N D6-branes wraps a different 3-cycle that intersects the first one at n points. Then the four-dimensional theory has $U(M) \times U(N)$ gauge symmetry, and it contains n bifundamental chiral matter multiplets transforming as $(M, \overline{N})$.

It is striking that constructions that are this simple can capture roughly the types of groups and representations that appear in the standard model. Moreover, here are generalizations involving orbifolds and orientifolds that are quite realistic. But, even then, there are some challenging problems that remain:

- Not easy to get the correct $U(1)$ of weak hypercharge and no other $U(1)$ groups at low energy. Some $U(1)$s are lifted by a four-dimensional version of the anomaly cancellation mechanism. However, this does not automatically guarantee that just the desired $U(1)$ group survives.

- There are many massless moduli. Localizing the branes at singularities eliminates some of them. However, it is hard to remove all of them, as is required. One approach to overcoming this problem is to have the brane intersections be localized at singularities of the geometry. As an example of this approach see [45].

- This type of scheme does not incorporate grand unification. It is not certain that we should, since the hints of grand unification could prove to be illusory. However, the more conservative guess is that it is not an illusion.

For reviews of this remarkable subject see [46, 47, 48].

3.2.4 Type II on a Calabi–Yau with Flux

Calabi–Yau compactification of Type II superstring theories

$$\mathcal{M}_{10} = \mathrm{CY}_6 \times M_{3,1}$$

give $\mathcal{N} = 2$ supersymmetry in four dimensions. Also, it does not give non-abelian gauge symmetry or chiral matter. So, without further measures, this is completely unrealistic. By including D-branes and/or flux one can overcome these difficulties. In fact, as discussed in the first lecture, the flux even helps to stabilize moduli. It also gives warping of the geometry, as was also discussed.

This approach allows one to embed Randall–Sundrum models in a string-theoretic setting. Recall that RS proposed that the warping could used to solve the hierarchy problem and that it could make quantum gravity effects accessible at LHC energies. In this setting one can break supersymmetry by adding anti-D3-branes. As we discussed, this approach leads to interesting string-theoretic models of inflation in which the origin of the big bang is described as a brane-antibrane collision. However, it has not yet yielded quasi-realistic particle physics models as far as I am aware.

Perhaps the absurdly large number of flux vacua that typically arise in flux compactifications has discouraged people from trying to construct viable particle physics models. In fact, this large number of vacua has motivated the suggestion that various parameters of Nature (such as the cosmological constant) should be studied statistically on the landscape. I don't really understand the logic of doing this, since this approach seems to assume implicitly that Nature corresponds to a more or less random vacuum. This in turn is motivated by some vague idea about how Universes are spawned in the Multiverse in a process of eternal inflation. Then the story gets even more entangled when the *anthropic principle* is brought into the discussion. Some people are enthusiastic about this approach, but I find it fundamentally defeatist. It is not the way I like to think about particle physics.

3.3 F-Theory Local Models

3.3.1 Local Versus Global Models

In *global models* one fully specifies the compactification geometry and all background fluxes and other fields. This completely determines a vacuum and hence the physics at all scales – up to and beyond the Planck scale. The heterotic string constructions that we have discussed are examples of global models. *Local models* are less ambitious. They are supposed to describe particle physics in a limit where gravity is turned off. More precisely, the ratio of the Planck scale to the unification scale, $M_{\rm Pl}/M_{\rm GUT}$ can be made arbitrarily large.

In certain classes of models in which matter fields are restricted to branes, the low-energy physics only depends on the geometry in the vicinity of the branes (in the extra dimensions). If one can decompactify the extra dimensions in directions transverse to the branes, this has the effect of increasing the four-dimensional Planck mass while keeping other scales fixed. Then one only needs to focus on the geometry in the immediate vicinity of the branes losing sensitivity to features in the geometry that are far away from the relevant branes. This approach is sometimes called *geometric engineering.*

Such local models are more amenable to a "bottom-up" construction. The basic idea is that one should first try to "engineer" a phenomenologically viable local model that correctly describes all nongravitational physics up to the unification scale to good accuracy. Then one should try to find a "global completion" of this local model that fully describes the geometry of the compact dimensions and brings gravity back into the mix. It is plausible (but not demonstrated, as far as I know) that the vast majority of local models do not have a global completion. Thus the existence of a global completion would be a confirmation that the original local model was on the right track. Furthermore, the global completion is likely to require that parameters (and other features) are correlated even though they were not correlated in the original construction of the local model. Altogether, one could imagine that the construction of a viable local model would be somewhat predictive, and that the construction of its global completion would be more predictive.

3.3.2 What is F-Theory?

Type IIB superstring theory [13] is the setting for F-theory. Type IIB superstring theory is also the framework where most of the recent progress in flux compactifications, moduli stabilization, and SUSY breaking has been made. F-theory, which utilizes nonperturbative type IIB vacua containing 7-branes, was introduced in 1996 by Vafa [49]. F-theory configurations that give four-dimensional models contain various 7-branes wrapping compact 4-cycles of the six-dimensional compact space B. They also fill the four-dimensional Minkowski spacetime.

The type IIB theory in ten-dimensional Minkowski spacetime contains two massless scalar fields, often called C_0 and ϕ, which are conveniently combined to form a complex scalar field

$$\tau = C_0 + ie^{-\phi}.$$

The field C_0 is a Ramond-Ramond zero-form, which couples electrically to D-instantons and magnetically to D7-branes. It behaves, at least in some respects, like a ten-dimensional analog of an axion. The field ϕ is called the *dilaton.* It has the special property that the value of e^ϕ is the string coupling strength, *i.e.*, $g_s = \langle e^\phi \rangle$.

Type IIB superstring theory has a *nonperturbative discrete gauge symmetry* given by [18]

$$\tau \to \tau' = \frac{a\tau + b}{c\tau + d}, \quad \text{where} \quad \begin{pmatrix} a & b \\ c & d \end{pmatrix} \in SL(2,\mathbb{Z}).$$

This means that a, b, c, d are integers satisfying $ad - bc = 1$. Other fields transform at the same time. A special case is the transformation $\tau \to -1/\tau$. For $C_0 = 0$ this maps $\phi \to -\phi$ and hence $g_s \to 1/g_s$, which is S-duality. The group $SL(2,\mathbb{Z})$ is also the modular group of the torus. As often happens in string theory, things that look the same actually are the same. In this context this means that τ can be interpreted as the complex structure of a torus. So the picture is that to every point $y \in B$ one associates a torus with the complex structure $\tau(y)$. As τ approaches certain special values the torus becomes singular. (Mathematicians say that the torus *degenerates.*) This happens generically on the 4-cycles where the 7-branes are located.

When one analytically continues $\tau(z)$ along a curve encircling a 7-brane it comes back to a transformed value given by an $SL(2,\mathbb{Z})$ transformation. The 7-branes are classified by these transformations (more precisely the conjugacy classes), so there are an infinite number of distinct types of 7-branes. They can be described by a pair of integers (p, q) without any common divisors. Type IIB superstring theory also has an infinite number of different types of strings with a similar classification scheme. In fact, a (p, q) 7-brane is one on which a (p, q) string can end. A special case of this statement is that a fundamental string can end on a D7-brane. The associated $SL(2,\mathbb{Z})$ transformation in this case is $\tau \to \tau + 1$. In this case the dilaton is constant and only C_0 varies, so it is possible to have weak coupling and use perturbative string techniques.

For a consistent F-theory solution, the varying field $\tau(z)$ and the compact six-dimensional space B are required to define an *elliptically fibred Calabi-Yau four-fold* (with section). Then, the low-energy effective four-dimensional theory has $\mathcal{N} = 1$ supersymmetry. One subtlety is that this construction makes it seem as if the compact space is eight-dimensional, since that is the dimension of a Calabi-Yau four-fold. If this were the case the entire spacetime dimension would be 12. However, this is definitely not the case. The tori that describe the fibres have a complex structure, given by $\tau(z)$, but they have zero area. This can be understood as a limit of tori with finite area, but a complete explanation requires invoking a duality relating F-theory to M-theory, which I will not present here. There had been a considerable amount of progress in understanding F-theory before this year [50, 51, 52], but it was not clear how to go about making realistic models.

3.3.3 Construction of Local F-Theory Models

The F-theory approach to phenomenology has been pursued this year by Beasley, Heckman, Vafa, Donagi, Wijnholt, Marsano, Saulina, and Schäfer-Nameki in various combinations [53] – [61]. The new proposal, which has given the subject a new lease on life, is to focus on models in which one can define a limit in which gravity is decoupled. The criterion is that it should be possible to make the dimensions transverse to the 4-cycles wrapped by the 7-branes arbitrarily large. Equivalently, it should be possible to contract the 4-cycles to points while holding the six-dimensional volume fixed. Such contractible 4-cycles must be positive curvature Kähler manifolds. These are fully classified and are given by manifolds called *del Pezzo manifolds* (or del Pezzo surfaces), which are denoted dP_n. The integer n takes the values $0 \leq n \leq 8$.[9] The del Pezzos have a close relationship with the exceptional Lie algebras E_n. The basic idea is that they contain 2-cycles whose intersections are characterized by the E_n Dynkin diagram. By this type of F-theory construction, one can construct an $SU(5)$ or $SO(10)$ SUSY-GUT model.

Constructions that involve 7-branes of various types are much more subtle – and also more interesting than ones that only involve D7-branes. D7-branes are mutually local. A stack of N of them gives $U(N)$ gauge symmetry. Matter fields at intersections (due to stretched open strings) are bifundamental. However, different kinds of 7-branes are mutually nonlocal. As a result, there are stacks (corresponding to the ADE classification of singularities) that can give $U(N)$, $SO(2N)$ or even E_N gauge symmetry. Exceptional gauge groups, spinor representations of $SO(10)$, and nonzero top-quark Yukawa couplings in $SU(5)$ are possible for F-theory vacua. In constructions that only involve D7-branes these phenomena can only arise nonperturbatively. Since the top quark Yukawa coupling must be quite large to account for the large top quark mass, the F-theory viewpoint seems preferable.

The crucial assumption of the local F-theory approach – the existence of a decoupling limit in which four-dimensional gravity can be turned off – may or may not be physically correct. The fact that $M_{\rm Pl}/M_{\rm GUT}$ is in the range 100–1000, depending on the precise definition of each of these mass scales, makes it plausible. There are an infinite variety of types of 4-cycles that 7-branes could wrap if one did not make this assumption. Therefore this principle picks out a tiny corner of the string theory landscape. In this way, the doom and gloom that the discovery of this landscape has generated is swept aside, at least if the basic assumption is correct.

In particular, it allows one to construct an $SU(5)$ or $SO(10)$ SUSY-GUT with three families. The $SU(5)$ case is more straightforward. There are three basic rules:

[9]There is also one other possibility, which is a product of two two-spheres.

- Each stack of 7-branes on a 4-cycle S determines a gauge group. One stack should give the GUT gauge group. Other "matter 7-branes" are also needed.
- Pairs of 4-cycles intersect on 2-cycles $S \cap S'$. Each such 2-cycle is a Riemann surface that has associated *chiral matter fields* given by zero-modes of the Dirac operator. The number of such zero modes determines the number of families.
- Three 2-cycles can intersect at points. Products of wave functions at these intersections determine the *Yukawa couplings*.

These three rules can be summarized by saying that the gauge fields live in eight dimensions, the matter fields live in six dimensions, and the interactions take place in four dimensions. Gravity, which is ignored in a local construction, lives in ten dimensions.

3.3.4 Gauge Symmetry Breaking

One of the strengths of the local F-theory approach to model building is the way that symmetry breaking can be incorporated. In conventional SUSY-GUTS based on four-dimensional quantum field theory, the breaking of the $SU(5)$ GUT symmetry to the $SU(3)\times SU(2)\times U(1)$ standard model gauge group requires the introduction of Higgs fields belonging to large representations of $SU(5)$. This is definitely ugly, and does not work very well. In string theory such representations are not present in the low mass spectrum, but there are other available mechanisms. In the case of the heterotic string constructions the standard method introduced in [16] is to associate nontrivial Wilson lines to noncontractible cycles of the Calabi–Yau geometry.

In the local F-theory models there are no noncontractible cycles and a different mechanism for breaking the $SU(5)$ gauge symmetry to the standard model gauge group is required. Fortunately, one is available. The proposal of Vafa and collaborators is introduce nonzero flux. In doing this one has to choose the $U(1)$ subgroup of $SU(5)$ that this flux corresponds to. By choosing it to correspond to the weak hypercharge $U(1)$ subgroup one gets exactly the desired symmetry breaking pattern. This flux is called *hyperflux* in the recent literature. One also has to choose the 2-cycle inside the del Pezzo surface that carries this flux. This is an elaborate and beautiful story that I don't intend to explain in detail here. One crucial point is that by making appropriate choices it is possible to arrange to obtain complete $SU(5)$ multiplets when one wants them and incomplete multiplets when that is desired. In the case of the quarks and leptons one wants complete $\bar{\mathbf{5}}$ and $\mathbf{10}$ multiplets. In the case of the Higgs bosons one wants only the doublet pieces of the $\mathbf{5}$ and $\bar{\mathbf{5}}$ multiplets, eliminating the triplets, which would give rapid proton decay. All this can be elegantly arranged.

One also needs to incorporate electroweak symmetry breaking. The standard Higgs mechanism is triggered by the breaking of supersymmetry. Supersymmetry requires that are two Higgs doublets, of course.

3.3.5 Supersymmetry Breaking, etc.

Local models are incompatible with gravity-mediated SUSY breaking. However, for *gauge-mediated SUSY breaking* both the visible and the hidden sectors, as well as their mediation, can be described in a single effective field theory decoupled from gravity. According to most experts, such models are favored for suppressing flavor-changing neutral currents. When the model is extended to incorporate a global completion and bring gravity back in the story, there will be a small admixture of gravity mediation. A proposal in the recent literature is that gauge-mediated supersymmetry breaking arises due to *stringy instantons* (Euclidean D3-branes that wrap the 4-cycle S) on an auxiliary 7-brane that has been dubbed a *Peccei-Quinn 7-brane.*

Beasley, Heckman, and Vafa [54, 55] have also proposed interesting mechanisms by which local F-theory models can address each of the following:

- R-parity conservation
- Peccei–Quinn symmetry and axion properties
- the μ and μ/B_μ problems
- the size of neutrino masses
- proton decay
- hierarchies of Yukawa couplings

There are many more details of this approach that I cannot go into here, including singularity enhancement, classification of fluxes, nonexistence of heterotic and M-theory duals, determination of superpotentials, threshold corrections, dark matter, sparticle spectrum, Higgsing the Peccei–Quinn symmetry, messenger fields, etc. This subject is still in its infancy, but it shows great promise. It will be very interesting to see what predictions can be made before the experimental results pour in and whether they turn out to be correct.

4 Conclusion

The understanding of string theory continues to progress on many fronts at an impressive rate. Many challenges remain, and it will undoubtedly require many decades to answer some of the deepest questions. However, there is no reason to be pessimistic. The subject is already proving useful for various purposes ranging from fundamental mathematics to particle physics phenomenology and cosmology. New experimental and observational results will be very helpful in stimulating further progress.

Acknowledgment

I am grateful to the Ettore Majorana Foundation and Centre for Scientific Culture for its hospitality during this delightful school. This work was supported in part by the U.S. Dept. of Energy under Grant No. DE-FG03-92-ER40701.

References

[1] M. B. Green, J. H. Schwarz, and E. Witten, *Superstring Theory* in 2 Volumes, Cambridge Univ. Press, 1987.

[2] J. Polchinski, *String Theory* in 2 Volumes, Cambridge Univ. Press, 1998.

[3] B. Zwiebach, *A First Course in String Theory*, Cambridge Univ. Press, 2004.

[4] K. Becker, M. Becker, and J. H. Schwarz, *String Theory and M-Theory: A Modern Introduction*, Cambridge Univ. Press, 2007.

[5] J. H. Schwarz, "The Early Years of String Theory: A Personal Perspective," arXiv:0708.1917 [hep-th].

[6] G. Veneziano, "Construction of a Crossing-Symmetric Regge-Behaved Amplitude for Linearly Rising Regge Trajectories," *Nuovo Cim.* **57A**, 190 (1968).

[7] P. Ramond, "Dual Theory for Free Fermions," *Phys. Rev.* **3**, 2415 (1971).

[8] A. Neveu and J. H. Schwarz, "Factorizable Dual Model of Pions," *Nucl. Phys.* B **31**, 86 (1971).

[9] J. Scherk and J. H. Schwarz, "Dual Models for Non-Hadrons," *Nucl. Phys.* B **81**, 118 (1974).

[10] T. Yoneya, "Quantum Gravity and the Zero Slope Limit of the Generalized Virasoro Model," *Nuovo Cim. Lett.* **8**, 951 (1973).

[11] T. Yoneya, "Connection of Dual Models to Electrodynamics and Gravidynamics," *Prog. Theor. Phys.* **51**, 1907 (1974).

[12] F. Gliozzi, J. Scherk, and D. Olive, "Supersymmetry, Supergravity Theories and the Dual Spinor Model," *Nucl. Phys.* B **122**, 253 (1977).

[13] M. B. Green and J. H. Schwarz, "Supersymmetrical String Theories," *Phys. Lett.* B **109**, 444 (1982).

[14] M. B. Green and J. H. Schwarz, "Anomaly Cancellation in Supersymmetric $D = 10$ Gauge Theory and Superstring Theory," *Phys. Lett.* B **149**, 117 (1984).

[15] D. J. Gross, J. A. Harvey, E. J. Martinec and R. Rohm, "The Heterotic String," *Phys. Rev. Lett.* **54**, 502 (1985).

[16] P. Candelas, G. T. Horowitz, A. Strominger and E. Witten, "Vacuum Configurations for Superstrings," *Nucl. Phys.* B **258**, 46 (1985).

[17] A. Sen, "Strong - Weak Coupling Duality in Four-Dimensional String Theory," Int. J. Mod. Phys. A **9**, 3707 (1994) [arXiv:hep-th/9402002].

[18] C. M. Hull and P. K. Townsend, "Unity of Superstring Dualities," Nucl. Phys. B **438**, 109 (1995) [arXiv:hep-th/9410167].

[19] E. Witten, "String Theory Dynamics in Various Dimensions," Nucl. Phys. B **443**, 85 (1995) [arXiv:hep-th/9503124].

[20] E. Cremmer, B. Julia and J. Scherk, "Supergravity Theory in 11 Dimensions," Phys. Lett. B **76**, 409 (1978).

[21] J. M. Maldacena, "The Large N Limit of Superconformal Field Theories and Supergravity," *Adv. Theor. Math. Phys.* **2**, 231 (1998) [arXiv:hep-th/9711200].

[22] S. S. Gubser, I. R. Klebanov and A. M. Polyakov, "Gauge theory correlators from non-critical string theory," Phys. Lett. B **428**, 105 (1998) [arXiv:hep-th/9802109].

[23] E. Witten, "Anti-de Sitter space and holography," Adv. Theor. Math. Phys. **2**, 253 (1998) [arXiv:hep-th/9802150].

[24] O. Aharony, O. Bergman, D. L. Jafferis and J. Maldacena, "N=6 Superconformal Chern-Simons-Matter Theories, M2-branes and their Gravity Duals," JHEP **0810**, 091 (2008) [arXiv:0806.1218 [hep-th]].

[25] T. Sakai and S. Sugimoto, "Low Energy Hadron Physics in Holographic QCD," Prog. Theor. Phys. **113**, 843 (2005) [arXiv:hep-th/0412141].

[26] A. M. Polyakov, "The Wall of the Cave," Int. J. Mod. Phys. A **14**, 645 (1999) [arXiv:hep-th/9809057].

[27] G. Policastro, D. T. Son and A. O. Starinets, "The Shear Viscosity of Strongly Coupled N = 4 Supersymmetric Yang-Mills Plasma," Phys. Rev. Lett. **87**, 081601 (2001) [arXiv:hep-th/0104066].

[28] P. Kovtun, D. T. Son and A. O. Starinets, "Viscosity in Strongly Interacting Quantum Field Theories from Black Hole Physics," Phys. Rev. Lett. **94**, 111601 (2005) [arXiv:hep-th/0405231].

[29] S. Sachdev and M. Mueller, "Quantum Criticality and Black Holes," arXiv:0810.3005 [cond-mat.str-el].

[30] K. Balasubramanian and J. McGreevy, "Gravity Duals for Non-relativistic CFTs," Phys. Rev. Lett. **101**, 061601 (2008) [arXiv:0804.4053 [hep-th]].

[31] D. T. Son, "Toward an AdS/Cold Atoms Correspondence: a Geometric Realization of the Schroedinger Symmetry," Phys. Rev. D **78**, 046003 (2008) [arXiv:0804.3972 [hep-th]].

[32] M. R. Douglas, "The Statistics of String/M Theory Vacua," JHEP **0305**, 046 (2003) [arXiv:hep-th/0303194].

[33] M. R. Douglas and S. Kachru, "Flux Compactification," Rev. Mod. Phys. **79**, 733 (2007) [arXiv:hep-th/0610102].

[34] L. Randall and R. Sundrum, "A Large Mass Hierarchy from a Small Extra Dimension," Phys. Rev. Lett. **83**, 3370 (1999) [arXiv:hep-ph/9905221].

[35] S. B. Giddings, S. Kachru and J. Polchinski, "Hierarchies from Fluxes in String Compactifications," Phys. Rev. D **66**, 106006 (2002) [arXiv:hep-th/0105097].

[36] S. Kachru, R. Kallosh, A. Linde and S. P. Trivedi, "De Sitter Vacua in String Theory," Phys. Rev. D **68**, 046005 (2003) [arXiv:hep-th/0301240].

[37] S. Kachru, R. Kallosh, A. Linde, J. M. Maldacena, L. P. McAllister and S. P. Trivedi, "Towards Inflation in String Theory," JCAP **0310**, 013 (2003) [arXiv:hep-th/0308055]

[38] J. Polchinski, "Cosmic String Loops and Gravitational Radiation," arXiv:0707.0888 [astro-ph].

[39] P. Hořava and E. Witten, "Heterotic and Type I String Dynamics from Eleven Dimensions," Nucl. Phys. B **460**, 506 (1996) [arXiv:hep-th/9510209].

[40] P. Hořava and E. Witten, "Eleven-Dimensional Supergravity on a Manifold with Boundary," Nucl. Phys. B **475**, 94 (1996) [arXiv:hep-th/9603142].

[41] E. Witten, "Strong Coupling Expansion of Calabi-Yau Compactification," Nucl. Phys. B **471**, 135 (1996) [arXiv:hep-th/9602070].

[42] V. Bouchard and R. Donagi, "An SU(5) Heterotic Standard Model," Phys. Lett. B **633**, 783 (2006) [arXiv:hep-th/0512149].

[43] V. Braun, Y. H. He, B. A. Ovrut and T. Pantev, "The Exact MSSM Spectrum from String Theory," JHEP **0605**, 043 (2006) [arXiv:hep-th/0512177].

[44] B. Acharya and E. Witten, "Chiral Fermions from Manifolds of G(2) Holonomy," arXiv:hep-th/0109152.

[45] H. Verlinde and M. Wijnholt, "Building the Standard Model on a D3-brane," JHEP **0701**, 106 (2007) [arXiv:hep-th/0508089].

[46] R. Blumenhagen, M. Cvetic, P. Langacker and G. Shiu, "Toward Realistic Intersecting D-Brane Models," Ann. Rev. Nucl. Part. Sci. **55**, 71 (2005) [arXiv:hep-th/0502005].

[47] R. Blumenhagen, B. Kors, D. Lust and S. Stieberger, "Four-Dimensional String Compactifications with D-Branes, Orientifolds and Fluxes," Phys. Rept. **445**, 1 (2007) [arXiv:hep-th/0610327].

[48] D. Malyshev and H. Verlinde, "D-branes at Singularities and String Phenomenology," Nucl. Phys. Proc. Suppl. **171**, 139 (2007) [arXiv:0711.2451 [hep-th]].

[49] C. Vafa, "Evidence for F-Theory," Nucl. Phys. B **469**, 403 (1996) [arXiv:hep-th/9602022].

[50] R. Friedman, J. Morgan and E. Witten, "Vector Bundles and F Theory," Commun. Math. Phys. **187**, 679 (1997) [arXiv:hep-th/9701162].

[51] M. Bershadsky, A. Johansen, T. Pantev and V. Sadov, "On Four-Dimensional Compactifications of F-Theory," Nucl. Phys. B **505**, 165 (1997) [arXiv:hep-th/9701165].

[52] B. Andreas and G. Curio, "On Discrete Twist and Four-Flux in N = 1 Heterotic/F-theory Compactifications," Adv. Theor. Math. Phys. **3**, 1325 (1999) [arXiv:hep-th/9908193].

[53] R. Donagi and M. Wijnholt, "Model Building with F-Theory," arXiv:0802.2969 [hep-th].

[54] C. Beasley, J. J. Heckman and C. Vafa, "GUTs and Exceptional Branes in F-theory - I," arXiv:0802.3391 [hep-th].

[55] C. Beasley, J. J. Heckman and C. Vafa, "GUTs and Exceptional Branes in F-theory - II: Experimental Predictions," arXiv:0806.0102 [hep-th].

[56] J. J. Heckman, J. Marsano, N. Saulina, S. Schafer-Nameki and C. Vafa, "Instantons and SUSY breaking in F-theory," arXiv:0808.1286 [hep-th].

[57] J. Marsano, N. Saulina and S. Schafer-Nameki, "Gauge Mediation in F-Theory GUT Models," arXiv:0808.1571 [hep-th].

[58] R. Donagi and M. Wijnholt, "Breaking GUT Groups in F-Theory," arXiv:0808.2223 [hep-th].

[59] J. Marsano, N. Saulina and S. Schafer-Nameki, "An Instanton Toolbox for F-Theory Model Building," arXiv:0808.2450 [hep-th].

[60] J. J. Heckman and C. Vafa, "F-theory, GUTs, and the Weak Scale," arXiv:0809.1098 [hep-th].

[61] J. J. Heckman and C. Vafa, "Flavor Hierarchy From F-theory," arXiv:0811.2417 [hep-th].

CHAIRMAN: J.H. SCHWARZ

Scientific Secretaries: A. Shcherbakov, A. Yeranyan

DISCUSSION I

- Zichichi:

Could you please say a few words on the "striking results" of AdS5/CFT4 duality for RHIC and LHC?

- Schwarz:

The quark-gluon plasma is supposed to be described by QCD at very high energy density and temperature, which are formed in collisions of heavy ions. One thing that we would like to have (but we don't have) is a 5-dimensional string theory that is dual to QCD. QCD is a 4-dimensional quantum field theory and it is plausible that there is a 5-dimensional string theory system that is its holographic realization. QCD is not a conformal field theory but this type of construction can be generalized. If we are optimistic enough we can use a theory with conformal invariance instead. So what people use is the maximally supersymmetrical case – N=4 super Yang-Mills. It is a particular conformal field theory for which the eta over s ratio is computed to be $1/4\pi$. There is a conjecture that this is the lowest value of this ratio that can be realized in any physical system. I'm not an expert on this stuff, but what people suspect is that even using a wrong theory could give the right answer since the result may be universal. There is no other approach that has been proposed to compute this quantity that agrees with observations.

- Klopot:

Does string theory unification inevitably require the existence of branes, or is there another way to unify the 5 types of string theory without the introduction of branes?

- Schwarz:

All the superstring theories have antisymmetric tensor fields, mentioned in my lecture, which can be represented by different forms. There are brane excitations in the theory that are charged with respect to these fields. So these branes exist as possible excitations. Whether they are present in the construction of the vacuum or not is another question. Roughly speaking, they are a very useful way to construct realistic vacua, particularly when you are working in Type I or Type II superstring theories. In the heterotic theory one is not invoking branes in the construction of the vacuum. There are no branes in the heterotic theory, besides the fundamental string and its magnetic dual, which is a 5-brane. They are not important in the construction of the vacuum, but nevertheless they exist as excitations.

- Carrasco:

Outsider question concerning calculation. What is the role of different formalisms – Nathan Berkovits's pure-spinor, Green-Schwarz, etc.

- Schwarz:

A major advantage of the RNS formalism is that it is very straightforward to quantize, but spacetime SUSY is not manifest. An advantage of the GS approach is in having space-time supersymmetry manifest. Unfortunately it is not possible to do quantization while maintaining Lorentz invariance. (This invariance still holds, but it is not manifest.) Berkovits's pure spinor formalism is undoubtedly correct, but it is not fully understood – details need to be clarified. It is not clear where the BRST operator comes from. I am sure that it is right, but there are mysterious aspects in his approach. In some cases it seems to combine the advantages of the other two formalisms.

- *Nicotri:*

If the Ads/CFT correspondence is true, can the weak coupling limit of string theory be considered the definition of a strongly coupled gauge theory and vice versa? If this is the case, can you please comment on the regime in which the two dual couplings are "intermediate"?

- *Schwarz:*

In the case of quantum field theory one has a possibility, in principle, of making a discrete approximation on a lattice and putting it on a computer as people do. So if you have something like N=4 super Yang-Mills theory I think it is fair to say that it's completely defined. The dual picture, where you have string theory on an AdS5 x S5 geometry, is really best understood in perturbation theory. I think if one speaks of completion of one theory by the other, then I would say that it is a string theory that is being completed by a field theory formulation. If you accept this notion that you don't have a complete understanding of one side then you cannot get a complete proof. All you can do is to argue that it works when you understand both sides well enough to test it in certain regimes. We can do that. Intermediate coupling is tough in any theory and in general you would have to resort to some numerical or other types of scheme to deal with it. One trick that we can sometimes exploit in string theory is to study systems with a lot of supersymmetry and compute certain protected quantities which you can continue from weak coupling to strong coupling maintaining control all the way through. But it plays a role only for very limited types of questions.

- *Nicotri:*

In string theory strings usually move in a flat spacetime. Is it possible to construct string theory in a curved or dynamical spacetime?

- *Schwarz:*

Sure, the reason that flat spacetime is nice to study is because the world-sheet theory becomes a free field theory. When you have a curved spacetime the world sheet theory becomes interacting and usually what you have to do is to consider two expansions at the same time. This forces you to look at the supergravity approximation to the superstring theory. For example, when you study AdS5/CFT4 by just looking at string theory in the tree approximation, this corresponds to the dual gauge theory in the planar approximation. And there has been a lot of work on both sides. You need to know the full dependence on the 't Hooft parameter, which then corresponds in the string theory picture to the full dependence on the ratio of the radius of compactification to the string length scale. That is already very difficult.

- *Yeranyan:*

Do we need to have a "confinement" mechanism for keeping interactions, except those of gravity, on the brane? If so, does there exist a realistic one?

- *Schwarz:*

The answer to this question is encoded in the definition of the D-branes. They are objects on which the open strings end. Matter fields are excitations of the open strings. On the other hand, closed strings, which correspond to gravity, are defined in the whole space. That is why we have a natural mechanism for confining matter fields to D-branes.

- *Di Mauro:*

Could you say something on the problem of the vacuum in the presence of the tachyons?

- *Schwarz:*

I briefly discussed the most difficult case, which is a tachyon of a closed string. So if we want to say something more, let's discuss the open string tachyon. In the bosonic string theory there are both open string and closed string tachyons. The open strings end on D-branes. So when we consider a 26-dimensional bosonic open string that means that you are looking at a vacuum that has spacetime filling D-branes, which are D25-branes. An observation of Ashoke Sen was that these branes are unstable. D25-branes do not carry conserved charges and consequently decay. So the transition to the true vacuum from the false one is believed to be due to the decay of a D25-brane. This picture has been confirmed by string field theory. The only string field theory that people have under control is the open bosonic string theory, exactly what you need for this particular problem. It has been shown that the change in the energy between the false and true vacuum is exactly equal to the tension of the D25 brane.

- *Di Mauro:*

On what basis can you say that string theory is UV finite? Does the length of the string act as a natural cutoff on the amplitudes?

- *Schwarz:*

Well, that's true.

- *Susskind:*

That is my fear, because we think that N=8 supergravity is finite now. So what you are saying is the way we said it 20 years ago.

- *Schwarz:*

Even if N=8 supergravity is perturbatively finite, I'm convinced that the nonperturbative completion of that theory will lead you to string theory in 10 dimension.

- *Slobodenyuk:*

String theory can tell us about the number of flavor quarks and leptons. Can it give us information about quark and lepton masses?

- *Schwarz:*

In my lecture tomorrow I am going to discuss this question. I will discuss a new scheme that has been proposed by Vafa and collaborators which looks particularly promising.

- *Shakirov:*

You have mentioned string field theory. Can you say a few words about it, in some more detail?

- *Schwarz:*

In ordinary particle field theory, the field is a function of the spacetime coordinates. But in string field theory, instead of a spacetime point you have a loop. Roughly speaking, the field is a functional of the loop and defining the action you have to do a functional integral.

- *Shakirov:*

So I should integrate over a loop space?

- *Schwarz:*

Yes.

- *Hoehn:*

There are 5 superstring theories related by dualities. Has it already been ruled out that there could be a 6th or 7th version of superstring theory and further dualities?

- *Schwarz:*

You can take one of these theories and compactify it on a torus. Then what you choose to call this theory – just a version of the old one or a new theory – is just a matter of taste. What is special about these 5 theories is that all of them are formulated with supersymmetry in 10-dimensional spacetime. Certainly, there are no other supersymmetric string theories in 10 dimensions.

- *Johansson:*

From the S-matrix point of view, what is the difference between the five superstring theories? What is the difference between the 2-, 3- etc. point functions?

- *Schwarz:*

Their form is pretty similar. Let's just talk about flat 10 dimensional Minkowski space. Four of them have oriented closed strings (the two heterotic string theories and the two Type II superstring theories). The fact that you have only an oriented closed string means that the perturbation expansion is particularly simple. An n-loop amplitude is given by a Riemann surface of genus n. So there is just one Feynman diagram in each order of the perturbation expansion. This is quite a remarkable property of the theory. The usual Feynman diagrams of point-particle theory correspond to different corners of the integration regions. In the case of the Type I theory the story is more complicated, because it contains both open and closed strings, and they are not oriented. So this means that you have to include worldsheets that are non-orientable, and you have to allow for boundaries. So they are much more complicated. The four particle amplitude in the tree approximation is always some ratio of gamma functions, but the details are different in each case.

- *Shcherbakov:*

How does the critical dimension of a superstring theory depend on the number of supersymmetries in the original theory?

- *Schwarz:*

The critical dimensions are determined by a world-sheet conformal anomaly cancellation. It may be shown in different formalisms. They all give critical dimension equal to 10 for the superstring theories. I would like to mention as well that there is another string theory that gives the critical dimension equal to four, but it has two time and two spatial directions, which is not what we want.

- *Ni:*

If supersymmetry is discovered at the LHC, how will the discovery impact the development of string or superstring theory? In other words, is string theory required for supersymmetry?

- *Schwarz:*

No, I think supersymmetry is required for gravity. We know that it exists. And string theory has supersymmetry. But we don't know for sure at what energy scale supersymmetry is broken. There are various arguments for which it is plausible that it should show up at the LHC, supersymmetric grand unification, dark matter, the hierarchy and other things. I think there's a very good chance it will show up. This won't prove that string theory is right. I think it is better to say that the existence of gravity proves that string theory is right.

- *Bonati:*

Do you think using AdS4/CFT3 it will be possible to obtain some exact results for three dimensional critical phenomena like with minimal models in CFT2?

- *Schwarz:*

In 3d conformal field theories the classical Lagrangians resemble N=4 super Yang-Mills in four dimensions more than they resemble two dimensional minimal models. There is no conformal anomaly.

- *Ciuffoli:*

You told us that there are a lot of different methods of compactifications. When does the difference between these various ways of compactification become relevant? Is there something like an energy scale?

- *Schwarz:*

The question of how to do compactifications is equivalent to the question how you're going to get a realistic particle physics phenomenology. I'm going to discuss this tomorrow. But I would like to mention one thing. If one does succeed in finding a scheme that works, it doesn't mean that the other schemes are wrong. There could be one or more dual descriptions of the same thing. These dual descriptions may refer to different coupling regimes so that there is no unique answer to this question.

CHAIRMAN: J.H. SCHWARZ

Scientific Secretaries: A. Shcherbakov, A. Yeranyan

DISCUSSION II

- *Samsonov:*

Professor, today you told us about F-theory, which is very promising for phenomenological applications. What does the letter "F" stand for?

- *Schwarz:*

This term was introduced by Vafa soon after the name "M-theory" was introduced. I think Vafa was under the impression that the "M" of M-theory stood for "mother". That motivated him to introduce this name.

- *Nath:*

It seems that in this theory the Higgs triplets and the X and Y gauge bosons should have masses around the GUT scale. The question is whether this is the case. If so, it would be very interesting to compute the $p \to e^+\pi^0$ and $p \to \nu K^+$ rates in this model, since they should be testable in the current and the next generation of proton decay experiments.

- *Schwarz:*

Yes, that is right. The second paper of Vafa and collaborators, which is 200 pages long, discusses these things. I didn't get through that section yet, and I don't know the details. I believe they identify approximate global symmetries that suppress dangerous couplings.

- *Dubinskiy:*

Could you please say a little bit more about how superstring theory could be applied in condensed matter physics? What problems might be solved?

- *Schwarz:*

You need a conformal symmetry to describe critical phenomena, so that's why you might want a conformal field theory. If it has an AdS dual, then strongly coupling calculations would be feasible. I think there are hopes of saying something about high T_c superconductors and quantum Hall systems, for example. As far as I know, nothing concrete has been achieved yet, but people are just starting to try.

- *Dubinskiy:*

At the beginning of today's lecture you said that we need to input some experimental facts and some assumptions to make superstring theory predictive. What exactly do you mean by these experimental facts?

- *Schwarz:*

We cannot explain everything from first principles, so we need to input some assumptions. To make a local construction in F-theory we have to decide how many families of quarks and leptons we want, what the gauge group should be, etc. The F-theory scenario that I described is engineered to contain these desired properties. It is nontrivial that this is possible, and interesting new possibilities for solving well-known problems have been discovered. An example is breaking the GUT symmetry to the standard model symmetry by means of a background flux on the 4-cycle rather than by a Higgs mechanism. Another is a natural mechanism for doublet-triplet splitting in the Higgs sector. Once one finds a local model that looks promising, it would be desirable to find a global completion. This could correlate parameters and restrict the possible scenarios. So there is a possibility of making predictions in this approach.

- *Marshall:*

You mentioned that this theory includes several nice features – R-parity conservation, and so on. Are there any particular unique or different decay chains or predictions that we might be able to test at the LHC?

- *Schwarz:*

Not yet – but once the community has had a chance to examine this approach more systematically, there may be. We hope it will be possible to make specific predictions before the LHC discovers the answers.

- *Nicotri:*

You have described procedures to obtain the Standard model starting from string theory. Could you please give us an idea of the parameters that these procedures introduce?

- *Schwarz:*

I'm not sure, but there may be some freedom in choosing the del Pezzo surfaces wrapped by the 7-branes. I think the space of del Pezzo surfaces is a space that requires some eight or nine parameters to describe it, and I'm not sure how these parameters affect the construction or the extent to which they can be determined. There also seems to be some freedom in the choice of fluxes and the configurations of the other 7-branes.

-*Akiba:*

Is there any guiding principle from the Planck scale to low energy? Is it possible that string/M-theory is so rich in scenarios that it can mimic the standard model or any theory?

- *Schwarz:*

The new principle that has been introduced in the F-theory construction is the notion of a local model in which gravity can be neglected. That forces the four dimensional cycles that are wrapped by the 7-branes to be a del Pezzo, which is not the only geometry that you might consider without this principle. This makes the scheme much more predictive than the general story of F-theory constructions. Without this principle you can probably make an enormous number of other F-theory constructions, but all of them might give too small a ratio of the Planck scale to the GUT scale. So the fact that there is a significant ratio between these two scales is really what is behind the justification of this assumption. One can imagine that these models will become even more predictive when we understand how to construct global completions.

- *Nagy:*

My question is about the Randall-Sundrum models (and other warped geometry models). Do they require string theory or can they be considered in a normal field theoretical setting?

- *Schwarz:*

Higher dimensional models are not really well-defined without string theory. String theorists focusing on AdS/CFT dualities were led to consider compactifications involving anti de Sitter geometries for spacetime. These geometries, when expressed in suitable coordinates, show an exponential dependence on a radial coordinate. I believe this motivated Randall and Sundrum to try to exploit this exponential dependence on a coordinate as a way of obtaining a possible new solution to the hierarchy problem. Their work has spawned a great deal of activity among phenomenologists. Kachru and collaborators have demonstrated specific schemes in which it seems possible to embed RS-like models in string theory.

- *Susskind:*

Does the F-theory construction get around Witten's unpleasant conclusion that $f_{axion} \sim 10^{16}$ GeV in string theory?

- *Schwarz:*

Yes. The number that they found is at the upper end of the allowed range, namely 10^{12} GeV.

- *Susskind:*

f there is a stack of M matter branes, what is the phenomenological meaning of M^2?

- *Schwarz:*

The examples constructed by Vafa et al. have some five or six nonparallel matter branes each of which just carries one U(1). So they do not introduce stacks of matter branes. The only nonabelian gauge group in their specific examples is on the GUT 7-brane.

- *Ferrara:*

It is usually believed that in string theory there are no continuous global symmetries. Is this still true in world-brane scenarios or in F-theories compactifications? In other words, what is the correct statement about continuous global symmetries?

- *Schwarz:*

I don't have any counterexamples to that assertion. I think it's also true for discrete symmetries – you cannot have an unbroken discrete global symmetry either. It is important that R-parity, which is a discrete symmetry, is not a global symmetry. In their construction it arises as an unbroken subgroup of a U(1) gauge symmetry, so even though it is discrete it can be a gauge symmetry.

- *Ferrara:*

Let us take, for example, N=4 super Yang-Mills theory. Isn't the SU(4) R-symmetry a continuous symmetry that is exact? It is a global symmetry on the world volume of the brane.

- *Schwarz + Susskind:*

Global symmetries are broken in string theory. There is a little confusion here as to what is a local and what is a global symmetry. Something like $SU(4)_R$ is a local gauge symmetry from the dual string theory viewpoint.

solution to the hierarchy problem. Then what [illegible]
phenomenologists. Kaluza and collaborators [illegible]
possible to embed ES-like models in string theory.

[illegible]

ERICE LECTURES 2008

Gerard 't Hooft

Institute for Theoretical Physics
Utrecht University

and

Spinoza Institute
Postbox 80.195
3508 TD Utrecht, the Netherlands

e-mail: g.thooft@uu.nl
internet: http://www.phys.uu.nl/~thooft/

Lecture II: Some Effects of Instantons in QCD

1 Lecture II: SOME EFFECTS OF INSTANTONS IN QCD

1.1 Construction of instantons

Consider a non-Abelian Yang-Mills gauge theory in 4 space-time dimensions, and the group or subgroup $SU(2)$ of gauge transformations. They may be space-time dependent, so we have a mapping from $\mathbb{R}4$ to the space Ω of $SU(2)$ transformations, to be indicated by the function

$$U(\vec{x},t) = a_0(\vec{x},t)\,\mathbf{1} + i\sum_{i=1}^{3} a_i(\vec{x},t)\,\sigma_i\ , \tag{1.1}$$

where σ_i are the Pauli matrices and the coefficients $a_0, \cdots,\ a_3$ are real and obey

$$\sum_{i=0}^{3} a_i^2 = 1\ . \tag{1.2}$$

Clearly, these coefficients (1.2) lie on a 3-sphere, $S3$.

Now the boundary of $\mathbb{R}4$ is also a 3-sphere, so suppose that we choose our mapping in such a way that the 3-sphere of the boundary is a 1-to-1 mapping onto the gauge-sphere (1.2). The mapping that we chose, as well as small deformations of it, is said to have winding number one, and is indicated as $U_1(\vec{x},t)$. We also have mappings of winding number n, by raising U_1 to the nth power:

$$U_n(\vec{x},t) = U_1(\vec{x},t)^n\ , \tag{1.3}$$

where n can be any integer.

Only if $n = 0$, however, this mapping can be extended to all of $\mathbb{R}4$ without any singularity anywhere. Such singularities are normally not admitted in a gauge transformation. When $n \neq 0$, our gauge transformation can only be well-defined on a space $\mathbb{R}4$ with a small region V cut out. The singularity is then assumed to lie inside V.

Suppose we apply this gauge transformation $U_1(\vec{x},t)$ to the vacuum field configuration that was defined as $A_\mu(\vec{x},t) = 0$. The transformed field,

$$A_\mu^{(1)}(\vec{x},t) = -\frac{i}{g}(\partial_\mu U_1)\cdot U_1^{-1}\ , \tag{1.4}$$

is non-vanishing now, but of course, outside the region V, it still describes the vacuum state.

Now the gauge function $U(\vec{x},t)$ could not be extended inside V, but there is no problem if one wishes to extend the field configuration (1.4) to also obtain values inside V. If V would be the sphere $\vec{x}^2+t^2 \le \varrho$, then we could take

$$A_\mu^{(1)}(\vec{x},t) = \frac{|(\vec{x},t)|}{\varrho} A_\mu^{(1)}\left(\frac{(\vec{x},t)}{|(\vec{x},t)|}\right)\ . \tag{1.5}$$

to hold inside V.

This is smooth and singularity-free everywhere inside V (mild singularities in the derivatives of A can be smeared away if so desired). However, inside V, this field $A_\mu^{(1)}$ does not describe the vacuum, since no function U can be found that gauge transforms it away. Indeed, one can compute the associated curvature fields $G_{\mu\nu}(\vec{x},t) = \partial_\mu A_\nu - \partial_\nu A_\mu - ig[A_\mu, A_\nu]$ to find that these do not vanish inside V. This is therefore a non-trivial field configuration, and since it only differs from the vacuum inside a region V that surrounds a point as well as an instant in time[1], it is called an *instanton.* If the winding number n is one, we call it a single instanton, if it is -1, we have an anti-instanton. For values of n greater than 1 (or lower than -1) we have multi-instantons.

1.2 Instantons and fermions

Instantons can exist either in Minkowski spacetime or in Euclidean spacetime, since no reference was made to any Lorentz invariant metric. Instantons are time-dependent, therefore fields such as fermion (quark) fields, can have modes where the energy before and after an instanton is different. In the case of massless fermions, chirality (or, equivalently, the helicity) of the fermion is preserved, since the gauge field interaction (which goes as $ig\gamma_4\gamma_\mu A_\mu(\vec{x},t)$) commutes with γ_5. Thus, left-handed fermions stay left-handed and right-handed ones stay right-handed. In the case of a single instanton, however, exactly one eigen mode of the fermionic Hamiltonian, for each chiral fermion flavor of positive helicity, is lifted from the Dirac sea to a positive energy state, see Fig. 1

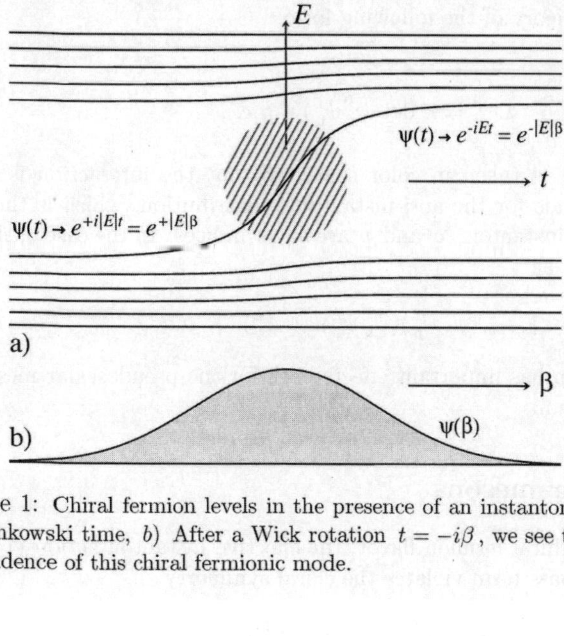

Figure 1: Chiral fermion levels in the presence of an instanton, *a)* as a function of Minkowski time, *b)* After a Wick rotation $t = -i\beta$, we see this Euclidean time dependence of this chiral fermionic mode.

The figure explains that this leads to one solution of the Dirac equation,

$$\gamma_\mu(\partial_\mu - igA_\mu)\psi = 0 \ , \tag{1.6}$$

in the presence of an instanton that rapidly approaches zero in all four Euclidean directions. [2][3]

Under a parity transformation,

$$\vec{x} \to -\vec{x} \ , \qquad t \to t \ , \tag{1.7}$$

the instanton transforms into an anti-instanton, since

$$(a_0\mathbf{1} + ia_i\sigma_i)^{-1} = a_0\mathbf{1} - ia_i\sigma_i \ , \qquad \text{if} \qquad \sum_{i=0}^{3} a_i^2 = 1 \ . \tag{1.8}$$

Therefore, left-helicity fermions do the opposite of right-helicity fermions: if a left-helicity fermion is created out of the Dirac sea, a right-handed helicity fermion disappears, and *vice versa.*

Since holes in the Dirac sea describe anti-fermions, one can extend the statement to include the possibility of the creation of a chiral antifermion in stead of the annihilation of a chiral fermion, and so on.

An instanton acts on all fermionic flavors the same way; therefore, an instanton effectively acts as a vertex where one unit of positive chirality of each flavor disappears and one unit of negative chirality appears. This vertex can be mimicked by an extra term in the Lagrangian of the theory of the following form:

$$\mathcal{L}^{\text{inst}} = \kappa \det_{a,p}(\overline{\psi}{}^R_a \psi^L_p) + h.c. \ , \tag{1.9}$$

where an inner product is taken in color space, so that the interaction is color-gauge invariant, and $h.c.$ stands for the anti-instanton's contribution, which is the hermitean conjugate of that of the instanton. a and p are flavor indices. In the case of three flavors, this is written explicitly as

$$\tfrac{1}{3!}\kappa\, \varepsilon_{abc}\varepsilon_{pqr}(\overline{\psi}{}^R_a\psi^L_p)(\overline{\psi}{}^R_b\psi^L_q)(\overline{\psi}{}^R_c\psi^L_r) \ + \ h.c. \ . \tag{1.10}$$

In QCD, this interaction has important effects both for the pseudoscalar mesons and for the scalar meson spectrum.

1.3 Pseudoscalar mesons

In the case of a single chiral fermion flavor, the effective instanton action (1.9) acts just as a mass term. This mass term violates the chiral symmetry

$$\psi^L \to U_L\psi^L \ , \qquad \psi^R \to U_R\psi^R \ , \tag{1.11}$$

if the complex numbers U_L and U_R are chosen to be independent. Global vector gauge invariance, corresponding to $U_L = U_R$ remains a good symmetry. This is exactly the result that one also obtains in perturbation theory, where the Adler-Bell-Jackiw anomaly[4] is found to break the symmetry in exactly the same way. This is not an accident[5].

In the case of more than one flavor, U_R and U_L are complex unitary matrices, and now one only needs to demand that

$$\det(U_R) = \det(U_L) . \tag{1.12}$$

This breaks the chiral symmetry

$$U(N_F)_L \otimes U(N_F)_R \tag{1.13}$$

into

$$SU(N_F)_L \otimes SU(N_F)_R \otimes U(1) , \tag{1.14}$$

again in agreement with the Adler-Bell-Jackiw anomaly in the perturbative theory.

Long before the advent of QCD, it was noted that the pseudoscalar particles π and K appear to act as the Goldstone bosons of the symmetry (1.14), which seems to be well obeyed by the interactions of these particles, while the symmetry (1.13) would require one more scalar meson, the η particle (in case of isospin $SU(2)$), or the η' (in the case of flavor $SU(3)$) to act as a Goldstone boson. But Goldstone bosons should be massless. The pion is much lighter than the η particle, and the kaons are much lighter than the η' (also sometimes called the *Beppo*[6]). We now understand that the excess masses of η and of η' are both due to the instantons in QCD.

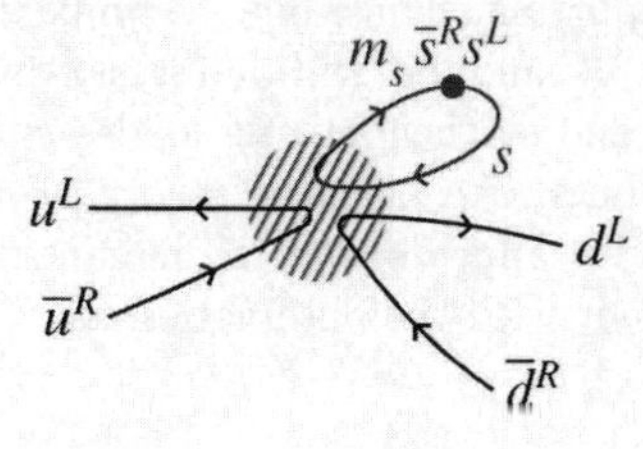

Figure 2: The strange quark s can form a closed loop in the effective quark vertex, an effect that scales with the strange quark mass m, since the left helicity strange quark must turn into a right-helicity one (or vice versa).

The fact that the interaction (1.10) for three flavors in $SU(3)$ turns into a four-fermion interaction for isospin, $SU(2)$, when the strange quark is assumed to be heavy, can be understood: the strange mass term explicitly mixes the chiral left handed strange quark with the righthanded one, so that $\langle \overline{\psi}_s^R \psi_s^L \rangle$ (where s stands for strangeness) can be seen to be proportional to the strange quark mass instead of a dynamical (pseudo)scalar field. In Figure 2 it is illustrated how the mass term for the strange quark s can turn an effective 6-quark vertex into a 4-quark vertex.

1.4 Scalar mesons

The effective instanton action, Eq. (1.9) or (1.10), not only causes shifts in the spectrum of pseusoscalar mesons; it also contributes in the scalar sector. At first sight, it seems to be more difficult to identify its effects there, since, in this channel, no violation takes place of any explicit symmetry pattern. There is, however, a more subtle way in which it manifests itself.

Effective QCD models showed a peculiar difficulty pertinent to the scalar mesons. The scalar mesons, like the pseudovectors, form $SU(3)$ octets or $U(3)$ nonets. One of the scalar spin zero states has sizeable branching ratios in the K - $\overline{K}$ sector, so that one would be tempted to describe as an $\overline{s}\,s$ bound state. However, it is about as light as the other scalar spin zero states that do not decay into strange mesons. The s quark is heavier than the u and d quarks. How do we explain this?

One possibility investigated in the literature is that the scalar mesons are systems containing two quarks and two antiquarks, the so-called *tetraquarks*. Realizing that a color-antisymmetric two-quark system transforms as a $\overline{\mathbf{3}}$ representation of color $SU(3)$, a teraquark is realized if, in a baryon, one quark is replaced by two antiquarks. Therefore, one might imagine that tetraquarks will be about as heavy as baryons are. If the scalar mesons were tetraquarks, their mass spectrum would look a bit better.

This, however, would still raise the question where the ordinary scalar biquark states are. Why would tetraquark states be more important than biquark ones? It is here that the instanton interaction (1.10) sheds new light. It couples tetraquark states such as $\overline{u}\overline{d}ud$ directly to the state $\overline{s}s$. The masses of these two states could be close together, and, as is well known from perturbation theories, when two states are close together, a relatively weak mixing term can give a relatively large splitting. Thus, one new mixed state arises that is relatively light. Biquark states such as $\overline{u}s$ and $\overline{u}d$ will be further away from the tetraquark states with which they are mixed, and so their mixing angles will be smaller and their masses are reduced by smaller amounts. An effective theory of scalar meson states can be constructed along these lines, and it appears that the instanton interaction (1.10) is just what is needed to obtain agreement with the observations.[7]

References

[1] A.A. Belavin, A.M. Polyakov, A.S. Schwartz and Yu.S. Tyupkin, *Phys. Lett.* **59B** (1975) 85.

[2] R. C.G. Callan Jr., R.F. Dashen and D.J. Gross, *Phys. Lett.* **63B** (1976) 334.

[3] G.'t Hooft, *Phys. Rev. Lett.* **37** (1976) 8; *Phys. Rev.* **D14** (1976) 3432.

[4] S.L. Adler, *Phys. Rev.* **177** (1969) 2426; J.S. Bell and R. Jackiw, *Nuovo Cim.* **60A** (1969) 47 S.L. Adler and W.A. Bardeen, *Phys. Rev.* **182** (1969) 1517; W.A. Bardeen, *Phys. Rev.* **184** (1969) 1848.

[5] G. 't Hooft, textitThe Physics of Instantons in the Pseudoscalar and Vector Meson Mixing, SPIN-1999/01, hep-th/9903189.

[6] A. Zichichi, *The Beppo Particle*, Int. Symp. "Highlights in Physics Today: One hundred Years after the Birth of Beppo Occhialini", Milan,February 16, 2007.

[7] G. 't Hooft, G. Isidori, L. Maiani, A.D. Polosa and V. Riquer, *Phys. Lett.* **B662** (2008) 424, arXiv:0801.2288.

CHAIRMAN: G. 't HOOFT

Scientific Secretaries: J. Dent, G. Stavenga

DISCUSSION II

- *Di Mauro*:

One of the physical evidences of instantons is that we can have a spin flip of fermions and this produces a chiral anomaly. So we have a very intriguing connection between topological non-trivial configuration and quantum anomalies. Can you comment more about the relation between the instanton and quantum anomalies connection with some examples?

- *'t Hooft:*

The way I describe instantons is that, say, a left-handed fermion pops out of the vacuum. Its energy level raises from the background sea into this set of particle states which correspond to visible particles, and the converse happens with the right-handed fermions. If a right-handed fermion was present, then it disappears into the sea, and if there was none then an anti-particle is produced – which is basically the same thing. The net effect of the instanton is that a right-handed particle disappears, and a left-handed particle appears. This is a notable violation of a conservation law. Before this observation, it was thought that left-handed particles were conserved and right-handed particles were conserved, that's helicity conservation. This is expressed as the conservation of helicity charge, helicity charge being, say, plus one for all negative helicities, or left-helicity particles, and minus one for all right-helicity particles. That charge, also called chiral charge, is apparently being destroyed by the instanton. Now, what does this have to do with the ABJ anomaly? According to ABJ, an anomaly occurs, if you write down a triangle diagram with two gauge particles going in, or say, one gauge particle going in and two gauge particles going out, in short, a triangle with three external gauge particles, two being the photons of some gauge fields, and one third being whatever the quantity is that you are looking at, which is usually chiral charge. Suppose that there was a gauge field coupled to chiral charge. Now, that can't be. Why can't that be? Well, we just saw that helicity charge is not conserved by instantons. So, if you try to couple a gauge field to a charge that is not conserved, then you are in for trouble. And, yes, you get trouble if you try to do this. In the triangle diagram, if you try to renormalize the triangle diagram, you discover you can't do it. You can't do it while conserving gauge invariance for those two external photons which are in a gauge theory, and also conserving chiral charge. And the result is that chiral charge is not conserved. In fact, the triangle diagram tells you about the non-conservation of chiral charge.

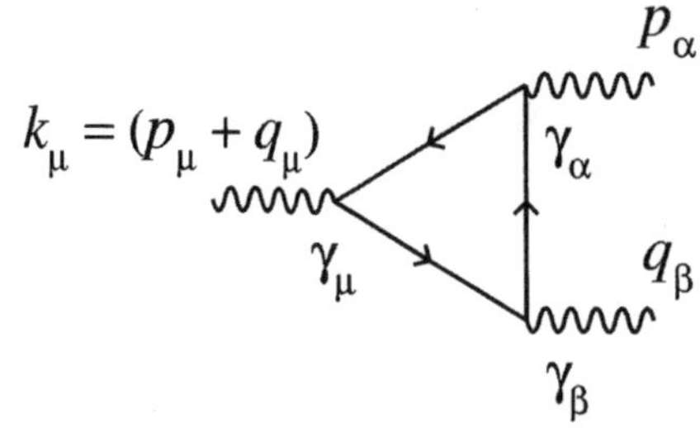

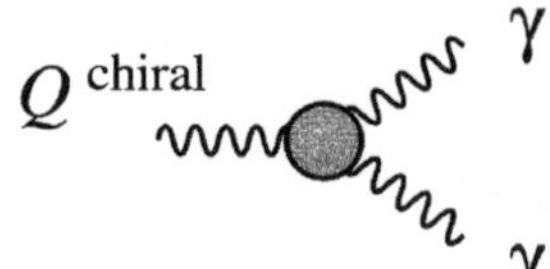

$$p_\alpha \Gamma_{\mu\alpha\beta}(p,q) = 0$$

$$\partial_\mu A_\mu^{\text{chiral}} = \frac{g^2}{16\pi^2} \tilde{F}_{\alpha\beta} F_{\alpha\beta}$$

$$q_\beta \Gamma_{\mu\alpha\beta} = 0$$

$$k_\mu \Gamma_{\mu\alpha\beta} \neq 0$$

$$\tfrac{1}{4}\int \tilde{F} F d^4x = \frac{8\pi^2}{g^2}$$

The gamma's here are the photons, with respect to which you must have gauge invariance, because these are gauge particles. This diagram must be gauge invariant for electromagnetism. Now this other thing is Q-chiral. So this is chiral charge, and now the statement is that there is a clash with respect to gauge invariance at these three vertices. Only on two of the three points you can demand gauge invariance, and if you do that, the divergence at the third vertex is fixed. So, this diagram has two possible counterterms, and those counterterms are fixed if you demand gauge invariance at vertex α and at vertex β. Once you fix those two counterterms, there is no third counterterm that you can use. So, you have to wait and see, and you discover, to your surprise perhaps, that gauge invariance is violated at the third vertex. For this amplitude Γ, gauge invariance means that at each of its vertices, labled with the indices, μ, α and β, an identity depending on the momenta k, p and q has to be obeyed. To be precise, if you multiply this amplitude Γ either with $p\alpha$ here or with $q\beta$ there, you find zero. That is gauge invariance. Now at the third vertex, the in-going momentum k is equal to p plus q. So if you multiply Γ with $k\mu$, then you find that this is not zero. Thus, if you impose gauge invariance at two of the vertices, then the third identity cannot be obeyed. That's the so-called Adler Bell Jackiw anomaly. You can compute its amplitude and you then find that it behaves as a diagram. The non-vanishing diagram is generated when you fix the associated counterterms. It is as if you have a local interaction, which causes that $\partial\mu A\mu$chiral, the divergence of the chiral vector current, is actually $F\alpha\beta$ $F\alpha\beta$ dual. These are the photon fields on the right hand side. This is the ABJ anomaly. There is a coefficient here, a combination of the coupling constant g and factors of π.

The remarkable thing is now that, the instanton, which is a gauge configuration that I described yesterday, which was discovered by four Russians, Belyavin, Schwarz, Tyupkin, and Polyakov, obeyes a special identity. They discovered that if you integrate F times F dual d4x you find a quantity that goes like 1 over the coupling constant squared (times some factors π), so the smaller the coupling constant, the bigger the instanton is because this is a non perturbative field configuration. This integral is very big when the g is very small. If you substitute this equation on the left hand side you get the integral over d4x of $\partial\mu A\mu$chiral. This actually is a calculation you have to do if you want to compute the chiral charge before and the chiral charge after the event. Because this is just a surface integral, it computes the integral of the total charge before, minus the integral of the total charge, after the instanton. Then you may simply substitute what you found out from the triangle diagram. You find that the coupling constant and the factors π cancel out. The net result of the whole calculation is that there are two units of chiral charge that disappear in the instanton. Two units because a right handed particle disappears while a left handed particle appears. A chiral charge of minus one disappears and a chiral charge of plus one appears so that gives a net two units of chiral charge appearing into the instanton. It all hangs together completely, mathematically. The triangle diagram is the diagram that tells you how much charge actually disappeared into the instanton or was created by the instanton. Of course the anti-instanton has the opposite effect. This tilde here means that you take the dual field, by multiplying with the coefficient $\varepsilon\alpha\beta\gamma\delta$, which is antisymmetric under parity. So if you change the parity of the configuration you get a minus sign, and the opposite happens with the chiral charge reappearing where it previously disappeared or vice-versa. So that's how an anti-instanton can undo the work of an instanton.

Admittedly, the argument boils down to be a very technical one, but this is how one can explain it in words.

- Di Mauro:

You mentioned weak instantons in your talk, and I would like to know how are instanton effects modified if gauge symmetries are realized in the Higgs model?

- 't Hooft:

The Higgs mechanism has very little to do with instantons directly, although I should add one thing: of course the Higgs effect gives mass to the gauge field and because of that it keeps the coupling constant small, so the one over g squared becomes large, and so we'll have a very large amount of action. If you compute the probability of the effect actually to occur, you have to exponentiate this quantity because it is e to the power of the action which is the amplitude of the event and you have to compute it in Euclidean space, then this quantity holds in Euclidean space. One finds e to the minus 8 π2 / g2 for the contribution of an instanton to the action. That is very small when you have the Higgs mechanism. The reason is that the Higgs mechanism breaks the symmetry, therefore it keeps the gauge field strength very tiny, the gauge fields become short range. Because it keeps the gauge fields tiny, it impedes the occurrence of instantons. So the Higgs effect tries to keep the instantons away. Consequently, a large amount of action is needed for instantons to occur, and that is translated into a very strong suppression, an exponential suppression in the amplitude. This is just another way of saying that now going from one state to the other state is like a tunneling effect in quantum mechanics. It is strongly suppressed by exponential factors. The net result is that in the weak interaction theory where instantons also occur, the instanton is very strongly suppressed by the Higgs mechanism, so strongly that you hardly ever see that instanton. That's a good thing because the charge which is violated by the electroweak instanton is actually baryon number and lepton number. The Standard Model does not conserve baryon number or lepton number, the instanton breaks it, but it is very strongly suppressed if the factor 8 π2 / g2 becomes of order 137 or something like that. Roughly speaking this is 137 with a minus sign that sits in the exponent, so e to the minus 137 is the probability that baryon number is spontaneously violated in the Standard Model. This is a property of the Standard Model, you may not have realized that the SM actually violates baryon number conservation by instantons.

- Stavenga:

If the instanton is considered on the lattice and if we want to go to the continuum limit, then I suppose that we have to take the lattice constant to zero whereas the instanton is a classical effect of the bare lagrangian so that means that the instanton is a solution when the coupling constant goes to zero. Is the lattice thereby pushing the instanton out of the spectrum completely in the continuum limit of QCD?

- *'t Hooft:*

That is fortunately not true. The point of an instanton on the lattice is a very subtle one because the lattice is discrete; the lattice contains only points and links. The instanton sits in four-space. Now the instanton could be a very tiny configuration in between the lattice sites. But, the lattice is discrete so there is nothing in between the lattice sites. So the instanton can actually disappear on the lattice. Now that is related to the difficulty of defining chiral fermions on the lattice. Remember, the instanton violates chiral charge, chiral charge is difficult to define on the lattice. It is difficult, but not impossible. We've seen lattice QCD talks by Kenway and others, so you can, with lot of effort, define chiral charge on the lattice, though it is difficult. The reason precisely is that the lattice is discrete while the instanton is something that is topologically defined, and these topological definitions appear to work only in a continuum. There is a conceptual clash here, and to get around it you have to do a certain amount of work. But it can be done, and the net result usually is that instantons work as expected in the continuum theory provided that they are large compared to the lattice. The instanton is a big thing, while the lattice spacing is very tiny, and in that case it is all right because the amount of gauge fields on every single lattice site is still very small. It is the entire integral of F F dual which has to be large but that integral now comprises so many lattice sites that on every single lattice element the field is very weak. Then you can study your instanton.

Now the instantons that are going to be negligible according to the arguments would be the instantons which sit on a single lattice site. Those are the very tiny instantons, and indeed in QCD the very tiny instantons are negligible. It is the large instantons that give large contributions, and that is because of the running coupling constant. The effect of the instanton is proportional to the quantity $e\text{-}8\pi 2/g2$, but g is a running coupling parameter so when you go to the single lattice site the running coupling parameter is small, and therefore this quantity is negligible. Lattice QCD lives from the fact that you may assume that the coupling constant is so small on every single lattice site that nothing very special happens between the lattice links. So this amplitude is ignorable there. However, g is a running coupling constant and in fact if the instanton is very large compared to the lattice length, then the accumulated effect of these different fields collaborating on the lattice becomes large because g is large at that scale. The instanton has a sizeable effect provided it is large compared to the lattice size.

- *Stavenga*:

So if only the large instantons contribute then they should be sensitive to the boundary conditions. Since the instanton is derived from boundary conditions at infinity you can take all kinds of boundary conditions on the lattice. For instance, periodic boundary conditions.

- *'t Hooft:*

That is of course true. It means that if you want to do a sensible lattice calculation you always have a difficulty in that you always want to choose the lattice length as small as possible, but you also want to choose the volume of space and time as large as possible. There is always this clash, there is nothing that can be done about that. The instantons must fit inside that volume. If you don't have those conditions right, then you do a bad lattice calculation.

- *Stavenga:*

May I ask one question to Richard (Kenway) about this? Do you see the instanton or the instanton configurations?

- *Kenway:*

Yes certainly. We certainly do, I mentioned actually in my talk that we checked that we have a sensible distribution of topological charge in the gauge configurations that we generate. Gerard is absolutely right. You need to have a reasonably small lattice spacing and a reasonably large volume in order to be able to do that.

- *Caffari:*

How should instantons interact with c quarks? Could they try to explain the resonances found near the open charm mesons threshold?

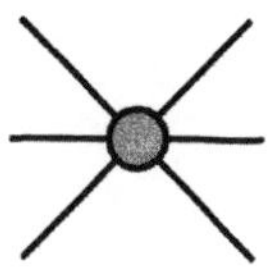

$$\kappa\ \overline{\psi}\,\overline{\psi}\,\overline{\psi}\ \psi\,\psi\,\psi\ + \text{h.c.}$$

- *'t Hooft:*

The instanton gives an effective interaction of this form. If you take three flavors for a moment, it is ψ bar, ψ bar, ψ bar, ψ, ψ, ψ, so this is an effective six point interaction. Because every fermion flavor eats up one chirality and produces the other chirality. This plus the Hermitian conjugate which does exactly the opposite thing. Now you ask what is the effect on c quarks...well one might expect a fourth psi bar and a fourth psi that search in the sea to take c quarks out to give them opposite chirality, and if those states were already occupied in the sea then the produced quarks would be physical quarks. But since the c quark has a large mass, and its mass term essentially hides this effect, the main contribution still comes from the six fermion effective interaction term, where some of these light quarks will be in the sea and some of these will be in the open, being, so to say, physical quarks. As usual in QFT the only difference from ordinary QFT is this six fermion effective interaction. Normally you have two, three, or four particles in the vertices of QFT, whereas here you have six.

- *Martin:*

What is the correct way to think about instantons at large number of color limit? Usual large N thinking as far as I know is that they are not relevant but I would think that as the number of colors gets large, the number of SU(2) embeddings gets bigger – you have many more instantons. So, what is the correct way to think about this?

- *'t Hooft:*

This is actually a very difficult question. Ed Witten, who has studied this question, claims that instanton effects do not go away very fast for large N, that they are suppressed by some mild factors such as 1/N, but not exponentially for large N as one might have expected. Because for large N the thing you want to keep constant is the coupling constant g2N, which is g tilde squared N. You keep that constant for large N, but the instanton action is 8 π2 / g2, not g2N. That is because the instanton sits in an SU(2) subgroup of the group and the instanton doesn't know about all these other colors. It doesn't care about them. Now you can think, ok this means that the instanton action is e to the minus 8 π2 N divided by g tilde squared. If you keep g tilde constant then the instanton is suppressed by an exponential factor e to the minus N times something. But then Witten asks for caution: this g is a running coupling constant. It goes all the way to infinity in the gauge theory. If it is infinity then this coefficient multiplying N in the exponent is zero so this amplitude is not exponentially suppressed at very large N. According to Witten it is 1/N or 1/N2. This is a very difficult argument, a very technical argument, and I'm not sure exactly why Witten arrives at this detailed result. But it is quite conceivable that instantons do not go away as you might expect naively because g tilde is a running coupling constant which runs all the way to infinity.

MARTIN

$$g^2 N = \tilde{g}^2$$

$$e^{-\frac{8\pi^2}{g^2}}$$

$$\boxed{e^{-8\pi^2 N/\tilde{g}^2}} \sim 1/N$$

ERICE LECTURES 2008

Gerard 't Hooft

Institute for Theoretical Physics
Utrecht University

and

Spinoza Institute
Postbox 80.195
3508 TD Utrecht, the Netherlands

e-mail: g.thooft@uu.nl
internet: http://www.phys.uu.nl/~thooft/

Lecture III: Crystalline Gravity

1 Lecture III: CRYSTALLINE GRAVITY

Abstract

Matter interacting classically with gravity in 3+1 dimensions usually gives rise to a continuum of degrees of freedom, so that, in any attempt to quantize the theory, ultraviolet divergences are nearly inevitable. Here, we investigate matter of a form that only displays a finite number of degrees of freedom in compact sections of space-time. In finite domains, one has only exact, analytic solutions. This is achieved by limiting ourselves to straight pieces of string, surrounded by locally flat sections of space-time. Next, we suggest replacing in the string holonomy group, the Lorentz group by a discrete subgroup, which turns spacetime into a 4-dimensional crystal with defects.

1.1 Introduction

One way of addressing the problem of quantizing gravity is first to search for its fundamental degrees of freedom. In all quantum systems that we know of, these degrees of freedom are denumerable, at least if one confines oneself to the data inside a box of finite dimensions. In gravity, one expects even more stringent constraints: according to the holographic principle, one expects the total number of degrees of freedom that reside inside a finite box not to exceed the number of microstates of a black hole that would just fit in there — in other words, the total number of degrees of freedom is finite!

Now imagine splitting the box into more and more finite subsystems. If space and time were to be described by coordinates consisting of real numbers, then, at least mathematically, splitting boxes into smaller ones can be done without limits. What this appears to imply then, is that at distance scales small compared to the Planck length, roughly 10^{-33} cm, there should be no degrees of freedom left at all. One way to accommodate for such a situation is to describe space and time there as to be absolutely flat. there is no physically non-trivial structure left. The most one can allow for is some defects when gluing together flat pieces of space-time at their boundaries. In other words: at the Planck scale, gravity must be replaced by some *topological* theory.

"Simplicial gravity", as described by R. Loll in her lectures at this school, is an example of just such a structure, but it is not quite the same as the theory to be discussed now. As will gradually become clear, "gravitational" and "matter" degrees of freedom will be indistinguishable, both being described as *defects* when gluing locally flat pieces of Minkowski space-time together. We assume the pieces not to have a cosmological constant, so, the bare cosmological constant $\Lambda = 0$.

The most conspicuous difference between the theory to be described here[1] and others is that, in essence, it is a classical theory. Its degrees of freedom, the defects, obey classical equations of motion. One can think of this in two ways: the most conventional attitude to take is to regard the theory as a precursor to a quantized version, as was done with many other theories, but we also have in mind a more radical idea, pursued by this lecturer for some time[2], which is called "prequantization", the idea that all states of the classical theory can directly be promoted to elements of a basis of Hilbert space, without first replacing Poisson brackets by Dirac brackets. We leave these options open for the time being.

The question first to address is then: what are the dynamical equations of motion for the defects? We have a kind of crystal in Minkowski space. The crystal has defects in it. Ordinary crystals only live in 3 dimensional Euclidean space, and their defects can be anything. The defects in our theory are 4 dimensional, and obey e.o.m. In the next chapter, we illustrate the idea by first studying gravity in only two spacelike dimensions.

1.2 Gravity in $2+1$ dimensions

The importance of the model of gravitating point particles in a locally flat 2+1 dimensional space-time, is still severely underestimated[3]—[5]. In 2+1 dimensions, pure gravity (gravity without matter in some small section of space-time) has no physical degrees of freedom at all (apart from boundary effects if spacetime is topologically non-trivial). This is because the Riemann curvature $R_{\alpha\beta\gamma\delta}$ can be rewritten in terms of a symmetric 3×3 matrix $Q^{\mu\nu}$ as follows:

$$R_{\alpha\beta\gamma\delta} = \varepsilon_{\alpha\beta\mu}\,\varepsilon_{\gamma\delta\nu}\,Q^{\mu\nu}\ , \tag{1.1}$$

so that the Ricci curvature is

$$R_{\alpha\gamma} = R_{\alpha\beta\gamma}{}^{\beta} = (g_{\mu\nu}g_{\alpha\gamma} - g_{\mu\gamma}g_{\alpha\nu})Q^{\mu\nu} = Q^{\mu}_{\mu}\,g_{\alpha\gamma} - Q_{\gamma\alpha}\ ; \tag{1.2}$$

$$\text{so that}\qquad Q_{\mu\nu} = \tfrac{1}{2}R^{\alpha}_{\alpha}\,g_{\mu\nu} - R_{\mu\nu}\ . \tag{1.3}$$

Clearly, if matter is absent, $R_{\mu\nu}$ vanishes, and therefore so do $Q_{\mu\nu}$ and $R_{\alpha\beta\gamma\delta}$. Conversely, a point particle represents point curvature. Thus, particles are point singularities surrounded by flat space-time. A particle at rest can be described as in Figure 1 a). The wedge is stitched closed, so that the points A and A' are identified. The defect angle can

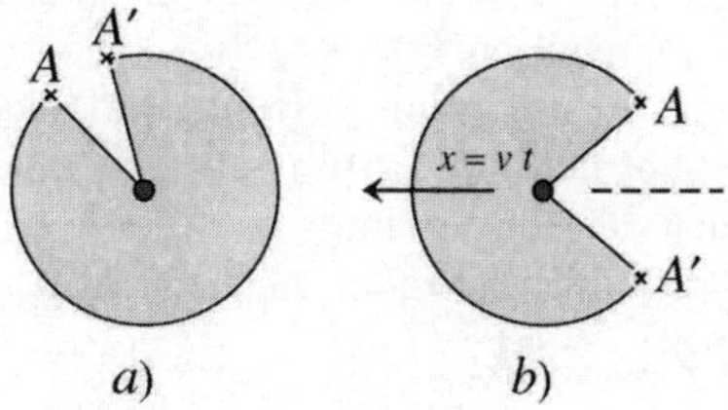

Figure 1: a) 2 dimensional space surrounding a point particle. Points A and A' are identified. b) Particle moving in the x-direction, after applying a Lorentz boost. Only if the defect is arranged in the way indicated, the identification of points A and A' is without a time shift.

directly be identified with the particle's rest mass. One can show that systems containing several point particles in motion, while the total momentum is kept zero, is surrounded by a conical space-time, of which the deficit angle can be identified with the total energy.

When a point particle is set in motion, the surrounding space-time is described by performing Lorentz transformations upon the stationary case, see Fig. 1 *b*). One then can study systems with N particles, by viewing space-time as a tessellation of locally flat triangles or polygons.[4] If all particles in a $2+1$ dimensional universe would be stationary or nearly stationary, the two-dimensional integrated scalar Ricci tensor, $\int \delta^2 x \sqrt{g}\, R$, would have to be positive, so with a sufficient number of particles the spacelike part of this universe would always close into an $S(2)$ geometry. Its timelike coordinate can form a compact dimension (featuring both a *bang* and a *crunch*), or a semi-infinite one, with either a bang or a crunch. With fast moving particles one can however also form 2-d surfaces with higher genus, without requiring negative mass particles.[5]

Quantization of this system is often carried out 'as usual',[6] but there are delicate problems, having to do with the fact that we are dealing with a strictly finite universe, so that the role of an 'observer' is questionable, and the statistical interpretation of the wave function is dubious because the finiteness of the universe prohibits infinite sequences of experiments to which a statistical analysis would apply. Carrying quantization out with care, one first observes that evidently, time is quantized into 'Planck time' units[7]. This is easily derived from the fact that the hamiltonian is an angle and it is bounded to the unit circle. Consequently, a quantum theory cannot be formulated using differential equations in time, but rather one should use evolution operators that bridge integral time segments. This indeed can be regarded as a first indication of some sort of space-time discreteness, which we will encounter later in a more concrete way. Moreover, a confrontation with foundational aspects of quantum mechanics will appear to be inevitable.

1.3 Three space dimensions: strings and joints

In the work presented here, the question is asked whether a similar "finite" theory can also be formulated in $3+1$ dimensions. This is far from obvious. One first notices that the absence of matter now no longer guarantees local flatness, since the Ricci curvature $R_{\mu\nu}$ can vanish without the total Riemann curvature $R^{\alpha}{}_{\beta\mu\nu}$ being zero. However, one still can decide to view space-time as a tessellation of locally flat pieces. The defects in such a construction again may represent not only matter, but also physical activities such as gravitational waves. The primary defects one finds are direct generalizations of the $2+1$ dimensional case. Take a particle-like defect in 2-space. In 3+1 dimensions, such defects manifest themselves as strings.

It could be that matter is always arranged in such a way that it can be regarded as defects in a locally perfectly flat space-time. As opposed to technical approaches towards solving General Relativity[8], we now regard matter of this form as elementary. The novelty in this idea is that, in spite of matter being distributed on subspaces of measure zero, we still insist that it obeys local laws of causal behavior. Let us see how this looks.

By simply adding the third space dimension as a spectator, orthogonal to the first two,

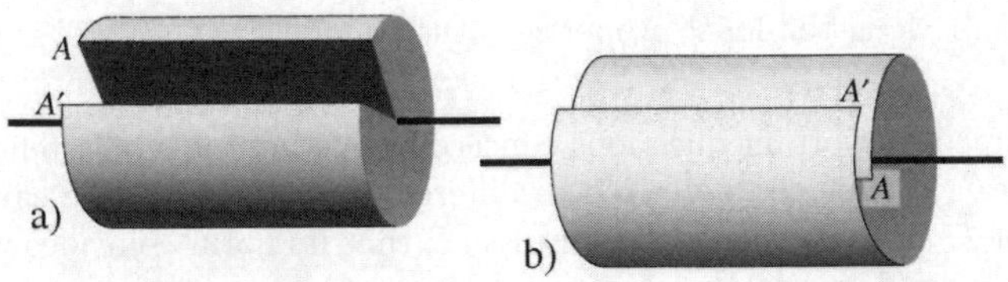

Figure 2: a) 3-space surrounding a positive string, with deficit angle $A\,A'$; b) 3-space surrounding a negative string, showing a surplus angle $A\,A'$.

we find flat 3 dimensional space-time surrounding a string. Indeed, the energy momentum tensor of a straight, infinite, static string pointing in the z-direction, is

$$T_{\mu\nu} = t_{\mu\nu}\,\delta^2(\tilde{x}) \,, \qquad t_{33} = -t_{00} = \varrho \,, \qquad \text{all other } \; t_{\alpha\beta} = 0 \,, \tag{1.4}$$

where ϱ is the string tension parameter.

Space-time surrounding a string is then seen to be as sketched in Fig. 2a. When the string constant ϱ is large, we *redefine* it to be the one generating exactly the deficit angle $\alpha = 8\pi G\varrho$. Note that a *positive* string constant leads to a *deficit* angle. A negative string constant would produce a *surplus* angle, see Fig. 2b. Normally, in string theory, a negative string constant would make a string highly unstable, as it contains positive pressure and negative energy. In our case, however, strings are constrained to form straight lines, and therefore this instability has no effect at small distances. We later may wish to include such negative strings in our models. We leave this open for the time being, but unless indicated specifically, we will usually be discussing strings with positive string constants and thus with positive deficit angles.

In addition to pure, straight stringlike defects, we will need pointlike defects where three or more strings join. If we take the point $\vec{x}$ where the strings join to be standing still, then all strings joining there will be standing still as well. Each string, (i), has a deficit angle that represents a rotation $\Omega_i = e^{2i\vec{\omega}_i\cdot L_i}$, where L_i are the generators of rotations in 3D, and $\omega_i = |\omega_i|$ are half the deficit angles. The condition that the strings match at a joint is that surrounding spacetime has no other deficits, so that

$$\Omega_1\,\Omega_2\,\Omega_3 = \mathbb{I} \,. \tag{1.5}$$

1.4 Moving strings

A description of a multitude of static strings is now straightforward, in principle. However, already here, one may expect considerable complications. The system of static strings is not unlike the $2+1$ dimensional universe with moving point particles, where the time coordinate is replaced by the coordinate z of the static string system. We know that the $2+1$ dimensional world has either a big bang singularity or a big crunch[4]. Similar "infrared" singularities might show up in the static string system. This question will not be further pursued here. It is the local properties of the model that we will investigate further.

Moving strings impose important questions concerning the internal consistency of the model. A moving string can be characterized in different ways. Firstly, one can specify the orientation vector $\vec{\omega}$ of the string, normalized such that its norm coincides with half the deficit angle, $\frac{1}{2}\alpha = 4\pi G\varrho$ when the string is at rest:

$$|\vec{\omega}| = \tfrac{1}{2}\alpha \, . \tag{1.6}$$

In addition then we specify the velocity vector $\vec{v}$ of the string. But, noticing that the string is invariant under boosts in the direction $\vec{\omega}$, only the component of $\vec{v}$ orthogonal to $\vec{\omega}$ matters, so we limit ourselves to the case

$$\vec{v} \cdot \vec{\omega} = 0 \, . \tag{1.7}$$

Finally, the position of the string at $t = 0$ should be specified. This requires another vector orthogonal to $\vec{\omega}$. All in all, we see that $3+2+2=7$ real parameters are needed, of which 5 are translationally invariant, and 2 can be set to zero by a spacelike translation in 3-space.

Alternatively, we can specify the string's characteristics by giving the element of the Poincaré group that describes the holonomy along a non-contractible cycle C around the string. For static strings through the origin of 3-space, this is just the pure rotation operator, which will be denoted as $U(\vec{\omega})$. For strings moving with velocity $\vec{v}$ through the origin, this is the element $B(\vec{v})\,U(\vec{\omega})\,B(-\vec{v})$ of the Lorentz group, where $B(\vec{v})$ is the element of $SO(3,1)$ that represents a pure Lorentz boost corresponding to the velocity $\vec{v}$. If the string does not move through the origin, we get a more general element of the Poincaré group.

Notice however, that an arbitrary element of the Lorentz group is specified by 6 parameters, not 5, and the elements of the Poincaré group by 10 parameters, not 7. This means that not all elements of the Poincaré group describe the holonomy of a string. Firstly, ignoring the translational part, the pure Lorentz transformation Q associated to the closed curve C has to obey one constraint:

There must be a Lorentz frame such that, in that frame, Q is a pure rotation in 3-space.

Since we plan to describe these Lorentz transformations in terms of their representations in $SL(2,C)$, we identify this as a constraint on the associated $SL(2,C)$ matrices. Write

$$Q = B(\vec{v})\,U(\vec{\omega})\,B(-\vec{v}) \, , \tag{1.8}$$

where $B(\vec{v})$ are 2×2 matrices representing boosts with velocity $\vec{v}$ (we'll see shortly that pure boosts are represented by *hermitean* 2×2 matrices), and U is a unitary matrix representing a pure rotation. When $\vec{\omega}$ points in the z direction, we have

$$U(\vec{\omega}) = \begin{pmatrix} e^{i\omega} & 0 \\ 0 & e^{-i\omega} \end{pmatrix} \, , \tag{1.9}$$

so that

$$\mathrm{Tr}\,(U) = 2\cos\omega\ . \tag{1.10}$$

Since the trace is invariant under rotations and the boosts (1.8), it follows quite generally that

$$\mathrm{Im}(\mathrm{Tr}\,(Q)) = 0\ ; \tag{1.11}$$

$$|\mathrm{Re}(\mathrm{Tr}\,(Q))| \leq 2\ . \tag{1.12}$$

Eq. (1.11) fixes one of the real variables of the Lorentz transformation Q. Inequality (1.12) is important in a different way. A generic $SL(2,C)$ matrix can de written in a basis where it is diagonal. Because the determinant is restricted to be 1, the diagonal form is then

$$Q = \begin{pmatrix} z & 0 \\ 0 & 1/z \end{pmatrix}\ , \tag{1.13}$$

where z can be any complex number. Imposing (1.11) leaves two options: either z is on the unit circle – in which case it represents a pure rotation in 3-space, or it is a positive or negative real number. In the latter case, Q is a pure Lorentz boost, and this is when Ineq. (1.12) is violated. It describes the holonomy of something that is quite different from a string.

An other restriction to be imposed on the holonomy of a physical string concerns the translational part of the element of the Poincaré group, which we ignore for the time being.

1.5 Collisions

When we were dealing with point particles, in the $2+1$ dimensional case, we could safely assume that the particles will never collide head-on. In general, they will miss one another, and consequently no further dynamical rules are needed to determine how an N particle system will evolve. This will not be true in higher dimensional spaces.[1] Strings in $3+1$ dimensional space-time will in general not be able to avoid one another. They will cross, and in doing so, two straight string sections will not be straight anymore after the collision.

Consider an initial state in which two strings are heading towards one another. We can always work in a Lorentz frame where one of the strings, call it A, is at rest. The conical 3-space surrounding it has a deficit angle $\alpha = 2\omega_A$. In the generic case, in this Lorentz frame, the second string does not have to be oriented orthogonally to the first one. Its string constant, β, does not have to be the same as α. Without loss of generality, we may assume that the velocity vector $\vec{v}$ of the second string is orthogonal to both A and B.

[1] Strings will in general not collide head-on in a space-time of more than 4 dimensions. However, the generalizations of the objects we discuss in this paper, in higher dimensions will be branes, not strings.

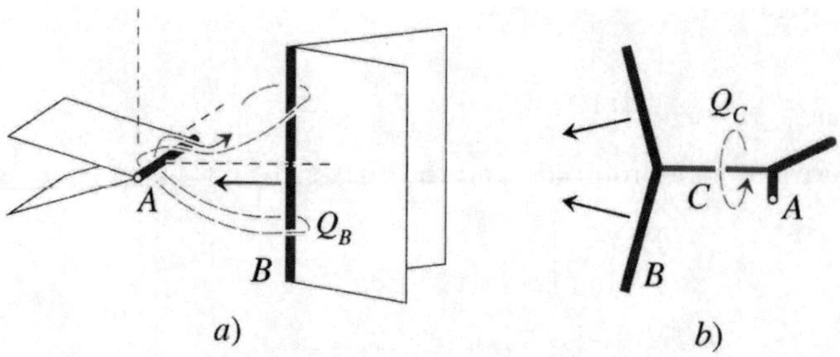

Figure 3: Orthogonal strings scattering. a) Initial state. String B moves towards string A (arrow). The cusps caused by their deficit angles are shown. Dashed lines are an orthogonal frame shown for reference. b) After the scattering, a new string C connecting the first two emerges. Strings A and B now both show a kink. The cusps in the last situation are not shown. The oriented dashed curves explain Eq. (1.14).

However, it is a physical limitation if we also assume A to be orthogonal to B. Just because it is special, we consider this case first. The collision event is sketched in Fig. 3. In $3a$, the two strings approach one another. They both drag a space-time cusp with them. Now, what happens when B hits A, is best understood by drawing the cusp of A in the opposite direction. The result of that, however, is that string B is seen to have a kink. The same thing happens to string A itself; it develops a kink due to the cusp of B. After the passage, the two kinks must be connected by a new string, C that stretches with A and B now moving away from one another.

Indeed, we see that, in general, the holonomy of string C is non-trivial; it is obtained from the holonomies Q_A and Q_B of strings A and B as follows (depending on sign conventions for Q_A, Q_B and Q_C):

$$Q_C = Q_B^{-1}\,Q_A^{-1}\,Q_B\,Q_A\,. \tag{1.14}$$

Q_A and Q_B do not commute because they represent rotations along two different axes. Clearly, upon crossing, two strings produce a third stretching between them. We see that the inevitability of string crossings forces us to generalize our model to string *segments*, ending in joints with other segments. All segments may have different string 'constants' ϱ. We will have to address this feature again later.

If the strings approach one another at an angle different from $90°$, a complication arises. In this case, one can convince oneself that no solution is possible with a single string stretching between the outgoing strings. This can be understood by studying the geometry, but we can also verify that, in general, the holonomy (1.14) is not of the string type: it violates Eq. (1.11).

To save the model, one can now propose the following. When two strings A and B collide at an angle $\varphi \neq 90°$, not one but two new strings appear[2], both stretching from

[2]Later, in this section, we will argue that even more than two new strings might emerge.

A to B. A single string cannot be associated with a holonomy of the form (1.14), but a pair of strings can. The question is now, whether the data of this pair of strings would be uniquely determined by the initial characteristics of A and B. To investigate this question, the author combined analytical arguments with computer calculations, just to see how things will work out. The topology is defined in Fig. 4.

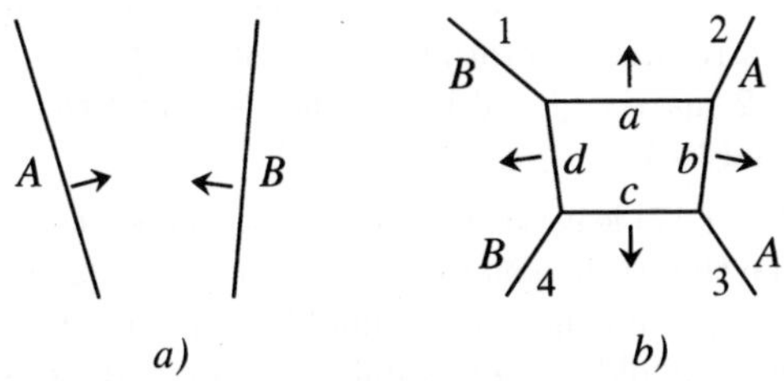

Figure 4: Scattering at an angle produces two new strands. a) A and B enter. b) Since A passes the cusp of B and *vice versa*, strands A_2 and A_3 form angles. These four pieces are labelled 1—4. Four new string pieces must be further specified, here labelled a—d.

At first sight, it seems that we have considerable freedom to define the orientations, strengths and velocities of the 'internal' strings a—d. However, if we fix one of these, all others are determined since the holonomies at a junction must obey Eq. (1.5). In addition, the strings must be properly attached to one another. As it turns out, the matching of the strings is guaranteed if Eq. (1.5) is obeyed at all junctions, and if in addition the *string conditions* (1.11) and (1.12) are obeyed by all four new holonomies, a—d.

The holonomy matrices at the 'external lines' 1—4 are fixed by the initial conditions. Originally, we had, in one conveniently chosen Lorentz frame, $Q_1 = Q_B$ and $Q_3 = Q_A^{-1}$. Here, the inverse sign arises if we decide to consider the holonomies with respect to observers looking towards the interaction point, and the holonomy curves C are chosen to go clockwise. Then,

$$\begin{aligned} Q_4 &= Q_3^{-1}\,Q_1^{-1}\,Q_3\,, \\ Q_2 &= Q_1^{-1}\,Q_3^{-1}\,Q_1\,. \end{aligned} \tag{1.15}$$

The last equation is actually the one required for consistency with the demand that

$$Q_1\,Q_2\,Q_3\,Q_4 = \mathbb{I}\,. \tag{1.16}$$

Given the holonomy Q_a of the string section a, the others can be defined as follows:

$$Q_b = Q_a\,Q_2\,; \quad Q_c = Q_b\,Q_3\,; \quad Q_d = Q_a\,Q_1^{-1}\,, \tag{1.17}$$

which is the most systematic definition, and consistency with Eq. (1.16) is ensured. We emphasize that there is always some ambiguity in defining the Lorentz frames for the holonomies Q_1—Q_4 and Q_a—Q_d, so we use Eqs. (1.15)—(1.17) also to specify these frames.

Let now all external holonomies Q_1—Q_4 be given. How much freedom is there for Q_a—Q_d? We do not need to consider the translation parameters in the Poincaré group; these will be taken care of automatically, since there is a point (0,0,0,0) where the colliding strings A and B first met. This point can be kept at the origin of our coordinate frame. Thus, we consider the equations for the elements of the Lorentz group, which are most conveniently described as $SL(2,C)$ matrices. Each element of the Lorentz group is characterized by 6 real variables: a rotation vector and a velocity vector, or alternatively the four complex numbers in a 2×2 matrix Q, subject to the constraint that the complex number $\det(Q)$ should be set equal to 1.

The junction equations (1.15)—(1.17) leave us the freedom to choose Q_a. This gives a space with 6 real parameters. Then the string equation (1.11) for Q_a—Q_d, together gives us 4 real constraints. The surviving 2-dimensional manifold is then further constrained by the demands (1.12). Thus, the manifold of all possibilities is a two-dimensional space.

To obtain somewhat more understanding of this manifold, let us consider the 8 dimensional set of all $L(2,C)$ matrices for Q_a, without the nonlinear constraint concerning the determinant. The conditions (1.11), $\mathrm{Im}\,\mathrm{Tr}\,(Q_{a,b,c,d})=0$, in combination with the junction equations (1.17), are 4 linear equations for the matrix elements of Q_a. This leaves us with a linear $8-4=4$ dimensional space. Then we have the inequalities (1.12), which for the matrix Q_a imply that

$$Q_a=\begin{pmatrix} a_1+ia_2 & b_1+ib_2 \\ c_1+ic_2 & d_1-ia_2 \end{pmatrix}\ ; \qquad |a_1+d_1|\le 2\ , \tag{1.18}$$

and similarly for the three other internal holonomies. Realizing that, in our 4 dimensional space, these conditions can be written as

$$|e_i\cdot x|\le 2\ , \qquad i=1,\,\ldots,4\ , \tag{1.19}$$

and assuming that, in general, the four vectors e_i will be independent, we see that, in the generic case, the surviving space is a compact one: a four dimensional hypercube. We can be sure that the inequalities (1.19) give us a *non empty* four dimensional space.

Next, however, we have the two constraints

$$\mathrm{Re}(\det(Q_a))=1\ , \qquad \mathrm{Im}(\det(Q_a))=0\ . \tag{1.20}$$

These two equations for Q_a ensure that the same equation will hold for Q_b—Q_d, because the determinant is preserved, and because $\det(Q_i)=1$ also for the external Q_1—Q_4. Now these are quadratic equations for the coefficients of Q_a, so the question whether these two equations are compatible with the inequalities (1.19) and with one another is a more delicate one.

We refer to Ref[1] for further details. What has been discovered is that in a fraction of all cases indeed a class of double-strand solution exists, but when the strings meet at relativistic speeds, and possibly in some other exceptional circumstances, this class is empty: the 'double-strand Ansatz' then fails. In those cases, we have to try something else. More complicated scattering diagrams were considered, but we were not yet able to prove that configurations with either a single internal quadrangle or multiple strands suffice to cover all eventualities, as the space of all possible external holonomies Q_1 — Q_4 is very large[3].

Strings crossing over is not the only kind of "events" that can take place in this model. We can also encounter the situation where a string segment, of the kind that results from collisions of the type described in the above, is reduced to zero length. It is bounded by two other junctions that herewith merge into one. Our first try should be whether the result could again be a one-strand, two-strand, or multiple strand final state, just like the ones described earlier. However, if we allow ourselves strings with negative string constants, then there is a simpler final state: the one where the original string gets a "negative length". This is really a string where the deficit angle has switched sign. Once it was decided to allow their presence, we could allow them here as well.

1.6 Crystalline gravity

The gravity model we arrived at is dynamical, finite and Lorentz invariant. It consists of straight string segments, attached to one another at string joints. Space-time surrounding the strings and the joints is locally flat and Minkowskian. The strings form cusp singularities with deficit angles equal to the string constants. These string constants must be allowed to be different for the different strings. Matching conditions at the joints automatically ensure that the tensile forces of these strings balance out. The dynamical rules are then as follows:

1. When two strings or string segments collide, new straight sting segments form, either in the from of a quadrangle (Fig. 4), or in a more complicated setting.
2. When a string segment shrinks to zero length, a similar rule will prescribe the newly formed string segments.
3. There is a considerable amount of freedom in choosing the details of these rules, which is comparable to the freedom one has in choosing the matter Lagrangian in a gravity theory.
4. There is no independent gravitational degree of freedom. "Gravitons" are best regarded as composites of matter. The smoothness of gravity fields and waves that we normally consider, can only be reproduced in a continuum limit of the model.
5. It has not been checked whether the string constants can be kept positive when collisions take place. Most likely, this is not the case, so that we do have negative energy configurations.
6. The model cannot be quantized in the usual fashion, since
   a) strings with a continuous spectrum of different string constants appear;

[3]Note added in proof: we now think that configurations where the joints connect 4 strings rather than 3, might provide complete solution sets.

b) there appears to be no time reversal (or CPT) symmetry: strings can be seen to break up but they do not tend to rejoin.

An entirely different picture however arises if we consider imposing one more constraint: let us assume that not all elements of the Poincaré group are admitted, but instead only a discrete subgroup. The easiest way to think of this is to assume space and time to form a lattice in Minkowski space, preferably a rectangular lattice. Our stringlike defects are then much more like crystal defects. This is why we call the resulting theory "crystalline gravity". It differs from a crystal however in three important ways: our crystal now exists not only in space but also in time, and furthermore, our crystal defects obey equations of motion, as in the above. And finally, our crystal is not imbedded in a flat background spacetime, but it itself defines spacetime, which is curved along the defects.

The defect angles of stationary strings are now restricted to multiples of $90°$. In an ordinary crystal, this would cause a lot of strain away from the defect, but that is because it would be imbedded in a flat background space, which is absent here. Unlike ordinary crystals, our spacetime crystal has an infinite rotation group in its crystal group: our set of discrete Lorentz transformations may contain the matrix

$$B = \begin{pmatrix} 0 & 1 & 1 & 1 \\ 1 & 0 & 1 & 1 \\ 1 & 1 & 0 & 1 \\ 1 & 1 & 1 & 2 \end{pmatrix} , \tag{1.21}$$

since it is a legitimate Lorentz transformation, mapping integral coordinates onto integral coordinates. Combining this with space rotations over $90°$, gives us the largest subgroup of the Lorentz group that leaves a cubical spacetime lattice invariant.[4]

One can find an infinity of further subgroups of this subgroup that could be considered, but we will not dwell on such features any further.

There are reasons why we will *not* advocate "quantization" of this model along the usual procedures. Perhaps, quantization can go by means of the "pre-quantization" procedure proposed earlier[2]. There are quite a few open questions apart from quantization. For instance, one might suspect an instability such as the formation of a black hole horizon, although precisely in this model one might also suspect the converse, that black holes cannot form. If all string constants are kept positive, localized matter configurations cannot exist, whereas all gravitational curvature must be associated with strings — there is no pure gravity in this model. So, if there is a black hole, strings will have to stick out from it.

The absence of pure gravity degrees of freedom is intriguing. In a sense, matter here is "unified" with gravity, not, as in many models, because gravity generates particle-like degrees of freedom, but the converse, because the matter degrees of freedom, here the string segments, carry around all the space-time curvature there is.

[4]Other discrete subgroups exist leaving different types of lattices invariant.

References

[1] G. 't Hooft, *Found. Phys.* **38** (2008) 733, ITP-UU-08/17, SPIN-08/16, arXiv:0804.0328.

[2] G. 't Hooft, *J. Phys: Conference Series* **67** (2007) 012015, quant-ph/0604008.

[3] S. Deser, R. Jackiw and G. 't Hooft, *Ann. Phys.* **152** (1984) 220.

[4] G. 't Hooft, *Nucl. Phys.* **B30** (Proc. Suppl.) (1993) 200; id., *Class. Quantum Grav.* **10** (1993) 1023.

[5] Z. Kadar, *Class. Quant. Grav.* **22** (2005) 809, e-Print: gr-qc/0410012.

[6] E. Witten, *Nucl. Phys.* **B311** (1988) 46; S. Carlip, *Canadian Gen.Rel.* 1993:0215-234 (QC6:C25:1993), gr-qc/9305020.

[7] G. 't Hooft, *Class. Quantum Grav.* **13** 1023, gr-qc/9607022.

[8] J.W. Barret et al, gr-qc/9411008, *Int.J.Theor.Phys.* **36** (1997) 815, and references therein.

[9] R. Brandenberger, H. Firouzjahi and J. Karouby, *Lensing and CMB Anisotropies by Cosmic Strings at a Junction*, arXiv:0710.1636 (gr-qc).

CHAIRMAN: G. 't HOOFT

Scientific Secretaries: S. Nicotri, R. Ziegler

DISCUSSION III

- *Dubynskiy:*

You were talking about the importance of instantons in explaining the masses of some scalar and pseudoscalar mesons. What could the instanton technique tell us about the branching ratios and decay constants of these states?

- *'t Hooft:*

I am not sure if I can answer that question. What I can say is that instantons give a quite noticeable six-quark effective interaction to be added to the perturbative interaction of QCD. That is a rather naïve statement because the QCD action itself automatically includes instanton effects and so it should include the six-quark effective interaction. It is as if you have an effective mesonic theory plus a six-quark effective action. If you have $\psi_L$barψ_R (referring to fig. 1 of my presentation), this roughly is a mesonic field φ_i^j (with two mesonic indices i and j), so you can mimic the determinant effect by writing a three meson effective interaction $\kappa\ \varphi\varphi\varphi$. If you include that into an effective meson theory, you can compute its coefficient κ by looking at the phenomenology of mesons. For instance, the eta mass is considerably bigger than the pion mass, so what you do is to write down an effective mesonic theory with all the kinetic terms of all the particles, assuming chiral symmetry to be spontaneously broken by the pion and sigma fields and then you ask what does it do to the eta particle. Well, the eta particle would also be massless like the pions in first approximation, but there is the instanton coefficient with the right quantum numbers to raise the mass of the eta and eta prime particles. So, eta and eta prime are considerably heavier than what you would have thought they were and the way to compute that in the above effective meson theory is to compare the coefficient κ in front of the three mesons term with the phenomenological observations, like the mass of the eta. In this way you find that it is a quite substantial coupling and then you can use it to compute other interaction strengths to compare experiments on decay widths and branching ratios with QCD-inspired effective theory.

- *Dubynskiy:*

Do you think that instanton effects can be important for the recently discovered states in the charmonium spectrum?

- *'t Hooft:*

Well, charmonium primarily requires a perturbative approach. The surprising thing when J/ψ was discovered was that it was a narrow state and it appeared to be very well described by perturbative QCD. In perturbative QCD the exponential instanton effect is very unimportant, so my guess is that instantons have a very suppressed effect on the charmonium and other heavy quark states. Since it is a three meson effective interaction, you should add the charmonium states as well, but then, since the J/ψ has a very large mass, which means that it couples left to right very strongly, it acts basically like a constant and it is usually included in the coefficient κ. Note added in proof: Instantons could play a role in the final state interactions among the decay products.

- *Klopot:*

Instantons are responsible of the mass of the eta prime meson. What is their role in the electromagnetic decay of this meson? Are there any quantitative calculations of the contribution of the instantons to this process?

- 't Hooft:

I have not done such calculations personally, but I presume that the answer is yes. The effective theories of mesons have been studied in lots of detail and the electromagnetic decay of the eta is a very important process that has been measured accurately and compared with theory. As far as I know there are no serious disagreements with the observations.

- Susskind:

You discussed the effect of the instanton on the mixing between single-mesons and two-mesons. You showed a level diagram and the effect of that mixing in shifting the energy levels. Can I ask you to just go through that again very quickly?

- 't Hooft:

Referring to fig. 2 (of my presentation): the top two levels are the tetraquark states and the lower two levels are the meson states. You expect the sbar-s meson to be heavier than ubar-u and you expect tetraquarks with two strange quarks in it to be heavier than tetraquarks without. But the two tetraquark states are strongly coupled and mixed with the single two-quark states. The instanton looks at every flavor once and it mixes ubar-u-dbar-d with sbar-s and ubar-u with dbar-d-sbar-s with comparable matrix elements.

- Susskind:

Then I have a hard time understanding how it can push so hard the two interior states that it can actually push them past the exterior ones.

- 't Hooft:

I think you are right. I was exaggerating a little bit in my figure.

- Liu:

We could measure the mixing between $a_0(980)$ and $f_0(980)$ to test the various models describing the lightest scalars. Can the quark model including instanton effects be experimentally distinguished from other models?

- 't Hooft:

QCD without instantons is probably an ill-defined theory. But, even if I drop the instanton term from the effective meson theory I just showed, an experimentalist would say that he sees an eta particle considerably heavier than the pion. The only term which can explain such an effect is the instanton term in QCD. This is a very strong indication that instantons indeed are extremely important and relevant in the description of strong interactions. For what concerns the tetraquark the story is a little bit subtle. From a phenomenological point of view, there has been some confusion about what the scalar mesons actually are. Are they two-quark or four-quark states? I am saying that, if you believe in instantons, they are mixtures. When quantitative calculations were done with these mixtures, they worked much better from an experimental point of view than without. So, also for scalar mesons the instantons explain some of the observations. Quantitatively these calculations seem to work very well.

- Korthals-Alte:

If you look at your six-quarks vertex it means that a normal quark can talk to a tetraquark. If you put a simple spectator quark, than the normal meson become two mesons and so you have a decay of a tetraquark in two normal mesons.

- *'t Hooft:*

(referring to fig.3of his presentation) As soon as the tetraquark is sufficiently heavy it will decay into other mesons. This also means that a tetraquark is strongly coupled to a single bi-quark state (sbar-s) which can decay into strange mesons. So the light tetraquark, which only contains up and down quarks, can decay into two strange objects. This is indeed a very important observation.

Strongly Coupled Gauge Theories

Richard Kenway
School of Physics & Astronomy, The University of Edinburgh,
Edinburgh EH9 3JZ, Scotland

Abstract

The theory of the strong interaction, QCD, explains how most of the mass we observe in the Universe is generated through gauge theory dynamics and lies far below the scale of any new physics, such as the Planck mass associated with quantum gravity. Lattice QCD simulations have made this explanation quantitatively precise and also hint at why some quarks are much lighter still. They are providing reliable predictions for much more besides, enabling precision tests of the Standard Model and imposing constraints on physics beyond. It is worth asking whether the mass scales associated with electroweak symmetry breaking and the heavier quarks could also be explained by a new gauge dynamics. Lattice simulations are now venturing to explore this theoretical territory, which is intrinsically non-perturbative.

1. The Dynamical Origins of Mass

In this year in which the LHC switches on and there are high hopes that the mystery surrounding the Higgs sector of the Standard Model (SM) will soon be resolved, attention is turning again to the question: what is mass? More specifically, what sets the different mass scales that we see in the SM? From the perspective of the SM, gravity lies outside and some yet-to-be-discovered theory of quantum gravity determines the physics of the Planck scale. We need to understand how the quarks and leptons can have much smaller masses than this and yet themselves have a range of masses that spans many orders of magnitude.

In this lecture, I will explore how far it may be possible to answer these questions using the dynamics of strongly-coupled gauge theories. The reason this might be a good starting point is that QCD generates 99% of the mass we experience in everyday life, through the binding energies of the quarks and gluons making up protons and neutrons. The technology of lattice QCD, developed over the past 35 years, is close to achieving precise quantitative control of quark confinement and the low-energy physics of hadrons. So we know with a high degree of confidence that QCD is the correct low-energy description. It tells us why the proton is light and it gives some pointers as to how chiral symmetry makes some fermions light. Recent results from the lattice are testing the region of validity of chiral perturbation theory and so tell us which quarks are light, in the sense that they are described correctly by this low-energy effective theory.

Precise control of lattice QCD also completes our arsenal of computational tools with which to constrain physics beyond the SM using low-energy experimental measurements. This is starting to be able to rule out some models of grand unification and, though less spectacular than direct discovery, the systematic build-up of these constraints could become important if the LHC results are too confusing.

Ideally, however, we would like some insights into the physics of what sets the mass scale of the electroweak interaction. Could it be some new gauge interaction like QCD – a well-worn idea, historically called technicolor? Although old fashioned, actually rather little is known about the dynamics of technicolor theories. There have been neither the tools nor the motivation to tackle the non-perturbative effects. This allowed room for speculation about dynamics very different from QCD, but now lattice techniques are starting to explore non-QCD theories systematically. The experience with QCD gives us confidence that this approach will also enable us to rule out many contenders for models of electroweak symmetry breaking.

2. Lattice Gauge Theories

Ken Wilson's formulation of Yang-Mills theories on a regular hypercubic space-time lattice in 1974 solved the problem of how to render gauge field theories finite without spoiling the local symmetry (at least at the lattice points). Computer simulation of a lattice gauge theory begins from the path integral in four-dimensional Euclidean space. This analytic continuation of Minkowski space-time to imaginary time converts the path integral into what looks like a rather complicated statistical mechanics problem. The advantage is that, when the Euclidean space is replaced by a finite hypercubic lattice (i.e., with a finite volume), the resulting path integral becomes a finite-dimensional integral over the fields at each of the lattice sites and, in the case of gauge theories, the lattice links, that may be evaluated numerically using Monte Carlo methods.

Michael Creutz recognised this and launched a huge computational effort in QCD, where quark confinement had seemed an insurmountable obstacle to a systematic approach at low energies. This led to both the development of new algorithms and new computer architectures, in order to achieve the performance required to simulate QCD on realistic lattices, i.e., with large enough volumes to contain the hadrons of interest, small enough lattice spacings to be close to the continuum, and QCD parameters close to their physical values.

Ensuring proximity of the bare parameters in the QCD action to their physical values is itself a significant computational task. These values are the inputs to the simulations. Determining them is logically the first stage in being able to use lattice QCD to compute quantities that can be useful for phenomenology. The cause of this difficulty is the confinement of quarks, which hides the fundamental degrees of freedom (the quark and gluon fields) from direct experimental measurements. Experiments measure the properties of hadrons and it is from these that we have to infer the values of the QCD parameters: the quark masses, the quark-gluon coupling and the vacuum angle.

In principle, we would do this by guessing the parameter values, performing a simulation to compute a set of hadronic quantities, comparing these with their experimental values, and then adjusting the parameters and repeating until we get agreement with experiment. We could then conclude that the input parameters have their physical values. In practice, this process is complicated by the lack of computer power to run simulations actually at the physical values, so that some extrapolation from unphysical values is required.

The hadronic quantities used to fix the bare masses of the various quark flavours are typically the masses of mesons whose valence quarks have those flavours. If there are N_f flavours, we need to input N_f meson masses from experiment (usually, we assume isospin symmetry so that the u and d quark masses are equal and fixed by the pion mass; the s quark mass is then fixed by the kaon mass). One additional dimensionful quantity is needed to set the lattice scale. Conventionally, this is used to define the lattice spacing in physical units, a, by demanding that, for $N_f + 1$ meson masses, $M^{\text{lattice}}(m_1,\ldots, m_{Nf}) = aM^{\text{experiment}}$, where m_i denotes the mass of quark flavour i.

If we could perform simulations at the physical quark masses on big enough lattices, this would be a direct iterative procedure. As it is today, this is still too expensive, so chiral perturbation theory is used to model the dependence of meson masses on quark masses and extrapolate the simulation results from the unphysically large u and d quark masses used in the simulations to their physical values. The lattice spacings available to today's computers correspond to a momentum cut-off comparable to the c quark mass, so that it is not possible simultaneously to simulate the c, b and t quarks without big discretisation effects. These flavours are only included as valence quarks (i.e., they don't appear in the sea). Even as valence quarks, additional techniques, such as improvement (see Section 4) and effective actions (see Section 7) are needed to control the discretisation effects for these flavours.

3. The Continuum Limit and Why the Proton is Light

The lattice spacing in physical units, a, tells us how close our simulation is to the continuum limit. If a is smaller than any scale relevant to the hadronic physics we are trying to compute, then that physics, expressed as dimensionless combinations of physical quantities, should become independent of a as it is made smaller. This is called 'scaling' and when a quantity 'scales' it is equal to its value in the continuum limit. The continuum limit is a critical point of the lattice theory – if physical lengths becomes independent of a, as a is decreased, then, relative to the lattice theory with fixed spacing, physical lengths diverge. Such a critical point is a fixed point of the renormalisation group (a zero of the β function).

The gauge or quark-gluon coupling, g, is the extra parameter in the QCD action (in addition to the quark masses) needed to vary the lattice spacing. More generally, non-abelian gauge theories without matter fields, or with a small enough number of fermion flavours in the fundamental representation of the gauge group, are asymptotically free, which means that they have a UV stable/IR unstable fixed point at $g = 0$. This fixed point lies in a critical surface of co-dimension one (the parameters which span the surface are irrelevant) and g parametrises the one unstable, i.e., relevant, direction – to reach the surface you have to tune g to zero, i.e., the lattice spacing vanishes with g.

Perturbation theory can be used to compute the β function near the origin, so we have analytical control over the functional dependence of the lattice spacing, a, on g near the continuum limit in asymptotically-free theories. Since this dependence guarantees that physical quantities scale, the net effect is that a physical length scale, Λ, emerges, which is invariant as g and a change according to the β function close to the fixed point and takes the place of the lattice spacing a in providing the dimension for physical quantities. Thus, in QCD, a hadron mass $M_i \rightarrow c_i \Lambda_{QCD}$ as $g \rightarrow 0$, where c_i is a computable number, and, in principle, dimensionless ratios of physical quantities become computable numbers. This phenomenon in which a physical scale replaces the artificially introduced cut-off scale (lattice spacing) is called dimensional transmutation, and it is the key to how mass scales are generated in asymptotically-free gauge theories.

Now we can answer the question of why the proton is light. In QCD, close to the continuum limit, from the particular form of the β function, it can be shown that the ratio of the hadron mass scale, Λ_{QCD}, to the cut-off scale, a^{-1}, (inverse lattice spacing) depends exponentially on the inverse of $g^2(a)$ at the cut-off scale. If we take the lattice spacing to be the scale of new physics, such as the Planck scale, then $g^2(M_{Planck}{}^{-1})$ is tiny and the ratio of mass scales can be enormous. The square of the ratio of the proton mass to the Planck mass is proportional to the ratio of the gravitational to the electrostatic force between two protons (10^{-36}). So the same argument explains why gravity is so weak.

Thus, if QCD is the origin of most of the mass of the proton, and recent lattice calculations confirm that it is, then the mystery of why the mass scale of ordinary matter is vastly smaller than the Planck scale is solved. It remains to (use lattice QCD to) compute the actual mass ratios of all the hadrons and any other dimensionless ratios of phenomenologically interesting quantities.

4. Universality

The continuum limit involves tuning g to a critical point of the lattice theory, where physical lengths in units of the lattice spacing diverge. Details of the physics on the scale of the lattice spacing should become 'washed out' as we approach a critical point, so that different lattice formulations may have the same continuum limit. This means we can add so-called 'irrelevant' interactions at the scale of the lattice spacing which affect the local physics, but which are suppressed by positive powers of the lattice spacing and vanish in the continuum limit. Lattice actions which differ only in this way, and hence have the same continuum limit, are said to belong to the same universality class. Generally, only the symmetries and the number of space-time dimensions of a lattice action dictate which universality class it belongs to.

In practice, we exploit this freedom to construct lattice actions in which irrelevant interactions are added to try to cancel the leading-order corrections in powers of the lattice spacing, in the hope that scaling can be achieved at larger values of g, i.e., larger lattice spacings, where simulations are less costly. Since taking the continuum limit involves demonstrating control over the extrapolation of results from simulations at several lattice spacings, a lot of effort is focused on these 'improved' lattice actions.

In four-dimensional gauge theories, it turns out that all the lattice interaction terms which break Lorentz symmetry are irrelevant and Lorentz invariance is recovered in the continuum limit 'for free'. Thus, Lorentz invariance is called an 'accidental symmetry' of the lattice theory. As I have mentioned, Wilson showed how to preserve local gauge invariance at non-zero lattice spacing. At strong coupling, i.e., large values of g, it is possible to prove that gauge charges are confined as a consequence of this gauge invariance. For SU(3) gauge theory with quarks in the fundamental representation of the colour group, all of the evidence from simulations suggests that there is no phase transition to a deconfined phase as g is reduced towards zero, so that confinement remains a property of the continuum limit (at zero temperature). A rigorous proof of this is yet to be found.

Demonstrating that the correct chiral and flavour symmetries of QCD are also recovered in the continuum limit has proved to be much more elusive. Early lattice formulations of fermion fields broke one or other of these symmetries explicitly, complicating renormalisation due to operator mixings allowed by the reduced symmetry, and calling into question whether the lattice actions really belong to the universality class of continuum QCD. Now, a class of lattice actions (which obey the so-called Ginsparg-Wilson condition) have been formulated that preserve both symmetries, and the rigorous lattice formulation of QCD is complete.

5. Why are Some Fermions Light?

The simple answer which is usually given to this question is that they are protected from acquiring a mass by chiral symmetry. This symmetry argument is normally based on perturbation theory and must certainly be scrutinised carefully in strongly-coupled gauge theories. It turns out that realising chiral symmetry non-perturbatively in four-dimensional space-time is not straightforward and may not even be possible in a local quantum field theory. So what makes the u and d quarks light?

The earliest formulations of lattice QCD encountered this problem when trying to define massless fermion fields on a lattice. A no-go theorem showed that, if the lattice fermion action is bilinear, local, translation invariant, Hermitian, and has continuous chiral symmetry, then the continuum limit contains fermions in opposite chirality pairs. Using a simple finite-difference approximation to the derivative in the Dirac action yields twice as many continuum fermion species as lattice species, per dimension of space-time, i.e., 16 continuum species in four dimensions correspond to a single lattice field. Nature is not like this and so, evidently, it does not realize light quarks this way.

Early on, Wilson proposed adding an irrelevant term to the lattice fermion action which gives 15 of these species masses of order the lattice cut-off so that they effectively decouple, but which explicitly breaks chiral symmetry. It should be possible to tune the fermion mass in the action so that, in the continuum limit, chiral symmetry and the correct number of flavours are restored, but the absence of chiral symmetry on the lattice allows additive renormalisation of the quark masses and the mixing of operators of different chirality. Both complicate the calculations of phenomenologically interesting matrix elements. It leaves open the question of whether there is a non-perturbative regularisation of QCD which preserves chiral symmetry.

Staggered quarks are another way of dealing with the problem. They exploit the fact that the finite-difference approximation to the Dirac action can be diagonalised to decouple the four spinor components of the lattice fermion field. If three of the four components are dropped, the resulting fermion doubling effect is reduced to four degenerate species in the continuum limit. This is still the wrong flavour structure for QCD and, to recognise this, the four species are called 'tastes', but there is a four-fold saving in computational cost coming from the reduction in the number of degrees of freedom. Each flavour corresponds to a different staggered lattice field and, hence, appears in the continuum with four tastes. Somewhat controversially, this is corrected in the numerical algorithm for simulating the staggered path integral by taking the fourth root of the fermion determinant that arises from integrating over the fermion fields. In perturbation theory, this gives the correct number of tastes, i.e., one. However, it cannot be described by a local field theory and there is no guarantee that it is in the QCD universality class. Current wisdom is that the correct continuum limit is obtained, provided the zero-quark-mass limit is not taken first. The price is considerable technical complexity for all but the simplest physical quantities in simultaneously controlling the extrapolation to the physical u and d quark masses and the continuum limit. Again a non-perturbative definition of chiral symmetry has evaded us.

In the early 1990's, it was realised that by suitably generalising the definition of chiral symmetry and using a lattice Dirac operator which obeys the Ginsparg-Wilson relation, full chiral symmetry could be realised at non-zero lattice spacing. Several such Dirac operators have been found. They evade the no-go theorem, because they are not ultra-local, i.e., they couple the fields at all lattice sites, but they are local (interactions fall exponentially with distance) when the gauge fields are sufficiently smooth. All share the feature that they can be approximated by a five-dimensional ultra-local lattice theory, the most widely used called 'domain wall fermions'.

This reflects a general result that chiral fermions may be trapped on four-dimensional domain walls, or branes, in five dimensions. There is no fermion doubling problem, because in five-dimensions γ_5 appears in the action and there is no chiral symmetry. Wilson's explicit dimension-5 term removes the doublers, so the standard Wilson lattice action works to produce a single light fermion in the continuum limit. A discontinuity in the fifth dimension traps a single chiral mode; in domain-wall fermions, the fifth dimension is of finite physical extent and zero modes of opposite chirality are trapped on the four-dimensional boundaries (where the four-dimensional gauge fields live – there are no gauge fields in the fifth direction and the same gauge fields occur on each four-dimensional slice of the five-dimensional lattice).

In the lattice theory, the chiral fields 'leak' into the fifth dimension, because there is only a finite number of steps between the two walls. This causes residual chiral symmetry breaking, which manifests itself as a non-zero fermion mass. However, this is easily controlled. The four-dimensional fermion spectrum comprises extended and localised states. The former are associated with non-zero eigenvalues of the logarithm of the transfer matrix in the fifth direction and are suppressed exponentially over a finite lattice distance, which therefore vanishes as the continuum limit is taken in the fifth dimension.

Localised states build up a non-zero density of zero modes of the logarithm of the transfer matrix, which are not exponentially suppressed and could cause trouble even in the continuum limit (each localised zero mode on one wall may propagate to the other wall, where it can have a non-zero overlap with another localised zero mode). However, the density of zero modes vanishes exponentially in the continuum limit of the four-dimensional theory.

These effects have been confirmed non-perturbatively by QCD simulations and show that chiral symmetry arises naturally in theories with a finite fifth dimension. Maybe, there is a hint here of extra dimensions? Practically, for QCD simulations this means we have the full continuum chiral and flavour symmetries to control operator mixing and renormalisation. However, there is a computational price for this – a factor roughly proportional to the size of the fifth dimension.

6. Lattice QCD is the Correct Theory of Hadrons

Apart from the obvious phenomenological importance of being able to carry out reliable and accurate lattice QCD calculations, establishing quantitative control of this non-perturbative theory can provide a solid base from which to explore whether other strongly coupled gauge theories could generate electroweak symmetry breaking.

Three factors have brought us close to establishing from first principles that QCD is the correct low-energy theory of hadrons. The first is a set of lattice actions that realise the continuum symmetries of QCD in different ways, thereby approaching the continuum limit from different directions, which can be compared to check that systematic effects are under control. The second is the Rational Hybrid Monte Carlo (RHMC) algorithm, which is exact for any number of flavours and is much faster than previous algorithms. The third is computers which are approaching the petaflop/s speeds needed for domain wall fermions, the most expensive and best controlled of the formulations.

By now, results have been obtained for many physically important quantities, but not everything is currently within easy reach. Hadron masses may be extracted from the exponential fall-off at large Euclidean time separations of expectation values of products of lattice fields at two times, and decay constants can be extracted from the amplitude of the exponential. When these results are combined with expectation values of products of fields at three Euclidean times, it is possible to extract matrix elements of operators between single-hadron states. These enable us to compute, for example, the mixing of neutral mesons and semileptonic decay form factors. Beyond these, greater ingenuity is required, because there is no general method to extract matrix elements involving multi-hadron final states.

On the computer, we are free to explore other worlds with different numbers of colours and flavours, and different quark masses from ours. In fact, even for QCD with 2+1 light flavours (degenerate u and d quarks, and an s quark), which is the current state-of-the-art, it is normal to vary the u and d quark mass, because the computational cost grows as their mass decreases. Some extrapolation to their physical mass value, using chiral perturbation theory, is usually employed, although we are on the verge of having enough computer power to simulate them directly. Famously, the Edinburgh Plot shows how the proton mass varies from the naïve quark model limit, corresponding to infinitely large *u* and *d* masses, down to their physical value.

As computationally the cheapest lattice formulation, staggered QCD, combined with improvement that helps decouple the unwanted tastes, has been hugely successful in computing some physical quantities with unprecedented precision. Where comparisons can be made, these results agree with experiment (with one interesting exception, see Section 8). This is certainly strong circumstantial evidence that QCD correctly describes strong-interaction physics. However, lingering unease about the theoretical foundation and the technical complexity of the calculations mean that, should a clear discrepancy between staggered QCD and experiment emerge, we would not be able to conclude today whether this was due to new physics or something wrong with the calculation. Other lattice QCD formulations are in hot pursuit and, with the first result from simulations 'at the physical point', i.e., at the physical *u* and *d* masses, all sources of systematic error are under serious investigation. Over the next couple of years, there will be enough evidence for most of us to conclude that QCD is the correct low-energy theory of hadrons.

7. Testing the Validity of Effective Theories

We can exploit the freedom to vary the input parameter values in simulations to test how well effective theories approximate QCD in particular parameter ranges. Historically, due to lack of computer power, simulations at the physical parameter values were prohibitively expensive, so faith in effective theories was needed to extrapolate the results to the physical point. Chiral perturbation theory was needed for the *u* and *d* quarks, heavy-quark effective theory or non-relativistic QCD for *b* quarks, and we were left rather unsure how best to treat *s* and *c* quarks. To further complicate matters, both discretisation errors and finite size effects vary with the quark masses, and neither additional source of error could be guaranteed to be small.

As computer power and algorithmic efficiency have grown, we have been able to perform reliable simulations over wider ranges of parameter values on large lattices and we are now testing the range of validity of effective theories. Ultimately, we will not need them to get accurate results for QCD if we are willing to pay for the computer power, but understanding how well effective theories work in this real case will give valuable insights into how well we can trust their use outside the Standard Model, e.g., for technicolor models.

Domain-wall QCD has enabled the most extensive tests so far of chiral perturbation theory in the case of 2+1 flavours. This effective theory, which is an expansion in the quark masses and momenta, can be applied to just the *u* and *d* quarks, with the *s* quark treated as heavy ($SU(2)_L \times SU(2)_R$ chiral symmetry), or assuming all three flavours are light ($SU(3)_L \times SU(3)_R$). The results of attempts to fit simultaneously the lattice data for the pseudoscalar meson masses and decay constants, using next-to-leading-order chiral perturbation theory, suggest that the $SU(3)_L \times SU(3)_R$ effective theory does not work at the physical *s* mass. A separate study using improved Wilson quarks very close to the physical point (with pseudoscalar meson masses as low as 156 MeV), but on rather small lattices where finite-size effects may be significant, supports this conclusion.

Hence, only the *u* and *d* quarks are truly light as a consequence of chiral symmetry. The fits to the lattice data determine the quark masses rather precisely. We find that the average of the *u* and *d* masses is $3.72(16)_{stat}(37)_{syst}$ MeV and the *s* quark mass is 107.3 $(4.4)_{stat}(10.8)_{syst}$ MeV.

8. Lattice QCD is a Precise and Predictive Tool

Already, lattice QCD has predicted results ahead of their experimental measurement. The mass of the B_c meson and the form factor for semileptonic decays of *D* mesons into kaons were both computed first (in 2004) and later (in 2005) confirmed by experiment. This is a crucial step in the development of the field, because it shows that lattice QCD can be an effective predictive tool and so it gives us confidence in using lattice QCD to search for new physics.

The main target for phenomenological applications of lattice QCD has been in flavour physics, to try to pin down the elements of the CKM quark mixing matrix. These parameters are tightly constrained if Nature has only three generations of quarks and leptons by the fact that it must be a unitary 3x3 matrix. The hope has been that sufficiently precise determinations of these parameters by different experiments might reveal inconsistencies that would be signals of missing new physics. However, Nature has not co-operated. There is still room for something new, because, in several cases, the largest uncertainty in these parameter determinations comes not from experiment, but from the calculation of QCD matrix elements. It has been an important challenge for lattice QCD to reduce these errors to below the size of the experimental errors.

The current generation of lattice simulations has achieved this for some key matrix elements. For example, the $\Delta S = 2$ matrix element that governs the mixing of neutral kaons, B_K, due to indirect *CP* violation in the kaon system has been computed in domain-wall QCD. The result for the renormalisation group invariant B_K is 0.72 ± 0.04 and this can be compared with the value for the same quantity that can be deduced from experiment alone, 0.78 ± 0.08. This shows that the theoretical uncertainty is being pushed down to the point where lattice QCD is able to improve our measurements of CKM matrix elements. Furthermore, similar computations in lattice QCD can obtain results for the hadronic matrix elements of all $\Delta S = 2$ four-fermion operators. These play an important role in constraining new physics, eg from GUTs or SUSY, which contributes to flavour-changing neutral currents (FCNCs) that are strictly limited by experiment.

A second example is the semileptonic decay of *B* mesons into *D* mesons, whose strength is determined by the CKM matrix element V_{cb}. Here, heavy-quark symmetry fixes the normalisation of the decay form factor at zero recoil (when the *D* meson is at rest in the *B*'s rest frame) to be 1. In reality, the *b* and *c* quarks are not infinitely massive and this modifies the normalisation. The deviation from 1 is vital for a precision determination of V_{cb} and can only be computed non-perturbatively. The result from lattice QCD is 0.921 ± 13 ± 20, significantly different from 1. Combined with the experimental measurement, this gives $|V_{cb}| = (38.7 \pm 0.9_{\text{expt}} \pm 1.0_{\text{theo}}) \times 10^{-3}$, which again shows that the theoretical uncertainty is becoming comparable to the experimental uncertainty.

Precision tests of the SM based on lattice QCD have revealed one discrepancy which may be significant. The decay constant for the D_s meson in the latest 2+1-flavour simulations with staggered quarks differs from experiment by 3.8σ, whereas those for the D, K and π mesons all agree with experiment. This result requires a 45-parameter fit to the lattice data in order to extrapolate the results to the physical quark masses and the continuum limit. This technical complexity is a feature of staggered quarks and, consequently, it is very important to check this result using a different lattice fermion formulation. Whether or not this turns out to be the first discovery of new physics using lattice QCD, it shows again how lattice QCD is now able to contribute to precision phenomenology.

9. Proton Decay Lifetime Bounds Rule out Some GUTs

Unification in whatever guise, be it SUSY or GUTs, means that quarks can decay into leptons and, in particular, that protons decay. The experimental lower bounds on various proton decay modes consequently impose constraints on models of unification. However, since the quark decay occurs inside a proton, there are QCD matrix elements which have to be computed in order to relate the quark-level decay in the unified theory to the proton lifetime. Lattice QCD is being used to compute the hadronic matrix elements of baryon-number violating operators that are allowed by SM symmetries in the low-energy effective theory.

Assuming that R parity holds, the lowest dimension SUSY GUT effective operator has dimension 5 and is due to colour triplet Higgsino exchange. The dominant proton decay mode is into kaons for which the lower bound on the lifetime from Super-Kamiokande is 2.3×10^{33} years. The most recent lattice QCD determination of the corresponding matrix element using 2+1 flavours of domain-wall quarks, combined with this lifetime bound, requires that the Higgsino mass in the minimal SU(5) SUSY GUT exceeds 3×10^{17} GeV. Other precision measurements from LEP cannot be accommodated if the mass is greater than 10^{16} GeV and, thus, this model is ruled out.

Conventional GUTs, both SUSY and non-SUSY, mediate proton decay via gauge boson exchange which leads to dimension 6 low energy effective operators. The lifetime bound of 8.2×10^{33} years on the resulting dominant proton decay mode to pions, together with a different lattice QCD matrix element determination, places a lower bound on the mass of the new gauge boson which, in a typical GUT, is around 10^{15} GeV. This is at a level which is becoming uncomfortable for some GUTs. As the lifetime bounds steadily increase, because Super-Kamiokande fails to see any decays, these mass bounds will steadily eliminate particular models. Lattice QCD, therefore, will play a crucial role in narrowing down the possibilities for beyond Standard Model (BSM) physics.

10. New Strong Dynamics as a Source of Electroweak Symmetry Breaking

It is an old idea that new strong dynamics may drive electroweak symmetry breaking (EWSB). The analogy is with QCD and the new gauge theory is called technicolor, although it was readily apparent that the new dynamics has to be very different from the asymptotic freedom/infrared slavery of QCD. If technicolor interactions, leading to a condensate of techniquarks, are responsible for the $W^{\pm}$ and Z masses, then the scale of this symmetry breaking must be of order 1 TeV. If an extended version of technicolor is also to generate the quark masses, its scale must be greater than 1000 TeV, because these interactions generate four-fermion effective interactions between normal quarks, as well as between quarks and techniquarks and amongst techniquarks themselves, and the former contribute to FCNCs. As mentioned earlier, for particular models, these contributions may be computed in lattice QCD, but the main issue here is that a large extended technicolor scale is required to ensure that FCNCs are within the experimental limits.

This large scale then suppresses the quark masses that can be generated through the techniquark condensate. Unless some different sort of dynamics operates between the extended technicolor scale and the scale of EWSB, which can provide an enhancement for the quark masses, extended technicolor cannot even explain the s quark mass. "Walking" technicolor is one possibility. The idea is that the extended technicolor coupling does not run between the extended technicolor scale and the EWSB scale, enhancing the expectation value of the techniquark condensate by a power of the ratio of these scales. The power is given by the techniquark mass anomalous dimension, which also has to be of order 1 for this to work. Then it is possible to generate quark masses of order a few GeV. Thus, walking technicolor could explain the s and c masses, and, at a push, even the b mass, but not the top quark mass.

Whether any gauge theory "walks" is, almost certainly, a non-perturbative question. It requires a near-zero of the β function at some non-zero coupling. The (universal) perturbative β function to two loops may have a zero at non-zero coupling for particular choices of the gauge group, fermion representation and number of flavours. The resulting theory built around this (infrared stable) fixed point is conformal and so cannot support chiral symmetry breaking or massive bound hadron-like states. Theories with such fixed points are said to be in the "conformal window". Their existence suggests there may be nearby theories, whose β functions have near-zeros. None has been found yet, but lattice theorists are beginning the search.

Recent lattice simulations for SU(3) gauge theories with 8 and 12 flavours of fermion in the fundamental representation, show that the former is QCD-like, with a coupling which grows in the infrared, whereas the latter has a conformal fixed point (the renormalized coupling tends to a constant value in the infrared). Walking might occur for some intermediate number of flavours. Simulations of SU(3) gauge theory with two flavours of sextet fermions also suggest this theory may have a conformal fixed point. This work is exploratory. It may not even bear any relation to reality, and we may learn so very rapidly once LHC data becomes available, but it is also possible that strong dynamics plays a key role in EWSB and lattice methods will become an essential tool in understanding LHC physics.

In conclusion, we now understand how the non-perturbative dynamics of gauge theories, made precise via lattice simulations of QCD, explains why the proton mass is far below the Planck scale and why the *u* and *d* quarks are so light, through the way in which chiral symmetry is realised. This makes the idea that all mass scales can be understood through strong dynamics very appealing. The success of lattice QCD as a precision tool for phenomenology, gives us confidence to embark on non-perturbative studies of other gauge theories. We therefore have a way forward to explore non-perturbatively the old idea of technicolor and, hence whether EWSB and the heavier quark masses might have their origin in some new strong dynamics.

CHAIRMAN: R. KENWAY

Scientific Secretaries: S. Nicotri, D. Reeb

DISCUSSION

- Musaev:

You have presented a model with the number of flavours equal to zero and you have shown us some results concerning the mass of the ψ meson and of the Υ meson. What is the quark content of those mesons in this theory? What does it mean that the number of flavours equals zero?

- Kenway:

It means that in the path integral the contribution from the sea quarks is ignored. If you think about sea quarks and valence quarks, then the states that I was showing results for were built out of valence quarks, but the background gauge configurations that are used in the path integral were purely gauge fields computed with the Yang-Mills action, without contributions coming from quark loops. The reason that this was even ever considered useful is because in the early days of lattice QCD simulations there was not sufficient computer power. This theory was computationally much cheaper to simulate than the one that includes the effects of fermion loops. So, it was studied as a sort of trial example of a theory that had some of the features of QCD. Remarkably, it turned out that for a quite lot of quantities the results you get from this zero-flavour, or quenched, theory only deviate from full QCD by roughly 10-20%, so it wasn't a bad model, although there are some quantities that one would get completely wrong.

- Weir:

How much more computationally demanding is it to add an extra quark flavour, compared with the zero-flavour case? Does it pay off?

- Kenway:

This is a time-dependent question! In the early days it was very expensive to add flavours, but, as I mentioned in the talk, we now have a new algorithm, the rational Hybrid Monte Carlo algorithm, that has been very highly optimized to do the Monte Carlo simulations with, and it means that adding flavours is a relatively cheap thing to do. That's also one of the reasons why we can now more easily look at other models, with different numbers of flavours, fermions in different representations, and so on. So, what really happened is that numerical technology has changed a lot over the last twenty or thirty years in which we have been trying to do the simulations. Today, putting in different flavour content is relatively cheap and straightforward.

- Carrasco:

I am very impressed by the amount of computation that these calculations require, so I spent most of my lunch break trying to understand where these Petaflops are going. This is my current naïve understanding: the goal is trying to model a 4D lattice of points where at each point we have one or more SU(3) vectors representing quark fields and SU(3) matrices between lattice sites representing the gluon field. We model this by actually having one or more SU(3) quarks at each site and SU(3) matrices on each link, and the Monte Carlo simulation involves multiple sweeps through the whole lattice. I am wondering if each point is really equal, deserving equal resources, or is there room for some sort of adaptive meshing here?

- *Kenway:*

This picture is correct. The problem with adaptive meshing is that you need to be able to pick regions in your lattice where there is a lot of physics happening. The action for the background vacuum configurations we simulate is translationally invariant, so there is no particular reason to expect activity in one part of a lattice and not in another.

- *Stavenga:*

In a non-asymptotically free theory we need a UV completion when we go to small distances. This can be seen in attempts to resum the Borel perturbation series. In QCD there is no problem for resummation, because the UV renormalons are not on the real axis. However, the resummation is obstructed by IR renormalons. Those introduce ambiguities coming from IR physics. Might it be the case that QCD needs an IR completion, so that lattice calculations are irrelevant?

- *Kenway:*

Well, we could have an indication of such IR effects if there were strong finite volume dependence in the results. But QCD appears to behave as I described previously, that is as an asymptotically free theory with its own scale, Λ_{QCD}, and once you take your lattice bigger than this scale, then there don't appear to be any IR effects. We do not observe any effect that may force us to conclude that QCD is IR incomplete.

- *Klopot:*

Am I right that in SU(2)×SU(2) chiral perturbation theory you fix the mass of the strange quark, while the masses of up and down quarks are considered to be small?

- *Kenway:*

This is correct.

- *Dubynskiy:*

How well can lattice QCD predict the position of the resonances in the charmonium spectrum? Could we describe molecular states, tetraquark states and other exotic states by lattice simulations? And why did the lattice not predict such states before their experimental discoveries?

- *Kenway:*

Except for flavor singlet states, which are challenging because it is hard to get a decent signal for them, it is quite straightforward to calculate the charmonium spectrum. Exotic multiquark states can in principle be calculated, even though they need more lattice volume since they have a larger number of quarks. The reason why we have not done it for all possible exotic states is that it would be a huge effort and it is hard to tell which ones are stable and which are very short-lived resonances. We have simply focussed our attention on what we thought were interesting problems.

- *Bonati:*

Is it possible to obtain theoretical limitations, using lattice methods, on the θ-parameter with precision comparable to the experimental ones?

- *Kenway:*

I think in principle we can get a handle on this parameter, which involves computing the electric dipole moment of the neutron on the lattice. The challenge with the calculation is that QCD with a θ-parameter has a complex action, so the Monte Carlo method does not work. Some very preliminary work has been done, for example, by assuming θ is small and expanding to leading order in it. The second challenge is that you have to properly sample from all the topological sectors to get the physics right, and that is very sensitive to the way in which you put in the light sea quarks. Today it is really an open question for simulations as to what it means to get the correct sampling of topologies. Since we now can handle chiral symmetry on the lattice much better, there is hope that the situation might improve soon.

- *Nicotri:*

Can you please comment more on infrared fixed points?

- *Kenway:*

Since the first two terms of the beta function expansion depend on the number of colors and on the number of flavours, you can choose some particular values of them such that those terms have opposite sign. If this happens at small values of the coupling constant, so that you can trust your perturbation expansion, then this might indicate the existence of a stable non-trivial IR fixed point. At the end of my talk I have shown some examples of how to study this issue of fixed points non-perurbatively. This is a relatively complicated calculation to do, hence we don't have complete coverage yet over the interesting range of the number of flavours, colors, representations, etc. I have also shown you a plot which indicates a zero in the discrete β-function in SU(3) Yang-Mills with two flavours of sextet fermions at the fairly small value of $g^2 = 2$, but a zero in the β-function does not automatically imply that there is a fixed point – a full renormalization group blocking analysis is required.

- *Ni:*

You mentioned that lattice QCD can be used to search for new physics, as has been done in the case of the decay constants of D mesons. Is there any other particular result that one should look into to search for "new physics"?

- *Kenway:*

There was a hope, a few years ago, of overconstraining the unitarity triangle to expose discrepancies between the Standard Model and experiments, particularly at B-factories. Unfortunately, history has shown that the quark mixing is well described within the Standard Model, although there still are possibilities to refine theoretical calculations to see some kind of discrepancies in the future. Examples like the $b \to s\gamma$ decays should receive contributions at leading order from beyond Standard Model physics, so, if you are able to precisely compute this decay theoretically and compare to experiment, you should be able to constrain BSM physics. A possibly productive direction in the future might be to calculate QCD matrix elements for low-energy effective interactions that are induced by physics beyond the Standard Model and constrain new models in that way.

- *Susskind:*

I assume that in the limit in which the fifth dimension becomes infinite (in lattice units) the pion becomes massless. What I would like to know is how fast the pion mass tends to zero. In particular, does it go to zero exponentially as you bring the lattice spacing to zero?

- *Kenway:*

Yes, the pion mass indeed goes to zero exponentially as the number of lattice points in the fifth dimension goes to infinity, and I have shown you a plot of this behaviour. This is crucial, because it shows that we have control over chiral symmetry breaking, which is exponentially small in our QCD simulations.

- *Arnowitt:*

One of the old problems in lattice gauge theory was why the pion is much lighter than the rho. This has now been solved. Can you give a physical explanation for this, i.e. is it due to the new treatment of chiral spinors?

- *Kenway:*

We are reaching the point where we understand SU(2)×SU(2) chiral perturbation theory with up and down quarks in the vicinity of the physical pion mass and we seem to be capturing the correct physics. Probably, we need to go a little bit below the physical light quark masses to convince ourselves that everything is really ok. We need to simulate non-degenerate up and down quarks, in particular to see if the up quark is massless, but that requires a much more complicated chiral perturbation theory analysis, and will be a challenge for the future.

- *Arnowitt:*

When Pran Nath and I worked on SU(3)×SU(3) chiral current algebra we knew it was a poorer approximation than SU(2)×SU(2). Can the lattice analysis tell now how bad it was? Also, it is very nice now seeing the proton decay matrix elements α and β so well calculated on the lattice!

- *Kenway:*

If we use SU(3)×SU(3) chiral perturbation theory to extrapolate the data for the pseudoscalar decay constant to the physical quark masses, adding the next-to-leading order term in chiral perturbation theory is a 60% effect, which is a strong indication that it is not working reliably in the vicinity of the strange quark mass.

- *Kopnin:*

The bag parameter B_k has now been calculated with an accuracy of around 5%, while a year ago it used to be around 10%. Is it possible that the situation with the bag parameters relevant to the B-Bbar and B_s-$B_s$bar oscillations will improve in the same way in the near future?

- *Kenway:*

Eventually yes, but here there are two difficulties. The improvement of this calculation for the kaon system is due to the use of domain wall fermions that preserve chiral symmetry. That is needed for the B system as well, of course, because there are light quarks involved, but you have got the additional problem that you also have to simulate the b quark itself, which is too heavy to fit on the lattices that we currently use for the light quarks. The cutoff due to the lattice spacing is just below 2 GeV, so there is no way to directly simulate the b quark and treat the light quarks correctly at the same time. Nevertheless, there are three possibilities to get the same precision for B physics. The first would be a huge computer which uses a very much finer lattice spacing, such that the cutoff is substantially above 5 GeV, so that you can simulate the b quark directly and also correctly simulate the light quarks. The second option is to use an effective theory for the heavy quarks, such as non-relativistic QCD, and match it onto the current simulations for light quark physics. That can become technically difficult to achieve high precision, because it means adding extra operators which you have to compute. The third possibility, which I think is the most feasible in terms of having real control, is to simulate the b quark on a fine enough lattice and then block down, using renormalization group techniques, to a coarser lattice, on which you can simulate the light quarks.

- *'t Hooft:*

I would like to ask a question related to the question by Professor Arnowitt about chiral symmetry. To me it seems that people do not use the 1/g expansion on the lattice, maybe it is too technically complicated. Concerning the question of Professor Arnowitt whether we understand the chiral vacuum, it seems that with ordinary Wilson fermions the 1/g expansion gives you a lighter pion than rho meson. So, if you can handle domain wall fermions in this 1/g expansion maybe you will see an exactly massless pion?

- *Kenway:*

The main reason why the 1/g expansion was abandoned were concerns about the roughening transition that it tends to suffer from as you try to take the continuum limit. It may be that in the five dimensional formulation, you are able to avoid this problem.

- *Susskind:*

Shouldn't you see the pion being massless on either side of this roughening transition if you get the chiral symmetry correct?

- *Kenway:*

I don't know. Do you think so?

- *Susskind:*

I don't see why not!

Strongly Interacting Matter at High Energy Density*

Larry McLerran

Riken Brookhaven Center and Physics Department

Brookhaven National Laboratory, Upton, NY 11973 USA

December 8, 2008

Abstract

This lecture concerns the properties of strongly interacting matter (which is described by Quantum Chromodynamics) at very high energy density. I review the properties of matter at high temperature, discussing the deconfinement phase transition . At high baryon density and low temperature, large N_c arguments are developed which suggest that high baryonic density matter is a third form of matter, Quarkyonic Matter, that is distinct from confined hadronic matter and deconfined matter. I finally discuss the Color Glass Condensate which controls the high energy limit of QCD, and forms the low x part of a hadron wavefunction. The Glasma is introduced as matter formed by the Color Glass Condensate which eventually thermalizes into a Quark Gluon Plasma.

1 Introduction

This lecture is about the structure of matter at very high energy density. Matter in thermal equilibrium is characterized by a temperature and various chemical potentials corresponding to conserved charges.[1]-[17] The only conserved charge I consider is baryon number. In addition to matter in thermal equilibrium, there is the matter which composes the dominant part of

*Lecture delivered at the International School of Subnuclear Physics, Erice, Italy 2008

the hadron wavefunction at high energy. This matter is a highly coherent configuration of gluons at large energy density and is called the Color Glass Condensate.[18]-[23] Such matter controls the interactions of strongly interacting particles in the high energy limit. In collisions, sheets of Colored Glass pass through one another, and in the first instants after the collisions, color electric and magnetic fields are produced which have a different topology from the Color Glass Condensate fields. The CGC fields are in the plane perpendicular to the direction of motion of the hadrons, but the Glasma fields are longitudinal and along the direction of motion. Such fields bear a close resemblance to the fields hypothesized within the Lund model. [24]-[28]

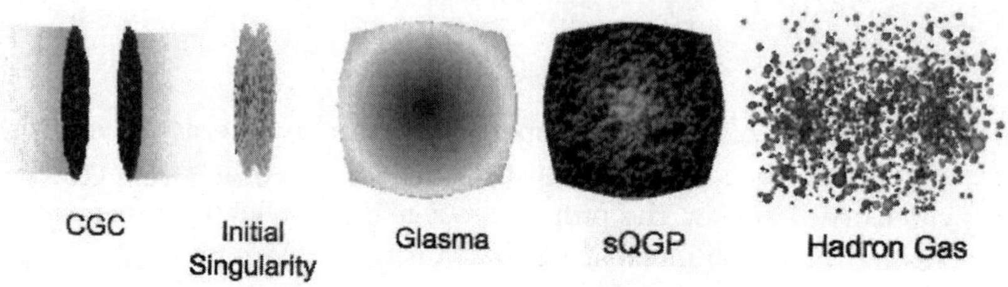

Figure 1: A schematic picture of the evolution of matter produced in the heavy ion collisions.

A cartoon of how such high energy density matter appears in high energy nuclear collisions in shown in Fig. 1. Sheets of Color Glass Condensate collide. When they pass through one another, the CGC fields rapidly change and a Glasma is formed. In the limit $E \to \infty$, there is an initial singularity corresponding to this rapid change. The decay products of the Glasma matter presumably thermalize, forming a strongly interacting Quark Gluon Plasma. The sQGP after some time hadronizes, and these hadrons make their way to detectors. The experimental study of such collisions involves knowledge of the properties of all of the forms of matter described above.

2 Matter at Finite Temperature

2.1 Confinement

The partition function is

$$Z = Tr\ e^{-\beta H + \beta \mu_B N_B} \tag{1}$$

where the temperature is $T = 1/\beta$ and N_B is the baryon number and μ_B is the baryon number chemical potential. Operator expectation values are

$$< O >= \frac{Tr\ O\ e^{-\beta H + \beta \mu_B N_B}}{Z} \tag{2}$$

Under the substitution $e^{-\beta H} \rightarrow e^{-itH}$, the partition function becomes the time evolution operator of QCD. Therefore, if we change $t \rightarrow it$,and redefine zeroth components of fields by appropriate factors of i, and introduce Euclidean gamma matrices with anti-commutation relations

$$\{\gamma^\mu, \gamma^\nu\} = -2\delta^{\mu\nu} \tag{3}$$

then for QCD, the partition function has the path integral representation

$$Z = \int\ [dA][d\overline{\psi}][d\psi]exp\left\{-\int_0^\beta\ d^4x\ \left(\frac{1}{4}F^2 + \overline{\psi}\left[\frac{1}{i}\gamma \cdot D + m + i\mu_Q\gamma^0\psi\right]\right)\right\} \tag{4}$$

Here the fermion field is a quark field so that the baryon number chemical potential is

$$\mu_Q = \frac{1}{N_c}\mu_B \tag{5}$$

This path integral is in Euclidean space and is computable using Monte Carlo methods when the quark chemical potential vanishes. If the quark chemical potential is non-zero, various contributions appear with different sign, and the Monte Carlo integrations are poorly convergent. Boundary conditions on the fields must be specified on account of the finite length of the integration in time. They are periodic for Bosons and anti-periodic for Fermions, and follow from the trace in the definition of the partition function.

A straightforward way to probe the confining properties of the QCD matter is to introduce a heavy test quark. If the free energy of the heavy test quark is infinite, then there is confinement, and if it is finite there is deconfinement. One can prove that the free energy of an quark added to the system is

$$e^{-\beta F_q} =< L > \tag{6}$$

where

$$L(\vec{x}) = \frac{1}{N_c} Tr\ P\ e^{i \int\ dt\ A^0(\vec{x},t)} \tag{7}$$

So confinement means $< L >= 0$ and deconfinement means that $< L >$ is finite. The path ordered phase integration which defines the line operator L is shown in Fig. 2 Such a path ordered phase is called a Polyakov loop or Wilson line.

Figure 2: The contour in the t plane which defines the Polyakov loop. The space is closed in time because of the periodic boundary conditions imposed by the definition of the partition function.

Exercise

Prove that the free energy of a heavy static quark added to the system is given by Eqn. 6. To do this, use the heavy quark action as

$$S_{HQ} = \int\ dt\ \overline{\psi}(\vec{x},t)\ \frac{1}{i}\gamma^0 D^0\ \psi(\vec{x},t). \tag{8}$$

Then show that the gluonic contribution to the action is invariant under gauge transformations that are periodic up to an element of the center of the gauge group. The center of the gauge group is a set of diagonal matrices matrix $Z_p = e^{2\pi i p/N}\overline{I}$ *where* $\overline{I}$ *is an identity matrix. Show that the quark contribution to the action is not invariant, and show that* $L \to Z_p L$ *under this transformation. Show that in a theory with only dynamical gluons, that the energy of a system of n quarks minus antiquarks is invariant under the center symmetry transformation only if n is an integer multiple of* N. *Therefore, when the center symmetry is realized, the only states of finite free energy are baryons plus color singlet mesons.*

In the exercise, you have shown that the realization of the center symmetry, $L \to Z_p L$ is equivalent to confinement. This symmetry is like the global rotational symmetry of a spin system, and it may be either realized or broken. At large separations, the correlation of a line and its adjoint, corresponding to a quark-antiquark pair is

$$lim_{r \to \infty} < L(r) L^\dagger(0) >= Ce^{-\kappa r} + < L(0) >< L^\dagger(0) > \tag{9}$$

since upon subtracting a mean field term, correlation functions should vanish exponentially. Since

$$e^{-\beta F_{q\overline{q}}(r)} =< L(r) L^\dagger(0) > \tag{10}$$

we see that in the confined phase, where $< L >= 0$, the potential is linear, but in the unconfined phase, where $< L >$ is non-zero, the potential goes to a constant at large separations.

The analogy with a spin system is useful. For the spin system corresponding to QCD without dynamical quarks, the partition function can be written as

$$Z = \int [dA] \, e^{-\frac{1}{g^2} S[A]} \tag{11}$$

The effective temperature of the spin system associated with the gluon fields is $T_{eff} \sim g^2$. By asymptotic freedom of the strong interactions, as real temperature gets larger, the effective temperature gets smaller. So at large real temperature (small effective temperature) we expect an ordered system, where the Z_N symmetry is broken, and there is deconfinement. For small real temperature corresponding to large effective temperature, there is disorder or confinement.

As you showed in the exercise, the presence of dynamical fermions breaks the Z_N symmetry. This is analogous to placing a spin system in an external magnetic field. There is no longer any symmetry associated with confinement, and the phase transition can disappear. This is what is believed to happen in QCD for physical masses of quarks. What was a first order phase transition for the theory in the absence of quarks becomes a continuous change in the properties of the matter for the theory with quarks.

Another way to think about the confinement-decofinement transition is a change in the number of degrees of freedom. At low temperatures, there are light meson degrees of freedom. Since these are confined, the number of degrees of freedom is of order one in the number of colors. In the unconfined world, there are $2(N_c^2 - 1)$ gluons, and $4N_cN_f$ fermions where N_f is the number of light mass fermion families. The energy density scaled by T^4 is a dimensionless number and directly proportional to the number of degrees of freedom. We expect it to have the property shown in Fig. 3 for pure QCD in the absence of quarks. The discontinuity at the deconfinement temperature, T_d is the latent heat of the phase transition.

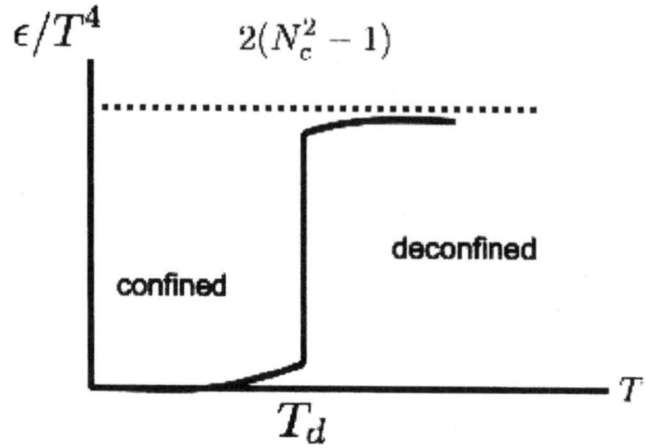

Figure 3: The energy density scaled by T^4 for QCD in the absence of dynamical quarks.

The energy density can be computed using lattice Monte Carlo methods. The result of such computation is shown in Fig. 4. The discontinuity present for the theory with no quarks becomes a rapid cross over when dynamical quarks are present.

The large N_c limit gives some insight into the properties of high temperature matter.[29]-[31] As $N_c \to \infty$, the energy density itself is an order parameter for the decofinement phase transition. Viewed from the hadronic world, there is an amount of energy density $\sim N_c^2$ which must be inserted to surpass the transition temperature. At infinite N_c this cannot happen, as this involves an inifnite amount of energy. There is a Hagedorn limiting temperature, which for finite N_c would have been the deconfinement temperature.

The Hagedorn limiting temperature can be understood from the viewpoint of the hadronic world as arising from an exponentially growing density of states. In a few paragraphs, we will argue that mesons and glueballs are very weakly interacting in the limit of large N_c. Therefore, the partition

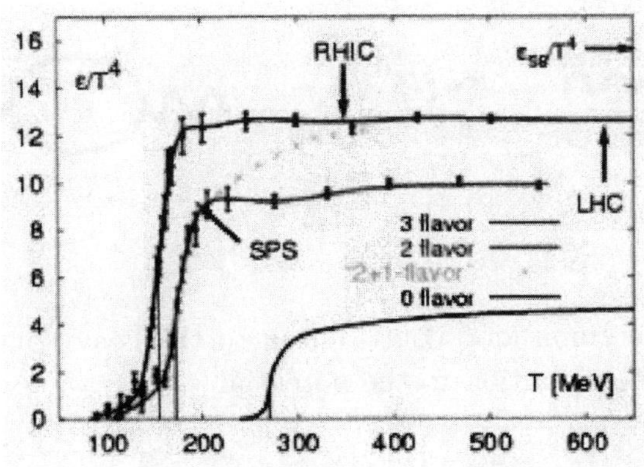

Figure 4: The energy density scaled by T^4 measured in QCD from lattice Monte Carlo simulation. Here there are quarks with realistic masses.

function is

$$Z = \int dm\ \rho(m) e^{-m/T} \tag{12}$$

Taking $\rho(m) \sim m^\alpha e^{\kappa m}$, so that

$$< m > \sim \frac{1}{1/T - \kappa} \tag{13}$$

diverges when $T \to 1/\kappa$

2.2 A Brief Review of the Large N_c Limit

The large N_c limit for an interacting theory takes $N_c \to \infty$ with the 't Hooft coupling $g^2_{'tHooft} = g^2 N_c$ finite. This approximation has the advantage that the interactions among quarks and gluons simplifies. For example, at finite temperature, the disappearance of confinement is associated with Debye screening by gluon loops, as shown in Fig. 5a. This diagram generates a screening mass of order $M^2_{screening} \sim g^2_{'tHooft} T^2$. On the other hand the quark loop contribution is smaller by a power of N_c and vanishes in the large N_c limit.

To understand interactions consider Fig. 6a. This corresponds to a mesonic current-current interaction through quarks. In powers of N_c, it is of

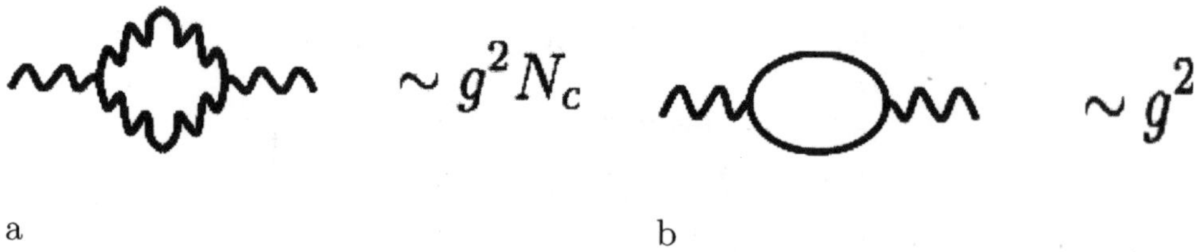

Figure 5: a: The gluon loop contribution to the heavy quark potential. b: The quark loop contribution to the potential

order N_c. Gluon interactions will not change this overall factor. The three current interaction is also of order N_C as shown in Fig. 6b. The three meson vertex, G which remains after amputating the external lines, is therefore of order $1/\sqrt{N_c}$. A similar argument shows that the four meson interaction is of order $1/N_c$.

Exercise

Show that the 3 glueball vertex is of order $1/N_c$ *and the four glueball interaction of order* $1/N_c^2$

These arguments show that QCD at large N_c becomes a theory of non-interacting mesons and glueballs. There are an infinite number of such states because excitations can never decay. In fact, the spectrum of mesons seen in nature does look to a fair approximation like non-interacting particles. Widths of resonances are typically of order $200MeV$, for resonances with masses up to several GeV.

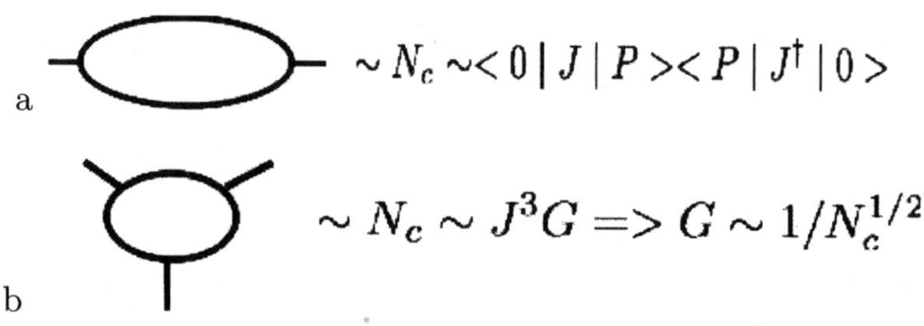

Figure 6: a: The quark loop corresponding to a current-current interaction. b: A quark loop corresponding to a three current interaction.

2.3 Mass Generation and Chiral Symmetry Breaking

QCD in the limit of zero quark masses has a $U(1) \times SU_L(2) \times SU_R(2)$ symmetry. (The $U_5(1)$ symmetry is explicitly broken due to the axial anomaly.) Since the pion field, $\overline{\psi}\tau^a\gamma_5\psi$ is generated by an $SU_{L-R}(2)$ transformation of the sigma field, $\overline{\psi}\psi$, the energy (or potential) in the space of the pion-sigma field is degenerate under this transformation. In nature, pions have anomalously low masses. This is believed to be a consequence of chiral symmetry breaking, where the σ field acquires an expectation value, and the pion fields are Goldstone bosons associated with the degeneracy of the potential under the chiral rotations.

Such symmetry breaking can occur if the energy of a particle-antiparticle pair is less than zero, as shown in Fig. 7. On the left of this figure is the naive vacuum where the negative energy states associated with quark are filed. The right hand side of the figure corresponds to a particle hole excitation, corresponding to a sigma meson. Remember that a hole in the negative energy sea corresponds to an antiparticle with the opposite momentum and energy. If the σ meson excitation has negative energy, the system is unstable with respect to forming a condensate of these mesons.

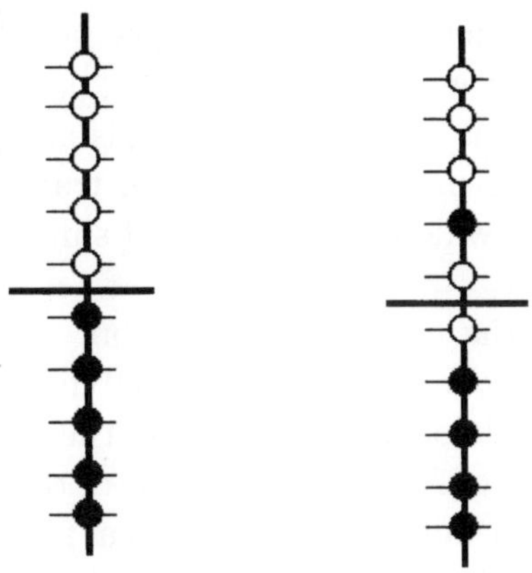

Figure 7: The energy levels of the Dirac equation. Unfilled states are open circles and filled states are solid circles. For the free Dirac equation, negative energy states are filled and positive energy states are unoccupied, as shown on the left hand side. A mesonic excitation corresponding to a particle hole pair is shown on the right hand side.

At sufficiently high temperature, the chiral condensate might melt. Indeed this occurs.[15] For QCD, the chiral and deconfinement phase transition occur at the same temperature. At a temperature of about $170 - 200 MeV$, both the linear potential disappears and chiral symmetry is restored. It is difficult to make a precise statement about the indentification of the chiral and deconfinement phase transitions, since as argued above, for QCD with quarks, there is not a real phase transition associated with deconfinement.[32]-[33] Also, when quarks have finite masses, as they do in nature, chiral symmetry is not an exact symmetry, and there need be no strict phase transition associated with its restoration. nevertheless, the cross over is quite rapid, and the point, and there are rapid changes in the both the potential and the sigma condensate $< \overline{\psi}\psi >$ at temperatures which are in a narrow range.

3 Finite Density

When the deconfinement phase transition was first discussed, the phase diagram of nuclear matter was postulated. This postulate is shown in Fig. 8. In such a diagram, the world is simply divided into that of a world of mesons and glueballs, corresponding to confinement, and to a deconfined world of quarks and gluons. There is a weak liquid gas phase transitions corresponding to nuclear matter, at $\mu_B \sim M_N$ and very low temperature. The deconfinement baryon chemical potential is assumed to be significantly larger than M_N. (There may be weak phase transitions corresponding to color superconductivity, which for purposes of simplicity, I shall not discuss in this lecture.[16])

The problem with the conventional wisdom about the phase diagram of QCD is that it is not consistent with our large N_c understanding.[17] As shown in Fig. 5, the quark loop contribution to the quark-antiquark potential is suppressed by a power of N_C relative to that of the gluons, and vanishes in the large N_c limit at fixed value of quark chemical potential, $\mu_Q \sim \Lambda_{QCD}$. It is only through such loops that the effects of finite baryon number density enter into the thermodynamic potential, or correct the quark-antiquark potential. The effects of quark loops on the quark-antiquark potential can only become important when $\mu_Q \sim \sqrt{N_c}\Lambda_{QCD}$. This occurs when the density of baryons is $\rho_B \sim N_C^{3/2}\Lambda_{QCD}^3$ which is parametrically large compared to the scale at which the coupling should become weak, $\rho_B \sim \Lambda_{QCD}^3$, and where we would have naively expected deconfinement to occur.

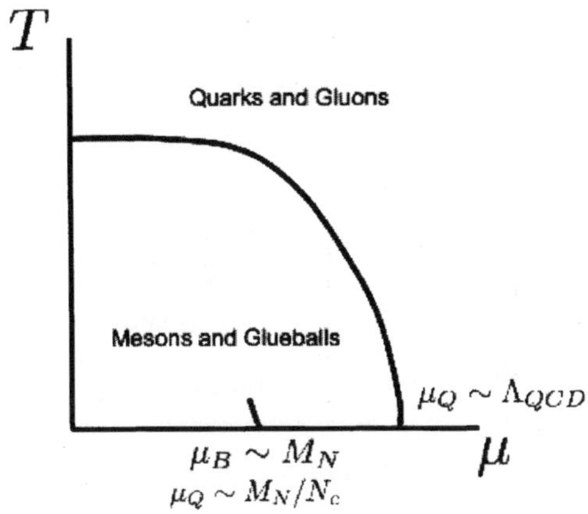

Figure 8: The popular folklore for the phase diagram of QCD.

At large N_C, there is another order parameter besides the free energy of a heavy test quark. The baryon number density itself is an order parameter. At finite chemical potential,

$$\rho_B \sim e^{\mu_B/T - M_B/T} \sim e^{-N_c\{M_B/N_cT - \mu_Q/T\}} \tag{14}$$

For $m_Q \leq M_B/N_c$, the baryon number is always zero in the large N_c limit. Above it, there is a finite baryon number density and confinement. Correspondingly, if one increases the temperature, at some temperature independent of μ_Q, there is deconfinement. Such a phase diagram is shown in Fig. 9

We call the third phase of matter, which has finite baryon number density but is confined, Quarkyonic. The name expresses the fact that the matter is composed of confined baryons yet behaves like quarks at high densities. At chemical potentials parametrically large compare to Λ_{QCD} but small compared to $\sqrt{N_c}\Lambda_{QCD}$, the bulk properties of the matter are computable in perturbation theory. Therefore the energy density should scale as N_c. The quarks should be approximately chirally symmetric. There may be non-perturbative effects associated with confinement and chiral symmetry restoration near the fermi surface, since there the interactions are sensitive to long distance effects, but the bulk properties should look like almost free quarks.

There is another order parameter for the system, the energy density itself.

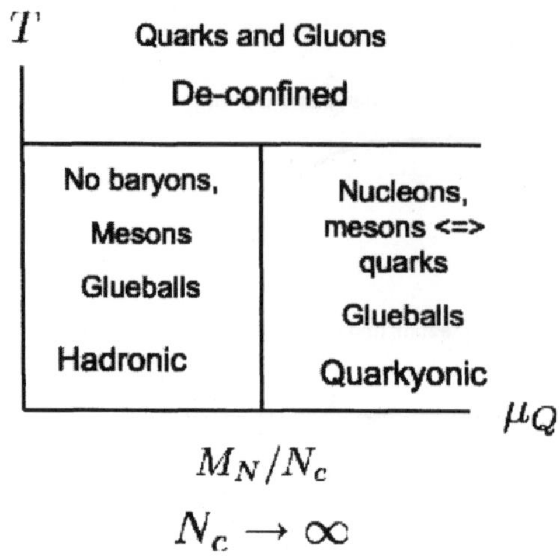

Figure 9: The phase diagram of QCD at large N_c.

In the hadronic phase, $\epsilon \sim O(1)$. In the de-confined phase, gluons dominate the number of degrees of freedom, $\epsilon \sim N_c^2$. The Quarkyonic phase is quarks plus thermal excitations of glueballs and mesons, so that $\epsilon \sim N_c$

The transition between hadronic matter and Quarkyoninc matter is in a narrow range of quark chemical potential. The typical Fermi momentum for the transition is $k_F \sim \Lambda_{QCD}$. On the other hand, the quark fermi energy is

$$\mu_Q \sim (M_N + k_F^2/2M_N)/N_c \sim \Lambda_{QCD}(1 + O(1/N_c^2)) \qquad (15)$$

Since the quarks are confined, there should be a dual description of the matter at large N_c in terms of either quarks or baryons. A baryonic description is provided by the Skyrme model.[34] In large N_c, the Skyrme action is of order N_c, which agrees parametrically with the dependence on N_c of a gas of weakly coupled quarks, and agrees with the results above.

In the discussion above, I have implicitly assumed that the limit $N_c \to \infty$ is taken while the number of quark flavors N_f is held fixed. If the number of flavors is large, then there there is never a linear potential between quarks because quark pairs can be made from the vacuum that short out the potential. In the fixed N_f limit, it is known from lattice Monte-Carlo studies that the linear potential provides a good description of the quark-antiquark interaction for distances between $.3\ Fm \leq r \leq 2\ Fm$. This suggests that the fixed N_f large N_c limit is valid for some range of distance scales.

For $N_F \sim N_c \to \infty$, there is a high degeneracy of low mass baryonic states. The number of lowest mass baryons with $I = J = 1/2$ can be computed with a result $e^{N_c F(N_f/N_c)}$. The baryon number density remains a good order parameter, and $\rho_B \sim e^{N_c F(N_f/N_c)+\mu_B/T-M_B/T}$. For $TF(N_f/N_c)+\mu_Q < M_B/N_c$ there is a baryon number zero phase and for greater values thee is a finite baryon number. The transition is caused by the Hagedorn spectrum of baryonic resonances.

Presumably for QCD, there are reasonably sharp cross overs or phase transitions that separate the hadronic, quarkyonic and confined worlds. The energy density of the confined world is $O(N_f)$, the Quarkyonic is $\sim 2N_cN_f$, and the deconfined is $\sim 4N_cN_f+2N_c^2$. Precisely how chiral symmetry restoration appears in this phase diagram is not yet known.

There is a hint from heavy ion collisions about Quarkyonic matter. If one measures the abundances of various species of particles produced in heavy ion collisions, one may fit these numbers to a temperature and baryon number chemical potential.[35] These temperatures and chemical potentials are interpreted as those for which the system freezes out or decouples. At some low temperature and density the matter is expanding so rapidly that the matter cannot maintain thermal equilibrium by rescattering. It is expected that such a freeze out temperature reflects the position of phase transition surfaces, since at a phase transition, the number of degrees of freedom change very rapidly. This results in a rapid lowering of the density as one goes across the transition surface. The lower density makes it harder for the system to maintain itself in thermal equilibrium. In numercal simulations of the properties of heavy ion collisions, this seems to be a reasonable first approximation.

In Fig. 10, the freeze out temperature and baryon chemical potential are plotted. At low baryon density, the curve follows expectations that the freeze out surface follows the phase boundary between the hadronic and confined world. At a chemical potential of about $\mu_B \sim 500\ MeV$, the decoupling line deviates from expectations for the confinement-deconfinement surface. This point is about where baryons begin to dominate the energy density. The line intersects the $T = 0$ axis at about $\mu_B = M_N$, which is what we expect for the Quarkyonic phase. If this interpretation is correct, then experiment has given us the rough parameters of the Quarkyonic-Hadronic boundary.

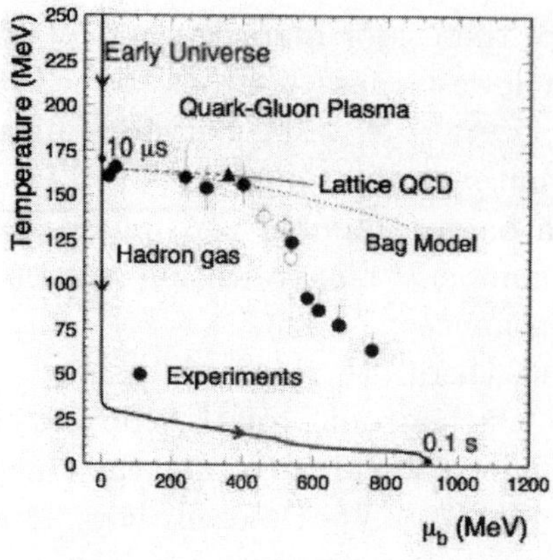

Figure 10: Freeze out temperature and baryon chemical potential determined by heavy ion collisions.

4 Color Glass Condensate

A remarkable experimental fact about the hadrons at high energy is that their size is very slowly increasing, but the number of gluons increases rapidly. This means that the high energy limit is the high gluon density limit. One would think that ultimately the density would saturate at high energy, since at some point the density would be so large that the repulsive gluon interactions would prevent further growth of the gluon number. This is almost true.[36]-[37] For gluons of a fixed size, the number stops growing rapidly, but one can then add in more gluons of smaller size, until the density of such gluon is high enough so that their density saturates. This can go on forever proceeding to saturate gluons of smaller and smaller sized as the typical energy scale goes to infinity. This is illustrated in Fig. 11.

This high energy density gluonic matter is the Color Glass Condensate.[18]-[23] The name color comes from the color of the gluons which compose it. The word condensate arises from the high phase space density of the gluons. Defining

$$\rho_{PS} = \frac{dN}{dyd^2p_Td^2x_T} \tag{16}$$

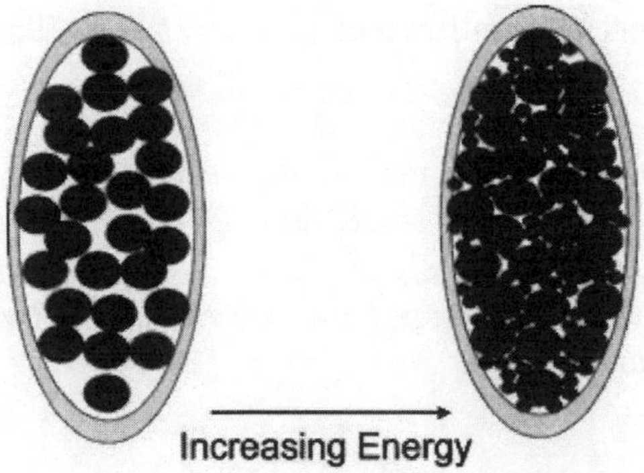

Figure 11: More gluons added to a hadron as the hadron energy increases..

where y is the gluon rapidity $y = \frac{1}{2}ln((E + P_z)/(E - P_z))$.

The energy of such a distribution of gluons is

$$E = -\kappa\rho_{PS} + \kappa'\alpha_S\rho_{PS}^2 \tag{17}$$

The first term in this expression is negative because the gluon density spontaneously generates itself, and the last term is positive because of repulsive gluon interactions (needed because the energy is bounded from below). This is extremized at $\rho_{PS} \sim 1/\alpha_S$, which is typical of condensation phenomena in Bose condensates and superconductivity. Such condensates are highly coherent, and the fields associated with them are classical, because of the high occupation number. Here $\alpha_S << 1$ because the transverse density of gluons is high compared to Λ_{QCD}^2.

The word glass is in the name because the partons which make the coherent CGC field are fast moving. Let us imagine we are in the infinite momentum frame of the hadron. The fast moving gluons produce the more slowly moving ones. Those at the lowest allowed value of momentum, which is of order Λ_{QCD} are slow moving, and their natural time scale of evolution would be $t \sim 1/\Lambda_{QCD}$. Because they arise from a coherent field produced from the fast partons, their time scale of evolution is Lorentz time dilated, and they evolve slowly compared to natural times scales. This is characteristic of glassy material.

The Color Glass Condensate is made of the Lorentz boosted Couomb fields of the fast moving gluons in the hadron. In light cone coordinates, the gluons are on a sheet of small extent in $x^- = t - z$. They evolve slowly in $x^+ = t + z$.

Exercise

Show that the because fields are rapidly varying in x^- and slowly varying in x^+, that F^{i+} is big, F^{i-} is small, and F^{ij} is of intermediated strength. Show that the large fields satisfy $\vec{E} \perp \vec{B} \perp \hat{z}$, so that the color electric and magnetic fields have the same Lorentz structure as the Lienard-Wiechart fields of electrodynamics.

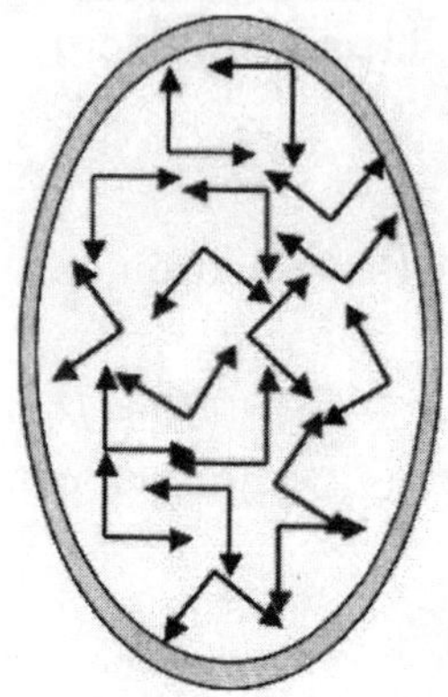

Figure 12: The color electric and color magnetic fields in the Color Glass Condensate.

The color electric and color magnetic fields of the CGC are shown in Fig. 12. These fields are of the form of the Lienard-Wiechart fields of electrodynamics. The color orientation is random. Their density is determined by the saturation momentum. Integrating the phase space density up to the highest momentum scale where the gluons are still highly coherent gives[38]

$$\frac{1}{\pi R^2}\frac{dN}{dy} = \frac{1}{\alpha_S} Q^2_{sat} \tag{18}$$

To determine the dependence of the saturation momentum scale on energy, one needs a theory of the CGC. This is given by QCD plus renormalization group in the gluon energy. I have argued that gluon of higher momentum acts as a source for those of low momentum. The scale which separates high energy gluons from soft gluons is arbitrary. This arbitrariness is resolved by renormalization group equations. Suppose we have the theory separated between sources and fields and there is one separation momentum scale P_1. We can move to a new scale P_2 by integrating out fields to produce sources at the new scale. This procedure can be done when the coupling is weak, even though the CGC fields are strong. It turns out that the solution to this renormalization group equation is universal, so that the high energy limit and the CGC are independent of hadron. The solution also has the property that the saturation momentum grows forever as the energy increases.

The saturation momentum scale acts as an infrared cutoff. For distance scales less than $1/Q_{sat}$, fields resolve individual sources of charge. For distance scales greater than $1/Q_{sat}$, the fields see multiple sources of charge which average to zero. Gluonic field strengths should be softer by one power of r at large distances or one power of p at small momentum than one would have ignoring coherence. For example, the gluon distribution itself, which is proportional to the gluon strength squared $< a^\dagger a >\sim A^2$, goes as

$$\begin{aligned} \frac{dN}{dyd^2p_T} &\sim \frac{1}{p_T^2} \qquad p_T \geq Q_{sat} \\ \frac{dN}{dyd^2p_T} &\sim lm(p_T) \quad p_T \leq Q_{sat} \end{aligned} \tag{19}$$

When applied to the computation of hadron-hardon scattering, the total multiplicity becomes computable because the $1/p_T^4$ singularity of perturbation theory becomes $\sim ln(p_T)$, which is integrable.

5 The Glasma

The collision of two hadrons may be visualized as that of two very thin sheets of Colored Glass.[25]-[28] The thickness of these sheets is determined by $\alpha_S \Delta y \sim 1$ since if one attempts a classical field description, quantum corrections to the classical field become important when this is satisfied. Using that

$$y = \frac{1}{2} ln((E+p_z)/(E-p_Z)) = ln((E+p_z)/M_T) \sim -ln(\Delta z Q_{sat}) \tag{20}$$

This gives the thickness of the sheet as $\Delta z \sim e^{-\kappa/\alpha_S}/Q_{sat}$. For small α_S this is much much smaller than the scale set by the saturation momentum.

In the time it takes the sheets to pass through one another, the topology of the fields change. As I shall argue in the following paragraphs, the fields which originally are embedded in the sheets of Colored Glass tsansform into longitudinal flux tubes of color electric and color magnetic fields, as shown in Fig. 13. This configuration of fields is called the Glasma because it is

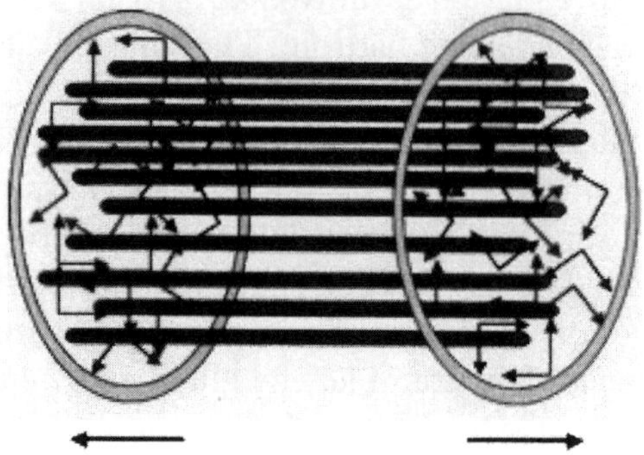

Figure 13: The color electric and color magnetic fields of the Glasma immediately after the collision of two hadrons.

created by the Color Glass Condensate and eventually decays into a Quark Gluon Plasma.

To understand the origin of the Glasma fields, we need to first understand a little more about the CGC fields. The CGC fields are generated by a source of field along the lightcone, $x^{\pm} = 0$. For definiteness, consider the sheet with $x^- = 0$. The Yang-Mills equations may be solved by a vector potential which is a two dimensional gauge transform of vacuum fields for either of the regions $x^- < 0$ or $x^- > 0$. We choose different gauge transforms of vacuum in either of these regions, and the discontinuity in Yang-Mills equations becomes the source of color charge on the sheets.

Now consider the scattering problem illustrated in Fig. 14. In addition to $x^{\pm}$, there are the space-time rapidity variable $\eta = \frac{1}{2}ln(x^+/x^-)$, and a proper time $\tau = \sqrt{x^+x^-}$.

Inside the backward lightcone, I use gauge freedom to choose the field $A^\mu = 0$. In the right hand and left hand side of the light cone I choose fields which are pure gauge transforms of vacuum. This guarantees that along the backwards lightcone, the Yang-Mills equations are solved. If I choose, $A^\mu = A_1^\mu + A_2^\mu$ infinitesimally close to the forward light cone, but inside it,

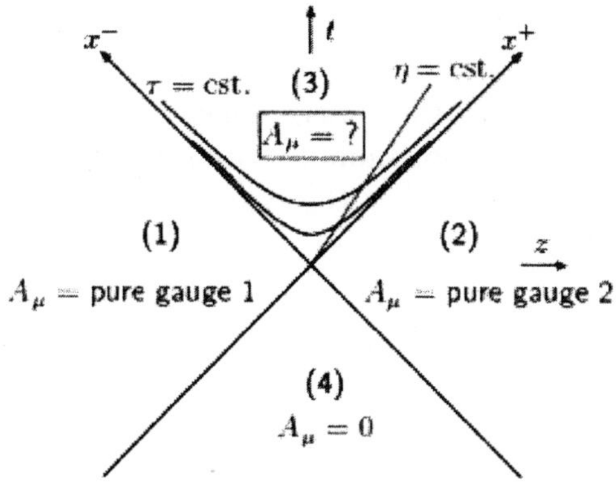

Figure 14: A space-time diagram for the scattering of two sheets of colored glass

then one gets the correct discontinuity to give the sources which sit on the forward light cone. Of course, this field configuration is no longer a gauge transformation of vacuum, and will evolve. It can be shown that this solution is independent of the longitudinal coordinate, the space-time rapidity.

The fields in the forward-lightcone are initially only longitudinal electric and magnetic fields. This is because in the forward light cone, there are induced sources for the color electric and magnetic fields:

$$\begin{aligned} \vec{\nabla} \cdot \vec{E}_{1,2} &= \vec{A}_{2,1} \cdot \vec{E}_{1,2} \\ \vec{\nabla} \cdot \vec{B}_{1,2} &= \vec{A}_{2,1} \cdot \vec{B}_{1,2} \end{aligned} \tag{21}$$

These fields are independent of longitudinal coordinate. They generate equal and opposite densities of color electric and color magnetic charge density on the sheets. The average energy density stored in the magnetic fields is the same as that in the electric fields because of the duality of the Yang-Mills equations under $E \leftrightarrow B$, and the symmetry of the initial conditions.

The classical fields have large strength, and decay by classical evolution. Presumably they evolve into an interacting Quark Gluon Plasma. There are in fact unstable modes of these solutions and given enough time may develop turbulence and might classically thermalize the fields.[39]-[44]

Note also that these classical fields carry a non-zero Chern-Simons charge, since $E \cdot B$ is non zero The integral of this charge vanishes, but on the transverse size scale of $1/Q_{sat}$ the charge density is non zero. It is expected that non-zero total Chern-Simons number density will be generated by the time evolution of fluctuations, as is the case in baryogenesis in electroweak theory.[45]-[48]

6 Summary

I have presented a description of some of the properties of high energy density matter which might be made in high energy collisions. The properties of matter in thermal equilibrium can be tested in heavy ion collisions. The Color Glass Condensate has various phenomenological tests in the theory of high energy lepton-nucleus scattering and hadron collisions. It has a successful phenomenology, which will be tested with greater precision in experiments at LHC. The concept of the Glasma is fairly new, and has qualitative features in accord with results from RHIC, and will also be tested in high energy heavy ion collisions at LHC.

7 Acknowledgments

I gratefully acknowledge my colleagues Yoshimasa Hidaka, Rob Piasarksi, and Raju Venugopalan for their many discussion with me concerning the ideas presented here.

This manuscript has been authorized under Contract No. DE-AC02-98CH0886 with the US Department of Energy.

References

[1] N. Itoh *Prog. Theor. Phys.* **44** 291 (1970).

[2] P. Carruthers *Coll. Phenom.* **1** 147 (1973)

[3] N. Cabibbo and G. Parisi, *Phys. Lett.* **B59** 67 (1975).

[4] J. C. Collins and M. J. Perry, *Phys. Rev. Lett.* **32** 1353 (1974).

[5] G. Baym and S. A. Chin, *Phys. Lett.* **B62** 241 (1976).

[6] B. A. Freedman and L. D. Mclerran, *Phys. Rev.* **D16**, 1130 (1977); 1147 (1977); 1169 (1977).

[7] E. V. Shuryak, *Sov. Phys. JETP* **47** 212 (1978); *Zh. Ekps. Teor. Fiz.* **74** 408 (1978).

[8] J. Kapusta, *Nucl. Phys.* **B148** 461 (1979)

[9] A. Polyakov, *Phys. Lett.* **B72** 477 (1978).

[10] L. Susskind, *Phys. Rev.* **D20** 2610 (1979).

[11] L. D. Mclerran and B. Svetitsky, *Phys. Lett.* **B98** 195 (1981); *Phys. Rev.* **D24** 450 (1981)

[12] J. Kuti, J. Polonyi and K. Szlachanyi, *Phys. Lett.* **B98** 199 (1981).

[13] D. Gross, R. Pisarski and L. Yaffe, *Rev. Mod. Phys.* **53** 43 (1981).

[14] J. Engels, F. Karsch, H. Satz and I. Montvay, *Phys. Lett.* **B101** 89 (1981)

[15] F. Karsch, *Lect. Notes. Phys.* **83** 209 (2002)

[16] For a review with references to original literature, see M. Alford, A. Schmidt, K. Rajagopal and T. Schafer, arXiv:0709.4365

[17] L. McLerran and R. Pisarski, *Nucl. Phys.* **A796** 83 (2007); Y. Hidaka, L. Mclerran and R. Pisarski, arXiv:0803.0279.

[18] L. D. McLerran and R. Venugopalan, *Phys. Rev.* **D49**, 2233(1994); 3352 (1994); **D50**, 2225 (1994).

[19] E. Iancu, A. Leonidov and L. D. McLerran, *Nucl. Phys.* **A692**, 583 (2001); E. Ferreiro E. Iancu, A. Leonidov and L. D. McLerran, *Nucl. Phys.* **A710**,373 (2002);J. Jalilian-Marian, *Phys. Rev,.* **C70** 027902 (2004) E. Iancu and L. McLerran, *Phys. Lett.* **B510**, 145 (2001); J. Jalilian-Marian, A. Kovner, L. McLerran and H. Weigert, *Phys. Rev.* **D55** (1997), 5414; J. Jalilian-Marian, A. Kovner, A. Leonidov and H. Weigert, *Nucl. Phys.* **B504** (1997), 415; *Phys. Rev.* **D59** (1999), 014014.

[20] I. Balitsky, *Nucl. Phys.* **B463** (1996), 99.

[21] Yu. V. Kovchegov, *Phys. Rev.* **D60** (1999), 034008; *ibid.* **D61** (2000), 074018.

[22] A. H. Mueller, *Phys.Lett.* **B523**, 243 (2001)

[23] E. Iancu and R. Venugopalan, Published in QGP3, Eds. R. C. Hwa and X. N. Wang, World Scientific. hep-ph/0303204

[24] B. Andersson, G. Gustafson, G. Ingelman and T. Sjostrand, *Phys. Rept.* 97 31 (1983).

[25] A. Kovner, L. McLerran and H. Weigert, *Phys. Rev.* **D52** 3809 (1995); **D52**, 6231 (1995).

[26] A. Krasnitz and R. Venugopalan, *Nucl. Phys.* **B557** 237 (1999); *Phys. Rev. Lett.* **84** 4309 (2000)

[27] T. Lappi, *Phys. Rev.* **C67** 054903 (2003).

[28] T. Lappi and L. McLerran, *Nucl. Phys.* **A772** 200 (2006).

[29] G. 't Hooft, *Nucl. Phys.* **B75**, 461 (1974).

[30] E. Witten, *Nucl. Phys.* **B160**, 57 (1979).

[31] C. B. Thorn, *Phys. Lett.* **B99**, 458 (1981).

[32] M. Cheng et. al. *Phys. Rev.* **D74**, 054507 (2006).

[33] Y. Aoki et. al. arXiv 0711.0024.

[34] G. Adkins, C. Nappi and E. Witten, *Nucl. Phys.* **B228**, 552 (1983).

[35] For a recent treatment of statistical models see P. Barun Muzinger, D. Magestro, K. Redlich and J. Stachel, *Phys. Lett.* **B518** 41 (2001).

[36] L. V. Gribov, E. M. Levin and M. G. Ryskin, *Phys. Rept.* **100**, 1 (1983).

[37] A. H. Mueller and Jian-wei Qiu, *Nucl. Phys.* **B268**, 427 (1986); J.-P. Blaizot and A. H. Mueller, *Nucl. Phys.* **B289**, 847 (1987).

[38] D. Kharzeev and M. Nardi, *Phys. Lett.* **B507** 121 (2001).

[39] S. Mroczynski, *Phys. Lett.* **B214** 587 (1988); **B314** 118 (1993); **B363**, 26 (1997); P. Romatschke and M. Strickland, *Phys. Rev.* **D68** 036004 (2003); **D70** 116006 (2004).

[40] P. Arnold, J. Lenaghan, and G. Moore, *JHEP* **0308** 002 (2003); P. Arnold, J. Lenaghan, G. Moore and L. Yaffe, *Phys. Rev. Lett.* **94** 072302 (2005).

[41] A. Dumitru and Y. Nara, *JHEP* **0509** 041 (2005); A. Dumitru, Y. Nara and M. Strickland, *Phys. Lett.* **B621** 89 (2005).

[42] P. Romatschke and R. Venugopalan, *Phys. Rev. Lett.* **96** 062302 (2006).

[43] H. Fujii and K. Itakura, arXiv:0806.1840; arXiv:0803.0410; A. Iwazaki, arXiv:0804.0188 [hep-ph]

[44] K. Fukushima, F. Gelis, and L. McLerran, *Nucl. Phys.* **A786** 107 (2007).

[45] D. Kharzeev, R. Pisrski and M. Tytgat, *Phys. Rev. Lett* **81**, 512 (1998).

[46] D. Kharzeev and R. Pisarski, *Phys. Rev.* **D61**, 111901 (2000).

[47] D. Kharzeev and A. Zhitnitsky, *Nucl. Phys.* **A797**, 67 (2007).

[48] D. Kharzeev, L. Mclerran and H. Warringa, *Nucl. Phys.* **A803** 227 (2008).

CHAIRMAN: L. McLERRAN

Scientific Secretaries: K.H. Khan, Y. Klopot

DISCUSSION

- *Fedosenko:*

What do you mean when you say "the size of gluons"? What are the small and large colored balls on the picture in slide number 17? Are these gluons?

- *McLerran:*

Suppose you start with a hadron at low energy. Then you would say the size of those color balls is of the size of the constituent quarks. This is some fraction of the size of hadron, maybe some scale-size of the order of $1/\Lambda_{QCD}$, maybe ½ fm, or some scale size like that. If you go to higher and higher energies, different states become a better approximation of a hadron. As you go to higher energy using the evolution equations or renormalization group equations which, for example, for structure functions are DGLAP eq. or BFKL eq., or whatever, you evolve the gluonic component of the original wave function. What happens is that you generate more and more gluons in the system, and those gluons begin to dominate the wave function of the system. So, imagine at some scale we have gluons, and those gluons have some typical size. I don't specify what size yet. As you go to higher energies you have more gluons, you have to pack them into the system. Repulsive interactions of the gluons will keep you from piling too many of them on top of one another. So, eventually, you have to pile gluons into the free spaces in-between those balls. So, how does the size came in? The size is a typical inverse Q (intrinsic momentum) scale in a wave function. If you go to the higher energies, you have to fit more gluons into the system, so the intrinsic size of the degrees of freedom that are to be fit into the system at high energies is smaller or the typical momentum scale is getting bigger. The point is that the renormalization group eq. takes you from one scale to the other scale. And as we go to higher and higher energy that just means we'll be fitting in smaller and smaller objects into the hadron wave function. You can ask what is the typical size of the gluons. It is the inverse of the typical momentum scale of the gluons at a typical energy scale. You can measure a typical saturation momentum associated with a size scale for a nucleus like gold. It is about 1-2 GeV. If you go up to CERN energies, LHC energies for heavy ions, numbers are 3-5 GeV. The size scale is typically of the order of 0.2 fm, at RHIC energy and about .02-.03 Fm at LHC.

- *Stavenga:*

Your talk seemed to indicate that chiral bag models are a good description. Can you comment on chiral bag model vs AdS/QCD models of baryons and why chiral bag model went out of fashion?

- *McLerran:*

AdS/CFT world is something that probably does describe qualitative aspects of a strong coupling limit of QCD. It provides a conceptual tool to understand a strong coupling limit. It is very interesting. You can use the AdS/CFT correspondence to describe saturation, and you'll get a picture which looks like a picture I described there for a color glass condensate, except the phase space density doesn't become big because the coupling is of order 1 (the phase space density is of order $1/\text{coupling}^2$. The phenomenology of such a theory probably best describes the low energy limit of scattering, not really the high energy limit. Now, when you go to QCD in the finite temperature and density world the story is perhaps different. There is some evidence there that maybe the thermal system made in high energy scattering is characterized by somewhat stronger coupling than one was expected. Now, the chiral bag theory is interesting because a chiral bag theory is an attempt to reconcile the dual description of a hadron which occurs in the Skyrme model. You can think about a nucleons made from the pion field. The topological quantum number associated with that pion field will correspond to baryon number. So, you can simultaneously think of a hadron as pion degrees of freedom and also as quark degrees of freedom. And when you decide to think about it in terms quarks degrees of freedom as opposed to meson degrees of freedom, again, it is somewhat arbitrary because of this dualism in the confining phase of the theory. So, the chiral bag model attempts to try to find some branching point when one description is somewhat better than the other description. To what degree it is realistic for the description for a single hadron, it's hard to say because the quark degrees of freedom as perturbative degrees of freedom are probably hard to sort out for the low energy properties of a hadron.

- *Susskind:*

This glass I would call "spaghetti". Electric and magnetic spaghetti. The spaghetti strength presumably is not stable. I would guess, that the electric ones can discharge by quark pairs and magnetic ones by gluons...

- *McLerran:*

It's even better (or worse). You have both electric and magnetic fields, these things can evolve classically.

- *Susskind:*

So you think the fields are large enough to evolve classically?

- *McLerran:*

Yes, the magnitude of the field strength is characterized by 1/alpha_s, and alpha_s is small when the density of gluons is large enough or the energy is high enough.

- *Susskind:*

Okay, what is the decay mechanism?

- *McLerran:*

The decay mechanism is a classical evolution of the fields. And the characteristic time-scale for fields decay is of order of the inverse saturation momentum. The time-scale in which this happens is a time-scale of the order of the inverse saturation momentum. This is much larger than the time it takes the sources to pass through one another. The sources are localized in a distance scale $\exp(-C/\alpha_s)/Q_s$

- *Susskind:*

Okay, what does the "inverse saturation momentum" mean?

- *McLerran:*

Saturation momentum is given by the inverse of the typical transverse separation scale between the gluons in a high energy nucleus.

- *Susskind:*

Is it energy dependent?

- *McLerran:*

- Yes, surely. As you go to the higher and higher energies the saturation momentum gets larger and larger, so the scale size gets smaller and smaller. This is because the number of gluons in a hadron grows more rapidly than the size of a hadron.

- *Korthals-Altes:*

You mentioned P and CP violating signatures in the QG plasma. At first sight this sounds like a contradiction in terms. QCD has a P and CP conserving ground state whatever the temperature. However it may be that during the formation of the plasma bubbles of metastable vacuum form that violation of P and CP. The question is then: how do these bubbles form? Some ten years ago Pysarski, Tytgat et al. came up with such a bubble scenario. What is the new mechanism?

- *McLerran:*

The new idea here is that if you have transitions that involve changes in topological charge, those transitions involve P and CP violation. Ordinarily we have a vacuum which violates P and CP, we have to bias it some way so it sits in some particular configuration. The QCD vacuum doesn't do that. We can still have transitions which violate P and CP and in the average, these things, you know, wash-out. But you can have large fluctuations during the collision process and so if you measure the system for some finite time there will be some fluctuations which will generate some P and CP violation on an event by event basis. Now how can you see that? The way you can see it is, if you have some instanton transitions, or sphaleron transitions at high temperature, what will happen is you align helicity preferentially in one direction compared to the other. Now, in order to select out helicity, you can apply an external ordinary electromagnetic field. Specifically, if you apply a magnetic field, you orient the magnetic moments of particles, and because the helicity is also aligned, you generate an electromagnetic current in the direction of magnetic field. So if you can apply magnetic field you can see how all theses helicities are aligned. In heavy ion collision that actually happens because the ions have the charges of valence quarks in the nuclei, and as they pass through one another a magnetic field is generated. It is the biggest at earliest times, but it has a tail. So then what you have to do, you have to convolute these transitions which involve instanton transitions for the strong interactions with the magnetic field which is applied, and you generate fluctuations in the flux of charged particles perpendicular to the reaction plane of the collisions Now, you have no net effect, because there is no net P and CP violations in QCD when averaged over all events. But nevertheless, you can calculate the fluctuation spectrum as a function of energy as function of charge, as a function of impact parameter of the collision to see the whether the underlying mechanism actually describes what happens. Together with Harmen Warringa and Dima Kharzeev we estimated this effect and in fact it describes qualitatively very, very preliminary data coming from RHIC. But there will be more on this from STAR experiment. We'll have more to say on this in the next few months.

- *Tannenbaum*:

You said that strings decay in time $1/Q_{satur}$, Q_{satur} =2 GeV , t=0.1 fm/c? The γ-factor of the colliding nuclei is 100 , 100x0.1=10 fm/c which seems reasonable – how do you take the γ-factor into account?

- *McLerran*:

The Lorenz contracted size of a nucleus at that energy is about 0.1 fm, so the separation is not particularly huge. If you have a sting, it begins to decay first in the middle. If you go to rapidities away from the central region the things which are decaying involve the gamma-factor, but the things which are produced right in the middle do not involve the gamma-factor. So, if you look at the things in the center of the mass frame, then it will first decay in the center, then it will take longer to decay if you go to edges. So, you have an extended longitudinal structure, where the time to decay depends on where you are on this longitudinal structure. The signature for this is very clear, however. You would generate long-range correlations in longitudinal variable and the rapidity associated with this structure. But, yes, you do have to take into account the gamma-factor in order to get the longitudinal structure.

- *Dubinsky*:

Is there any way to make a prediction of hydrodynamic properties of the QGP by other techniques which are different from AdS/QCD approach?

- *McLerran*:

There are at least 2 things that are interesting. One is the observation that you could understand the deviation of the energy density and structure for an ideal gas values by string calculations. The (energy density)/T^4 at high T is about .75 or .8 rather then ideal gas value. And it happens naturally in a strong coupling scenario. The other piece of information that got people a little bit excited was stuff by Kovtun, Son and Starinets and others. They said that there is a lower bound on the viscosity/entropy ratio. That lower bound you can sort of understand as an uncertainly principle argument because you are really measuring basically a mean free path divided by the thermal wave length of the system When η/s is $1/(4\pi)$ it turns out that also the value for a mean free path is of the order of thermal wave length and that's where transport is no longer a classical transport , it is a quantum transport. The reason why people got excited is because in descriptions of RHIC data, the viscosity free hydrodynamic model works quite well. It actually describes incredible numbers of features of the data. The η/s values which you do get are the η/s values which are close to this bound. But on the other hand, if I calculate it in terms of the mean free path I get the value of about 0.1 - 0.2 which corresponds to the mean free path of about ½ fm, so I am not so sure I need AdSCFT/QCD to explain the ratio. A value of ½ fm does not seem so small for a high density strongly interacting system.

- *Korthals-Altes*:

There is a remark: the lower limit $1/(4\pi)$ on the ratio η/s in N=4 SUSY is violated in N=2 conformal Sp(N) theory.

The nature and the mass of neutrinos. Majorana vs. Dirac

A. Bettini
Dipartimento di Fisica G. Galilei dell'Università and INFN, Padova
Laboratorio Subterráneo de Canfranc, Spain

Abstract

Spin ½ particles are described by the Dirac equation, which, in particular, predicts antiparticles as different from particles. Charged particles are distinguished from their antiparticles by electric charge. Neutrinos and antineutrinos are distinguished by the lepton number only and, if this is not absolutely conserved, may be the same particle. Notice that there are no experimental proofs that neutrinos follow the Dirac equation, rather then the Majorana equation.
After a summary on the present status of neutrino physics, I shall analyse the predictions of the Majorana equation and show that the experimental search on the Majorana vs. Dirac nature of neutrinos is possible only in a physical process, the neutrino-less double beta decay. I shall review the status of this search and conclude by mentioning a recently proposed different approach based on atomic physics.

1. The present status of neutrino physics

Contrarily to the assumptions of the Standard Model, 1. neutrinos are massive, 2. their stationary states, namely the mass eigenstates, are not the states of definite flavour, 3. flavour lepton numbers are not conserved.

The experimental evidence for oscillations and matter conversion implies that the flavour states are not the mass eigenstates. We call the latter ν_1, ν_2 and ν_3, and m_1, m_2 and m_3 their masses. Flavour states are linear combinations of the eigenstates. The mixing matrix can be written as the product of three rotations including a phase factor, as in the case of quarks, and a fourth diagonal matrix with two more phases. The latter, called Majorana phases, can be absorbed in the wave functions only if neutrinos and antineutrinos are different (Dirac) particles. With $c_{ij}=\cos\theta_{ij}$, and $s_{ij}=\sin\theta_{ij}$ we have

$$\begin{pmatrix} \nu_e \\ \nu_\mu \\ \nu_\tau \end{pmatrix} = \begin{pmatrix} 1 & 0 & 0 \\ 0 & c_{23} & s_{23} \\ 0 & -s_{23} & c_{23} \end{pmatrix} \begin{pmatrix} c_{13} & 0 & s_{13}e^{i\delta} \\ 0 & 1 & 0 \\ -s_{13}e^{-i\delta} & 0 & c_{13} \end{pmatrix} \begin{pmatrix} c_{12} & -s_{12} & 0 \\ s_{12} & c_{12} & 0 \\ 0 & 0 & 1 \end{pmatrix} \begin{pmatrix} 1 & 0 & 0 \\ 0 & e^{i\alpha} & 0 \\ 0 & 0 & e^{i\beta} \end{pmatrix} \begin{pmatrix} \nu_1 \\ \nu_2 \\ \nu_3 \end{pmatrix} \qquad (1)$$

In total there are nine quantities to measure, three masses, three mixing angles and the three phases. The latter, if $\neq 0$ and $\neq \pi$, give *CP* violation effects in the lepton sector. Majorana phases are irrelevant in oscillation and matter conversion phenomena; they appear only (for any practical purpose) in double beta decay.

Different groups have performed global fits including all the available data, providing values for the three mixing angles (only an upper limit for θ_{13}) and two squared mass differences. Results can be summarised as follows[1] (2 σ uncertainties):

$$\begin{aligned} &\left|\Delta m^2\right| \equiv \left|m_3^2 - \left(m_2^2 + m_1^2\right)/2\right| = 2380 \pm 270 \text{ meV}^2 \qquad \delta m^2 \equiv m_2^2 - m_1^2 = 76.6 \pm 3.5 \text{ meV}^2 \\ &\sin^2\theta_{12} = 0.326^{+0.05}_{-0.04} \qquad \sin^2\theta_{23} = 0.45^{+0.16}_{-0.09} \qquad \sin^2\theta_{13} < 0.032\,. \end{aligned} \qquad (2)$$

Information on the absolute value of neutrino masses comes from: the measurement of the end point energy and shape of the beta decay spectrum, the search for neutrino-less double beta decay, the subject of this lecture, and the measurement of large structures spectrum in the Universe. Each of them has provided only upper limits, on different quantities but each of them might not be far from discovery.

The measurement of the electron spectrum in the beta decay of a nucleus near its end-point gives a direct determination of the quantity

$$m_{\nu_e}^2 = \sum_{i=1}^{3} |U_{ei}|^2 m_i^2 = c_{13}^2 c_{12}^2 m_1^2 + c_{13}^2 s_{12}^2 m_2^2 + s_{13}^2 m_3^2 \quad (3)$$

The present upper limit is $m_{\nu e} < 2.3$ eV from the Mainz[2] and Troitz[3] experiments on ^{3}H decay. Tritium is used due to its extremely low endpoint energy (18.6 keV). The next step is the KATRIN experiment under construction at Karlsruhe aiming to increase by an order of magnitude the sensitivity, down to 0.2 eV (detection limit at 5σ=0.35 eV).

If neutrinos are such, an extremely rare weak decay can happen, the neutrino-less double beta decay ($0\nu2\beta$), $(A,Z) \to (A,Z+2) + 2e^{\S}$. The inverse lifetime is proportional to the square of the quantity, the "Majorana mass"

$$M_{ee} = \left| \sum_{i=1}^{3} U_{ei}^{\ 2} m_i \right| = c_{13}^2 c_{12}^2 m_1 + c_{13}^2 s_{12}^2 m_2 e^{i\alpha} + s_{13}^2 m_3 e^{i\beta} \quad (4)$$

Notice that the phase factors can induce cancellations, which can be almost complete in case $\Delta m^2 > 0$.

Cosmology provides an indirect but very sensitive limit on the sum of neutrino masses $\sum m_i$. All the structures present in the Universe have been seeded by quantum fluctuations that took place when the Universe was extremely small. These initial fluctuations grown during the evolution of the Universe, due to the gravitational instability. Neutrinos, due to their very small masses, had speeds larger than the escape velocity from the smaller structures. Consequently they could 'free stream' out of those structures diminishing their total mass. As a net result, neutrinos tend to erase the structures at scales smaller than a certain dimension, which is a decreasing function of the sum of the neutrino masses.

The most conservative limit include only the cosmic microwaves spectrum (WMAP and other experiments), the Supernovae data and the absolute distance indicator for recent epochs, the Baryon Acoustic Oscillation peak, giving[4]

$$\sum m_i \leq 600 \text{ meV} \qquad \Rightarrow m_i < 200 \text{ meV} \quad (5)$$

Notice the smallness of neutrino masses, 12 orders of magnitude smaller than the top.

2. The Majorana equation

In the Standard Model neutrinos are assumed to be Dirac particles, but we do not have experimental evidence for that. In 1937 E. Majorana published on Il Nuovo Cimento the fundamental note "Teoria simmetrica dell'elettrone e del positrone"[5].

The abstract reads: "*We show that it is possible to achieve complete formal symmetrization in the electron and positron quantum theory by means of a new quantization process. The meaning of Dirac equations is somewhat modified and it is no more necessary to speak of negative-energy states; nor to assume, for any other type of particles, especially neutral ones, the existence of antiparticles, corresponding to the "holes" of negative energy*".

The starting point of Majorana is the unpleasant asymmetry in the Dirac theory between electrons and positrons. He writes *"The interpretation of the so called "negative energy states" proposed by Dirac* (in 1924) *leads, as it is well known, to a substantially symmetric description of electron and positrons"*. At the end the two partners are perfectly symmetric, but at the beginning of the Dirac argument they are not. Majorana states not to be completely satisfied and to have tried another way leading directly to the goal.

"From this (new way) *we can expect only a formal progress for electrons and positrons; but it looks to us important, in view of possible extensions, that the notion itself of negative energy*

state be abandoned. Indeed, we shall see that to be perfectly possible to build, in the most natural way, a theory of the neutral elementary particles without negative states."

He continues by stating that the new approach allows to "*not only to give a symmetric form to the electron-positron theory, but also to build a substantially novel theory for the particles deprived of electric charge (neutrons and hypothetical neutrinos)*".

He concludes that it is probably "*not yet possible to ask to the experience to decide between this new theory and the simple extension of the Dirac equations to the neutral particles*"

The neutrino had been introduced by W. Pauli in 1930, but it was not known to exist. Majorana observes that the new theory allows one to introduce "*in this poorly explored field the minimum possible number of new objects*". Namely, only two states, instead of the four states of the Dirac particles.

Is it possible now, 70 years later, to ask to the experience to decide between the two theories? The answer is: yes. Do we have already that answer? The answer is: no. When a definite answer will come? The answer is: may be soon or may me be in many years of experimental efforts.

Let us start by recalling some basic concepts. The Dirac equation is

$$\left(\gamma_\mu p^\mu - m\right)\psi = 0 \tag{6}$$

where ψ is the Dirac bi-spinor. It is useful to identify its two spinor components that we call φ and χ, namely $\psi = \begin{pmatrix}\varphi \\ \chi\end{pmatrix}$.

They represent the particle and the antiparticle. The two elements of each have a definite spin projection on the given quantization axis. If the particle is free and not at rest, the natural quantization axis is the particle velocity. The spin projection operator on this oriented axis is called helicity $h = \frac{\vec{\sigma}\cdot\vec{p}}{|p|}$, where σ_i are the Pauli matrices. Notice that helicity is a property of the spinors, not of the bi-spinors, and that it should not be confused with chirality, which is a property of the bi-spinors. The states of defined chirality are the eigenstates of the γ_5 matrix and are called left (L) if its eigenvalue is –1, right (R) if it is +1. They are obtained via the chiral projections

$$\psi_L = \frac{1}{2}\left(1-\gamma_5\right)\psi, \qquad \gamma_5\psi_L = -\psi_L; \qquad \psi_R = \frac{1}{2}\left(1+\gamma_5\right)\psi, \qquad \gamma_5\psi_R = +\psi_R \tag{7}$$

and for the conjugate states

$$\bar{\psi}_L = \bar{\psi}\frac{1}{2}\left(1+\gamma_5\right), \qquad \bar{\psi}_R = \bar{\psi}\frac{1}{2}\left(1-\gamma_5\right) \tag{8}$$

As it well known chirality is important because only negative chirality fields have weak charge current interactions, i. e. couple to the *W*. However, chirality is not observable, because γ_5 does not commute with the mass term in the (free) Hamiltonian.

Chirality would be an observable for mass-less spin ½ particles, which do not exist in nature. We measure the helicity instead. Let us have a closer look. We chose the representation of the Dirac matrices in which γ_5 is diagonal, and we have

$$\gamma^0 = \begin{pmatrix}0 & 1\\ 1 & 0\end{pmatrix}, \quad \gamma^i = \begin{pmatrix}0 & -\sigma^i\\ \sigma^i & 0\end{pmatrix}, \quad \gamma^5 = \begin{pmatrix}1 & 0\\ 0 & -1\end{pmatrix} \tag{9}$$

The chiral projections are

$$\psi_L = \frac{1}{2}\left(1-\gamma_5\right)\psi = \begin{pmatrix}0 & 0\\ 0 & 1\end{pmatrix}\begin{pmatrix}\varphi\\ \chi\end{pmatrix} = \begin{pmatrix}0\\ \chi\end{pmatrix}; \quad \psi_R = \frac{1}{2}\left(1+\gamma_5\right)\psi = \begin{pmatrix}0 & 1\\ 0 & 0\end{pmatrix}\begin{pmatrix}\varphi\\ \chi\end{pmatrix} = \begin{pmatrix}\varphi\\ 0\end{pmatrix} \tag{10}$$

Notice that both the left and the right spinor have only two non zero components corresponding respectively to the χ and φ spinors.

We can now write the Dirac equation as

$$\begin{aligned}(E+\vec{p}\cdot\vec{\sigma})\chi-m\varphi=0\\(E-\vec{p}\cdot\vec{\sigma})\varphi-m\chi=0\end{aligned}\tag{11}$$

Notice that the two equations are coupled, in the general case $m\neq 0$. Let us examine the helicity content of the spinor χ of the left bispinor. Taking the z axis along the velocity, we have

$$\chi=\frac{1}{2}\left(\frac{E-p_z}{m}\right)\chi^{+1/2}+\frac{1}{2}\left(\frac{E+p_z}{m}\right)\chi^{-1/2}\approx\frac{m}{E}\chi^{+1/2}+\frac{1}{2}\left(\frac{E+p_z}{m}\right)\chi^{-1/2}\tag{12}$$

We see that the spinor of the left bi-spinor contains components of both helicities. However, the positive helicity component is small at energies much larger than the mass. It is rigorously zero for m=0.

In the Dirac theory the two φ and χ spinor components of the bispinor have different transformation properties under Lorentz transformations. The question posed by Majorana, in modern language, is: is it possible to find a spinor φ_c constructed with the components of φ only, and hence without further degrees of freedom, which transforms like a χ instead than a φ and that can consequently take its place in the Dirac equation? And similarly for χ? The Majorana answer is: yes, they are

$$\varphi_c=i\sigma_2\chi_M^*;\quad \chi_c=i\sigma_2\varphi_M^*\tag{13}$$

We can then write the Majorana equation in the form corresponding to the Dirac equation (11)

$$\begin{aligned}(E+\vec{p}\cdot\vec{\sigma})\chi-im_a\sigma_2\chi^*=0\\(E-\vec{p}\cdot\vec{\sigma})\varphi-im_b\sigma_2\varphi^*=0\end{aligned}\tag{14}$$

We see that the two equations are now decoupled. Consequently, the mass can have a different value in the two of them, as we indicated.

Notice that for massless particles both in the Dirac and in the Majorana case the two equations decouple and are identical in the two cases. Consequently, there is no difference between the theories. However, nature has chosen to give a mass to all the spin ½ particles and the two theories give different predictions, which we might be able to observe, as we will see. More precisely, the Dirac (4-component) bispinor field is equivalent to two degenerate (m_a=m_b) Majorana fields with, say

$$\varphi_D=i\sigma_2\chi_M^*\qquad\qquad \chi_D=i\sigma_2\varphi_M^*\tag{15}$$

We can now write the 4-component Majorana bi-spinor as

$$\psi_M=\begin{pmatrix}\varphi\\ i\sigma_2\varphi^*\end{pmatrix}\tag{16}$$

Its charge conjugated is

$$\psi_M^C=i\gamma_2\psi_M=\begin{pmatrix}0&-i\sigma_2\\ i\sigma_2&0\end{pmatrix}\begin{pmatrix}\varphi\\ i\sigma_2\varphi^*\end{pmatrix}=\begin{pmatrix}\varphi\\ i\sigma_2\varphi^*\end{pmatrix}=\psi_M\tag{17}$$

We see that the charge conjugate field is equal to the field. Majorana particles are their own antiparticles.

In general, particle and antiparticle have all the "charges" with opposite values and consequently can be distinguished as different objects. Neutrinos and antineutrinos are possibly distinguished by a unique "charge", the lepton number. The latter has not the same theoretical status as the electric charge. Electric charge conservation corresponds to a local symmetry $U(1)$, which governs the dynamics. Lepton number conservation is a global symmetry, which does not govern the dynamics, rather it is a consequence of the dynamic and of the field composition of the Standard Model. If lepton number is not conserved nothing distinguishes neutrino from antineutrino. Neutrino may be its own antiparticle.

As we have seen, neutrino masses are a fraction of an electron-volt. Consequently, in all practical circumstances neutrinos are ultra-relativistic, namely their energy $E>>m$. In such conditions an approximated distinction between Majorana "neutrinos" and "antineutrinos" is possible.

The charged current $\bar{\nu}_{lL}\gamma^{\mu}l_L$ creates a negative chirality ν_l neutrino. If it is mass-less, as assumed in the Standard Model, it is also a pure negative helicity state. If on the contrary it is massive, it is a superposition of a predominant negative helicity component and a small fraction, with amplitude m/E, of positive helicity. Similarly, $\bar{l}_L\gamma^{\mu}\nu_{lL}$ creates a $\bar{\nu}_l$ of positive helicity if massless, with a dominant component of positive helicity and a small component, with amplitude m/E, of negative helicity if it has a mass. We then define the Majorana neutrino as the state of negative helicity and the antineutrino as the state of positive helicity. They interact respectively almost as Dirac neutrinos and antineutrinos respectively (if $E>>m$).

The difference between a Majorana and a Dirac neutrino is the following. The Dirac ν_l neutrino, whether massive or not, when interacting via CC with matter will always produce a charged lepton l. On the contrary, a Majorana ν_l neutrino will produce almost always a charged lepton but, in a fraction $(m/E)^2$ of the times, an anti-lepton $\bar{l}$. However the difference at the usual energies is really extremely small. Consider for example a Majorana neutrino of E=1 GeV and, say, m=100 meV. Interacting via charged currents it will produce an e^-, almost always as if it was a Dirac particle. Differently from a Dirac particle, it will produce an e^+ in a fraction $(m/E)^2=10^{-20}$ of the collisions. These effects are so small that cannot be detected presently.

In conclusion, the Majorana neutrino even if it is not rigorously an eigenstate of the lepton number has, at all practical effects, an effective lepton number $L_{\text{eff}}=+1$. Notice however, that the difference between Dirac and Majorana neutrinos would be spectacular at energies comparable with the mass, below or even around the electron volt, namely energies characteristic of atomic physics.

Notice that a meson can be its own antiparticle both if it is massive, like the π^0, and if it is massless, like the γ, a spin ½ particle needs to be massive.

The fraction of antineutrino in the amplitude (m/E) is not Lorentz-invariant. The same particle that is neutrino ($L_{\text{eff}}=+1$) in a frame in which $E\approx+p_z>>m$, is mainly antineutrino in a frame in which $E\approx-p_z>>m$, hence its effective lepton number is $L_{\text{eff}}=-1$. The fact of L_{eff} not being Lorentz invariant does not mean however that the same collisions will produce mainly electrons for an observer and mainly positrons for another one, because what matters is the neutrino velocity relative to the target, which is independent on the observer.

Consider now the neutrino mass term in the Lagrangian. In the Dirac case we can have the terms $m\bar{\psi}\psi$ and $m\bar{\psi}^C\psi^C$, in the Majorana case the term $\bar{\psi}\psi^C+\bar{\psi}^C\psi$. The Dirac terms are invariant under a global phase transformation, namely $\psi\Rightarrow e^{i\varphi}\psi$, $\psi^C\Rightarrow e^{-i\varphi}\psi^C$, which can be identified as corresponding to lepton number conservation, the Majorana term is not. In conclusion, and as obvious, Majorana particles violate lepton number.

Corresponding, there are physical processes with $\Delta L=\pm2$. One example is the scattering of a negative muon on a nucleus producing a positron, with the matrix element

$$\mu^- + (Z,A) \to e^+ + (Z-2,A) \tag{18}$$

As anticipated the cross section is too small for the process to be observable at present. We are left with the decays, better one decay, the neutrino-less double beta decay. We shall discuss only the principal topics here, referring to a recent review of Avignone, Elliot and Engel[6] for more information.

As it is well known, beta decay is energetically forbidden for even-even nuclei. The open channel is a second order weak process, the two-neutrino double beta ($2\nu 2\beta$) decay, in which two nucleons experience beta decay contemporarily

$$(Z,A) \rightarrow 2e^- + 2\overline{\nu}_e + (Z+2,A) \tag{19}$$

or similar into positrons. The underlying process at nucleon level is

$$2n \rightarrow 2e^- + 2\overline{\nu}_e + 2p \tag{20}$$

The corresponding lowest order diagram at the quark level is shown in Fig. 1a).

Notwithstanding their extremely long values, the lifetimes have been measured for several nuclei, mainly by the NEMO3 experiment[7]. For example, the $2\nu 2\beta$ half-life of ^{76}Ge is 150±10 10^{19} yr, of ^{100}Mo is 0.71±0.04 10^{19} yr, of ^{130}Te is 90±10 10^{19} yr, while for ^{136}Xe only the limit >81 10^{19} yr exists.

If neutrinos are Majorana particles, namely their own antiparticle, the quark level process shown in Fig. 1b) is possible (and similar with positrons), the neutrino-less double beta ($0\nu 2\beta$) decay. The process violates the lepton number by two units.

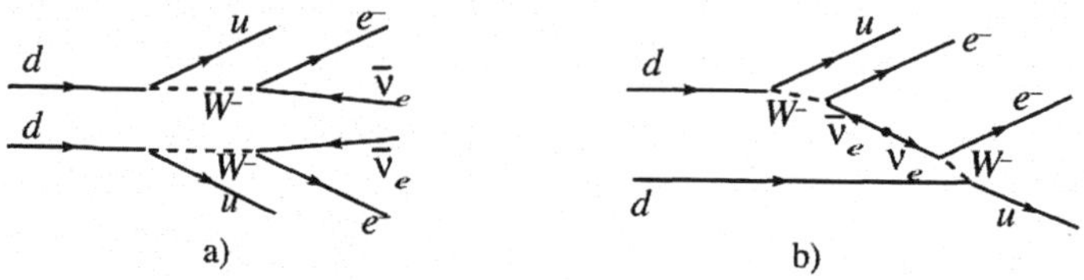

Fig. 1. Lowest order diagrams at quark level for: a) the ($2\nu 2\beta$) decay; the ($0\nu 2\beta$) decay

We can only observe the process at the nuclear level, namely

$$(Z,A) \rightarrow 2e^- + (Z+2,A) \tag{21}$$

The matrix element in Fig. 1a can be reliably calculated in term of a parameter with the dimensions of a mass, called Majorana mass M_{ee} given in eq. (4). Very difficult is on the contrary to calculate the corresponding half-life for the process (21), due to the nuclear physics effects. However, the difference between the initial and final nuclear levels has been accurately measured. Consequently, the sum of the electrons energies, called $Q_{\beta\beta}$, (which is typically a few MeV) is known, usually within a fraction of a keV.

The measured quantity is the half-life. It is related to the Majorana mass by the equation

$$T_{1/2}^{-1} = G_Z\left(Q_{\beta\beta}\right)\left|M^{0\nu}\right|^2 M_{ee}^2 \tag{22}$$

where G_Z is the phase space volume, depending on the nucleus, which is easily calculated and $M^{0\nu}$ is the "nuclear matrix element".

To calculate $M^{0\nu}$, and the similarly defined $M^{2\nu}$, one needs to evaluate the ground state wave functions of the initial and final nuclei and then construct the complete set of states of intermediate nucleus. Approximations and truncations are necessary. Two are the main theoretical techniques: the "Quasi Particle Random Phase Approximation" (QRPA) and the "Shell Model". In the first case the parameter called "particle-particle" coupling g_{pp} is fixed by comparing the similarly calculated $M^{2\nu}$ with the $2\nu 2\beta$ data. Notice however that $2\nu 2\beta$ is a long distance phenomenon with momentum transfer smaller than a few MeV, while $2\nu 2\beta$ happens at 2-3 fm distances (momentum transfers of 100-200 MeV). Moreover, cancellations between large terms take place in the calculation.

Theoretical work has progressed considerably in recent years, the reasons for the different results obtained with different methods have been understood and convergence is increasing, even if not on all the nuclides. In particular, Rodin at al.[8] have published calculations of $M^{0\nu}$ for the relevant nuclides with quoted uncertainties of 20-50%. We shall use these values in the following, if not differently stated. Just to quote four results relevant for the following discussion, for M_{ee}=50 meV, the half-lives are: for ^{56}Ge: 8.6 $\times 10^{26}$ yr, for ^{82}Se: 2.4$\times 10^{26}$ yr, for ^{130}Te: 2.2$\times 10^{26}$ yr, for ^{130}Xe: 4.6$\times 10^{26}$ yr.

With the above quoted numbers, it is immediate to calculate that for M_{ee}=50 meV, one expects in 1 t detector of ^{76}Ge enriched at 86% (the natural one is only 8%) 4 events per year, in 1 t of natural TeO_2, 3 events per year and in 1 t of ^{130}Xe enriched at 85%, 4.3 events per year. According to eq. (22) the count rate scales with the square of M_{ee}.

Fig. 2 shows the Majorana mass, calculated with the known values of the mixing angles, as a function of the lightest neutrino mass. The experiments under construction and design aim to explore the region down to about M_{ee}=100 meV.

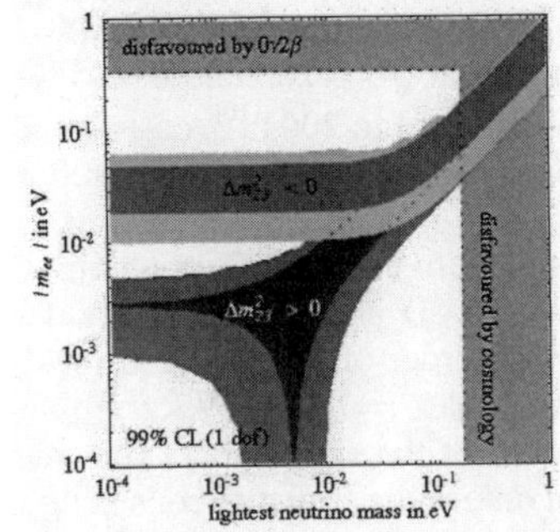

Fig. 2. Majorana neutrino mass vs. lightest neutrino mass [9]

3. The experiments

The sensitivity of an experiment to the lifetime τ in presence of a background rate b is proportional to the factor

$$F_\tau \propto \left(\frac{MT}{b\Delta E}\right)^{1/2} \tag{23}$$

where M is the sensitive mass, T the exposure time and ΔE the energy resolution. Consequently, the sensitivity to M_{ee}, which is inversely proportional to the square root of the lifetime, scales as the fourth root of the sensitive mass in presence of constant background. This implies that the background must be reduced substantially to zero in the region of interest, namely an energy interval of the order of the energy resolution (FWHM) around the known position of the signal. In these conditions the sensitivity to M_{ee} scales as the square root of the exposure, $F_M \propto \sqrt{MT}$.

Schematically one can divide the experiments in two classes: calorimetric and tracking experiments. In a calorimetric experiment source and detector are the same object, in practice Ge diodes (as in GERDA and MAJORANA experiments) or bolometers, as the TeO_2 of CUORE. The total energy of the two electrons is measured and a peak at $Q_{\beta\beta}$ is searched for. Advantages are that the detector materials can be made extremely clean and that the energy resolution is extremely good, a few keV FWHM. Tracking of the two electrons gives an additional handle for background rejection, which is however paid with an inferior energy resolution. Both gas (NEMO3, superNEMO and NEXT) and liquid (EXO) can be used as tracking media.

Even after eliminating all the backgrounds external and internal to the detector the irreducible background due to the $2\nu 2\beta$ decay remains. Near the end-point of the two-electron kinetic energy K_{ee} the $2\nu 2\beta$ decay spectrum is proportional to $\left(Q_{\beta\beta}-K_{ee}\right)^5$ and consequently the fraction of $2\nu 2\beta$ events in the signal region, namely a ΔE after the end-point, is

$$F \simeq \frac{7(\Delta E)^6}{m_e Q_{\beta\beta}^5} \tag{24}$$

and the signal to background

$$\frac{S}{B} \simeq \frac{m_e Q_{\beta\beta}^5}{7(\Delta E)^6} \frac{T_{1/2}^{2\nu}}{T_{1/2}^{0\nu}} \tag{25}$$

No compromise on energy resolution is tolerable.

The most sensitive experiment so far has been the Heidelberg-Moscow (HM) experiment at the INFN LNGS Laboratory. Using enriched Ge diodes, the experiment reached the lowest background index, b=0.11/(keV kg yr) in an exposure of MT=71.7 kg yr and energy resolution ΔE=3.27 keV. The signal is expected at Q_{bb} = 2039.006±0.05 keV. A subset of the collaboration reported a positive evidence at 4.2σ level in 2004[10], corresponding to 100 meV<M_{ee}<900 meV, depending on the assumed nuclear matrix element (about 700 meV with that of Rodin et al.[8]). IGEX[11], of a little smaller sensitivity, gives the upper limit M_{ee}<330-1350 meV.

There are presently two projects with enriched Germanium, GERDA[12] under construction at LNGS and MAJORANA[13] in its R&D phase in the USA. I shall concentrate on the former. Aiming to a final sensitivity of the order of one hundred meV, GERDA needs an exposure of about 100 kg yr and a background index b=10^{-3}/(keV kg yr), i.e. two orders of magnitude better than now. Fig. 3a) is an artist view of the experiment. The shielding of external environmental backgrounds is realized using an onion-like structure consisting in, from outside to inside: hyper-pure water in a 10 m diameter S/S tank shown open in Fig. 3, a S/S cryostat with internal lining of hyper pure copper and ultra-pure liquid Ar. The Ge crystals will be immersed naked in the liquid Ar[14].

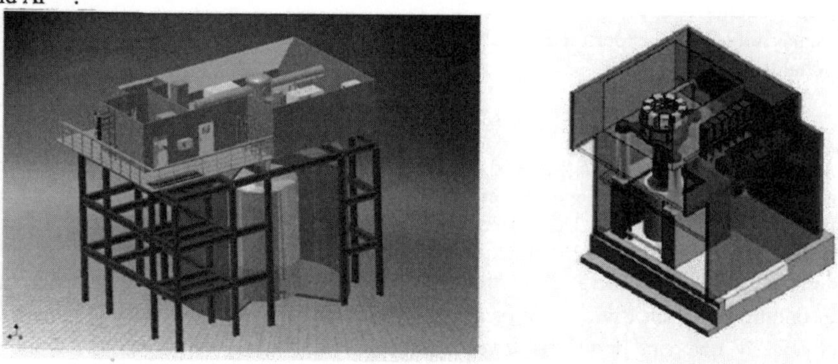

a) b)

Figure 3. Artist view of a) the GERDA experiment, b) the CUORE experiment, both at LNGS

The experiment is foreseen in two phases. In phase-I the existing detectors of HM and IGEX (about 15 kg) will be used. They are expected to contribute with about 10^{-2} events/(kg keV yr), due mainly to their intrinsic cosmogenic ^{60}Co contamination. This will dominate the background budget. However, this level will be enough to reach in one year exposure the sensitivity of $T_{1/2}$=2

$\times\ 10^{25}$ y, corresponding to $M_{ee}\approx$ 400 meV and confirm or refute the Klapdor et al. evidence. Notice that this check needs to be performed on the same isotope, due to the uncertainty on the nuclear matrix elements.

In phase-II about 20 kg of new, segmented, detectors will be added aiming to a background index b=10^{-3}/(keV kg yr). If this is achieved, an exposure of 100 kg yr will give a sensitivity of $M_{ee}\approx$ 140 meV. To reach the required background index, no component of its budget should be larger than, say, b=10^{-4}/(keV kg yr). The method followed by GERDA, as by other similar experiments, to this scope is the following. All background sources in every component of the apparatus are calculated by analytical methods (for example the cosmogenic activation of ^{60}Co and ^{68}Ge during the detector production) or Monte Carlo simulations. These calculations provide the allowed specific activity for each component, depending on its location relative to the detectors. μ-induced prompt signals are eliminated with two vetoes, a plastic scintillator roof and the water used as Cherenkov medium. All materials and components (of the suspensions, the front-end electronics, the cabling, etc.) are screened, by measuring by γ-spectroscopy and by Inductively Coupled Plasma Mass Spectroscopy. Mechanical design is finalised to minimum mass suspension and contact system for the Ge detectors. Special effort is dedicated to, low mass, ASIC and front-end electronics, taking into account its relatively large contribution to the background budget, presently of a few 10^{-4} /(keV kg yr). Geometrical localization and pulse shape analysis will be used to discriminate the backgrounds intrinsic to the Ge diodes. Indeed, the two electrons of the signal deposit their energy within a few millimetres, whilst the gammas, which constitute the principal background, do that with a few Compton scatterings separated by a few centimetres. Consequently, the signal events are "single site" (SS); background events are "multiple site" (MS). SS can be distinguished from MS with pulse shape analysis and anticoincidence between diodes and between segments of the detectors.

The bolometric techniques, proposed by E. Fiorini and T. Niinikovsky in 1984[15] extract the energy deposited by the two electrons in a crystal as heat by measuring the corresponding temperature increase Δt. The crystal is weakly coupled to a heat bath, whose temperature must be very low to reduce the specific heat that is proportional to $(T/T_D)^3$. Measurable Δt of a few millikelvin can obtained. The technique has been used by CUORICINO, which has just finished, and will be used by CUORE, both at LNGS with TeO_2 crystals. The natural abundance of the interesting isotope, ^{130}Te is 34% that is large enough to allow working with natural Tellurium. The high $Q_{\beta\beta}$=2529 keV is an additional advantage.

CUORICINO was built as a "tower" of several planes, each made of 4 detectors of 5×5×5 cm^3 and a mass of 790 g, for a total of 15.53 kg of ^{130}Te. Running under stable conditions from April 2003 to July 2008 achieved the energy resolution (FWHM) ΔE=8 keV and a background index b = 0.18±0.02 ev/(kg keV yr). No signal was detected in its 11.8 kg yr exposure providing the limit on the half-life $T_{1/2}$>3.1 10^{24} yr at 90% c. l. corresponding to M_{ee}<200-980 meV, where the large uncertainty depends on the nuclear matrix element[16].

CUORICINO is also the prototype of CUORE (Fig. 3b), which will consist on 988 detectors arranged in 19 CUORICINO-like towers (13 planes 4 detectors each) for a total mass of 741 kg, of which 203 kg of ^{130}Te. By reducing the background index to b=10^{-2} ev/(kg keV yr) and with an energy resolution ΔE=5 keV CUORE claims a sensitivity of M_{ee}=20-100 meV (depending on the nuclear matrix element) in five year running.

As in all the double beta experiments the crucial issue is the background. A Monte Carlo model has been developed on the basis of the experimentally tested CUORICINO background model. The contributions to the background budget due to radioactive traces in the detector materials, in the shielding and other components of the set-up, the neutrons from the rocks and those induced by muon interactions are all below 10^{-3} ev/(kg keV yr), according to the

Montecarlo calculations. These shows also that the background due to contamination of the crystal surfaces should be < 3 $\times 10^{-3}$ ev/(kg keV yr). The main problems come from the surfaces of the hyper-pure Cu structure facing the detectors, giving a contribution of 20-30 $\times 10^{-3}$ ev/(kg keV yr). The background model developed by CUORE interprets that as due to 100 pg/g of ^{238}U and ^{210}Pb on 10 μm depth of the Cu surface. There is here a diagnostic problem, to reach the sensitivity necessary to measure such very low level of radioactive traces. Both passive (mechanical polishing, electro-polishing, chemical etching, sputtering in UHV, etc.) and active (surface sensitive bolometers, measure heat and light) strategies are being developed to control the surface backgrounds.

We notice that the surface contribution to the background budget are irrelevant for the Ge diodes, but dominant for the bolometers. This fact is understood by noticing that the mass of one of the former detectors is about ten times larger than that (with reference to ^{230}Te) of one of the latter. This is an example of the extreme importance of the surface/bulk ratio.

The most sensitive experiment using the tracking technique is the already mentioned NEMO3 running since February 2004 at the Laboratoire Subterrain de Modane (LSM). The sources of the relevant isotopes are foils of square-metre order area. The two electrons are detected and tracked in drift gas chambers, operated in Geiger mode, at the two sides of the foil. Their energy is measured by plastic calorimeters at the end of the tracking volumes. A magnetic field of a few mT allows determining the charge of the track. An advantage of the technique is the possible use of a variety of source isotopes, which is exploited by installing 6.9 kg of ^{100}Mo ($Q_{\beta\beta}$=3034 keV), 0.9 kg of ^{182}Se ($Q_{\beta\beta}$=2995 keV) and several grams of ^{116}Cd, ^{96}Zr, ^{150}Nd and ^{48}Ca.

The technique, once the Rn contamination has been controlled, allows to reduce the background in the $2\beta 2\nu$ signal almost to zero. Consequently, NEMO3 produced beautiful results for this decay. The sensitivity to $2\beta 0\nu$ is limited by the $2\beta 2\nu$ which is relevant due to the poor energy resolution, ΔE=140-170 keV at 1 MeV. Two are the principal contributions to ΔE: the thickness of the foil and the corresponding degradation of the electrons energy and the resolution of the calorimeter.

Based on the NEMO3 experience, the superNEMO detector is presently in its R&D phase. It is conceived as a modular structure with 20 modules each containing a 5 kg source foil of ^{82}Se (or ^{150}Nd, if the enrichment of the latter will be practically feasible). The ratio $T_{1/2}^{2\nu} / T_{1/2}^{0\nu}$ is large for these isotopes. The expected efficiency for $2\beta 0\nu$ is limited to 20%. The experiment will be very large and will need a dedicated hall of 15×30 m^2.

The contamination of the very large surfaces should be drastically reduced: for ^{208}Tl from the NEMO3 level of < 20 μBq/kg to <2 μBq/kg and for ^{214}Bi from < 300 μBq/kg to <10 μBq/kg. A special detector, called BiPo, is being developed to measure these levels that are beyond the sensitivity of the standard techniques. Reaching these unprecedented limits of surface cleanliness will however not be sufficient. The energy resolution must also be substantially improved to a design figure is $\Delta E/E$=4% at 3 MeV. To this aim the thickness of the foils will be reduced to 40 mg/cm^2 and the calorimetry will be substantially improved. The best figure obtained till now is ΔE=8% at 1 MeV. At the present level of energy resolution, even in presence of the only $2\beta 0\nu$ background the limiting sensitivity of the experiment is M_{ee}=100-150 meV.

Differently from NEMO3 and superNEMO, in which source, tracking and calorimeter are separate elements, the use of time projection chamber (TPC) techniques may allow building a single device covering all these functions. I will discuss the EXO and the NEXT projects based on liquid and gas Xe TPC respectively. As TPCs reconstruct the events in three dimensions, it is possible to define a central cleaner fiducial volume and a surrounding shielding volume, possibly of the same fluid, by geometric cuts at the analysis level. The absence of any physical surface in between avoids the dangerous background surface sources. The energy deposit of charged particles in the TPC fluid is a complex process that must be completely understood to control the energy resolution[18]. We only mention here that only part of the energy is deposited as ionisation and moreover a fraction of the ion pairs, depending on the applied electric field, recombine in the column. At the end, a fraction of the energy appears as scintillation light. Especially at the low energies of the double beta decay there are large fluctuations event by event in the fraction of energy going to ionisation charge and to scintillation light. Consequently, both should be measured.

EXO[19] is a phased programme aiming to 1 to 10 t enriched (80%) ^{130}Xe sensitive mass. An energy resolution ΔE=3.3 % FWHM has been obtained in prototype detectors by measuring ionisation and light. 200 kg of enriched ^{136}Xe are already available to the Collaboration and a liquid TPC, EXO-200, is under construction in the WIPP underground facility in the USA. The engineering run with natural Xe, is planned for 2009. A physics run of two years will follow. With 70% efficiency and with 40 expected background events the sensitivity of M_{ee}>170 meV can be reached.

A potentially powerful element that has not been employed so far is the tagging of some characteristic element of the final state. Both EXO and NEXT have such a perspective in their R&D programs.

The idea that EXO is developing since several years is the Ba-grabbing & tagging technique, proposed by Moe[20] in 1991 and discussed by Danilov et al. in 2000[21]. The double beta decay of the Xe leads to a Ba doubly ionised atom

$$^{136}\text{Xe} \rightarrow ^{136}\text{Ba}^{++} + e^- + e^- + 2\nu_e \tag{26}$$

The process is immediately followed in the liquid by the capture of one electron

$$^{136}\text{Ba}^{++} + e^- \rightarrow ^{136}\text{Ba}^{+} \tag{27}$$

The singly ionised Ba is stable and one can try to extract it from the liquid into the gas phase, which is above the liquid, by using a suitable electrode. Its relevant atomic level structure is shown in Fig. 4. With an adjusted LASER beam one can optically pump the ion to the excited level $^2P_{1/2}$, from which it will decay to $^4D_{3/2}$. The atom is pumped back to the $^2P_{1/2}$ with a tuned LASER beam. The ion decays again, and again is pumped up. The process can continue again and again, obtaining up to 10^7 photons per second, which can be easily, in principle, detected.

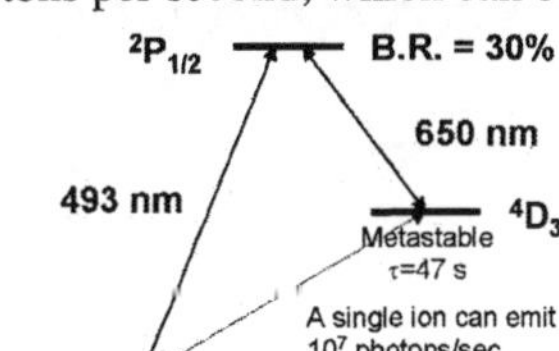

Figure 4. The energy levels of Ba$^+$

More recently a complementary approach has been developed in a proposal to the Laboratorio Subterráneo de Canfranc (LSC). The NEXT[22] proposal is based on a TPC containing an high pressure enriched Xe gas (an option that is being studied also within the EXO collaboration). The tag in this case is the topology, geometry and ionisation, of the tracks. The simulated example in Fig. 5 shows in a colour code the specific ionisation along the tracks, with two characteristic blobs at their ends. A rejection power against the dominant background of Compton electrons of two orders of magnitudes is expected by this topological signature.

The NEXT proposal has been approved by LSC and funded by the Spanish CONSOLIDER programme. Is presently in its initial R&D phase. The design criteria are the following:

- Avoid charged background from surfaces by eliminating them, base the design on 100% active, completely closed virtual fiducial surface.
- Obtain fine topological information.
- Obtain 1% energy resolution FWHM. This a crucial demand in order to suppress the $2\nu 2\beta$ background. Different alternatives are presently under study.
- Evaluate by Montecarlo the tolerable radioactive contaminant in the materials and screening

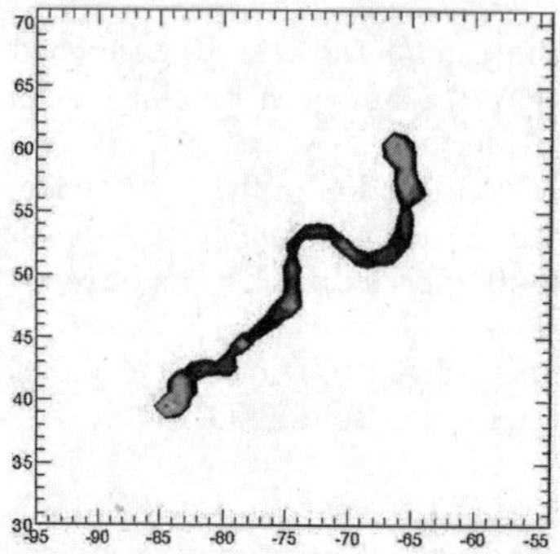

Figure 5. Simulation of two electrons tracks in NEXT

Before closing, it is instructive to try to answer to the following questions. Why is neutrino spectroscopy so difficult? Why establishing the nature of neutrinos is so demanding? The answer to both, apparently uncorrelated, questions is the same. The effects are extremely small because we work at neutrino energies much larger than their masses. They should be much more important at sub-electronvolt or even at electronvolt energy scales. This is the domain of atomic physics. Recently, along these lines, Y. Yoshimura[23] has proposed the "LASER Irradiated Neutrino Pair Decay" technique, illustrated in Fig. 6

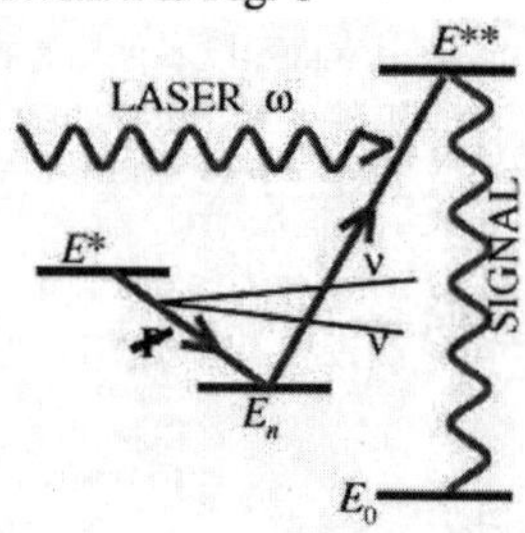

Figure 6. Two neutrino assisted

The metastable ($\tau \approx 1$ s) energy level I^* of energy E^* is initially populated by pumping. The electromagnetic transition to the level I_n of energy E_n is forbidden by parity conservation. It may happen via weak interactions with emission of two neutrinos (if Majorana or a neutrino-antineutrino pair if Dirac). The inverse spontaneous life is extremely long, namely

$$\Gamma_{\nu\nu} \propto \frac{G_F^2 (\Delta E)^5}{15\pi^3} : \ 3.3\times 10^{-34} \ \text{s}^{-1}\left[\Delta E / \text{eV}\right]^5 \qquad (28)$$

It can be strongly enhanced however by forcing the transition from the level I^* to the level I^{**} (the level of energy E^{**}) with an intense electromagnetic field, a LASER beam, in resonance.

Calling ω the LASER frequency, $\Delta\omega$ its band width, P its power and σ the absorption cross section, the rate of the transition $\gamma_{LASER} + I^* \rightarrow I^{**} + \nu_i + \nu_j$ is

$$\Gamma_{I^*I^{**}} = \Gamma_{\nu\nu} \times \frac{\omega}{\Delta\omega} \times \tau \times \omega P \times \sigma \tag{29}$$

Let us evaluate the orders of magnitudes. With P=1 W mm^{-2}, $\Delta\omega/\omega$ =10^{-9}, ω= 1 eV, τ = 1s we have $\Gamma_{I^{**}I^*} \approx 10^{-21}$ s^{-1} and hence a rate of 1 event/day. This can be enhanced by a factor 10^3-10^4 by employing a resonant cavity. We see that the idea is not obviously impossible. There are however many important elements to consider before claiming this as a viable technique. But it is certainly worth to try: if it works one can measure directly the absolute neutrino mass spectrum by searching for the energy thresholds for production of the different ν_i ν_j pairs and determine the nature of neutrinos from the behaviour of the excitation function as a function of frequency just above threshold.

References

1. G. L. Fogli et al. hep-ph/0805.2517 and references therein
2. C. Kraus et al. *Eur. Phys. J.* **C40** (2005) 447
3. V. M. Lobashev et al.; *Nucl. Phys. B Proc. Suppl.* **91** (2001) 280
4. E. Komatsu et al.; arXiv:0803.0547v2 [astro-ph]
5. E. Majorana; *Nuovo Cimento*, **5** (1937) 171
6. F. Avignone, S.Elliott and J.Engel, Rev. Mod. Phys. **80** (2008) 481
7. R. Arnold *et al.*, Phys. Rev. Lett. **95** (2005) 182302.
8. V. A. Rodin, et al.; Nucl. Phys. **A 766** (2006) 107; **A 793** (2007)
9. F. Feruglio, A. Strumia, F. Vissani; *Nucl. Phys.* **B637** (2002) 345
10. H. V. Klapdor-Kleingrothaus et al.; *Phys. Lett.* **B586** (2004) 198 and references therein
11. C. E. Aalseth et al. Phys. Rev. **D** 73 (2006) 053004
12. I. Abt et al. LNGS P38/04; hep-ex /04040390 and http//www.mpi–hd.mpg.de/GERDA/proposal.pdf
13. R. Gaitskell *et al.* (2003) nucl-ex/0311013.A
14. G. Heusser et al., Ann. Rev. NPS **45** (1995) 54; H.V. Klapdor-Kleingrothaus et al, GENIUS-Proposal, LNGS-LOI 9/97 1997
15. E. Fiorini and T.Niinikovsky; Nucl. Instrum. and Methods **224** (1984) 2139.
16. C. Arnaboldi et al. (CUORICINO) arXiv:0802.3439v1 [hep-ex]
17. A. S. Barabash; Phys. At. Nucl. **67** (2004)1984
18. D. Nygren "0ν double beta decay, WIMPs and Xenon. Will the search converge?" talk at NEUTRINO 2008
19. R. Neilson *et al*; *J. Phys.: Conf. Ser.* **136** (2008) 042052 and references therein
20. M. K. Moe; Phys. Rev. **C 44** (1991) R931
21. M. Danilov et al.; Phys. Rev. **C 62** (2000) 044501; Phys. Lett. **B 480** (2000) 12.
22. F. Nova et al. (2008) EOI-08 NEXT
23. M. Yoshimura hep-ph/0611362

CHAIRMAN: A. BETTINI

Scientific Secretaries: P. Hoehn, C. Maiano

DISCUSSION

- Wen:

You have mentioned that 0νββ decay may be the only way to answer the question of whether a neutrino is Dirac or Majorana. But as I understand, a neutrino magnetic momentum measurement could also verify if a neutrino is a Majorana or Dirac particle. Many 0νββ decay experiments were mentioned in the talk, but could you also comment on neutrino magnetic momentum measurement?

- Bettini:

Right now I do not really know how big the contribution of the Majorana term is to the magnetic moment of the neutrinos, however it is too small to be measurable. [After the discussion I checked that it is expected to be $\mu_\nu \sim 3\times10^{-20}\mu_B\left(m_\nu / 100 \text{ meV}\right)$, where μ_B is the Bohr magneton. This is 9 orders of magnitude below the present bounds. Notice that other "new physics" may give much larger contributions; see A. Strumia and F. Vissani hep-ph/0606.054 and references therein]
In general there are many processes which in principle could establish if neutrino is a Majorana particle or not. In practice this depends on the expected size of the observable, decay rate, branching ratio, cross section, etc. For example, the Majorana contribution to the branching ratio for the decay $\mu \to e+\gamma$ mentioned this morning is expected to be of the order of $\left(m_\mu M^M_{e\mu} / m^2_W\right)^2 \sim 10^{-50}$, 35 orders of magnitude lower then the sensitivity of the next generation of experiments, like the one mentioned by Prof. Oddone. Unfortunately, with the possible exception of the atomic physics experiments, neutrinoless double beta decay is the only process that can be experimentally reached.

- Schwarz:

Based on what you told us, it seems to me that detection of neutrinoless double beta decay requires an inverted mass hierarchy. Is that correct?

- Bettini:

The experiments being built now will be sensitive to 50-100 meV. As such, they will be able to reach almost the lower limits of the inverted mass hierarchy range. If no signal is found, the inverted hierarchy will be excluded. However, if we are lucky and an effect is established, this does not distinguish between the two hierarchies.

- Ni:

How can an experiment for the measurement of the neutrino mass by laser atomic levels be sensitive to a neutrino mass of 10 meV if the atomic level difference is around 1 eV?

- Bettini:

First of all let me repeat that I am not completely convinced that this class of experiments is feasible. It is a very interesting perspective that should be further studied. These are two examples (a transparency was shown) where you see that the levels are of the order of less than 1eV, one or two orders of magnitudes larger than neutrino masses. This factor may look large, but it is enormously smaller than the 7-8 orders of magnitude of the double beta decay. From the quantitative point of view, one can estimate rates of the order of one event per day, with the available LASER powers in the processes I mentioned this morning. It appears that 3 or 4 orders of magnitude can be gained by enclosing the target atoms in an optical cavity, tuned on the LASER frequency. In this way one obtains an effective multiple pass of the LASER beam.

- *Pacholek:*

Is it possible to construct a Majorana mass term for the neutrino, which has a non-zero weak hypercharge?

- *Bettini:*

This is a very theoretical question, I do not know. You probably have in mind some new model. What do you mean exactly?

- *Pacholek:*

I know there is the possibility of the so-called sterile neutrinos which do not have any charge so it is not a problem to construct a Majorana mass term for those neutrinos. But what if we would like to construct a Majorana mass term for non-sterile neutrinos?

- *Bettini*:

Sterile neutrinos are a theoretical hypothesis. We do not know if they exist or not. The existent neutrinos might be Majorana particles or Dirac particles. We do not know. If you build a theory you need to represent what you see in nature. When you build the SM you get to give the neutrino and the other particles the quantum numbers that are correct to interpret the phenomena you see. But perhaps you can build a novel theory in which the answer is yes.

- *Baldini:*

You presented many experiments which look for $\beta\beta0\nu$ decay. Which, according to you, is the most promising one?

- *Bettini*:

By the way, I did not present all the experiments and proposals. They are in different stages of evolution. The closest to being operational are GERDA and CUORE at Gran Sasso and EXO. Cuoricino, which was the most sensitive experiment, was just finished. It is also a prototype of CUORE, which is presently under construction. The construction of GERDA will be completed by the end of this year. The sensitivity of GERDA will be around 140 meV, that of CUORE will be somewhat better, 100 meV or even less, due to the larger source mass, provided the design background level is reached. In the USA, EXO is constructing its engineering set-up. The collaboration has already the 200 kg of enriched LXe needed for the physics phase. Its sensitivity should be similar to that of CUORE. Notice that, due to the uncertainty in the nuclear matrix elements, searches with different isotopes are mandatory. SUPERNEMO is a proposal which is still in a R&D phase, aiming to the same level of sensitivity.

- *Maiano:*

1. In your opinion, why should the neutrino be the only Majorana particle among the Standard Model elementary particles? Do you think this is only due to the fact that the neutrino is the only electrically neutral elementary particle?

2. If $\beta\beta0\nu$ experiments will not find evidence for a 0ν peak, how can one understand if this is due to cancellations in the phases or if it is due to a Dirac nature of ν ?

- *Bettini*:

1. Neutrinos are the only particles in the SM which have the opportunity to be Majorana particles, because they are neutral and possess spin ½. Outside the SM, the neutralinos of SUSY are Majorana particles.

2. Unfortunately, as you have seen this morning, the cancellations due to the phase factors may be complete or almost so in the case of "normal" mass hierarchy. Consequently, if the experiments will discover neutrino-less double beta decay, the Majorana nature of neutrinos will be established. Viceversa, if nothing will be found, we will not know whether they are Majorana or Dirac. There might be another way via atomic physics, however.

- *Aprile:*

But we have the KATRIN experiment starting in 2010. What if it finds the signal?

- *Bettini*:

KATRIN will be very important because it is a laboratory experiment that will provide a direct measurement as opposed to cosmology, which gives us indirect information. However, the 200 meV KATRIN sensitivity in the "electron neutrino mass" parameter is already excluded by cosmology, even in the must conservative hypothesis.

- *Aprile:*

Cosmology is only an upper limit, they are improving the sensitivity which according to your presentation is currently 600 meV.

- *Bettini*:

Yes it is 600 meV on the sum or, equivalently, 200 meV on each mass. One can extract from cosmology limits that are more or less stringent depending on the number of pieces of information used. Obviously, more stringent limits are also more model dependent. The limit I quoted is the most conservative, being based only on the CMB spectrum from WMAP5 and the Baryon Acoustic Oscillations peak. As you say, the sensitivity to neutrino mass of cosmology is improving. I add that it might be close to find a positive evidence. In this case, the present generation of experiments may well be able to find the double beta decay signal. However, this is not completely granted if cancellations are maximal.

A. DESHPANDE

"The Anomalous Spin Distributions in the Nucleon"

The Slides of the Lecture can be found at:

http://www.ccsem.infn.it/issp2008/

CHAIRMAN: A. DESHPANDE

Scientific Secretaries: S. Dubynskiy, S. Shakirov

DISCUSSION

- *Tannenbaum:*

On your slide 18, you had no Q squared in any of your formulas. Why not?

- *Deshpande:*

Oh, it should be there. That is assumed.

- *Tannenbaum:*

Do you have any measurements of this fact, as a function of Q squared?

- *Deshpande:*

Delta sigma, as a function of Q squared? Yes, there are measurements, and there are also fits to that data. So, if you look at the 1998-1999 papers by the experimentalists as well as the theorists, they surely give you evolution of delta sigma as a function of Q squared. It is really similar to alpha-s evolution.

- *Tannenbaum:*

And then I had another question. You have shown us these beautiful cross-sections measurements, thank you, and then you have said they agree with the Standard Model. So we can believe the asymmetry measurements. However, it is even more than that, because one can do the same for the DIS measurements.

- *Deshpande:*

Oh, absolutely, that's basically what we've tried to convince all collaborations to do. Precisely because of the fact, that we don't believe that the DIS cross-sections have no higher-order corrections.

- *Tannenbaum:*

No, but you can not do this with the DIS, because you are measuring the structure functions with DIS. The beauty of the hardronic measurements is that they prove, that the structure functions that you measure in DIS do work -- and that is what makes you believe the DIS measurements. You understand? That is the point!

- *Deshpande:*

Ok. That is fine. If you wanted to make that point to me, you have made it. But I think, the present difficulty is that simply the DIS collaborations have never done this. They have assumed that what they are measuring is correct, the asymmetry requires no other basis. It is only when the asymmetry measurements are being interpreted using the perturbative QCD, that is where we get in to this problem of establishing that pQCD works in that kinematic region.

- *Nicotri:*

May be this is not a right question to ask at this moment, but can you please explain, what is the link between the magnetic levitating top that you have shown at the beginning of the lecture and the rest of the lecture, please?

- *Deshpande:*

Oh, I simply wanted to get your attention. The spinning, the rotation of the top, is non-trivial although it appears trivial but indeed it is unexpected.

- *Ni:*

This is just a simple technical question. You have shown that you have polarized the beam with like 40% - 60% of efficiency. How did you actually maintain the polarization efficiency?

- *Deshpande:*

Oh, that is a good question. It is a very technical question, it should be answered in terms of accelerator physics. We start with the source, which is a polarized hydrogen source. We have a polarization which is, in fact, of order of 85%. You have to start with the highest possible polarization and try as hard as you can to maintain it. At the end of the linac we believe we have measured the first step at 200 MeV, where the polarization drops to 82%, starting from 85%. We know, how to preserve it, that is really nicely understood. When you go into the post-circular accelerator that is called the booster that makes it from 200 MeV to 2 GeV, there are no depolarizing resonances. What is a depolarizing resonance? In a collider, that is this vertical magnetic field (holding field), if the precession frequency becomes integer multiple of the rotation frequency, then there is a resonance. And this resonance destroys polarization. So you have to somehow maintain polarization through these resonances. It turns out, that up to 2 GeV there are no serious resonances for a proton. A resonance frequency depends on the mass and the energy that you go -- at 2 GeV there are not too much. Therefore, at the end of a booster, we still see about 75% or 80% of polarization. Then, from 2 GeV to 24 GeV, in the AGS is our maximum loss of polarization. Now, I have told you about these Siberian Snake magnets. Those are the magnets that help us to maintain polarization. Those magnets, ideally what they are -- imagine a dipole magnet then hold one side and twist it. The proton goes through on one side, pointing upwards, and comes out from the other end 180 degrees rotated. That is the concept of a Siberian snake magnet. This is a full Siberian snake magnet, not all snakes are full. To have a full snake, you need long length, depending on how much energy you have, because you turn/rotate a proton-spin all the way. Enough straight sections are needed in the accelerator ring, to put these Siberian snakes which we absolutely need to maintain polarization. At the end, at 24 GeV, we have a decrease from about 75% to about 65-67% polarization. Then we transfer the beam to RHIC, where from 24 GeV to 100 GeV we essentially see a few percent loss. As you see, the point is not in creating polarization, the point is not loosing it.

- *Patterson:*

You have mentioned the quasi-simultaneous detector scheme with polarized target. What about the systematics, which comes from the sign-dependent beam effects? Is there any significant beam-related polarization dependent systematics?

- *Deshpande:*

The one most important thing that is beam-related, and that affects our measurement, is not what you have mentioned. The accelerator does not change the beam spot from bunch to bunch to have a significant effect, however, what changes in the bunch structure of the beam is the flux, the number of protons from bunch number one to bunch number two can vary up to about 10 - 20% and we have seen that. We have 120 bunches or so in the RHIC, and bunch number one rom one ring collides to bunch number one of the other, bunch number two to bunch number two.... And so on. This is where the advantage of our bunch trains with our alternating spin orientations comes into picture. At the end of the experiment, on the average, you should not see any large difference. It is good to have as small as possible deviation in the population from bunch to bunch, that is true, but you can not completely get rid of it at this level. So 10 - 20% is what we see, and then we try to lower it by rotating them often. And then there is one more technique, which is a slightly different accelerator technique, called spin flipping. Suppose you have beam number one, with plus-minus-plus-minus-plus-minus, the other has a double structure like plus-plus-minus-minus-plus-plus. Bunch number one collides with bunch number one, ... And this goes on at the experiment. Now what I do is to flip all bunches in the beam number one, which means that what was plus-plus-minus-minus- becomes minus-minus-plus-plus. If I can do that, only on one beam, then I am automatically addressing your issue. Now you, instead of the same two bunches, with (say) 50 and 49 protons, this was plus and this was minus, this becomes plus and this becomes plus (hand-gesture). Now I can do it once more, this becomes minus and this becomes plus (more hand-gestures). And I can do it one more time, this becomes minus and this becomes minus. This is called spin-flipping, and this is another difficult technique, we are going to try it next year. We have tried it in the past, but in some other more complicated way, where at the interaction point it tried to change the orientation of both beams together, and it did not work. What it will do to polarization - we do not do yet. Our assumption, or hope, is that we will be able to overcome even that technique. As you see, that is the biggest thing that we worry about. That is called relative luminosity uncertainty, that comes from the fact that the bunches are not equally populated. That is the largest spin uncertainty in the systematic that we know of. It is now of the order of ten to the power minus four.

- *Nikiforov:*

For the future experiment, you have said that there are some plans to use low energy to have really low four-momentum transfer to measure just the orbital momentum of the quarks, not to have the influence of the sea and the gluons. Why is it not possible to use the measurement of very high x, where both the sea and the gluons do not give a contribution, so that we have just valence quarks.

- *Deshpande:*

As long as you don't break the proton, you are fine. What you should want to do is really not to break the proton, that is the crucial thing. The final state has the scattered electron from the initial state, the direct photon or meson, and the proton which remains intact, ok? Only under those conditions this physics is valid. If you do it at high x, this is possible and fine, and people do that. This is perfectly fine to do. As long as the proton is detected in the final state. It is important. But it is difficult - the cross section will be very small.

- *Susskind:*

Are you not going to explain us how has the toy worked?

- *Deshpande:*

Let's have some wine this night and then we will discuss that.

- *Susskind:*

Oh, I can explain that even without blackboard.

- *Zichichi:*

We have a very old system, which uses an electromagnetic blackboard if you want to write something.

- *Deshpande:*

Go ahead...Come here.

- *Susskind:*

Ok. Do I have to write down an explanation? Or part of it?

- *Zichichi:*

Of course.

- *Susskind:*

I have no idea if this is the way the thing actually works or not. At some point during the talk I have been coupled for couple of minutes trying to figure out how it might work. Are we talking now about something which is levitating? Here is the way you could make a levitator, I think. Let me start with a disk. I have a disk, from material, and in the center of the disk it is polarized, so at the center of the disk I have a plus charge and on the outer boundary of the disk I have a bunch of minus charges. Obviously the thing was spinning, I saw it was spinning, have you?

- *Deshpande:*

The top or the bottom? It was the box and then it was the top which was spinning.

- *Susskind:*

The top is spinning. Minus charges on the outside plus charges on the inside spinning around fast. And there is a magnetic field. So let's assume the magnetic field has components outward like that. Than what happen? There is ....because of B field v x B force. Let's look at the point over here. There is no force on the plus charge, because it is not moving, it is in the center.

- *Deshpande:*

Excuse me. The magnetic field lines are vertical.

- *Susskind:*

Magnetic lines of what? There is a magnet in the bottom and we'll come to it in a moment. So this thing is spinning and if the outward components of the magnetic field are over here then this moving negative charges I made the sign wrong but moving negative charge in a magnetic field exerts a force which is either vertically upward or downward and if I have my right hand rule I don't remember. So let's make it upward. So that thing is spinning, and if there is such a magnetic field than the thing can levitate. OK, so how do you make such a field? Only what you have to do to put a magnetic monopole over here. Magnetic monopole has lines of force, which have components that go outward. Of course if you do not want to put magnetic monopole over there you may put a magnetic bar with North and South poles. Now I am assigning this as a problem for the students to figure out if this thing is stable or not. What happens if you do not swap the axes? Is it stable? Does it have good pull back? Or does it get outward? That is the problem for the students. I worked it out.

- *Zichichi:*

Do not take the microphone out, Leonardo, keep it there is a question here.

- *t'Hooft:*

I have such a top at home and I know that this explanation is totally false.

- *Susskind:*

Ha-ha-ha. So maybe you will explain?

- *t'Hooft:*

Well, I am not sure I can explain it but the top itself is also magnetic. And if the north poles point to each other and the south poles point away from each other so they exert force.

- *Susskind:*

Then why is it spinning?

- *t'Hooft:*

Well the spin is to stabilize because if you do not have spin then the thing flips.

- *Susskind:*

So you do not need magnetic monopole?

- *t'Hooft:*

No, those poles are quite sufficient. It is only better if you have monopole but I could not get it in the shop.

- *Susskind:*

But you think you could make one like this?

- *t'Hooft:*

No, because electro field is much too weak. This is a relativistic effect we talk about. You need a relativistic top. But the thing he was showing did not rotate with a relativistic speed. But magnets are just fine. They repel each other if they point the same poles toward each other and the only thing you need is a special kind of magnetic field with a slight minimum over the potential at the center. So you need, actually, more complicated magnetic settling in the disk to make sure that there is a maximum around it and there is a local minimum at the center, so it is stable against sideward movement and then we stable it against flipping which is what you need a rotation for.

- *Susskind:*

Ok, thank you.

- *Deshpande:*

Have you opened the box?

- *'t Hooft:*

Oh, I did not open actually the box. Did you? Is there a single magnet in the box?

- *Deshpande:*

Yes there is just a bar magnet, a very strong one.

Results from PHENIX at RHIC

M. J. Tannenbaum *
Physics Department, 510c,
Brookhaven National Laboratory,
Upton, NY 11973-5000, USA
mjt@bnl.gov

December 10, 2008

1 Introduction

In nucleus-nucleus collisions at large c.m. energy or baryon density, a phase transition is expected from a state of nucleons containing confined quarks and gluons to a state of "deconfined" (from their individual nucleons) quarks and gluons, in chemical and thermal equilibrium, covering a volume that is many units of the confining length scale. This state of nuclear matter was originally given the name Quark Gluon Plasma(QGP) [1], a plasma being an ionized gas. However the results at RHIC [2] indicated that instead of behaving like a gas of free quarks and gluons, the matter created in heavy ion collisions at nucleon-nucleon c.m. energy $\sqrt{s_{NN}} = 200$ GeV appears to be more like a *liquid*. This matter interacts much more strongly than originally expected, as elaborated in recent peer reviewed articles by the 4 RHIC experiments [3], which inspired the theorists [4] to give it the new name "sQGP" (strongly interacting QGP).

*Research supported by U.S. Department of Energy, DE-AC02-98CH10886.

2 Issues in Relativistic Heavy Ion Physics

Since 1986, the 'gold-plated' signature of deconfinement was thought to be J/Ψ suppression. Matsui and Satz [5] proposed that J/Ψ production in A+A collisions will be suppressed by Debye screening of the quark color charge in the QGP which would dissolve the $c, \bar{c}$ bound state. "Anomalous suppression" of J/Ψ was found in Pb+Pb collisions at the CERN SpS $\sqrt{s_{NN}} = 17.2$ GeV [6]. The search for J/Ψ suppression and thermal photon/dilepton radiation from the QGP drove the design of the RHIC experiments.

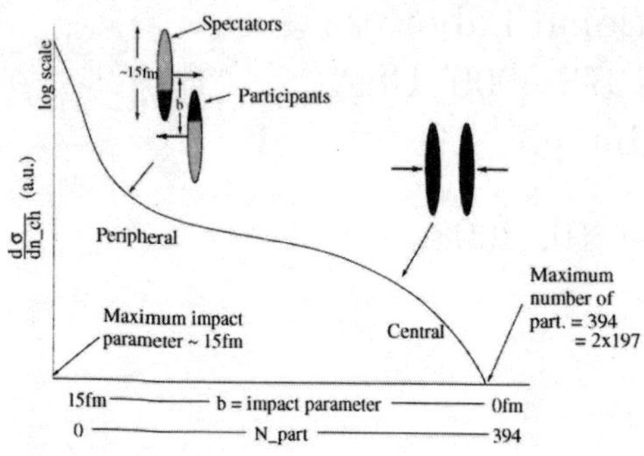

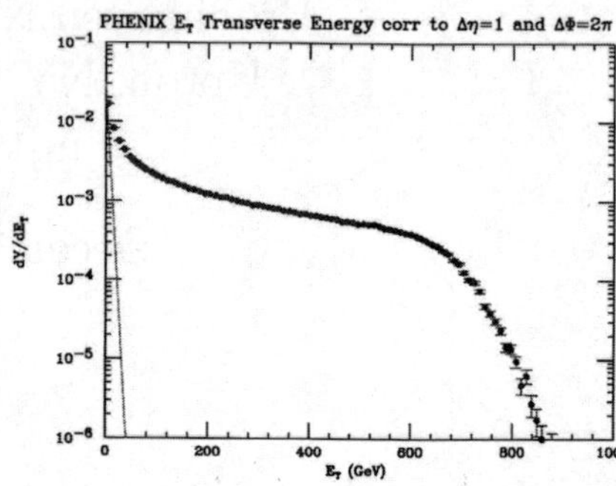

Figure 1: a) (left) Schematic of collision of two nuclei with impact parameter b. The curve with the ordinate labeled $d\sigma/dn_{\rm ch}$ represents the relative probability of charged particle multiplicity $n_{\rm ch}$ which is directly proportional to the number of participating nucleons, $N_{\rm part}$. b)(right) Transverse energy (E_T) distribution in Au+Au and p-p collisions at $\sqrt{s_{NN}} = 200$ GeV from PHENIX [7].

Another main concern of experimental design in RHI collisions is the huge multiplicity in A+A central collisions compared to p-p collisions. A schematic drawing of a collision of two relativistic Au nuclei is shown in Fig. 1. In the center of mass system of the nucleus-nucleus collision, the two Lorentz-contracted nuclei of radius R approach each other with impact parameter b. In the region of overlap, the "participating" nucleons interact with each other, while in the non-overlap region, the "spectator" nucleons simply continue on their original trajectories. The degree of overlap is called the centrality of the collision. The energy of the inelastic collision is predominantly dissipated by multiple particle production, where $n_{\rm ch}$, the number of charged particles produced, is directly proportional to the number of participating nucleons ($N_{\rm part}$) as sketched on Fig. 1. Thus, $n_{\rm ch}$ or the total transverse energy E_T in central Au+Au collisions is roughly A times larger than in a p-p collision, as shown in the measured transverse energy spectrum in the PHENIX detector for Au+Au compared to p-p (Fig. 1b) [7] and in actual events from the STAR and PHENIX detectors at RHIC in Fig. 2.

As it is a daunting task to reconstruct all the particles produced in such events, the initial experiments at RHIC concentrated on the measurement of single-particle or multi-particle inclusive variables to analyze RHI collisions, with inspiration from the CERN ISR which emphasized those techniques before the era of jet reconstruction. The two major detectors presently in operation at RHIC (Fig. 2), are STAR, a conventional TPC, and PHENIX, a very high granularity high resolution special purpose detector covering a smaller solid angle designed to measure and trigger on rare processes involving leptons, photons and identified hadrons at the highest luminosity.

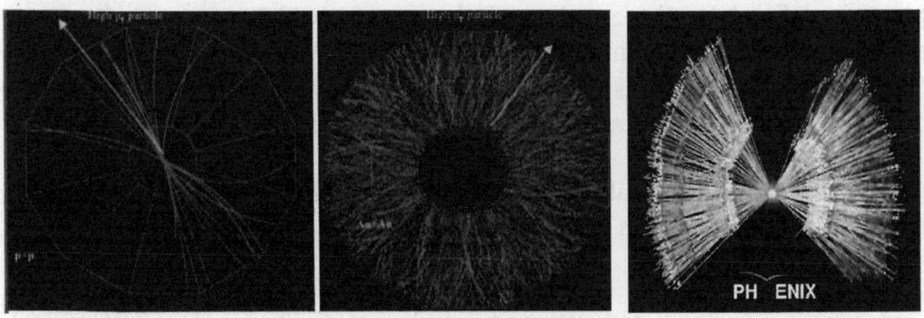

Figure 2: a) (left) A p-p collision in the STAR detector viewed along the collision axis; b) (center) Au+Au central collision at $\sqrt{s_{NN}} = 200$ GeV in the STAR detector; c) (right) Au+Au central collision at $\sqrt{s_{NN}} = 200$ GeV in the PHENIX detector.

In addition to the large multiplicity, there are two other issues in A+A physics which are different from p-p physics: i) space-time issues, both in momentum space and coordinate space—for instance what is the spatial extent of fragmentation? is there a formation time/distance?; ii) huge azimuthal anisotropies of particle production in non-central collisions (colloquially collective flow) which are interesting in their own right but can be troublesome.

Collective flow, or simply flow, is a collective effect which can not be obtained from a superposition of independent N-N collisions. The almond shaped overlap region of the A+A collision causes the particles to be emitted more favorably in the reaction plane (see Fig. 3). The semi-inclusive single

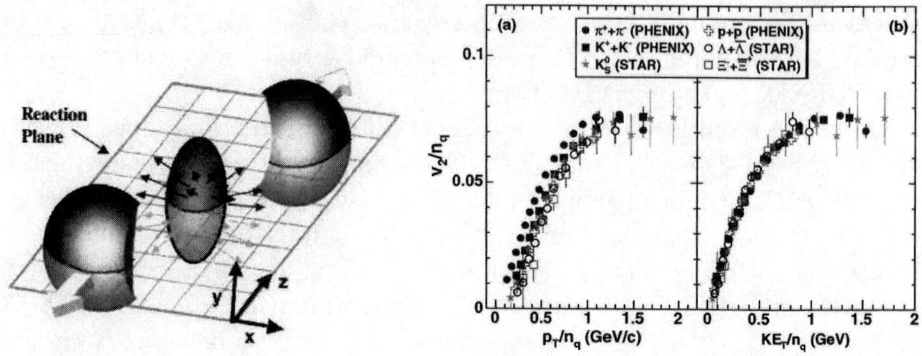

Figure 3: (left) Almond shaped overlap zone generated just after an A+A collision where the incident nuclei are moving along the $\pm z$ axis. The reaction plane is defined by the z axis and the x axis (which contains the impact parameter vector and defines $\Phi_R = 0$). (right) Measurements of elliptical-flow (v_2) for identified hadrons plotted as v_2 divided by the number of constituent quarks n_q in the hadron as a function of (a) p_T/n_q, (b) KE_T/n_q [8].

particle spectrum is modified by an expansion in harmonics of the azimuthal angle of the particle with respect to the reaction plane, $\phi - \Phi_R$:

$$\frac{Ed^3N}{dp^3} = \frac{d^3N}{p_T dp_T dy d\phi} = \frac{d^3N}{2\pi\, p_T dp_T dy}\left[1 + \sum_n 2v_n \cos n(\phi - \Phi_R)\right]. \qquad (1)$$

The expansion parameter v_2, called elliptical flow, is predominant at mid-rapidity and shows scaling rules which imply that it is the constituent quarks rather than the final hadrons that flow (Fig. 3b) [8].

3 Measurements in proton-proton collisions

In order to understand whether an effect observed in A+A collisions exhibits a sensitivity to collective effects or to the presence of a medium such as the QGP it is important to establish a precise baseline measurement in p-p collisions at the same value of nucleon-nucleon c.m. energy $\sqrt{s_{NN}}$. PHENIX measurements of the invariant cross section, $Ed^3\sigma/dp^3$, for π^0 and direct-single-γ production in p-p collisions at $\sqrt{s} = 200$ GeV are shown in Fig. 4a

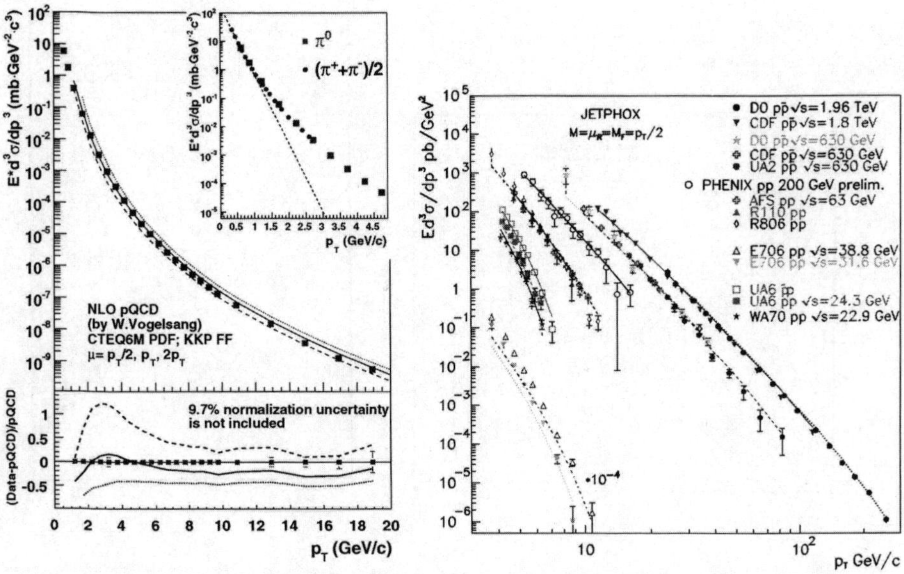

Figure 4: a) (left) PHENIX measurement of invariant cross section of π^0 vs. p_T at mid-rapidity in p-p collisons at $\sqrt{s} = 200$ GeV. [9]. b) (right) PHENIX measurement of inclusive direct-single γ in p-p collisons at $\sqrt{s} = 200$ GeV ($\circ$), together with all previous data compared to the theory. [10]

and Fig. 4b, respectively. The inset on Fig. 4a shows that the π^0 cross section is exponential $\sim e^{-6p_T}$ for $p_T < 2$ GeV/c, which is the region of soft-multiparticle physics. For $p_T > 2$ GeV/c the spectrum is a power law which is indicative of the hard-scattering of the quark and gluon constituents of the proton. The excellent agreement of the measurements with theory is rewarding, although not surprising, because the production of π^0 at large transverse momentum was discovered at the CERN-ISR in 1972 and proved that the partons of deeply inelastic scattering (DIS) interacted strongly with each other, which was explained by QCD in 1978 [11].

The ISR discovery [12] (Fig. 5a) showed that the e^{-6p_T} dependence at low p_T breaks to a power law with characteristic $\sqrt{s}$ dependence for $p_T > 2$ GeV/c, which is more evident from the log-log plot of subsequent data [13] (Fig. 5b) as a function of $x_T = 2p_T/\sqrt{s}$. Further ISR measurements utilizing inclusive single or pairs of hadrons established that high p_T particles in p-p

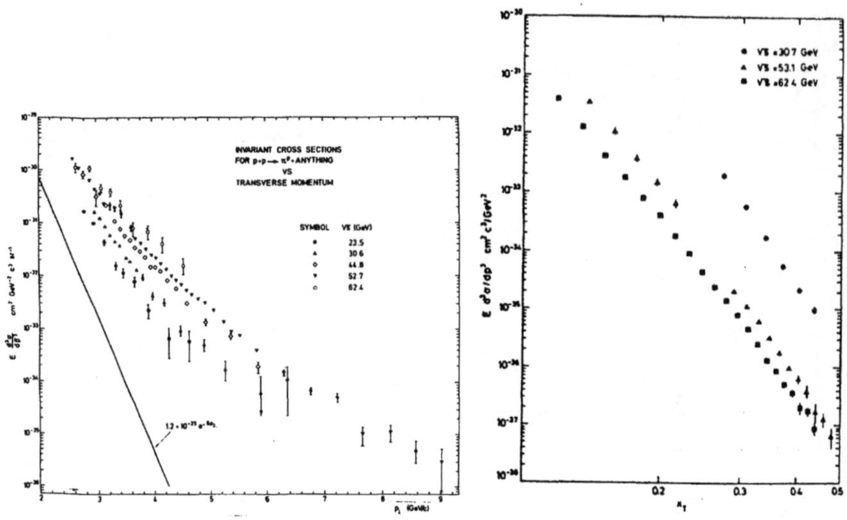

Figure 5: a) (left) CCR [12] measurement of the invariant cross section of π^0 vs. p_T at mid-rapidity in p-p collisons for 5 values of $\sqrt{s}$. b) (right) Later ISR measurement of invariant cross section of π^0 vs. $x_T = 2p_T/\sqrt{s}$ at mid-rapidity in p-p collisons for 3 values of $\sqrt{s}$ [13].

collisions are produced from states with two roughly back-to-back jets which are the result of scattering of constituents of the nucleons as described by Quantum Chromodynamics (QCD), which was developed during the course of those measurements [11]. These techniques have been used extensively and further developed at RHIC since they are the only practical method to study hard-scattering and jet phenomena in Au+Au central collisions at RHIC energies.

The di-jet structure of events triggered by a high p_T π^0, measured via two-particle correlations, is shown in Fig 6. The peaks on both the same side (Fig. 6a) as the trigger π^0 and opposite in azimuth (Fig. 6b) are due to the correlated charged particles from jets. The integrated (in $\Delta\phi$) yield of the away side-particles as a function of the variable $x_E \equiv -p_{T_a}\cos(\Delta\phi)/p_{T_t} \approx z_a/z_t$, where $z_t \approx p_{T_t}/\hat{p}_{T_t}$ is the fragmentation variable of the trigger jet (with $\hat{p}_{T_t}$) and $z_a \approx p_{T_a}/\hat{p}_{T_a}$ is the fragmentation variable of the away jet (with $\hat{p}_{T_a}$), was thought in the ISR era to measure the fragmentation function of the away jet (Fig. 6c) but was found at RHIC to be sensitive, instead, to the ratio of

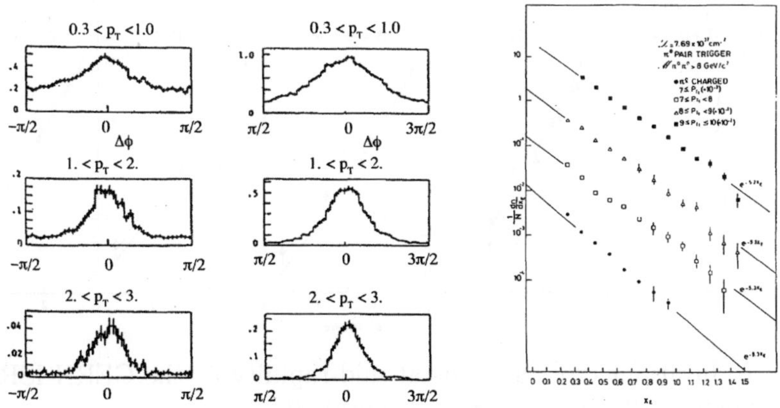

Figure 6: a,b) Distributions of azimuthal angle ($\Delta\phi$) of associated charged particles of transverse momentum p_{T_a}, with respect to a trigger π^0 with $p_{T_t} \geq 7$ GeV/c, for 3 intervals of $p_{T_{(a)}}$: a) (left-most panel) for $\Delta\phi = \pm\pi/2$ rad about the trigger particle, and b) (middle panel) for $\Delta\phi = \pm\pi/2$ about π radians (i.e. directly opposite in azimuth) to the trigger. The trigger particle is restricted to $|\eta| < 0.4$, while the associated charged particles are in the range $|\eta| \leq 0.7$. c) (right panel) x_E distributions (see text) corresponding to the data of the center panel.

the transverse momenta of the away-jet to the trigger jet, $\hat{x}_h \equiv \hat{p}_{T_a}/\hat{p}_{T_t}$ [14].

Two other ISR discoveries, direct single-γ production and direct-single $e^{\pm}$ production, and one near miss, J/Ψ production, are important components of physics at RHIC. Direct single-γ production via the inverse QCD-compton process $g + q \to \gamma + q$ is an important probe in A+A collisions because the γ is a direct participant in the reaction at the constituent level, emerges from the medium without interacting and can be measured precisely [15]. Direct single-$e^{\pm}$ at a level of $e^{\pm}/\pi^{\pm} \approx 10^4$ for all values of $\sqrt{s}$ at the CERN-ISR were discovered before either the J/Ψ or open-charm [16] (Fig. 7). After the discovery of the J/Ψ in 1974, it was demonstrated that the J/Ψ was not the source of the direct single-$e^{\pm}$ [16] and two years later, when open charm was discovered, it was shown that the direct single-$e^{\pm}$ were due to the semi-leptonic decay of charm mesons [17].

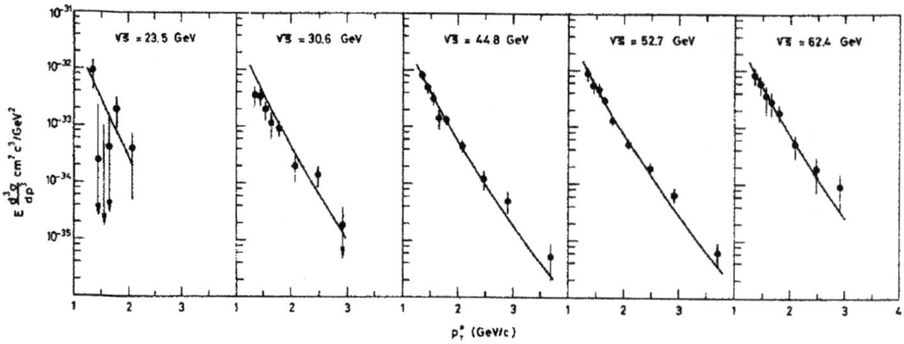

Figure 7: Invariant cross sections at mid-rapidity at the CERN-ISR: $(e^+ + e^-)/2$ (points); $10^{-4} \times (\pi^+ + \pi^-)/2$ (lines) [16].

4 From ISR p-p to RHIC A+A physics

The discovery, at RHIC, that π^0 are suppressed by roughly a factor of 5 compared to point-like scaling of hard-scattering in central Au+Au collisions is arguably *the* major discovery in Relativistic Heavy Ion Physics. Since hard-scattering at high $p_T > 2$ GeV/c is point-like, with distance scale $1/p_T < 0.1$ fm, the cross section in p+A (A+A) collisions, compared to p-p, should be larger by the relative number of possible point-like encounters, a factor of A (A^2) for p+A (A+A) minimum bias collisions. When the impact parameter or centrality of the collision is defined, the proportionality factor becomes $\langle T_{AA}\rangle$, the average overlap integral of the nuclear thickness functions.

In Fig. 8a), the PHENIX measurement of $Ed^3\sigma/dp^3$ for π^0 production in p-p collisions at $\sqrt{s} = 62.4$ is in excellent agreement with the ISR data and the PHENIX π^0 data follow the same trend as the lower energy data, with a pure power law, $Ed^3\sigma/dp^3 \propto p_T^{-8.1\pm0.1}$ for $p_T > 3$ GeV/c at $\sqrt{s} =$ 200 GeV. In Fig. 8b), the 200 GeV p-p data, multiplied by the point-like scaling factor $\langle T_{AA}\rangle$ for (0-10%) central Au+Au collisions are compared to the semi-inclusive invariant π^0 yield in central (0-10%) Au+Au collisions at $\sqrt{s_{NN}} = 200$ GeV and, amazingly, the Au+Au data follow the same power-law as the p-p data but are suppressed from the point-like scaled p-p data by a factor of ~ 5. The suppression is represented quantitatively by the "nuclear modification factor", $R_{AA}(p_T)$, the ratio of the measured semi-inclusive yield

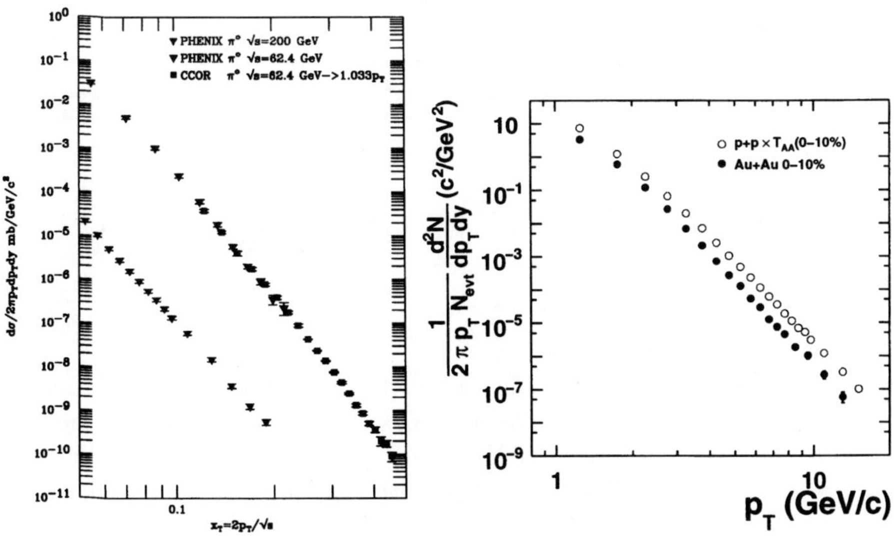

Figure 8: a) (left) $Ed^3\sigma/dp^3$ vs. x_T for PHENIX mid-rapidity π^0 at $\sqrt{s} = 200$ GeV in p-p collisions [9] plus PHENIX and CCOR-ISR [13] measurements at $\sqrt{s} = 62.4$ GeV, where the absolute p_T scale of the ISR measurement has been corrected upwards by 3% to agree with the PHENIX data. b) (right) π^0 p-p data vs. p_T at $\sqrt{s} = 200$ GeV from (a) multiplied by $\langle T_{AA} \rangle$ of Au+Au central (0-10%) collisions compared to semi-inclusive π^0 invariant yield in Au+Au central (0-10%) collisions at $\sqrt{s_{NN}} = 200$ GeV.

in A+A collisions to the point-like scaled p-p cross section at the same p_T:

$$R_{AA}(p_T) = \frac{d^2N^{\pi}_{AA}/dp_T dy N_{AA}}{\langle T_{AA} \rangle d^2\sigma^{\pi}_{pp}/dp_T dy} \quad . \tag{2}$$

In Fig. 9a, $R_{AA}(p_T)$ is shown for π^0, η mesons and direct-γ for $\sqrt{s_{NN}} =$ 200 GeV Au+Au central (0-10%) collisions [18]. The π^0 and η mesons, which are fragments of jets from outgoing partons are suppressed by the same amount while the direct-γ which do not interact in the medium are not suppressed. This indicates a strong medium effect on outgoing partons.

Fig. 9b shows that R_{AA} for central (0-10%) Cu+Cu collisions is comparable at $\sqrt{s_{NN}} = 62.4$ and 200 GeV, but that there is no suppression, actually a Cronin enhancement, at $\sqrt{s_{NN}} = 22.4$ GeV [19]. This indicates that the medium which suppresses jets is produced somewhere between $\sqrt{s_{NN}} = 22.4$ GeV, the SpS Fixed Target highest c.m. energy, and 62.4 GeV.

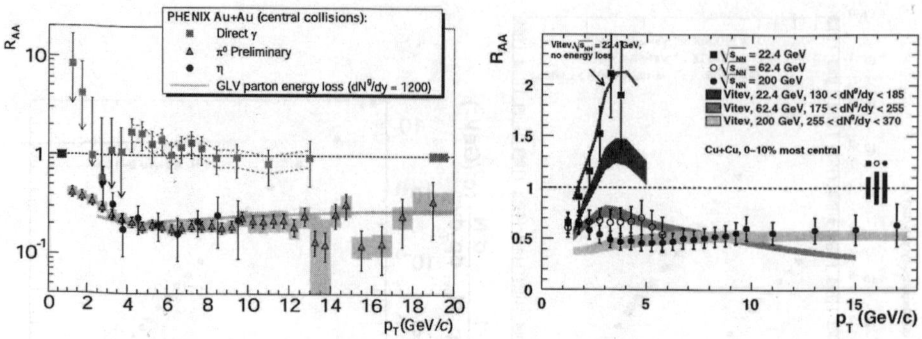

Figure 9: a) (left) Nuclear modification factor, R_{AA} for direct-γ, π^0 and η in Au+Au (0-10%) central collisions at $\sqrt{s_{NN}} = 200$ GeV [18], together with GLV theory curve [20]. b) PHENIX R_{AA} for π^0 in Cu+Cu central collisions at $\sqrt{s_{NN}} = 200$, 62.4 and 22.4 GeV [19], together with Vitev theory curves [21].

Remarkably, the dramatic difference in π^0 suppression from SpS to RHIC is not reflected in J/Ψ suppression, which is nearly identical at mid-rapidity at RHIC and the SpS, casting some doubt on the value of J/Ψ suppression as a probe of deconfinement [2]. One possible explanation is that c and $\bar{c}$ quarks in the QGP recombine to regenerate J/Ψ, miraculously making the observed R_{AA} equal at SpS and RHIC c.m. energies. The good news is that such models predict J/Ψ enhancement ($R_{AA} > 1$) at LHC energies, which would be spectacular, if observed.

Returning to the jet suppression shown in Fig. 9, the measurements seem to be in excellent agreement with the theoretical curves [20, 21]. The suppression can be explained by the energy loss of the outgoing partons in the dense color-charged medium due to coherent Landau-Pomeranchuk-Migdal radiation of gluons, predicted in QCD [22], which is sensitive properties of the medium. Measurements of two-particle correlations confirm the loss of energy of the away-jet relative to the trigger jet in Au+Au central collisions compared to p-p collisions. In analogy to Fig. 6 (above), the two-particle correlations in Au+Au collisions (Fig. 10a) show clear di-jet structure in both peripheral and central collisions. The away-side correlation in central Au+Au collisions is much wider than in peripheral Au+Au and p-p collisions and is further complicated by the large multiparticle background which is a modulated in azimuth by the v_2 collective flow of a comparable width to the jet correlation. After the v_2 correction, a double peak structure $\sim \pm 1$ radian

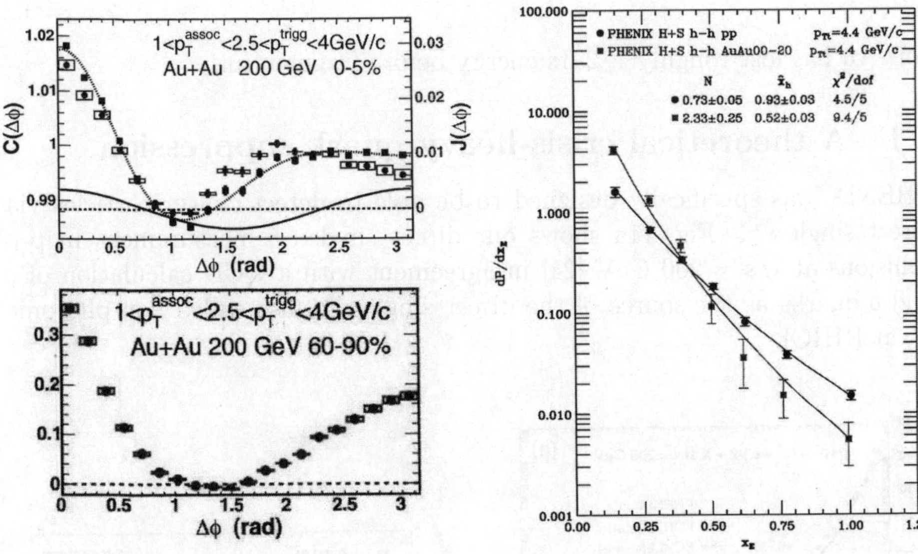

Figure 10: a) (left) Conditional yield azimuthal correlation function, $C(\Delta\phi)$ (black squares), flow background (solid line) and Jet function $J(\Delta\phi)$ (red dots) after flow subtraction, per trigger $h^{\pm}$ with $2.5 < p_{T_t} < 4$ GeV/c for associated $h^{\pm}$ of $1.0 < p_{T_a} < 2.5$ GeV/c from PHENIX [23]. b) (right) $x_E \approx p_{T_a}/p_{T_t}$ distribution for the Au+Au-central data compared to p-p.

from π is evident, with a dip at π radians. This may indicate a reaction of the medium to a passing parton in analogy to a "sonic-boom" and is under active study both theoretically and experimentally. The energy loss of the away-parton is indicated by the fact that the x_E distribution in Au+Au central collisions (Fig. 10b) is steeper than that from p-p collisions. As noted above, we found in PHENIX that the x_E distribution did not measure the fragmentation function of the away-jet but is sensitive instead to $\hat{x}_h$, the ratio of the transverse momentum of the away-parton to that of the trigger parton, specifically [14]:

$$\left.\frac{dP}{dx_E}\right|_{p_{T_t}} = N(n-1)\frac{1}{\hat{x}_h}\frac{1}{(1+x_E/\hat{x}_h)^n} \tag{3}$$

where N is a normalization factor, and n (=8.1 at 200 GeV) is the power of the inclusive invariant p_{T_t} distribution. The value of $\hat{x}_h = 0.52$ in Au+Au collisions compared to 0.93 in p-p collisions shows that the away-parton in

Au+Au has lost roughly 1/2 its energy before fragmenting.

4.1 A theoretical crisis-heavy quark suppression

PHENIX was specifically designed to be able to detect charm particles via direct single-$e^{\pm}$. Fig. 11a shows our direct single-$e^{\pm}$ measurement in p-p collisions at $\sqrt{s} = 200$ GeV [24] in agreement with a QCD calculation of c and b quarks as the source of the direct single-$e^{\pm}$ (also called non-photonic $e^{\pm}$ at RHIC).

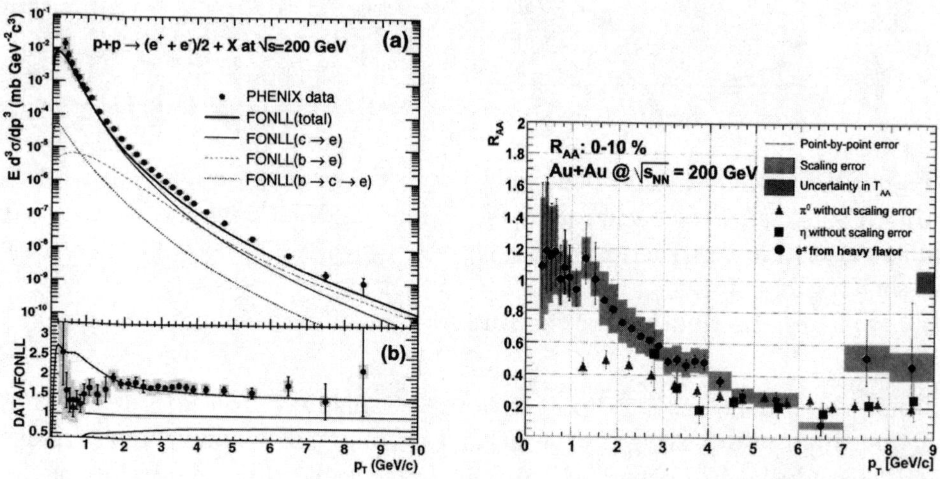

Figure 11: a) (left) Invariant cross section of direct single-$e^{\pm}$ in p-p collisions [24] compared to theoretical predictions from c and b quark semileptonic decay. b) (right) R_{AA} as a function of p_T for direct single-$e^{\pm}$ [25], π^0 and η in Au+Au central (0-10%) collisions at $\sqrt{s_{NN}} = 200$ GeV.

In Au+Au collisions, a totally unexpected result was observed. The direct single-$e^{\pm}$ from heavy quarks are suppressed the same as the π^0 and η from light quarks (and gluons) in the range $4 \leq p_T \leq 9$ GeV/c where b and c contributions are roughly equal (Fig. 11b) [25]. This strongly disfavors the QCD energy-loss explanation of jet-quenching because, naively, heavy quarks should radiate much less than light quarks and gluons in the medium; but opens up a whole range of new possibilities including string theory [26].

5 Zichichi to the rescue?

In September 2007, I read an article by Nino, "Yukawa's gold mine", in the CERN Courier taken from his talk at the 2007 International Nuclear Physics meeting in Tokyo, Japan, in which he proposed: "the reason why the top quark appears to be so heavy (around 200 GeV) could be the result of some, so far unknown, condition related to the fact that the final state must be QCD-colourless. We know that confinement produces masses of the order of a giga-electron-volt. Therefore, according to our present understanding, the QCD colourless condition cannot explain the heavy quark mass. However, since the origin of the quark masses is still not known, it cannot be excluded that in a QCD coloured world, the six quarks are all nearly massless and that the colourless condition is 'flavour' dependent."

Nino's idea really excited me even though, or perhaps because, it appeared to overturn two of the major tenets of the Standard Model since it seemed to imply that: QCD isn't flavor blind; the masses of quarks aren't given by the Higgs mechanism. Massless b and c quarks in a color-charged medium would be the simplest way to explain the apparent equality of gluon, light quark and heavy quark suppression indicated by the equality of R_{AA} for π^0 and R_{AA} of direct single-$e^{\pm}$ in regions where both c and b quarks dominate. Furthermore RHIC and LHC-Ions are the only place in the Universe to test this idea.

It may seem surprising that I would be so quick to take Nino's idea so seriously. This confidence dates from my graduate student days when I checked the proceedings of the 12th ICHEP in Dubna, Russia in 1964 to see how my thesis results were reported and I found several interesting questions and comments by an "A. Zichichi" printed in the proceedings. One comment about how to find the W boson in p+p collisions deserves a verbatim quote because it was exactly how the W was discovered at CERN 19 years later: "We would observe the μ's from W-decays. By measuring the angular and momentum distribution at large angles of K and π's, we can predict the corresponding μ-spectrum. We then see if the μ's found at large angles agree with or exceed the expected numbers."

Nino's idea seems much more reasonable to me than the string theory explanations of heavy-quark suppression (especially since they can't explain light-quark suppression). Nevertheless, just to be safe, I asked some distinguished theorists what they thought, with these results: "Oh, you mean the Higgs field can't penetrate the QGP" (Stan Brodsky); "You mean that the propagation of heavy and light quarks through the medium is the same" (Rob Pisarski); "The Higgs coupling to vector bosons γ, W, Z is specified in the standard model and is a fundamental issue. One big question to be answered by the LHC is whether the Higgs gives mass to fermions or only to gauge bosons. The Yukawa couplings to fermions are put in by hand and are not required" "What sets fermion masses, mixings?" (Chris Quigg-Moriond2008); "No change in the t-quark, W, Higgs mass relationship if there is no Yukawa coupling: but there could be other changes" (Bill Marciano).

Nino proposed to test his idea by shooting a proton beam through a QGP formed in a Pb+Pb collision at the LHC and seeing the proton 'dissolved' by the QGP. My idea is to use the new PHENIX vertex detector, to be installed in 2010, to map out, on an event-by-event basis, the di-hadron correlations from identified $b, \bar{b}$ di-jets, identified $c, \bar{c}$ di-jets, which do not originate from the vertex, and light quark and gluon di-jets, which originate from the vertex and can be measured with π^0-hadron correlations. A steepening of the slope of the x_E distribution of heavy-quark correlations as in Fig. 10b will confirm in detail (or falsify) whether the different flavors of quarks behave as if they have the same mass in a color-charged medium. If Nino's proposed effect is true, that the masses of fermions are not given by the Higgs, and we can confirm the effect at RHIC or LHC-Ions, this would be a case where Relativistic Heavy Ion Physics may have something unique to contribute at the most fundamental level to the Standard Model, which would constitute a "transformational discovery." Of course the LHC could falsify this idea by finding the Higgs decay to $b, \bar{b}$ at the expected rate in p-p collisions. Clearly, there are exciting years ahead of us!

References

[1] E. V. Shuryak, Phys. Repts. **61** (1980) 71–158.

[2] For a more detailed discussion, see M. J. Tannenbaum, Rep. Prog. Phys. **69** (2006) 2005–2059.

[3] Nucl. Phys. A**757** (2005) 1–283.

[4] D. Rischke and G. Levin, eds., Nucl. Phys. A**750** (2005) 1–171.

[5] T. Matsui and H. Satz, Phys. Lett. B**178** (1986) 416.

[6] M. C. Abreu, *et al.* NA50 Collab., Phys. Lett. B**450** (1999) 456–466.

[7] S. S. Adler, *et al.* PHENIX Collab., Phys. Rev. C**71** (2005) 034908.

[8] A. Adare, *et al.* PHENIX Collab., Phys. Rev. Lett. **98** (2007) 162301.

[9] S. S. Adler, *et al.* PHENIX Collab., Phys. Rev. D**76** (2007) 051006(R).

[10] P. Aurenche, *et al.*, Phys. Rev. D**73** (2006) 094007.

[11] See M. J. Tannenbaum, Nucl. Phys. A**749** (2005) 219c for references.

[12] F. W. Büsser, *et al.* (CCR), Phys. Lett. B**46** (1973) 471.

[13] A. L. Angelis, *et al.* (CCOR), Phys. Lett. B**79** (1978) 505.

[14] S. S. Adler, *et al.* (PHENIX Collab.), Phys. Rev. D**74** (2006) 072002.

[15] e.g. see A. L. S. Angelis, *et al.* (CMOR), Nucl. Phys. B**327** (1989) 541.

[16] F. W. Büsser, et al. (CCRS), Nucl. Phys. B**113** (1976) 189-245.

[17] I. Hinchliffe and C. H. Llewellyn Smith, Nucl. Phys. B**114** (1976) 45-60.

[18] Y. Akiba *et al.* (PHENIX Collab.), Nucl. Phys. A**774** (2006) 403–408.

[19] A. Adare, *et al.* (PHENIX Collab.), Phys. Rev. Lett. **101** (2008) 162301.

[20] M. Gyulassy, P. Levai and I. Vitev, Phys. Rev. Lett. **85** (2000) 5535-5538; I. Vitev and M. Gyulassy, Phys. Rev. Lett. **89** (2002) 252301.

[21] I. Vitev, Phys. Lett. B**639** (2006) 38–45, and private communication to Ref. [19].

[22] See R. Baier, D. Schiff and B. G. Zakharov, Ann. Rev. Nucl. Part. Sci. **50** (2000) 37–69, and references therein.

[23] A.Adare, *et al.* (PHENIX Collab.), Phys. Rev. Lett. **98** (2007) 232302.

[24] A. Adare, *et al.* PHENIX Collab., Phys. Rev. Lett. **97** (2006) 252002.

[25] A. Adare, *et al.* (PHENIX Collab.), Phys. Rev. Lett. **98** (2007) 172301.

[26] e.g. see Ref. [25] for a list of references.

CHAIRMAN: M.J. TANNENBAUM

Scientific Secretaries: B.Guerzoni, A. Silenzi

DISCUSSION

- *Nikiforov:*

Do you remember what is the cell size of the electromagnetic calorimeter in the PHENIX experiment?

- *Tannenbaum:*

Yes, I do!

- *Nikiforov:*

And what is it?

- *Tannenbaum:*

(laughs)

- *Nikiforov:*

and can you tell me how much it is compared to the Moliere radius?

- *Tannenbaum:*

Oh, compared to the Moliere radius the cell is larger, it's 0.01 x 0.01 radian and the cell 5.5 x5.5 cm^2 at 5 m and the moliere radius order of 1cm. See NIMA 499 (2003) 521—536.

- *Nikiforov:*

But, how can you measure π^0 up to 25 GeV as you stated in the talk?

- *Tannenbaum:*

They merge at 25 GeV, they are one cell apart. And right now our main problem is that at 18 GeV they start merging, but we have 2 calorimeters one is lead glass and one is lead scintillator with different segmentation so they can be compared. The π^0 decay that gives you problem is when they decay symmetrically because you will have for a 20 GeV π0 two 10 GeV photons; but you also have decays with 1 photon with 19 GeV and 1 photon with 1 GeV which are far apart so they can be resolved, but I think direct photons has more problems with the merged π0 background.

- *Johansson:*

You say that QCD might not be flavour blind, as an explanation of heavy quark suppression as you observe, does this mean that hadrons with 2nd and 3rd generation of quarks can decay through a flavour changing strong process, rather than a weak process?

- *Tannenbaum:*

Well, I'm an experimentalist, not a theorist. What I said referred to what happens in the color charged medium, but the quarks decay outside the medium. We are talking of a high pT di-quark embedded in a volume the size of a nucleus with quark and gluons deconfined, in principle according to what Nino says the quark masses can be equal, and I think that they both must be light because there is no other way of explaining light quark radiative energy loss.. The second thing is that we could have the same cross section for . production of light and heavy quarks in the medium if they both have the same mass. You have a lot of b quarks in the medium but you detect them outside of it, unless they decay inside the medium, but they don't. When the b quarks get outside the medium they return to their correct mass before they decay and that selects, due to quantum mechanics, only a small sample of the b quarks that could have been produced in the medium with low mass. That's why, in my opinion, Nino's idea on quark masses works. Serban Protopopescu of BNL told me that top quarks decay in the medium, which would mean a 10^{-24} s lifetime. I don't know if it has been measured but the theory is good and I saw this value in the particle data book. So at LHC there would be a large number of zero mass top quarks decaying semi-leptonically in the medium, and therefore a huge number of 90 GeV/c leptons coming out, depending on the semi-leptonic branching ratio. This could be quite an unexpected discovery at the LHC-Ions if Nino's idea about all quarks having the same mass in the QGCW are correct.

- *Klopot:*

Could you please comment on the causes of the large azimuthal anisotropy in heavy ion collisions in contrast to anisotropy in pp collisions?

- *Tannenbaum:*

Well, it's easy! There is no way we have thought of yet for finding anisotropy in the pp collisions. The only thing you can say is that the proton is composite. The multiplicity in pp collisions is so small that there is no obvious way of measuring the reaction plane to high precision. Maybe LHC could do that, but I don't know...

- *Nicotri:*

Do you think it is possible to study the color superconductive phase of QCD at RHIC? Or will it be possible to study it in the near future?

- *Tannenbaum:*

I don't think we can make it in a laboratory. You see the problem in the phase diagram: the superconductive phase proposed is at very low temperature, very high baryon chemical potential, μ_B. Maybe some astronomer can see them, but they are not accessible in lab.

- *Gazis:*

In the last part of your presentation. it was shown the specific role QGCW and the specific experimental technique you have shown to confirm it. My question is if in that object proposed (QGCW) the QCD study effects for the quarkonia have been taken into account. (i.e. quark internal motion, glueballs, color octet states)

- *Tannenbaum:*

No, that's a theorist's job. In this field we find a new phenomenon and five minutes later twenty theorists claim they have the right answer. The point is that if you predict something I have seen and also something I didn't see yet, I will go and measure it. If we find the effect that the b-quark energy loss is the same as the light quark energy loss we can then worry if this means that their masses are low or whether there is some other explanation other than radiative for energy loss in a medium.

- *Baldini:*

Long time ago I heard that in heavy ion collisions J/ψ should be suppressed and the ϕ enhanced. Is it still true?

- *Akiba:*

comparing the binary scaling the ϕ is suppressed at intermediate pt.

- *Tannenbaum:*

We measured J/ψ suppression the same as at CERN fixed target energy. But the phi is suppressed nearly the same as a π0 , for pT>5 GeV/c, a factor of 5. At lower pT the phi suppression is a bit less than π0 similar to Kaons.

- *Marshall:*

I have been thinking about Prof. Zichichi's experiment of firing a beam through the QGP, so I ask how long does the QGP live? And how could it be lengthened.

- *Tannenbaum:*

It typically lives 10 fm/c so it's 0.3 10^{-22}s. So it will be very difficult to time that beam. If Nino thinks he can do it, good luck to him. I have no idea how to do it. You can clearly send in a pulse of protons of a ns length but you will have one interaction in a Zillion. In principle if you throw enough protons one of them will hit, but I can't imagine that it can hit the A+A collision within a time of 10^{-23} second.

- *Marshall:*

and there is no way to lengthen the lifetime?

- *Tannenbaum:*

I wish we could control that. Maybe with a Penning trap with Chromo Electric and Chromagnetic fields (Ha,Ha), it would be wonderful.

- *Akiba:*

I wish to comment. We cannot shoot in the collision area, but we can actually study event by event the reaction frame, i.e. elliptic flow I have shown in my lecture.

- *Tannenbaum:*

b b-bar pairs are born as the medium is formed and pass through the medium on the way out. All the hard scatterring occurs in the first moment of the reaction and the di-partons are distributed through the medium So we can study this effect with hard scattered parton-pairs.

- *Cifarelli:*

The basic idea is to synchronize with the plasma and hit this object before hit has expanded, hadronized, avoiding examining the latter part of the reaction using pre-shower synchronization. I can tell they are working hard on this problem.

- *Tannenbaum:*

I think that the hard scattered partons can penetrate the plasma and it happens very quickly. A 2 GeV π can not be resolved until 20 fm which is outside the plasma while the energy loss of its parent parton has taken place inside the plasma. The soft stuff fragmenting takes place well after that time.

- *Ni :*

Is there any advantage by using Pb-Pb at LHC than using Au-Au collision like at RHIC?

- *Tannenbaum:*

When I was a particle physicist I used 5 x 10 x 20 cm lead bricks for a heavy element target [see PRL 26 (1971) 46] while nuclear physicists used gold targets that looked like gold coins. That's why I became a nuclear physicist. No, the reason is that when you make an ion source you can use any gold wire which is isotopically pure, while the standard lead brick is not isotopically pure (but of course all has the same Z). The difference between gold and lead is no big deal, Gold is 197 nucleons Lead is only 11 nucleons more. We will try to use uranium in the future, it's football shaped, so something interesting can come out.

- *Wen:*

How to decide to run the beam on the central collision case on under an off-central case? Is that based on some special physics purpose? What is the running-time fraction for different detector configurations?

- *Tannenbaum:*

We cannot select how the collision will take place. We squeeze the beam at the mm level but the ion is fm in size, so the centrality of the collision is just random.

- *Silenzi:*

You have shown that the particle yield in momentum with HI collisions decreases more steeply than with pp collisions. Is it possible to state from this feature that partons with higher energy are stopped more easily within the QGP?

- *Tannenbaum:*

Well, the energy dependence of the parton energy loss in a medium is one of the key issues of the field. In the 2 particle correlation we don't know how the energy loss by a parton in the QGP depends on its energy, although in principle we could map it out with enough data. In my opinion, if dE/E is constant we would have a constant R_{AA}, just on general principles, and that's what we see. I hope that LHC will sort that out. A colleague from Brookhaven, Dima Kharzeev found the way to have energy loss ΔE proportional to E^2, you know that synchrotron radiation energy loss is proportional to E^4.

- *Silenzi:*

So you confirm that there is a clear energy dependence in the energy loss during QGP crossing?

- *Tannenbaum:*

Actually from a more experimental point of view it's very difficult to say anything. If you want to trigger on a jet by selecting a high pT particle you have a trigger bias: you are most likely to get a low energy jet that fragments high than a high energy jet that fragments soft because the jet spectrum falls like $1/p_t^8$ and the fragmentation function is exponential. Similarly when you look at a particle coming out of the medium: instead of having a 100 GeV jet that has lost 50 GeV in crossing the medium you will have a 50 GeV jet from the surface that have not lost much, so this tremendous surface effect make this 2-particle correlations much more complicated to understand in A+Au collisions compared to p-p collisions as an experimental channel. We have people studying 3-particle correlation and 4 particle correlations, it's alot of fun.

- *Bonati:*

You spoke about deconfinement in the phase transition but usually deconfinement is associated also with restoration of chiral symmetry. Could this have some experimental consequences?

- *Tannenbaum:*

Do not ask me this, I am an experimental physicist. I think that chiral symmetry gets restored in the first instant of the collision. I do not know what happens when the particles hadronize.

- *Dubynskiy:*

On the cartoon of strongly interacting matter as a nuclear matter and quark matter at higher density the value of 1.6 fm is related to the size of a nucleon But where does the value for 0.8 fm for quark case come from?

- *Tannenbaum:*

The answer is that if we look at the formula for gluon radiation in the color-charged medium, the minimum of momentum transfer is μ^2. This μ is the same as the debye mass, and it 0.5 GeV/c, that is 0.4 of a fm. That's why I think 0.8 fm is the radius we study.

Highlight from RHIC

Yasuyuki Akiba
RIKEN Nishina Center for Accelerator Based Science,
Wako, Saitama 351-0198, Japan

November 30, 2008

Abstract

Collision of heavy nuclei at very high energy provides possibility to study nuclear matter at very high density. It is established that new form of very dense matter is formed in Au+Au collisions at RHIC. At such high density, it is believed that quarks and gluons are no longer confined in hadrons, but become constitutes of a quark-gluon plasma (QGP). The matter is very dense and opaque, and it has almost no viscosity and behaves like perfect fluid. These conclusions are based on two key discoveries, namely, strong suppression of high transverse momentum particles and strong elliptic flow. To understand the properties of the matter, it is important to measure many observables and relate them together. Penetrating probes such as J/ψ, high p_T jets, heavy quarks, lepton pairs and direct photons are particularly important. Recent experimental results from RHIC are presented.

1 Introduction

RHIC is the first collider of heavy ion beams. It is the only heavy ion collider until LHC start colliding Pb beam. Since its start of operation in year 2000, RHIC has collided variety of beams from polarized $p+p$ to d+Au to Cu+Cu to Au+Au for various energies ranging from about $\sqrt{s_{NN}} \simeq 20$ GeV to 200 GeV. A large amount of experimental results are obtained in the last eight years of RHIC operation.

Collision of heavy nuclei at very high energy provides possibility to study nuclear matter at very high density. The colliding nuclei at RHIC energies would melt protons and neutrons itno a collections of quarks and gluons. Lattice QCD calculations show that at high energy density and temperature ($\epsilon > 1$ GeV/fm^3, $T > Tc \simeq 170$ MeV) the QCD matter is no longer in hadronic phase but in the quark gluon plasma (QGP) phase. The phase transition from QGP phase to hadroic phase took place in early Universe about 10 μ second after the Big Bang. It should be noted that this is the only phase transiton that occured in the early Universe that can be recreated in the laboratory.

2 RHIC's two major discoveries

It is now well established that high density partonic matter is formed in central Au+Au collisions at RHIC[1, 2]. This conclusion is primarily based the following two major discoveries

- Strong suppression of high p_T hadrons in central Au+Au collisions (jet quenching)[3]
- Strong elliptic flow[4]

Based on these two major disoveries and other experimental evidence RHIC experiments concluded that state of dense partonic matter is formed in $A+A$ collisions at RHIC. I explain what those two discoveries are and how these discoveries lead to the conclusion of formation of high density matter at RHIC.

2.1 High p_T Suppression

One of the major discovery at RHIC is the strong suppression of high p_T hadrons in central Au+Au collisions relative to the scaled $p+p$ collision[3]. High p_T hadrons are produced from the fragmentation of high p_T scattered partons (quark and gluons) from hard scattering at initial stage. As the scattered partons propagate through the dense matter, they serve as probes of the medium.

The suppression is measured as nuclear suppression factor R_{AA} which is defined as

$$R_{AA}(p_T) = \frac{d^2N_{AA}(p_T)/dp_Tdy}{T_{AA}d^2\sigma_{pp}(p_T)/dp_Tdy}, \tag{1}$$

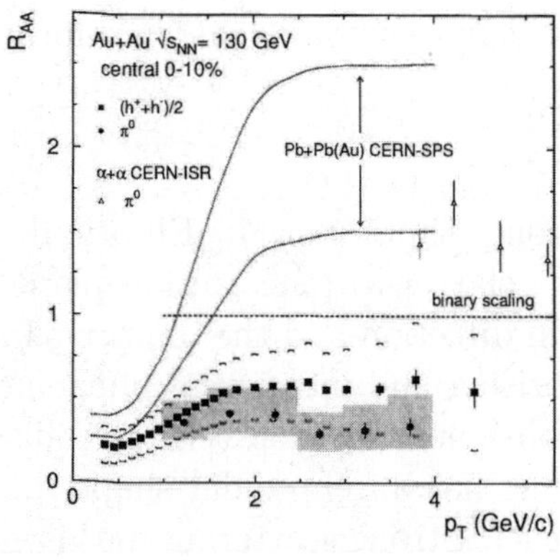

Figure 1: The nuclear suppression factor R_{AA} for charged hadrons and neutral pions in central Au+Au collisions at $\sqrt{s_{NN}} = 130$ GeV measured by PHENIX[3].

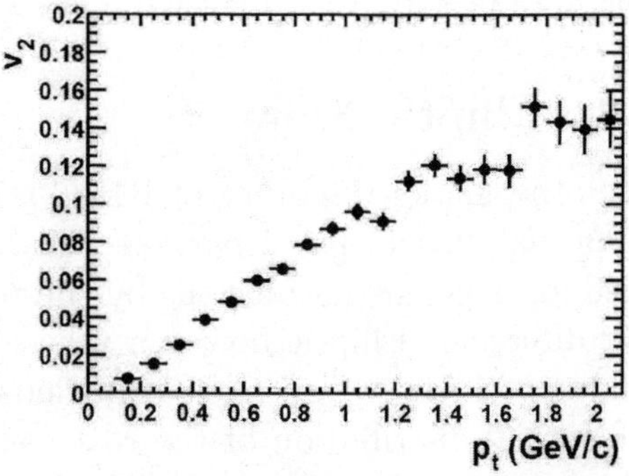

Figure 2: Elliptic flow strength v_2 of charged particles as a function of transverse momentum for minimum bias Au+Au collisions at $\sqrt{s_{NN}} = 130$ GeV measured by STAR[4].

where $T_{AA} =< N_{coll} > /\sigma_{pp}$ is nuclear thickness function calculated from a Glauber model. $< N_{coll} >$ is the number of binary $N + N$ collisions, and σ_{pp} is total inelastic $p + p$ collisions. If hadron production in $A + A$ is simple superposition of independent $N + N$ collision, $R_{AA} = 1$.

Figure 1 shows the first measurement of nuclear modification factor R_{AA} in Au+Au collision at $\sqrt{s_{NN}} = 130$ GeV[3]. The suppression is quite strong and the data points are well below unity expected for simple superposition of $N + N$ collisions. The suppression factor reaches approximately a factor of five ($R_{AA} \simeq 0.2$) at $p_T \simeq 4$ GeV/c, where most hadrons originate from parton fragmentaion (jets). This is the first evidence that very dense matter is formed in nuclear collisions at RHIC and high p_T scattered parton suffered large energy loss traversing through the medium.

The lack of such suppression of π^0 and other hadrons in d+Au collision[5] as well as the lack of suppression in high p_T direct photon yield in Au+Au[6] demonstrate that the suppression is caused by the dense medium formed in Au+Au collision. The common suppression pattern of π^0 and η at high p_T[7] demonstrates that the suppression occurs at the parton level before fragmentation. These experimental results support that the suppression is caused by the energy loss of hard scattered partons in the dense medium.

The data provide the direct evidence that dense partonic matter is formed.

2.2 Elliptic Flow

The other major discovery at RHIC is strong elliptic flow[4]. Elliptic flow is azimuthal anisotropy of particle emission relative to the reaction plane defined for non-central collisions by the beam direction and the impact parameter direction. Elliptic flow is a very surprising and interesting phenomena, unique to high energy nucleus-nucleus collsions. For non-central collision, the spatial distribution of the reaction zone has an elipsoidal shape due to incomplete overlap of the two colliding nuclei. Strong scattering in this early stage converts this spatial anisotropy into a momentum anisotropy which is observable as an elliptic flow of emitted hadrons.

The strength of the elliptic flow is characterized by the 2nd Fourier coefficient $v_2(p_T)$ of the azimuthal distribution,

$$\frac{d^2N}{d\phi dp_T} = N_0\Big[1 + \Sigma_{n=1}^{\infty} 2v_n(p_T)\cos n\phi\Big], \tag{2}$$

where ϕ is the emission angle of particle relative to the reaction plane. Figure 2 shows the first measurement of elliptic flow strength v_2 as function of transvserve momentum p_T[4]. The strength of v_2 is quite strong and it increases with increasing p_T.

Elliptic flow can be readily understoond in thermodynamics limit. If strong scattering is sufficient to establish local thermal equilibrium, the pressure gradient is largest in the shortest direction of ellipsoide i.e. the direction of the event plane. The gradient produces stronger expanding flow in that direction, resulting in anisotropic emission of particles.

In low p_T ($p_T < 2$ GeV/c), the elliptic flow data are well described by relativisitc hydrodynamical calculations assuming zero viscosity to entropy ratio ($\eta/s = 0$), i.e. perfect fluid. In particular, the hydrodynamical models can successfully reproduce the hadron mass dependence of v_2 strength[8, 9].

It should be noted that very early thermalization is required in these hydrodynamical calculations to reproduce the strong elliptic flow observed at RHIC. If thermalization is not established at early stage, the spatial eccentricity of the reaction zone is quickly reduced due to expansion. This in turn reduce the pressue gradients that drives the development of elliptic flow. The comparison between the data and the hydrodynamical calculation suggests that the thermalization time of less than 1 fm/c is required. Thus the large

v_2 observed at RHIC provide strong evidence for ealry local thermalization. How thermalization can be achived so quicky is one of the puzzle at RHIC.

The data and the theory comparion also shows that the viscosity to entropty ratio η/s is very small for the matter produced at RHIC. Even a small value of η/s can strongly suppress the elliptic flow strength v_2.

Since initial discovery of strong elliptic flow, large amount of data of v_2 of many kinds of particle have been collected at RHIC.

3 Highlights from more recent RHIC results

The main focus of heavy ion research at RHIC now is to understand the properties of the high density matter discovered. For this purpose, it is important to measure many observables and relate them together. Penetrating probes such as J/ψ, high p_T jets, heavy quarks, lepton pairs and direct photons are particularly important. In the following, I present some of the highlights of the recent experimental results from RHIC.

3.1 J/ψ Suppression

J/ψ has been considered as the probe of deconfinement. Matsui and Satz predicted J/ψ suppression due to Debye screening in QGP more than 20 years ago.

The top panel of figure 3 shows[10] R_{AA} of J/ψ as a function of number of participating nucleons. In both of mid and forward rapidities, R_{AA} decrease with increasing N_{part}. In most central collisions, approximately a factor of five suppression is observed. Surprisingly, R_{AA} is smaller for the forward direction, as shown in the bottom panel of the figure. This is contrary to a naıve expectation that the suppression should be stronger in mid-rapidity since the energy density is higher at mid-rapidity. Recombination models can give a possible explanation. Since a large number of $c\bar{c}$ pairs ($\simeq 20$) are produced in a single event of central Au+Au collisions at RHIC, a significant yield of J/ψ can be produced by recombination of these c and $\bar{c}$ quarks. The effect of recombination is expected to be stronger at mid-rapidity as the density of c and $\bar{c}$ is higher. Therefore, if the contribution from recombination is sufficiently strong, it can overcome stronger suppression of initially produced J/ψ at mid-rapidity.

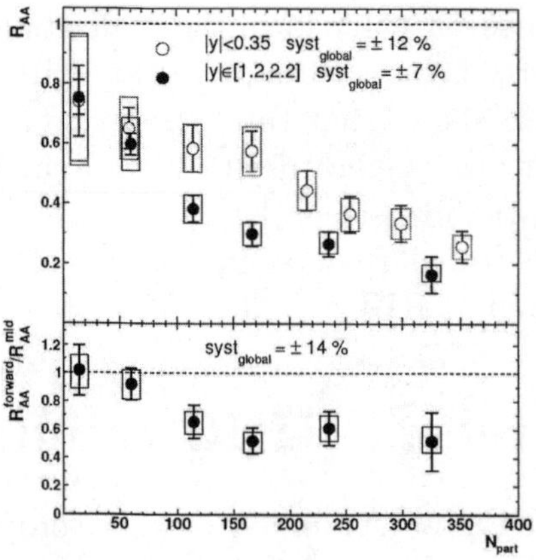

Figure 3: Top: R_{AA} of J/ψ vs N_{part} in Au+Au collisions at $\sqrt{s_{NN}} = 200$ GeV in mid ($|y| < 0.35$) and forward ($1.2 < |y| < 2.2$) rapidities[10] vs Bottom: Ratio of forward/mid-rapidity R_{AA} vs N_{part}.

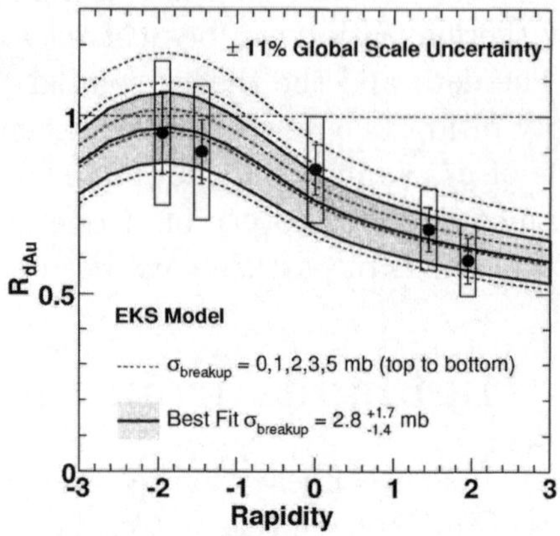

Figure 4: R_{dAu} of J/ψ in d+Au compared with theory prediction curves which includes EKS shadowing and J/ψ breakup cross section.

It is important to study the cold nuclear matter (CNM) effect on J/ψ production to interpret the Au+Au data. PHENIX has studied CNM effect on J/ψ in d+Au data set of RUN3 (2003). Figure 4 shows R_{dAu} from RUN3 d+Au data[11]. The data shows stronger suppression of J/ψ for $y > 0$, which is the deuteron going direction and corresponds to small x of partons in Au nucleus. The band in the figure represents a theoretical calculation of R_{dAu} that includes the effects of gluon shadowing parameterized by EKS and a small J/ψ breakup cross section. The calculation well describes the data. From the comparison of the data and the calculation, the breakup cross section is determined as $\sigma_{breakup} = 2.8^{+1.7}_{-1.4}$ mb, which is consistent with the value, 4.2 ± 0.5 mb, determined at CERN-SPS at $\sqrt{s_{NN}} \simeq 20$ GeV.

Due to large uncertainties of CNM effects from the available d+Au data, one can not conclude that the suppression beyond CNM effect is stronger in forward rapidity than in mid-rapidity. More precise d+Au data is needed to make definitive conclusion. In 2008 PHENIX collected very high statistic d+Au data, approximately 30 times of d+Au data in RUN3. The new data will give sufficient constraint on CNM effect.

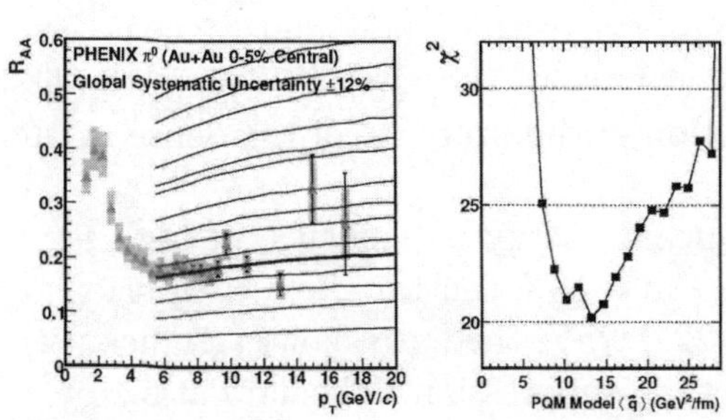

Figure 5: Left: π^0 R_{AA} for 0-5% central Au+Au collisions at 200 GeV with PQM model calculations for different values of $< \hat{q} >$. Right: χ^2 distribution as a function of $\hat{q}$[12].

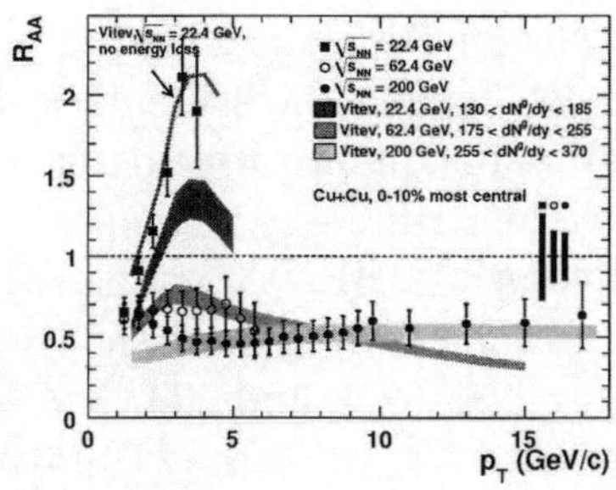

Figure 6: π^0 $R_{AA}(0_T)$ in central (0-10%) Cu+Cu collisions at $\sqrt{s_{NN}}$=22.4, 62.4, and 200 GeV. The theory curves by Vitev[16] are compared with the data.

3.2 Hadron suppression at higher p_T

The measured p_T range of R_{AA} of π^0 in central Au+Au collisions increases as more data are accumulated. The p_T range in the measurement in RUN1 (2000) is only up to 4 GeV/c[3]. It is then extended to 10 GeV/c at in RHIC RUN2 (2001). The latest results from high statistic RUN4 (2004) Au+Au data set have R_{AA} up to 20 GeV/c[12]. R_{AA} is approximately 0.2 in the top 10% centrality for $p_T > 5$ GeV/c with little p_T dependence. Since the p_T spectrum is well described by a power-law ($\propto p_T^{-n}$) at high p_T, this almost constant behavior of R_{AA} implies that high p_T partons suffer nearly constant fractional energy loss in the dense matter.

The high statistics data test the energy loss models and give constraint on the model parameters. PHENIX has performed a quantitative comparison between the R_{AA} data and various energy loss models[12, 13]. Figure 5[12] shows an example of such a comparison. Here R_{AA} data of π^0 in the 0-5% central Au+Au collision are compared with PQM model[14]. Each of the theory curves corresponds to different value of transport coefficient $< \hat{q} >$, the key parameter that controls the energy loss of partons in the medium. For each curve, the χ^2 value that takes into account statistical and systematical uncertainty of the data is calculated. The χ^2 value as a function of $< \hat{q} >$ thus calculated is shown in the right panel of the figure. In this way, the value of $< \hat{q} >$ is determined as $13.2^{+2.1}_{-3.2}$ GeV^2/fm. We should note that the value of $\hat{q}$ is determined in the context of the model and that the error does not include the systematic uncertainty of the model. This is the first step towards quantitative determination of the property of the dense matter formed at RHIC.

Recently, PHENIX studied the beam energy dependence of high p_T π^0 suppression[15]. Figure 6 shows R_{AA} in 0-10% central Cu+Cu collisions at $\sqrt{s_{NN}}$=22.4, 62.4, and 200 GeV. At 22.4 GeV, the data shows significant enhancement ($R_{AA} = 1.5 - 2.0$) rather than suppression. The data at 62.4 GeV and 200 GeV shows suppression. The calculations by an energy loss model by Vitev[16] are compared with the data. The model takes into account the Cronin enhancement, parton energy loss in cold nuclear matter, and final state parton energy loss in the dense medium. The model can reproduce the Cronin enhancement data in $p + A$ at fixed target experiments. While the model calculations, shown as bands in Fig. 6, agree with the data at 62.4 GeV and 200 GeV, it underestimate the data at 22.4 GeV. If the final state parton energy loss in the model is turned off, the model can reproduce the 22.4 GeV data. This suggests that parton energy loss starts to prevail over the Cronin enhancement in Cu+Cu collisions between $\sqrt{s_{NN}}$ = 22.4 GeV and 62.4 GeV.

3.3 Modification of Jet correlation

The strong suppression of high p_T π^0 in central Au+Au collision shows that the parton lose significant fraction of its energy while it propagates through the dense medium. The energy loss can modify the fragmentation of the parton into jets of hadrons. In heavy-ion collisions at RHIC, jets are not directly reconstructed due to very large soft background. They are, however, observed in correlation between two high p_T hadrons. The medium effects on the jets are then observed as modification of jet-induced two particle correlation.

Figure 7 shows the jet-pair distribution for two high p_T hadrons A ($2.5 < p_T < 4.0$ GeV/c) and B ($1.0 < p_T < 2.5$)[17]. The correlation due to the elliptic flow is subtracted from the data using ZYAM ("zero yield at minimum") assumption. The distribution is normalized as per trigger particle A. In the most peripheral collisions (60-90%) the data shows two back-to-back peak as expected for normal di-jet events. For more central events, the shape of the "away-side" jet is dramatically modified: the distribution becomes much broader, and there seems to be a minimum of the distribution at

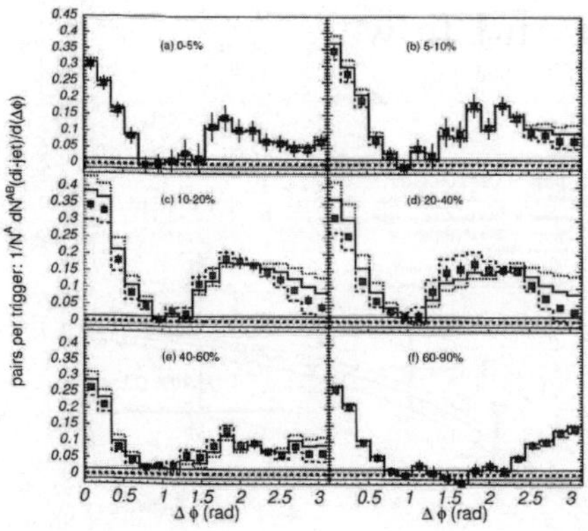

Figure 7: Jet-pair distribution in Au+Au collisions at $\sqrt{s_{NN}}$=200 GeV in various centralities[17].

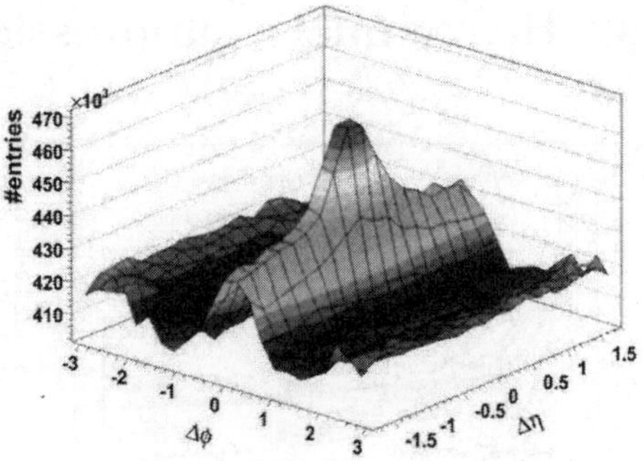

Figure 8: dihadron correlation function in $\Delta\eta \times \Delta\phi$ space in central AAu collisions reported by STAR, showing the "ridge" along η direction at $\Delta\phi = 0$

$\Delta\phi = \pi$. Similar effects are reported by STAR. These quite surprising results triggered many theoretical works to explain the effect, but so far there is no clear consensus on the interpretation of the data. A very interesting interpretation of the effect is that it is caused by Mach cone in the medium. The scattered parton travels faster thant he speed of sound the in the medium, causing a shock wave. If this is the case, the opening angle can be related to teh speed of sound in the medium.

Recently, another very interesting phenomena, "ridge" is reported by STAR[18]. Figure8 shows the dihadron correlation function in 2D ϕ and η space reported by STAR in Au+Au collision. Here a long $\Delta\eta$ correlation is seen at the same azimuthal direction ($\Delta\phi \simeq 0$) of the high p_T trigger hadron. This is the latest surprise in the jet correlation study in Au+Au collisions and it becomes one of the hottest topics. Similar effects are also reported by PHOBOS and PHENIX. This phenomenon is not well understood. The correlation length in the η direction is 2 to 4 units of pseudo-rapidity. It is difficult to understand how such a long range $\Delta\eta$ correlation can be generated. It might be caused by some non-linear interplay between jets and v_2.

3.4 Heavy quark suppression and flow

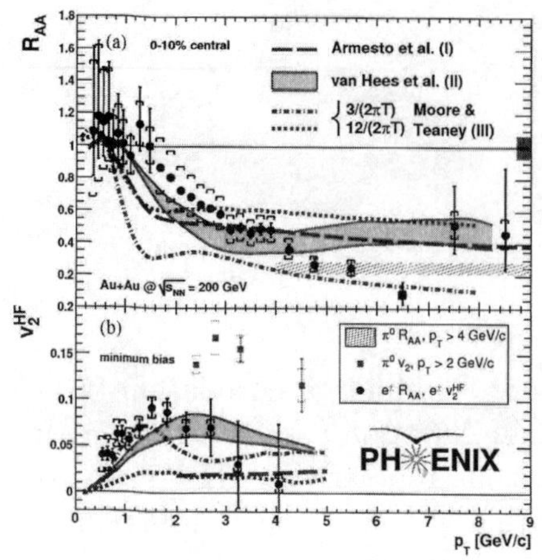

Figure 9: R_{AA} and v_2 of single electrons from heavy flavor decay in 0-10% central Au+Au collisions.

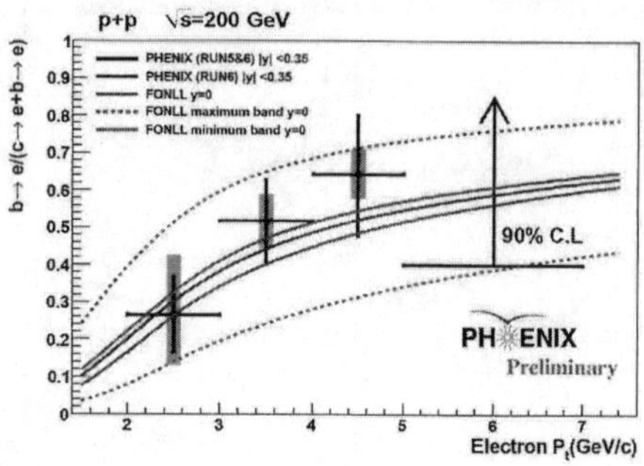

Figure 10: The fraction of $b \to e$ contribution in heavy flavor decay electrons as a function of electron p_T. The curves are prediction by FONLL calculation and its uncertainties.

Study of heavy quark (charm and bottom) production in heavy-ion collision can provide further insight on the property of the dense medium. Due to its large mass, heavy quarks are produced only at the initial energetic parton collisions. It is expected that energy loss via gluon radiation is small for heavy quarks, thus increasing their thermalization time. Consequently, weaker high p_T suppression and smaller v_2 is expected for heavy quarks.

PHENIX measures heavy quark production indirectly through measurements of single electrons from heavy quark decays. Figure 9 shows[19] R_{AA} and v_2 of single electrons form heavy flavor decays in Au+Au collisions at $\sqrt{s_{NN}} = 200$ GeV. A large suppression and strong v_2 is observed. The data show that heavy quarks strongly interact with the medium. A short relaxation time and/or a small diffusion coefficient for heavy quarks are required by the data. Two models (the models by van Hees and Rapp (II) and the models by Moore and Teaney (III)) qualitatively describe both of R_{AA} and v_2. From model comparison, the viscosity to entropy ratio η/s of the medium can be estimated as $\eta/s = (4/3 \sim 2)/4\pi$[19]. The estimated value is very small and is close to the conjectured quantum lower bound $1/4\pi$ from AdS/CFT.

A limitation of the heavy quark data shown in Fig. 9 is that we cannot distinguish b and c. To overcome this limitation, PHENIX has made the first attempt to disentangle $b \to e$ and $c \to e$ in $p+p$ collisions. Figure 10 shows[20] the fraction of $(b \to e)$ in the inclusive heavy-flavor electron from the analysis of electron-hadron correlation. The contribution from b becomes dominant for $p_T > 4$ GeV/c, as predicted by FONLL calculations. From the bottom electron spectrum, the total bottom production cross section in $p+p$ at 200 GeV is obtained as $\sigma_{b\bar{b}} = 4.61 \pm 1.31^{+2.57}_{-2.22}$ μb.

4 Photons and dileptons

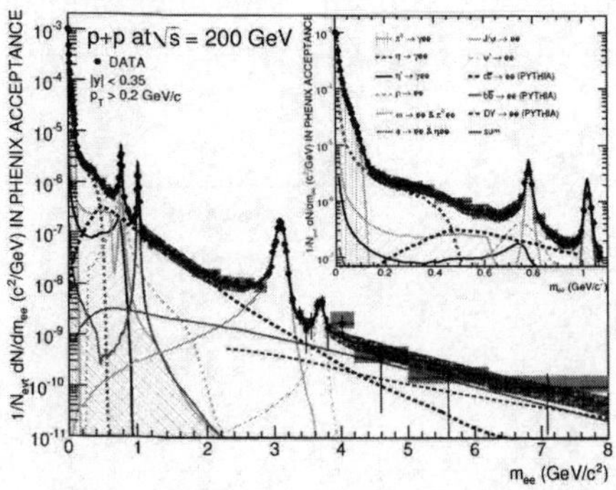

Figure 11: Low mass ee continuum in pp

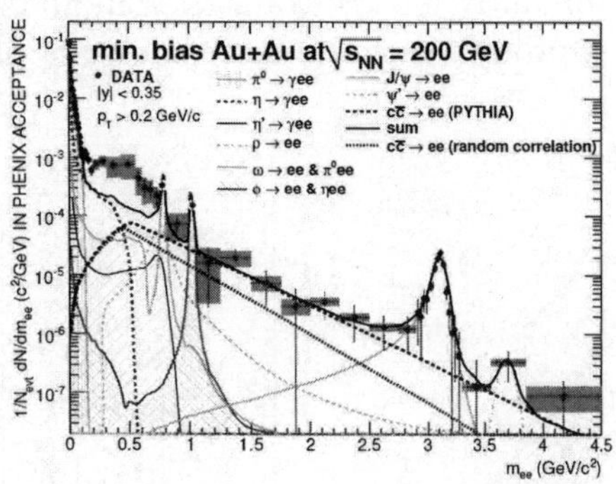

Figure 12: Low mass ee continuum in Au+Au

Electromagnetic probes, i.e. lepton pairs and direct photons, are ideal probes of dense medium. Since they have no strong interaction, they leave the medium without final state interaction, carrying the information deep inside of the medium.

Figure 11 shows[21] PHENIX measurement of low mass electron pair continuum in $p+p$. The data are compared with the expected mass spectrum from hadron decays. The hadron decay cocktail calculation is based on measured cross sections of light hadrons. It also includes the contribution from charm based on PYTHIA event generator. The total charm cross section and a few other parameters of PYTHIA is tuned so that it reproduce the single electron spectrum from heavy quark decay measured by PHENIX reasonably well for $p_T < 2$ GeV/c. The calculated mass spectrum agrees very well with the data. Charm is the dominant contributor for the mass region between ϕ

and J/ψ. From the comparison with the data and the calculation in this mass range, total charm cross section is obtained as $\sigma_{c\bar{c}} = 518 \pm 47 \pm 135 \pm 190$ μb. This value is consistent with that from single electron measurement. Similarly, b cross section is obtained from the mass region above J/ψ as $\sigma_{b\bar{b}} = 3.9 \pm 2.5^{+3}_{-2}$ μb, which is in agreement with the b cross section obtained from the electron-hadron correlation analysis discussed in the previous section.

Figure 12 shows[22] electron pair continuum in Au+Au. Unlike the $p+p$ data, the Au+Au data shows a large enhancement over the hadronic cocktail in low mass region. Integrating over the mass range $150 < M_{ee} < 750$ MeV/c^2, the enhancement is a factor of $3.4 \pm 0.2 \pm 1.3 \pm 0.7$. These excess is dominantly for low transverse momentum $p_T < 1$ GeV/c. This suggests that it is produced in the later hadronic phase of the reaction[23].

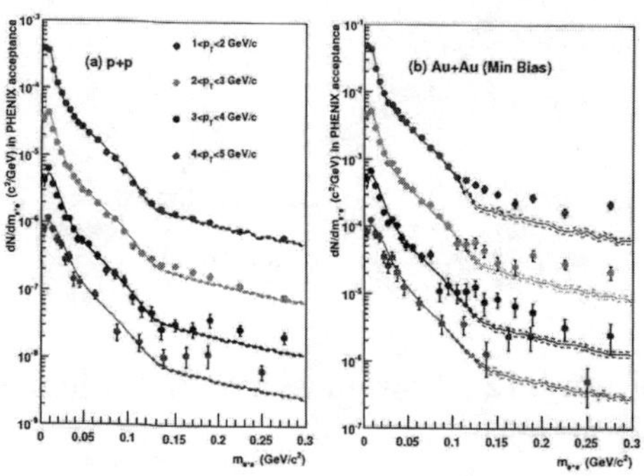

Figure 13: Low mass ee continuum in pp and Au+Au collisions compared with the expected mass distribution from hadron decays.

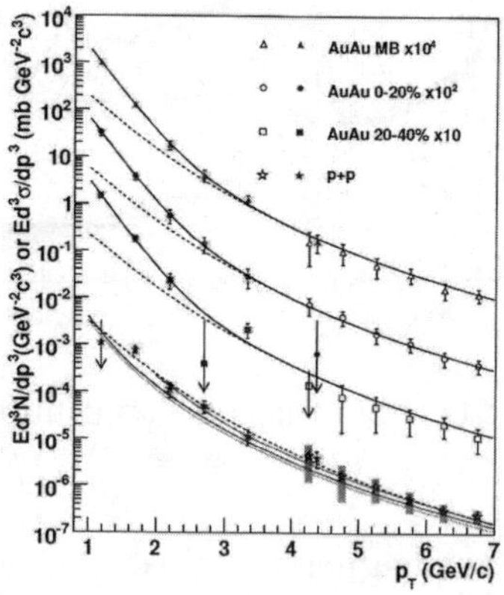

Figure 14: Direct photon spectra in $p+p$ and Au+Au collisions.

If the dense partonic matter formed in early stage of Au+Au collisions thermalized, it should emit thermal photon. If the thermal photon spectrum is measured with sufficient accuracy, it would provide a significant constraint on the initial temperature of the dense matter. It is predicted that thermal photons from the partonic phase is the dominant source of direct photons for $1 < p_T < 3$ GeV/c.

The measurement of real direct photon in such a low p_T range is, however, very difficult due to large background of hadron decay photons. To overcome this problem, PHENIX instead measured low mass, high p_T electron pairs, i.e. quasi-real virtual photons. By measuring the electron pairs just above π^0 mass, the background contribution from π^0 decays is removed, and the background is reduced by a factor of five.

Figure 13 shows[24] electron pair mass spectra in $p+p$ and Au+Au collisions for $m_{ee} < 300$ MeV/c^2 and $p_T > 1$ GeV/c in various p_T ranges. The Au+Au data shows a significant e^+e^- excess above the hadronic cocktail. A small but significant excess is also observed in $p+p$. The shape of the excess is found to be consistent with that expected from internal conversion of direct photon. Treating the excess as photon internal conversion, the invariant yield of direct photon is deduced.

Figure 14 shows[24] p_T distribution of direct photon from this analysis, together with the direct photon data at higher p_T from real photon measurements using electromagnetic calorimeter[6, 25]. The direct photon spectra in $p+p$ agrees well with pQCD calculation. The direct photon spectra in Au+Au shows significant excess over the binary collision scaled $p+p$ spectrum. The Au+Au spectrum is fit to an exponential plus the binary scaled $p+p$ spectrum. The fit results are shown as solid curves in the figure. In central Au+Au collisions, the inverse slope T obtained from the fit is $T = 221 \pm 23 \pm 18$ MeV. Several hydrodynamical models[26] can reproduce the central Au+Au data within a factor of two. These models assume formation of a hot system with initial temperature T_{init} = 300 MeV to 600 MeV.

References

[1] K. Adcox et al., Nucl. Phys. **A**757 (2005) 184.

[2] J. Adams et al., Nucl. Phys. **A**757 (2005) 102.

[3] K. Adcox et al., Phys. Rev. Lett. 88 (2002) 022301.

[4] K.H. Ackermann et al., Phys. Rev. Lett. 86 (2001) 402.

[5] K. Adcox et al., Phys. Rev. Lett. 91 (2004) 072303.

[6] S.S. Adler et al., Phys. Rev. Lett. 94 (2005) 232301.

Cherenkov, if the phyics is just what we are thinking it is , and we can apply the cuts in the water as ideally as you think it might work. With liquid argon there is no question, you actually see the interaction, you know exactly what's happening, it's almost like a bubble-chamber. Ultimately, it would be great to have liquid argon available as a detector; the caverns that you have to build will be considerably smaller, by a factor of 3 or 4, relative to the water Cherenkov counter. The other aspect for liquid argon is that it's sensitive in principle to different proton decays tha water Cerenkov counters. It's important to do it in water for some of the modes, and in liquid argon for other modes. Of course, super-K has worked for a number of years, so for a water Cherenkov counter to compete it in terms of proton decay, you need to make a much greater mass. By the time one starts this detector, super-K will have integrated 20 years of running. In order to catch up relatively quickly, you need a factor of 5-10 in mass.

- *Bernhard:*

How have the event yields for the rare K-decays (K^+ and K^0) been estimated?

- *Oddone:*

What people have looked at, was an experiment proposed at Brookhaven for measuring those rates. Typically, they would have had a handfull of events. With this number of events those ratios are known to 40% accuracy. We looked at the various factors that would come in. Once you will have a beam of the kinds of intensity produced with project X. One of the advantages, for example in terms of the K^0 to $\pi^0\nu\nu$mode of the K_{long}, that in the Brookhaven experiment a ribbon beam would have been used, in order to get enough beam to the detector. Therefore, there was not really a good 4π coverage, for example for the calorimeter veto. Whereas, with project X we can produce a real pencil beam going into the detectors. We multiplied all those factors and it looked like what was 10 events in Brookhaven, could easily be 1000 events here, and that would give you a measurement with ~3% accuracy.

CHAIRMAN: Y. AKIBA

Scientific Secretaries: F. Bellini, D. Caffarri

DISCUSSION

- *Caffarri:*

How is centrality measured in heavy ion collisions at Relativistic Heavy Ion Collider (RHIC)?

- *Akiba:*

For the centrality measurement with PHENIX detector we combine signals from the forward Beam-Beam Counter (BBC) and the Zero-Degree Calorimeter (ZDC). The BBC counts the charged particles produced with pseudorapidity belonging to range $3<\eta<4$ by means of a Cherenkov light detector. The signal is then proportional to the number of participants, roughly. Higher centrality means higher multiplicity. We also have a ZDC which measures the number of forward neutrons, which is proportional to the number of spectator nucleons. Once you set the correlation between the BBC and ZDC signals, you can select the class centrality of the collision.

- *Caffarri:*

If I'm not wrong, RHIC will be upgraded in the next months. How will it be pursued?

- *Akiba:*

We want to increase luminosity, so the idea is to use stochastic cooling to get luminosity of several times more. RHIC is upgrading its injector to EBIS. When this will be completed, RHIC will be able to accelerate heavier nuclei, up to uranium. But this is not a next months thing. I think it will take a few years.

NEWS FROM THE GRAN SASSO UNDERGROUND LABORATORY

Eugenio COCCIA
INFN Gran Sasso National Laboratory and U. of Rome Tor Vergata, Italy
E-mail: eugenio.coccia@lngs.infn.it

ABSTRACT

Experiments in underground laboratories have opened a new window in neutrino physics, beyond the standard model, and are today at the frontier in the investigation of neutrino features and in the study of dark matter nature. The future of underground science appears challenging and rich. This is true in particular for the INFN Gran Sasso Laboratory whose main characteristics, present scientific activity and future perspectives are presented.

1. Underground physics

Underground laboratories provide the necessary low radioactive background environment to investigate fundamental physics processes: very high energy phenomena, such as p-decay and neutrinoless $\beta\beta$ decay, happen spontaneously, but at extremely low rates; the study of neutrino properties from natural and artificial sources and the detection of dark matter candidates requires capability of detecting extremely weak effects. Underground laboratories appear complementary to those with accelerators in the basic research of the elementary constituents of matter, of their interactions and symmetries. It is remarkable that the first clear evidences for physics beyond the standard model (SM) comes from underground experiments.

In general one can consider two classes of motivations to push beyond the SM. There are "particle physics" reasons: the SM does not truly unify the elementary interactions, it leaves the problem of fermion masses and mixings unsolved and it exhibits the gauge hierarchy problem in the scalar sector. The second class of reasons finds its origin in "astroparticle and underground physics" issues: the problems of the solar and atmospheric neutrinos deficits, Majorana or Dirac neutrinos, baryogenesis, dark matter. These issues can be faced by contemporary experimental physics and constitute a formidable motivation for the underground laboratories activity.

In underground physics the struggle to advance the high energy frontier and to go beyond the standard model is the struggle for background control and reduction. Environment is the principal source of background. The environmental backgrounds of a laboratory depend on the depth and on the nature of the surrounding rocks and, as a consequence, may differ in the different facilities.

The relevance of the different background sources clearly depends on the experiment. For example, the energy directly deposited by the muons in the detector is more relevant for neutrinoless $\beta\beta$ decay experiments than for dark matter searches, while the neutron backgrounds and cosmogenesis are important for both; in any case a depth of 1000 m or a little more is enough. Even atmospheric neutrinos, interesting by themselves as they are, are a background; the principal one already at shallow depth, for proton decay experiments, which, as a consequence, need not to be too deep.

Frontier experiments do not require to the laboratory only a low background environment, but also technological support, easy and safe access and general as well as specific support structures.

I review here the present activity and future perspectives of the INFN Laboratori Nazionali del Gran Sasso. More information can be found on our WEB-site http://www.lngs.infn.it.

2. The Gran Sasso National Laboratory

The INFN Laboratori Nazionali del Gran Sasso (LNGS) is the largest underground laboratory in the world. It is located besides a freeway tunnel under the Apennines, in central Italy near the town of L'Aquila, 120 km from Rome.

The mission of the Laboratory is to host experiments that require a low background environment in the field of particle and astroparticle physics and nuclear astrophysics and other disciplines that can profit of its characteristics and of its infrastructures.

The geographical location (inside the National Park of Gran Sasso - Monti della Laga) and the special operating conditions (underground, near a highway tunnel and in close proximity to a large water basin) demand that special attention is paid to the safety and environmental aspects of the activities.

The realization of a large, high technology, underground laboratory was proposed in 1979 by A. Zichichi, then the President of INFN, and approved by the Italian Parliament in 1982. Since the original project the orientation of the halls was towards CERN, in order to host detectors to study neutrino physics on a beam produced at that laboratory.
Civil engineering works, under the responsibility of ANAS, the Italian Road Department, started in autumn 1982 and were completed by 1987.

The horizontal access allows easy transportation and installation of large pieces of apparatus. The underground facilities consist of three experimental halls and a set of connecting tunnels and service areas, for a total surface area of 18000 m^2. The three halls are approximately 100x18x18 m^3. An almost angle-independent 1400m rock overburden provides a μ flux attenuation of 10^{-6}. The neutron flux is $= 3.7 \pm 0.3 \times 10^{-2} m^{-2} s^{-1}$.

The infrastructures of the laboratory are completed by the surface site in Assergi, nearby the western entrance of the tunnel, hosting the directorate, offices, laboratories, workshops, library, canteen, etc. General services include administrative and

Figure 1: The underground facilities and the Gran Sasso massif geology

user support offices, the prevention, safety and environmental protection services, maintenance and, in direct support to the experiments, low activity techniques (including very low background facilities with Ge detectors underground), cryogenics, mechanics, chemistry, electronics, computing and networking. Last but not least, the laboratory has an important activity in the outreach, in the diffusion of the scientific culture and in the training of young people, with the active support of the Abruzzo Region.

The permanent staff is composed by 76 persons. Presently, the user are about 800 in number, from 25 different countries.

After almost 20 years of operational life, several first generation experiments have been completed and decommissioned, like MACRO and Gallex-GNO. A programme for the second phase has been implemented.

Main research topics of the present scientific programme are: neutrino physics with neutrinos naturally produced in the Sun and in Supernova explosion and neutrino oscillations with a beam from CERN (CNGS program), search for neutrino mass in neutrinoless $\beta\beta$ decays, dark matter search, studies of nuclear reactions of astrophysical interest.

Let me summarise the recent activity of the main research lines.

3. Neutrinos and stellar physics

Solar neutrino physics is one of the traditional research sectors of the laboratory. After the glorious life of GALLEX and GNO, the focus is now on Borexino, which is dedicated mainly to the measurement of the Be line component of the solar neutrino spectrum. The data taking started in May 2007. The excellent radio-purity reached by the Borexino experiment, in some cases at the level of about two orders of magnitude better than what expected, open a new window in the solar neutrino physics. In particular, recent results published by the Borexino Collaboration on the measurement of the solar neutrino flux from ^{7}Be, demonstrates that, with enough statistics, systematic errors will dominate on the statistical ones. The actual statistical error of 10% will be reduced to below 1% after 1 year of lifetime and the systematics at 3-5% after calibrations. Such precise measurement would represent a strong constrain for the actual Solar Model, providing a double test of our knowledge of the energy production mechanisms in the Sun and of the LMA-MSW flavour oscillation model.

The solar models are based on data and extrapolations; in particular the thermonuclear cross sections of the involved reactions are not measured in the relevant energy range but rather extrapolated from higher energies. The direct measurements are made very difficult by the very low values of the cross sections. Using the new 400 kV accelerator, LUNA is continuing his activity for the measurements of the cross section of nuclear reaction of astrophysical interest.

Stars substantially more massive than the Sun end their life through spectacular stellar explosions known as core collapse supernovae. Supernovae are among the most violent phenomena in the universe. The huge amount of the binding energy of the collapsed core is emitted in the form of neutrinos and anti-neutrinos of all flavours. The observation of neutrinos released in the various stages of the explosion will provide direct information about the dynamics at the center of the Supernova and therefore of the explosion mechanism. Neutrinos are in fact a unique tool to study the central region of the star, and hence the nuclear matter in the extreme state of temperature and density encountered in the Supernova core. Neutrinos originating from Supernovae explosions are continuously monitored, with 100% live time, by the LVD detector, constituted by 1000 tons of liquid scintillator.

LVD is part of the SNEWS (Supernovae Early Warning System), which involves an international collaboration of experimenters representing several of the current supernova neutrino detectors in the world. The goal of SNEWS is to provide the astronomical community with a prompt alert of the occurrence of a Galactic core collapse event and enable optical observation of the early phases of the Supernova.

LVD is also monitoring the neutrino beam from CERN.

4. Neutrino mass and nature

Elementary particles are different from their antiparticles because their charges - not only the electric one, but all of them - are opposite. The standard model assumes that neutrinos have only one charge, the lepton number. But, if this charge is not conserved, neutrinos and antineutrinos can be two states of the same particle. In this case well-specified nuclides would decay through the neutrinoless $\beta\beta$ channel. The Laboratory hosts today experiments searching for these very rare decays, employing different and complementary techniques, and is preparing new important activities. The Heidelberg-Moscow experiment, which with a sensitive mass of 11 kg of enriched ^{76}Ge was the most sensitive experiment in the world for several years and accumulated data for a 75 kg y exposure, was terminated in 2003, claiming a positive detection. In the same year CUORICINO, which employs TeO_2 bolometers for a total of 41 kg, started taking data and is the most sensitive running experiment today. The Laboratory approved two new experiments: CUORE, which brings the CUORICINO technique to a mass of more than 400 kg of TeO_2 bolometers, and GERDA planning to employ 500 kg of enriched ^{76}Ge. These activities represent the state of the art of the field. They will start taking data, respectively, in 2009 and in 2010.

One of the major commitments of the Gran Sasso laboratory in the next years will be the search of tau neutrino appearance on the artificial neutrino beam being built at CERN in Geneva: the CERN Neutrinos to Gran Sasso (CNGS) project. A beam of protons, with an energy of 400 GeV, produced in the SPS accelerator of CERN, collides with a target of graphite to produce a secondary beam consisting partially of pions (π) and kaons (K), charged particles with short lifetimes. After having been focused and directed towards the Gran Sasso, the π^+ and K^+ decay, giving origin to neutrinos that maintain practically the same direction of the particles that generated them. The neutrino beam contains almost exclusively muon neutrinos with an average energy of 17.4 GeV and a contamination on the level of a few percent of anti-muon neutrinos, electron and anti-electron neutrinos. The beam is directed through the Earth crust to Gran Sasso at 732 km distance and reaches the Gran Sasso Laboratory with a flux of tens of billions per day. The first neutrinos arrived in august 2006, and the beams is working regularly since this year 2008.

The OPERA experiment is designed for the direct observation of tau neutrinos resulting from oscillations of the muon neutrino of the beam. This search requires both micrometer scale resolution, obtained with modern emulsion techniques and large sensitive mass (1300 t) obtained with Pb sheets interleaved by emulsion layers. The experimental apparatus is made up of two supermodules, each formed by a target followed by a muon spectrometer. The entire target consists of 150,000 bricks, where the neutrino interactions occur. The brick walls are interleaved with layers of Target

Trackers, detectors that electronically track the event, permitting the determination in real time of the position. The measurement of the position of the neutrino interaction allows the identification of the brick in which the interaction occurred. That brick can be extracted and, after developing the photographic emulsions, can be analyzed using automated microscopes. While the events concerning muon neutrinos are relatively frequent (about 30 per day), only a few tau neutrino events are foreseen each year. The production and installation of the 150 000 bricks containing the Pb sheets and the emulsion layers is terminated and the experiment is in full data taking.

ICARUS is a general-purpose detector, with a broad physics programme, based on the novel concept of the liquid Argon time-projection chamber. After the development of the "definitive project" and of the risk analysis, completed in 2003, the 600 ton module was transported from Pavia to the underground Laboratory in december 2004 and is now at the end of the phase of installation and completion. The experiment should be ready for the beginning of 2009.

5. Dark Matter

There are many evidences today that most of the matter in the Universe does not significantly emit or absorb electromagnetic radiation, but manifests itself only through its gravitational interaction with luminous matter (stars and interstellar gas).

The conclusion from cosmological background observations, in particular the ones of WMAP, and nucleosynthesis arguments is that within an apparently flat universe at least 85% of the matter in the universe is in the form of non-baryonic, non-luminous matter.

Models of galaxy formation and simulations indicate that most of this dark matter component must be *cold*, i.e. consisting of particles that were non-relativistic at the time the universe became matter dominated and that are weakly interacting. Hence the denomination for the elusive particles of this dark matter as Weakly Interacting Massive Particles, or WIMPs. Galaxy rotation curves measurements lead to consider a WIMP halo of local density of 0.3 GeV/cm, a maxwellian velocity distribution with rms value of 220 km/s, and a relative velocity of the sun within the halo of 230 km/s. One of the most natural WIMP candidates is provided by supersymmetric theories: If R-parity is conserved, the lightest supersymmetric particle is stable and leads to a relic density compatible with the missing mass component when the interaction cross-section is of the order of the weak interaction cross-section. A natural candidate is the neutralino. Its predicted mass range is bound at low masses by LEP to 45 GeV, and at high masses to a few TeV, which gives it a non-negligible chance of being produced at LHC in the coming years. An indirect method to detect such particles is looking for products of WIMP annihilations in the Universe. The direct method is looking directly for WIMP interactions within a target material. The signal expected from a WIMP contribution have several distinctive signatures. WIMPs, like neutrons, scatter elastically off a nucleus in the target material, while gammas and betas scatter off the electronic cloud. If one can discriminate in the data a nuclear recoil from an electronic recoil, a large part of the background will have thus been eliminated. Moreover, for given WIMP characteristics (mass and cross-section) and for a given target material, a recoil energy spectrum is predicted, which can ideally be identified to the data, or at least allow to set an upper limit on the mass and cross-section. Last but not least, due to the relative speed distributions between this halo and an observer on Earth, both an annual and diurnal modulation of the WIMP signal flux and directionality, respectively, are expected. The amplitude of the annual modulation can only amount for as high as a few percent of the total WIMP signal.

The principles behind direct detection comes down to measuring the energy deposited by a WIMP interaction, which is expected to be in the keV to tens of keV range. There are three ways of measuring this energy, each choice leading in turn to a choice of target material, and which can eventually be combined:

- light signal, produced by a scintillation process, in pure crystals or in a liquid noble gas (fast).

- ionization signal from semi-conductor materials like Si, Ge, or CdTe (slower but very efficient). It is present also in liquefied noble gases.

- heat signal from a target material cooled at low temperature (slow but also very efficient).

Because of the very small interaction cross-section of WIMPs with ordinary matter, an extremely low event rate of events is expected typically, less than a few events per kg of target material and per year for preferred SUSY models. Background suppression is therefore a vital issue and operation deep underground is mandatory. It is also very important that the level of radio-purity of each component is kept to a minimum, in particular the parts closer to the detectors. Taking into account all these considerations has led to a number of experiments set in various underground sites all over the world and to the design of a new generation of large mass detectors, up to onc ton.

At Gran Sasso several experiments, using different techniques, are active and new experiments have been approved and started operating in the last years.

DAMA/LIBRA employs NaI crystals to detect the WIMPs by means of the flash of light produced in the detector by an Iodine nucleus recoiling after having been hit by a WIMP. To distinguish these events from the background, DAMA searches for an annual modulation of the rate, a behaviour that has several aspects that are peculiar of the searched effect and not of the main backgrounds. With about 100 kg sensitive mass DAMA was the only experiment worldwide sensitive to the annual modulation signature. After the conclusion of the experiment, results were published showing a signal of annual modulation. The larger experiment LIBRA, with 250 kg sensitive mass, published this year results from 4 years of data, confirming the annual modulation signal to better than 8 sigma.

CRESST searches for WIMPs with a cryogenic technique, looking for a very tiny temperature increase in the detector, due to the energy deposited by nuclei hit by the WIMPs. After the final design, the CRESST2 $CaWO_2$ detector, looking for a simultaneous temperature increase and light emission, has been installed and is operating.

Two new dark matter experiments, WARP and XENON have started their operation during 2006. They are based on the simultaneous detection of the ionization and scintillation signals in, respectively, liquid argon and liquid xenon. WARP performed preliminary measurements underground with a small (2.3 l) prototype and XENON took data with a 10 l apparatus. Both experiments are now preparing their 100 liter size detectors and will provide data in the next year.

6. More on Gran Sasso and european underground labs

The main activity of the theory group, staff and visitor scientists, has been focused on astroparticle physics, including solar and Supernova neutrinos, massive neutrinos, ultra high energy cosmic rays, topological defects and relativistic astrophysics. Important activity took place also in particle phenomenology and computer simulations of Lattice Field Theories. Finally it must me mentioned that a number of experiments in different disciplines, mainly geology and biology are profiting of the peculiar conditions of the laboratory.

The directors of the Western Europe underground laboratories (Gran Sasso in Italy, Modane in France, Canfranc in Spain and Boulby in UK) initiated actions to co-ordinate activities and to optimise the use of the available resources, taking into account the different characteristics of the infrastructures.

In the EU 6th "framework programme". The ILIAS project (Integrated Large Infrastructures for Astroparticle Science) was submitted successfully by ApPEC, the European inter-agency committee for astroparticle physics.
ILIAS contains several elements: the co-ordination of a number of activities of the laboratories (such as environmental background measurement and control, safety procedures, outreach activities, etc.), the funding for access of new users and common R&D in neutrinoless $\beta\beta$ decay and dark matter search. Detectors have evolved toward much larger masses, experimental setups and analysis tools have became more complex, the experimental collaborations, used to work as rather small groups, most evolve toward larger organisations. ILIAS represented a good opportunity to foster the necessary convergence. The continuation of this program, and in particular the coordination among the underground labs, is now being discussed in the community.

7. Conclusions

Experiments in underground laboratories have opened a new window in neutrino physics, beyond the standard model, and are at the frontier on the study of dark matter.

The activity of the Gran Sasso Laboratory gives an idea of how exciting are the perspectives for underground physics in the next decade. The challenges are the following:

- Investigate, by direct detection, the nature of dark matter.

- Determine neutrinos absolute masses and nature (Majorana vs. Dirac), searching for neutrinoless $\beta\beta$ decay.

- Solar Be and pp neutrino fluxes should be measured with flavour sensitivity and high (few percent) precision. Weak Interactions may be still richer than what we know today (remember lessons of J. Bahcall and of neutrino oscillations).

- Measuring or limiting $|\theta_{13}|^2$. This step is necessary before seriously launching a project for a neutrino factory.

- Design a new p-decay experiment with Megaton scale Cerenkov detector or tens of kiloton liquid argon TPC.

There is an increasing interest in realizing a large particle detector based on liquid Argon technology to afford exciting searches on neutrino physics with artificial sources, investigating CP violation in the leptonic sector, and preparing the technology for a proton decay experiment. Such perspectives could drive the realization of a new shallow depth laboratory in the Gran Sasso area and exploit again a dedicated CERN to Gran Sasso neutrino beam.

CHAIRMAN: E. COCCIA

Scientific Secretaries: I. Gnesi, K. Ni

DISCUSSION

- Silenzi:

LUNA is an accelerator experiment undergoing in Gran Sasso lab. Why is it needed to go in a quiet background environment instead of raising the luminosity of the accelerator?

- Coccia:

LUNA studies the nuclear reactions at the low energies existing inside the stars. If you want to study the effect of these reactions at low energy, as a matter of fact natural radioactivity is an important source of background and must be reduced. Going underground this background is reduced by 6 orders of magnitude. A gain which cannot be compensated by an increase of luminosity. Before LUNA, the cross section of these reactions were known only at higher energies and were extrapolated at the lower energies of the stars. To perform these measurements at low energies you have to pay a price which is to work underground.

- Kaminska:

1) What can we learn about supernova from supernova neutrino detection?
2) Do you expect that we will ever be able to detect relic neutrinos?

- Coccia:

The answer to the second question is: very hard to detect relic neutrinos. We can only, with the present technology, detect low energy neutrinos in certain cases. For instance, the detection of solar neutrinos below the MeV, that is what Borexino is able to measure now, required an extremely hard work for many years for background reduction. Relic neutrinos are out of present experimental possibilities.

About what we can learn by detecting neutrinos from supernova, well, we can learn a lot. You have to consider that neutrinos come out as a result of the collapse of the stars, 10 to 20 solar masses. The collapse forms a neutron star, and the binding energy is emitted mainly as neutrinos of all types, maybe even more than the 3 the types we know now. The supernova is really a peculiar laboratory that cannot be reproduced on earth. You can learn about the dynamics of the explosion and about what really is left after such a kind of event. The energy distribution of the neutrinos and their time of arrival on Earth, compared with the time of arrival of the photons and of the gravitational wave burst emitted by the supernova, allow the neutrino masses to be evaluated.

- Cappello:

About the results of DAMA experiment: it seems that modulation indicates a region in the Sigma vs mass plane for dark matter that is excluded by other experiments (like CDMS). How this discrepancy can be explained? Is it still possible to save the "dark matter explanation" for the modulation?

- Coccia:

This question is debated as a hot topic. It is possible that this experiment is measuring dark matter candidates. The reason why the comparison between different experiment results is not trivial at all and it's difficult to state one experiment is excluding any other is that the experiments are basically different. The target recoil nuclei are different among CDMS,

DAMA, XENON, WARP: you don't know actually the features of the WIMP particle, you don't know if the interaction is spin dependent or not, you don't know the coupling. It is really a story which has not yet ended.

- *Patterson:*

Is DAMA group actively pursuing alternative explanations to their modulation?

- *Coccia*:

It is an interesting question. They considered other possibilities of explanation; you have to consider that there is not only the annual modulation, but also the right phase. This is to say that the maximum and the minimum fall in the right days, the 2 of June and the 2 of December. Then you have to consider that the modulation occurs only in single hit events: the background of multiple hit events, are not modulated. There are several effects, including natural radioactivity, that are annually modulated, thus DAMA has studied many effects and made checks looking at all the logical possibilities, and they exclude that the source of the modulation can be anything different from dark matter. Of course what a some people think is that, even if they have excluded several local sources, it might exists a source no one has yet thought about. In my opinion it is absolutely necessary to continue this search with this and with different techniques and it would be an interesting idea to perform an identical DAMA experiment in some other laboratory, perhaps located in the other hemisphere, where the phase of many of the annual disturbances is opposite, the temperature, the pressure, the cosmic rays induced background.

- *Patterson*:

Is anyone pursuing a repeat of the DAMA style experiment?

- *Coccia*:

What I know is that the DAMA collaboration doesn't want to go to the other hemisphere. It is not simple to find another underground laboratory there. It would be simpler in this hemisphere, in Europe. Of course there are other laboratories in the USA and in Japan, but the attitude of the collaboration is that they don't see why they should go to another laboratory since they have done all the checks and the controls, but maybe someone in the community could repeat the experiment.

- *Patterson*:

Do the OPERA event counts include selection efficiencies, and what are the selection efficiencies, particularly for the non-muonic channels?

- *Coccia*:

The numbers I showed you included the efficiencies. My main worry for stating that these numbers are the right ones, as a laboratory director, is they are reported for 5 years of nominal beam intensity. But the nominal beam intensity has not yet been reached. This year (2008) we have an intensity half the nominal.

- *Patterson*:

What drives the complicated shape of typical dark matter theoretical regions on the usual σ vs M plots?

- *Coccia*:

Unfortunately I don't see here Elena Aprile, who could perhaps give a better answer to this question but I think that the reason is that if you take into account all the experimental and theoretical constraints, and the possible models, you get those peculiar shapes that you see in the cross section versus mass plot.

- *Klopot:*

A question concerning dark matter search experiments. How can one separate the events, caused by neutrino interaction from events caused by WIMP interaction?

- *Coccia*:

Most of these experiments are looking for nuclear recoils, and the consequent generation of heat or light or ionization or all of them. This is well beyond the possibilities of the neutrino. This could be possible only for masses of hundreds of GeV.

- *Wen:*

Which one is better for tau appearance measurement, OPERA or ICARUS?

- *Coccia*:

For detecting tau neutrino appearances, OPERA is better: it has explicitly been designed for these kind of detections. ICARUS has a wider research program, also devoted to astroparticle neutrinos. The size of ICARUS, which is now 600 tons, is not enough for reaching the sensitivity of OPERA.

- *Wen:*

How long will it take for OPERA to reach a θ_{13} sensitivity of 0.06? What is the biggest effect on its θ_{13} sensitivity?

- *Coccia*:

In principle you have to wait 5 years. It's the size, which is limited to the one that is there now. For having best performances you should have a much larger detector.

- *Nikiforov:*

What are the means to control geometrical deformations of emulsion in OPERA experiment due to gravity or temperature inhomogeneity, as well as the way to control relative alignment of different emulsion layers?

- *Coccia:*

There has been a lot of efforts for building emulsions and sheet of lead of the right size. They need to prepare the bricks automatically by a robot, and for doing that you need precise sizes for all these things. Once they are part of the detector, the work is already done. The only thing you have to see is the signature of the tau particle at the automatic imaging system of the emulsions, but for this part I think that gravity and other mechanical effects are not important. Gravity and mechanical distortions of the structure have been taken into account in such a way the precision is enough for allowing the detection.

- *Arnowitt:*

You showed the DAMA data for 2 keV – 5 keV, 2 keV – 6 keV with ~ 8σ effect. Do you know what the data gives out of these ranges? One should expect it to go down, but how much?

- *Coccia*:

Below 2 KeV there is not enough sensitivity and above 6 kev there is no more modulation.

- *'t Hooft:*

You mentioned the mass determination of neutrino. This would be one way to measure mass without measuring Δm^2, measuring flavor m^2 absolute, as it would be for supernova neutrino. If you had a very high time resolution you could even think to measure it from CERN neutrinos. Then the question is what is your time resolution. If CERN sends highly modulated neutrino beam, having a very high time resolution you can think even to measure the mass from that.

- *Coccia:*

I'm afraid it's not possible.

- *'t Hooft:*

What is your time resolution?

- *Coccia:*

The time resolution in the Opera experiment at GRAN SASSO is of the order of 100 picoseconds.

- *'t Hooft:*

So if you could do that at very low energy, and if you manage to send very low energy neutrino to Gran Sasso....

- *Coccia:*

...but not in the present experiment. You have to modulate strongly the energy of the beam.

- *'t Hooft:*

Yes you have to modulate...

- *Coccia:*

Thanks for the question. I will think about it.

HIGHLIGHTS FROM TRIUMF

Nigel S. Lockyer

ABSTRACT

We present highlights from the major research areas at TRIUMF, Canada's National Laboratory for Particle and Nuclear Physics. These include particle physics, nuclear physics, molecular and material sciences, and nuclear medicine. Examples from each of these areas and future plans are presented.

INTRODUCTION

TRIUMF is Canada's national laboratory for particle and nuclear physics and is owned and operated by a consortium of seven Canadian universities as a Joint Venture. An additional seven Canadian universities are associate members and participate in the laboratory governance through observer status on the Board of Management. TRIUMF is located on the campus of the University of British Columbia, in Vancouver, Canada. The laboratory site, with English Bay and Vancouver in the background, is shown in an aerial view in Figure 1.

Figure 1: The TRIUMF laboratory in Vancouver British Columbia, Canada.

All science at TRIUMF derives from accelerators and TRIUMF is the steward of accelerator science in Canada. The TRIUMF Rare Isotope Beam (RIB) program is driven by the 500 MeV H^- cyclotron, shown in Figure 2. There are also four small cyclotrons on site used mostly for medical isotope production, supporting both clinical and research applications. In addition there are several heavy ion linacs, the 35 MHz Split-ring RFQ in ISAC-I (Isotope Separation Accelerator Complex), the ISAC-I 106 MHz separated function DTL, and the superconducting radiofrequency (SRF) heavy ion linac in ISAC-II. A prototype copper ¼ wavelength cavity is shown in Figure 3. The prototype cavity is a precision mechanical copper model of a solid niobium high-beta superconducting cavity planned for an extension of the present SRF heavy ion linac. Twenty cavities will be required. The RFQ accelerates ions with $A/q < 30$ from 2 KeV/u to 150 KeV/u. The DTL accelerates ions with $A/q<6$ to final energies fully variable from $0.15<E<1.8$ MeV/u. The SRF heavy ion linac boosts ion energy to 20 MeV to provide RIBs above the Coulomb barrier. A new superconducting electron accelerator (e-linac [50 MeV, 500 kW]) is proposed for rare isotope production (SRF). The combined ability to produce RIB with a high power proton source and a high flux of photons would be unique in the world.

Figure 2: The TRIUMF original 500 MeV cyclotron, the world's largest, is the workhorse for the rare isotope beam (RIB) program and material science investigations.

The TRIUMF accelerator division is developing significant SRF infrastructure over the next few years. In collaboration with PAVAC Inc., located in Richmond, B.C., TRIUMF is constructing 20 beta< 1 cavities at PAVAC. Buffered Chemical etching or BCP is performed at TRIUMF. PAVAC is a precision machining and electron beam welder company. Collaborations with Fermilab, Jefferson Laboratory, and the Global Design Group (GDE) for the International Linear Collider are just beginning.

Figure 3: Mechanical prototype copper ¼ wavelength cavity. Twenty such solid niobium SRF cavities will be built by PAVAC, a local company. This cavity was designed at TRIUMF.

TRIUMF has significant programs in particle physics (mostly "offshore"), nuclear physics, nuclear medicine, and materials and molecular science. There are about 500 scientists, students, and staff on the TRIUMF campus, about 90 of whom are associated with MDS Nordion Inc., one of the world's largest producers of medical isotopes. The laboratory has about 1000 users and about 2000 visitors per year.

PARTICLE PHYSICS

The two largest projects at TRIUMF in particle physics revolve around investigating phenomena at the high energy frontier using the ATLAS detector at the Large Hadron Collider (LHC) at CERN and the T2K neutrino experiment at JPARC.

TRIUMF led the Canadian effort to join the LHC by constructing a number of components for the accelerator. These include 54 quadrupole magnets, shown in Figure 4, that act as a cleaning insertion for the beam before it enters the main LHC ring. Canada also designed the pulse forming networks for the kicker magnets. The Canadian particle physics community designed and constructed the hadronic endcap (HEC) for ATLAS. The HEC project was centered at TRIUMF. In addition, the forward hadronic calorimeter was constructed at the Universities of Toronto and Carleton with collaboration from TRIUMF. Canadian scientists make up about 4% of the ATLAS collaboration. TRIUMF also hosts a LHC GRID computing (ATLAS) Tier-1 centre. This centre is one of 11 around the world.

Figure 4: LHC cleaning insertion quadrupole magnets built in Canada.

The T2K experiment sends a beam of ν_μ from Tokai to Kamiokande. The far detector is the 50,000 ton SuperK water Cherenkov detector. Canada plays a major role in the near detector, the purpose of which is to measure the initial neutrino flux.

The T2K physics goal is to make a precise measurement of ν_μ disappearance and to discover $\nu_\mu \rightarrow \nu_e$ appearance, and thereby measuring θ_{13}. Canadian physicists are building the TPC and the fine grained scintillation detectors at TRIUMF. This detector system will constitute about 2/3 of the near detector. In addition, Canada has delivered the OTR, or optical transition radiation detectors, which measure the proton beam size and position that produces the neutrinos. In addition, TRIUMF has designed a part of the remote handling system associated with the target area. A major challenge of T2K will be how to design and operate a target when beam powers rises above a megawatt. This will be new territory and significant research will be required before it can be successful. Data taking will commence in 2009.

TRIUMF has a small internal particle physics program aimed at the $\pi \rightarrow e\ \nu$ decay chain. The experiment will measure the branching ratio of $\pi \rightarrow e\ \nu$ with a precision of better than 0.05%. The experiment will improve the sensitivity to new pseudoscalar interactions which are not helicity suppressed. The mass of the hypothetical particle probed (contact term) will be > 1000 TeV. This experiment will be able to set a limit on the mass of the charged Higgs boson of about 200 TeV.

RARE ISOTOPE PHYSICS

The major questions of nuclear physics are:

1) What binds protons and neutrons into stable nuclei and rare isotopes and where are the limits of existence?
2) When and how did the elements from iron to uranium originate?
3) What causes stars to explode? What are the details of core collapse supernovae?

The rare isotope beam (RIB) at TRIUMF is primarily for Z<82. TRIUMF is the highest power ISOL (Isotope Separation OnLine) facility in the world (50 kW of beam power on target) and consequently delivers the most intense beams of certain species. The power is limited by target technology not the beam driver. Halo nuclei are the most extreme rare isotopes, with a core nucleus and extra neutrons or proton on very large, classically forbidden orbits. This provides excellent testing ground for state of the art nuclear theory at extreme environments. The most famous halo nucleus is Li-11, which can be produced and studied in record quantities. There are a dozen world class experiments located in the three experimental areas (low, medium, and high energy). The TRIUMF RIB facility is a single user facility which is a major limiting factor to the amount of science performed. A further limitation to delivering science is the need to develop more varied targets that can function at high incident beam power. A layout of the low and medium energy RIB facility is shown in Figure 5.

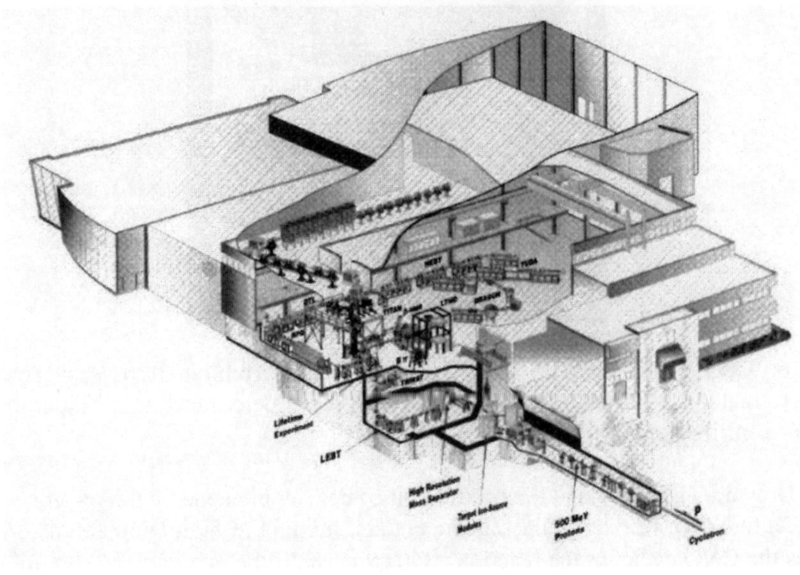

Figure 5: A layout of the RIB beam facility at TRIUMF showing the low and medium energy areas. The beam from the cyclotron enters from the lower left and strikes a target. The rare isotopes are then transported up one floor level to the experimental areas. The RFQ and DTL accelerators, designed and built at TRIUMF are shown. The high energy area is not shown.

NUCLEAR ASTROPHYSICS

What is the source of Al-26 in the galaxy? Satellites are able to detect a 1.809 MeV gamma ray from the decay process of Al-26. The COMPTEL satellite map, displayed in Figure 6, shows Al-26 concentrated in the spiral arms and the core of the galaxy. The source of Al-26 could be either from core collapse supernovae or from novae, although the satellite data suggests that most is made in supernovae. The DRAGON collaboration performed an experiment using a beam of Al-26 in which a weak resonance (one interaction every 3 trillion Al-26 particles) was searched for in $^{26}Al(p,\gamma)^{27}Si$. This information allows models to predict the supernova rate of 1.9 +/- 1.1 per century and 7.5 new stars born per year. DRAGON did observe a very weak resonance, indicating a small novae contribution and therefore a much larger core collapse supernovae contribution to the creation of Al-26 in our galaxy.

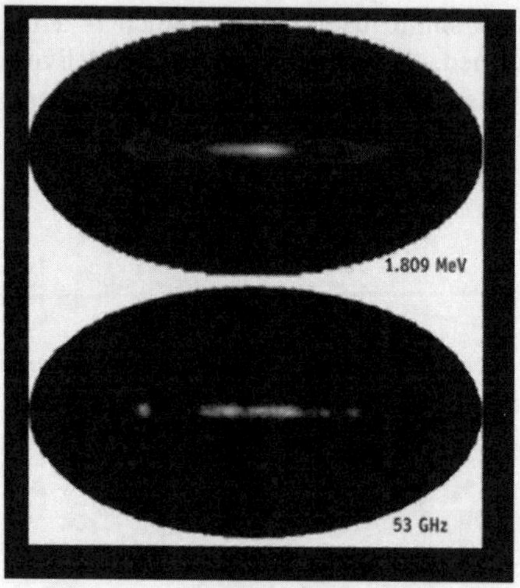

Figure 6: COMPTEL satellite data showing where Aluminum-26 is detected (and therefore created) in our galaxy. The primary source is determined to be supernovae and a smaller contribution from novae.

How do we understand the production of heavier elements in the rp-process (rapid proton capture)? In heavier main sequence stars, burning at high temperatures (0.5 GK) involves the CNO cycle. In the reaction $^{15}O(\alpha,\gamma)^{19}Ne$, if the rate of alpha capture to O-15 β^+ is high in hot explosive environments, large amounts of Ne-19 is formed.If Ne-19 captures a proton and beta decays to Ne-20, breakout from the hot CNO (HCNO) cycle into the rp-process occurs and synthesis up to Se-68 and beyond occurs.

It is also important to know composition of neutron star crust for understanding x-ray bursts and super x-ray bursts. The $^{15}O+\alpha \rightarrow {}^{19}Ne+\gamma$ reaction rate must exceed a threshold value or accreted matter burns stably without x-ray bursts. This rate is determined by an experiment at TRIUMF by measuring the lifetime of $^{15}O+\alpha \rightarrow {}^{19}Ne+\gamma$ and the branching ratio of Ne-19 states. The reaction rate is dominated by a single resonance corresponding to 4.03 MeV state in Ne-19. The result yielded 6.9+/-1.5(stat)+/-(sys) fs., which is ~4 times more precise than previous measurements and implies the reaction rate is twice as large as thought at x-ray burst temperatures.

NUCLEAR STRUCTURE

Halo nuclei are of great interest because they stretch the models trying to describe their properties. In 1985, Tanihata et al. used a beam of light nuclei impinging on beryllium, carbon, and aluminum targets. They found the Li-11 to be much larger than its neighboring nuclei, in fact the same radius size as Pb-208. Referred to as a halo nucleus and shown in Figure 7, the two neutrons "orbit" the central core Li-9 nucleus. The Li-9 plus one neutron is unstable, hence the name Borromean nucleus, and Li-11 has the lowest binding energy of all nuclear systems. The discovery by Tanihata was not predicted. The charge radius has been measured at TRIUMF by the ToPliS collaboration by measuring laser frequency shifts for the lithium isotopes. More recently, the Li-mass measurement by TITAN (TRIUMF Ion Trap) has set a new scale for precision of mass measurements of unstable nuclei. Shown in Fig. 8 is the most recent data indicating mass measurements of Li-8-9-11. The Li-11 half life of 8.8 ms is the shortest-lived isotope trapped in a penning trap. Until this measurement, the world uncertainty was δm=19 keV, and now this single measurement is 0.55 keV, confirming the net mass shift observed by CERN from the world average of about 70 keV. The two neutron separation energy displaying the world measurements are shown in Figure 8, including the most precise from TITAN. The Maya experiment at ISAC II has measured the halo neutron correlation using a two-neutron transfer reaction. The data suggests a large s-wave component.

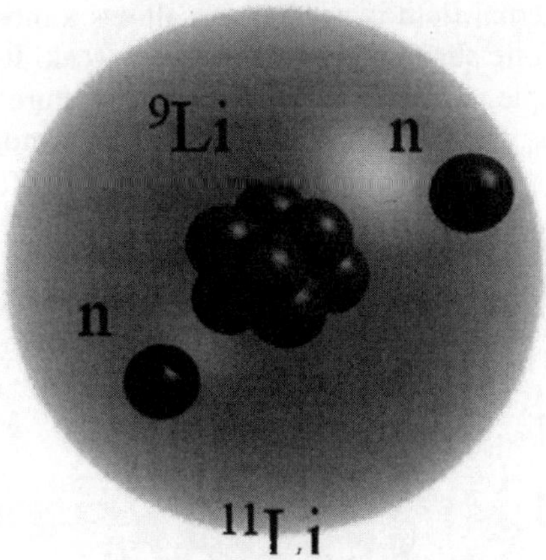

Figure 7: Artist's rendition of the Lithium-11 nucleus. There is a Li-9 core with two neutrons "orbiting" the core. Li-10 is unstable, indicating a three-body force is at play. The radius of Li-11 is similar to that of Pb-208.

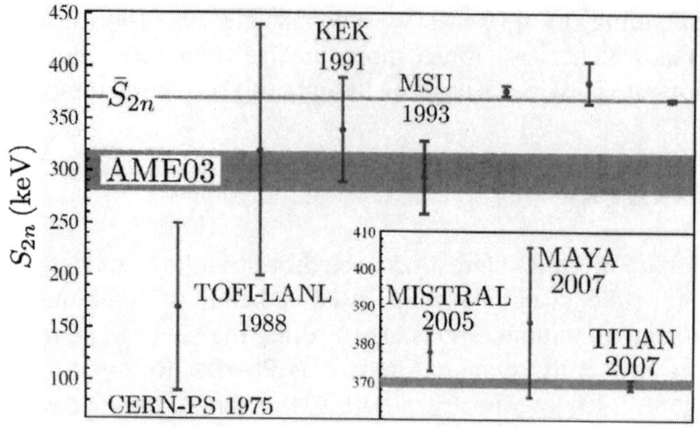

Figure 8: World data displaying two neutron separation energy in Li-11. The TITAN and MAYA results are from TRIUMF. Note the shift in world average and high precision of the MISTRAL and TITAN results.

MATERIAL SCIENCE

The experimental technique of muon spin relaxation (μSR), shown in Figure 9, is performed at four laboratories: ISIS at Rutherford in the UK, JPARC in Japan, Paul Scherrer Institute in Switzerland, and TRIUMF. At TRIUMF, the 500 MeV cyclotron proton beam strikes a target and produces charged pions (lifetime 26 ns) that decay at rest to a muon neutrino and a fully polarized 4.1 MeV muons from production. The muon acts as a magnetometer that is sensitive to fractions of a Gauss. The fact it is already polarized is important since no external field is needed and allows a more sensitive probe. This is ideal because fields in exotic superconductors are very weak. It is possible to deposit one muon at a time in a sample, which includes high-temperature superconductors, ordered magnets, and spin glasses. The muon can also replace the proton in hydrogen and behave as muonium (μ^+, e^-). It is then possible to study quantum diffusion, hydrogen in metals and semiconductors and chemical kinematics.

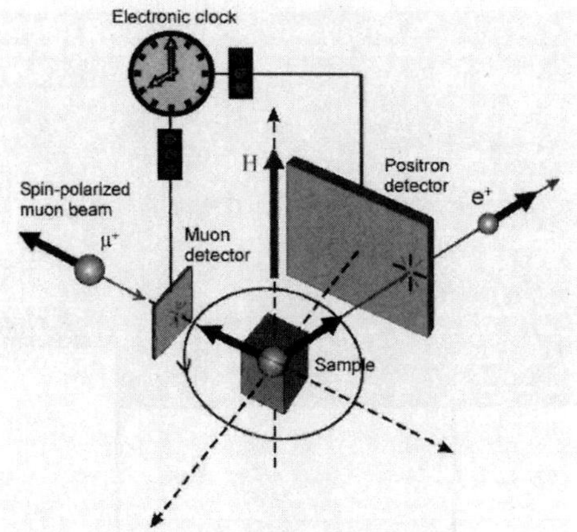

Figure 9: A simple cartoon explaining the μSr technique.

NUCLEAR MEDICINE

TRIUMF uses a small TR-13 (TRIUMF designed) 13 MeV cyclotron to produce a range of medical isotopes. TRIUMF provides the Pacific Parkinson's Research Centre with the short lived isotopes (C-11, F-18) for their program. These include a number of radiotracers for studies of dopamine in the brain. This program is now expanding into Alzheimer's and addiction. TRIUMF also provides all the fluorodeoxyglucose (FDG) for the British Columbia Cancer Agency (BCCA) for oncology staging. A PET/CT image is shown in Figure 10. This image uses another radiotracer EF5-F-18, more specific than FDG. The direction of nuclear medicine and cancer imaging is to use more specific radiotracers, an area of intense research.

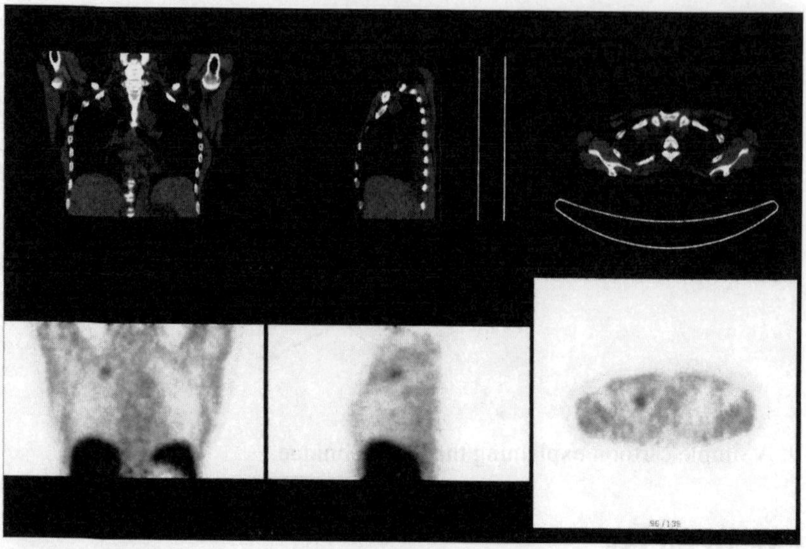

Figure 10: A PET/CT scan of a patient using the radiotracer EF5-F-18. The "hot spot" in the lower figure indicates the tumor is hypoxic and will likely be resistant to radiation therapy and certain types of chemo-therapy. Knowing the biochemical details of a tumor is important information for the doctor prescribing the treatment. This radiotracer was developed by researchers at the University of Pennsylvania and the labeling chemistry was performed by TRIUMF chemists.

TRIUMF has just celebrated a 30-year manufacturing partnership with MDS Nordion. TRIUMF operates three small cyclotrons for MDS Nordion. Located on the TRIUMF site, MDS Nordion distributes 2.5 million patient doses of isotopes per year around the world or about 15% of Canadian isotope production.

FUTURE

TRIUMF is planning two major initiatives to be implemented over the next 5-year funding period. First, an expansion of the nuclear medicine program is being developed with private sector and academic partners. The long history of collaboration in brain research will continue in addition to a new initiative in oncology. The new oncology program will focus on radio-chemistry and the development of new radio-tracers for positron emission tomography (PET) imaging in collaboration with private sector partners and BCCA.

TRIUMF is also proposing to build a new electron linear accelerator (50 MeV, 0.5 MW) to produce rare neutron rich isotopes. The electron accelerator (e-linac) will operate in CW (continuous wave) mode and will exploit superconducting radiofrequency

(SRF) technology. The isotopes produced will require actinide targets and will allow studies that attempt to approach the neutron drip-line and therefore understand the limits of stability. This technology is the basis for the design for the International Linear Collider. By developing the know-how with this technology, TRIUMF, and therefore Canada, will be able to participate in this global project. The e-linac when coupled to a beryllium target, will also allow production of Li-8 for materials studies. One of the primary goals of the materials program is to understand the magnetic properties of high-temperature superconductors, including the cuprates and the newly discovered iron (LaOFeAs) based.

Finally, the TRIUMF particle physics program will help coordinate the Canadian analysis teams in order to fully realize the benefits to Canada from engaging in the great opportunity for discovery at both the LHC facility at CERN and the T2K neutrino program in Japan.

ACKNOWLEDGEMENTS

Thank you to the organizers of the Erice 2008 International School of Subnuclear Physics. A special thank you goes to Professor Zichichi. Thanks are also extended to Jens Dilling, Achim Schwenk, Tom Ruth and all my TRIUMF colleagues.

CHAIRMAN: N. LOCKYER

Scientific Secrataries: S. Kawasaki, Z. Marshall

DISCUSSION

- Ni:

I found it quite interesting to see measured the radius of a Lithium isotope. How exactly do you measure the radius of the Lithium-11 halo?

- Lockyer:

You measure two things: the matter radius, which you can imagine as a scattering experiment, and you measure the charge radius. You measure the charge radius by looking at the energy levels of the electrons in the atom. It is a very precise laser experiment that helps you determine the charge radius. The combination of those two is one method. Another method is to just measure the mass. If you measure the mass of the whole atom, $E=mc^2$ tells you how tightly it's bound, and you're able to determine the radius from that also.

- Ni:

From the mass measurement you can tell the radius?

- Lockyer:

You can tell how tightly bound it is, which is part of the information you need.

- Patterson:

About the beta NMR - how small a B-field can you measure, and what limits that?

- Lockyer:

The fields that are measured in these superconductors are half a Gauss - that is the size of the field that is spontaneously created inside of the crystals, so that gives you a ballpark number.

- Patterson:

What about stray fields in the cryostat - is that an issue, and how is it handled?

- Lockyer:

Stray fields are always an issue, but they do these experiments in a zero-field environment. You can polarize it with a laser in advance, without a field, so the Lithium-8 comes in polarized.

- Patterson:

Do they then just cancel the field with coils?

- *Lockyer:*

They can create fields that are static and larger than what they think they are measuring and study the effect as a function of the additional field. You can cancel them, but in general you just try and see how your measurement varies with the external field. They are sitting inside of a magnet if they want to turn it on.

- *Patterson:*

Could TRIUMF be a facility for a low-energy beta beam neutrino source? Are the luminosities high enough to do something like that there?

- *Lockyer:*

It has been suggested. Carlo Rubbia, one of the beta-beam people, came. There is something called a Rubbia Ring, which you would store your favorite isotope in - neon is popular, helium is popular. It is a pretty complicated setup, and I think you would have to devote the laboratory to doing it. It's a big deal, and I think we have to spend a fair amount of time to understand what it is that you're trying to accomplish in the test, since it would only be a test to try and demonstrate the principle. But as far as production goes, I think people can look at production and say, "Yeah, you could do it."

- *Baudis:*

You mentioned the pioneering work of TRIUMF in PET. As far as I know, PET is often not used because of the fear of false positives - why is that, and can you improve it with a better detection technique, for example? I know that it is often not clear to doctors how the data should be interpreted, but I don't know why that is.

- *Lockyer:*

I can only speculate. I know there are false positives and false negatives as well. There is an active community that spends its time interpreting these images. So, the first thing to remember is that what you're looking at is radiation inside the body, and how that connects to the biology of the body is not completely clear. That's a challenge for the person looking at and interpreting the picture. What it means from the physics standpoint is that you're seeing radiation. Some of it may be related to the biochemistry of how it collects the FDG, which is a labeled sugar. Different cells have different "uptake values" - a term that describes how much a healthy cell takes up relative to a tumor cell. That ratio varies from 5 to 20, and since you don't know that *a priori* in a given patient, I can imagine that when you look at a signal there is a hidden constant in what you're looking at. That has to be a part of your interpretation.

- *Baudis:*

You also mentioned programs outside of TRIUMF, but you didn't mention the PICASSO experiment. Was that because it is not a high priority, or did you just not get to it?

- *Lockyer:*

I don't mention all Canadian experiments, or all the things that are going on. There's a certain bias in Canada toward wanting PICASSO to succeed, and so it's reasonably well supported. But, it's a technically challenging experiment. It has to demonstrate that it's working, and then it will be well supported. It's in SNOLAB now, as you know, and it is taking data.

- *Carrasco:*

You mentioned Niobium superconducting resonators that you've built. I would imagine they have a high Q - what are they used for?

- *Lockyer:*

You can use a copper cavity or a superconducting cavity. You shape the cavity in such a way that the electric field is pointing in a convenient direction, you poke a hole in it, and you can put a beam through it and accelerate the beam. The cavities that I was talking about are used for accelerating beams. They can be used for accelerating electrons, protons, and heavy ions. The issue with these cavities that makes them attractive is that the power dissipation is low, because as you go to higher and higher currents you can imagine that losses in the wall are a factor. There are technologies in accelerators now called "energy recovery linacs," for example, where you can take a beam of electrons, you can accelerate it (particularly as a light source), wiggle it, get your x-rays out of it, and still have most of the energy of the beam there. So you can bend it back and bring it back through the accelerator and charge up the cavity. That's been demonstrated at Jefferson Lab to be efficient at the 99% level at low current. That technology is something that a lot of people are looking at for next generation accelerators. It's a property of the superconducting cavity, and that's one of the reasons that Niobium is popular.

- *Tannenbaum:*

It seems that you have a very strong radioactive beam program. How does that compare to RIBF, proposed in the US?

- *Lockyer:*

There are two techniques for generating radioactive beams in the world: one is called fragmentation, and one is called ISOL. I described an ISOL facility, wherein you take a beam, put it on a thick target, and heat the target up to a few thousand degrees. There is a hole in the target, and a diffusion of these longer-lived isotopes, which have to get out of the target. Then you ionize them and accelerate them. The shortest-lived isotopes have a lifetime of a few milliseconds. Lithium 11, for example, lives 8 milliseconds. That's fairly short, and you still have to transport it somewhere and do something with it. If you want intense beams of something, that's the way to go, but you're limited by the lifetime. The fragmentation facilities, like FAIR (GSI) and RIKEN, accelerate a heavy ion of some kind and put it through a thin target to break it apart. In the process you might pick up a few extra nucleons in the way, and then you can accelerate those and do experiments with them. They can deal with things with much shorter lifetimes. If you want to get to the neutron drip line or the R-process, where this interesting physics is for understanding the generation of the elements and so on, then as you get to a very unstable nucleus, when you add a neutron, it won't stay there very long. The further you push out, the shorter lived these elements will be, and you will need this other technique. FAIR is a billion euro facility, and RIKEN is half a billion dollars equivalent. They're really pursuing different things, and so I would say they are complimentary.

- *Musaev:*

You mentioned x-ray bursts are explosions. Could you explain the current status of the gamma-ray burst problem?

- *Lockyer:*

I'm not sure - I only know about x-ray bursts and super-bursts - is there anyone in the audience who knows about gamma ray bursts?

- *'t Hooft:*

They come from high-Z things, far away in the universe.

- *Bettini:*

These are very powerful, very energetic astrophysical phenomena that are around in the universe. There was an example for the order of magnitude that I remember: the energy released is the same as colliding a sun and an anti-sun.

- *Lockyer:*

There's another phenomenon that people think may be the origin or location of the R-process that I was talking about, which is neutron stars merging. They may generate gamma rays. I hope that was a reasonable answer to your question.

- *Baldini:*

Among the various activities in TRIUMF, I saw the measurement of the electric dipole moment of Radon. Could you tell us about that?

- *Lockyer:*

I just mentioned it at the end - the challenge, experimentally, is to have Actinide targets. Actinide targets are often Uranium targets, and a number of elements are created that are alpha emitters. You have safety issues for the people who work there, how you deal with the targets, and so on. It has taken a little over a year to get a change in our license to do a test with Actinides. The idea is to measure the yields of Francium and Radon from this test, so that we would know how many we made. The experimenters have run a number of tests with Radon-equivalent but stable atoms to demonstrate that they can polarize them, they can precess them and so on. You have to collect the Radon-223 first, and measure its distortions. It's assumed to be a nucleus that has a very non-spherical shape. The first stage of their program is to measure the shape. The present best limit on the electric dipole moment of an atom comes from Mercury. If the shape of Radon-223 is sufficiently distorted, this could be a factor of 30 or so better than Mercury. It is a challenging experiment, because you have to trap them with lasers and polarize them, but atomic physicists are very good at that kind of thing. I think the issue is our getting the beam to them. Most people think it shouldn't be too hard. It's really a safety issue. I'm guessing the earliest we can begin the full experiment is around 2013 - it will take that long to setup to do it.

- Nikiforov:

To produce an ion you accelerate a proton and hit the appropriate target, then you ionize it and accelerate it. In your view of the future, wherein each hospital would have a "coffee machine" where you could push a button and have a dose of ions, how do you imagine all the machinery will go - it takes a big space to accelerate, and so on - into a single box?

- Lockyer:

Now is my chance to comment on engineering. The word “engineering” has been used by the theorists in a certain way, and I have a very different view of engineering. Engineering is something you do when you know what you're doing and you're ready to build something, and then the engineers take over and they build it right. This problem comes in two pieces. The medical cyclotrons that are out there now are all of the warm technology type, the low energy one. The first thing you would do is to build a superconducting machine, and I'm guessing the machine would be about a meter long when you finished. The technology exists to do that now. The second thing, which is the R&D aspect of the idea, is called microfluidics. You would have some kind of target cartridge inside the cyclotron, and you would choose the appropriate type of target to use with the beam. You would then generate the necessary isotopes and flow them into a "chip." The chip is silicon technology, the same technology use for optical transmission on a chip, which is like little pipes or wave guides that you can put fluids into. You would then mix the relevant chemicals together on this chip to produce your FDG. That technology exists, there have been some prototypes, though it's not fully developed. But, you would have to integrate that technology with the next-generation medical cyclotron. People are thinking about it already. I think it will happen, and the issue will be the exact energy, size, and quality. You could probably build the whole thing for $1.5M, when you're finished.

- Nikiforov:

I haven't seen any radiation protection in any of your pictures - are the energies just too low to require it?

- Lockyer:

The small cyclotrons come in two varieties: they come in a type that is already shielded,or you can put them in a room, and let the room act as shielding. For about 20 MeV, you need shielding about 20 cm thick - which might give you some sense of how much space one needs. It's an important consideration if you're going to put it in every hospital. But it's closer to 20 cm than 2 meters.

- Dubynskiy:

You spoke about results in medicine - do you know what kind of side effects these methods could have, and how well they have been studied? And how good are the statistics for curing patients - was there only one result, or many?

- *Lockyer:*

There are two issues: imaging and therapy. As to therapy: let me give you an older example - Iodine. If you want to treat your thyroid, you know that Iodine goes to your thyroid, and you can treat it quite successfully. That's a simple example that is out there, everyone uses, and works well. As for these more modern things I'm talking about, using alpha emitters: the clinic studies with people have only just begun. Duke University has just published one study on treating brain cancer with alpha emitters, where they connect the Astitine-211 to an antibody, and the antibody goes to the tumor. Those studies were two years long with 11 patients. The paper says that they are very encouraged, but it is a huge effort for the medical community. The results are promising for the idea of being able to target where the drug goes. All the patients in this case die - they were terminal patients, and the definition of success is how much longer they live - but they each lived about a factor of three longer than was expected. It's something.

- *Dubynskiy:*

Could this method be applied to curing any type of cancer?

- *Lockyer:*

I can't say. What I never mentioned is a very hot topic called apoptosis and the way that a cell decides to kill itself. There are various receptors on cells that you can attach to in order to understand the detailed mechanism by which the cell makes that decision. If the cell kills itself, it's over. If it decides not to kill itself, that could be a cancerous cell. The ability to look at the biochemical pathways with various isotopes, with imaging, which is where we are today and where we need to be tomorrow, is where the field is going.

- *Weir:*

You mentioned that Canadian labs must work collaboratively rather than running competing experiments. Is that a funding condition, and do you believe it benefits Canadian science?

- *Lockyer:*

Absolutely. Canada couldn't build the LHC by itself. There are two points. You have to have enough internal strength - either technical or scientific - that you can add value to a project and people want you to be a part of it. In the case of Canada, we aren't a part of CERN, so we had to be able to bring something to the table. I mentioned the kicker magnets, but that was a major component. We had the world's kicker magnet expert at TRIUMF, and that person has played a big role. My model of how particle physics will work in the future is this: there will be a series of small labs around the world that will have to collaborate with the big labs like Fermilab and CERN and help them out. They'll also have their own internal program as well. So, collaboration is essential. And it works in both directions. We deliver something to them, but we have 150 Canadians, of which 70 or so are students, so the students get to participate in what may be the most exciting scientific project out there right now. It's absolutely essential.

- *Weir:*

So you see TRIUMF as a center of excellence that consults on these projects in the future?

- *Lockyer:*

Yes. We try to carry our weight and contribute in some way to the project. I think, based on my sense of how CERN has receives our contribution, that this is a model that works. Convincing the government of that is another story.

- *Weir:*

To go back to those large nuclear charge radius ions - Lithium-11 is a sort of Rydberg nucleus, if you wish. Rydberg atoms have applications in quantum optics. Do you see any other applications for these large-charge-radius nuclei?

- *Lockyer:*

Where I see the excitement in the nuclear physics community is in trying to understand the three-body force in nuclear physics. I've seen applications of that to trying to understand neutron stars. I know it's not an application, but it's another are of science that this knowledge leaks into. Outside of nuclear physics I haven't heard of anything, but that is an interesting suggestion.

Super-KamiokaNDE (Universe, Mankind and Neutrinos)

2008-09-05 at Erice

Abstract:
The recently known properties of the neutrinos will be summarized. Their relation to the Mankind as well as to the Universe will be discussed. Future possibilities with the neutrinos will also be discussed.

M. Koshiba

(1) We have 3 families of elementary particles each consisting of 16 members. See next slide.

Each of these has its anti-particle, and this makes the total number of elementary particles equal to 96. Large number indeed.

Most of them are electrically charged and only the neutrinos are neutral and this makes them extremely difficult to observe. They are very penetrating and makes it possible to observe the stellar interior; X-rays make the bones visible.

(2) We have three topics in this lecture. Namely, the supernova neutrinos, the solar neutrinos and the neutrino oscillation.

Before discussing these neutrino topics, we spend sometime explaining what we know about the neutrinos.

The 92 chemical elements to form our bodies: where and how were they created?

Can we see the real beginning of Big-Bang?

$u^L, u^L, u^L,$ $d^L, d^L, d^L,$ e^L ν_e^L	$c^L, c^L, c^L,$ $s^L, s^L, s^L,$ μ^L ν_μ^L	$t^L, t^L, t^L,$ $b^L, b^L, b^L,$ τ^L ν_τ^L	**4 Interactions and 4 mediating Bosons** Strong Interaction on colored Quarks and Gluons, **g.**
$u^R, u^R, u^R,$ $d^R, d^R, d^R,$ e^R ν_e^R	$c^R, c^R, c^R,$ $s^R, s^R, s^R,$ μ^P ν_μ^R	$t^R, t^R, t^R,$ $b^R, b^R, b^R,$ τ^R ν_τ^R	E-M interaction on charged particles and Photon, γ. Weak Interaction and **Z^0,W^{+-}.**
1st Family or e-Family	2nd Family or μ-Family	3rd Family or τ-Family	Gravitational Interaction on all and **Gravitons.**

W.Pauli

Introduction of neutrino

W.Pauli; Letter to L.Meitner and her colleagues (letter open to the participants of the conference in Tubingen) (1930).

Majorana neutrino

Ettore Majorana
Nuovo Cim. 14 (1937) 171-184.

The possibility raised.

neutrino=anti-neutrino

Still unsetteled problem.

Relevant experiment is

ββdecay.

F.Reines and C.Cowan

Detectors by Reines and Cowan

Observation of anti-ν_e from the nuclear reactor.

F.Reines and C.L.Cowan, Phys. Rev. 92 (1953) 830.

F.Reines and C.L.Cowan, Phys. Rev. 113 (1959) 273.

3 events observed

B.Pontecorvo

Neutrino oscillation hypothesis

Бруно Понтекорво

B.Pontecorvo,
Zh. Eksp. Teor. Fiz. 53 (1967) 1717
[Sov. Phys. JETP 26 (1968) 984].

JOINT INSTITUTE FOR NUCLEAR RESEARCH

Maki, Nakagawa, Sakata

Possibility of neutrino oscillation pointed out.

Z.Maki, M.Nakagawa, and S.Sakata, Prog. Theo. Phys. 28 (1962) 870.

Z.Maki

M.Nakagawa

S.Sakata

Sakata Memorial Archival Library, Nagoya University (photo by Kenichi Aoki)
Research Center for Neutrino Science, Tohoku University

J.Steinberger, M.Schwartz and L.Lederman

Discovery of ν_μ different from ν_e

Discovery of muon neutrinos, ν_μ

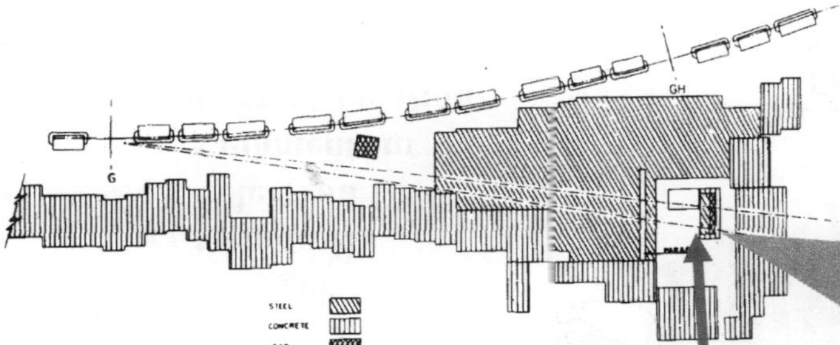

Neutrino
detector

G.Danby et al.,

Phys. Rev. Lett. 9 (1962) 36.

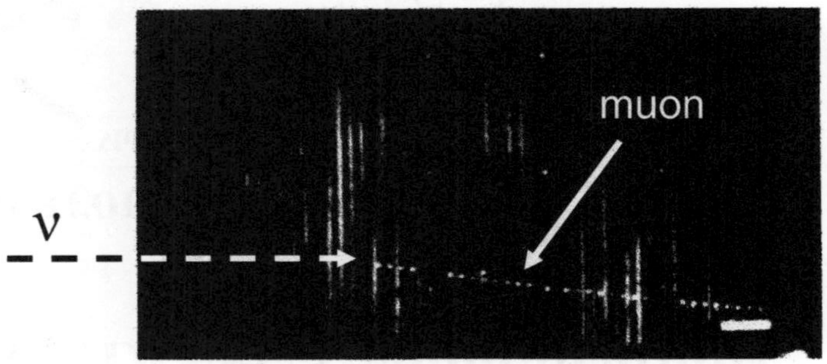

Observed event

Brookhaven National Laboratory

Observation of ν_τ (DONUT collaboration)

Spectrometer

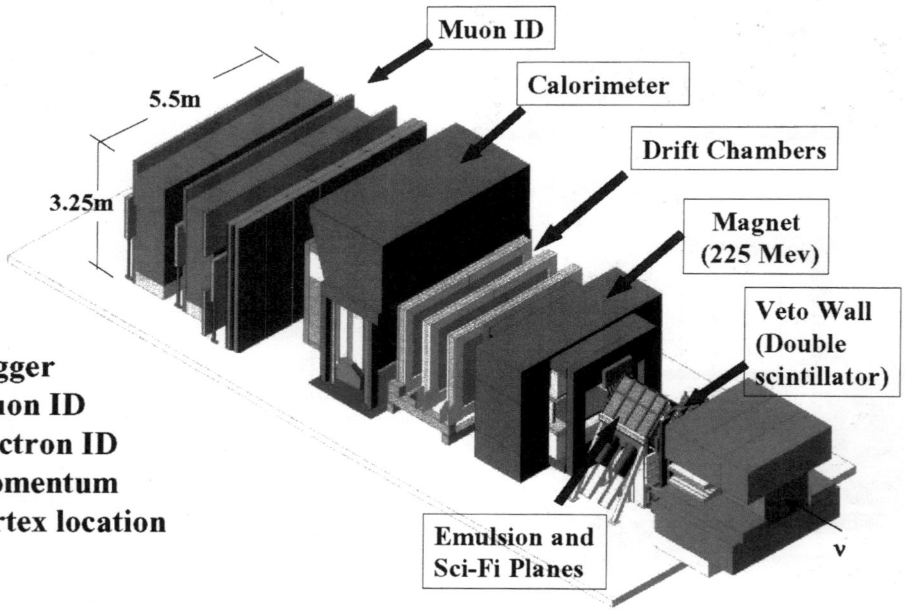

- trigger
- muon ID
- electron ID
- momentum
- vertex location

Observation of ν_τ

K.Kodama et al.,

Phys. Lett. B 504 (2001) 218.

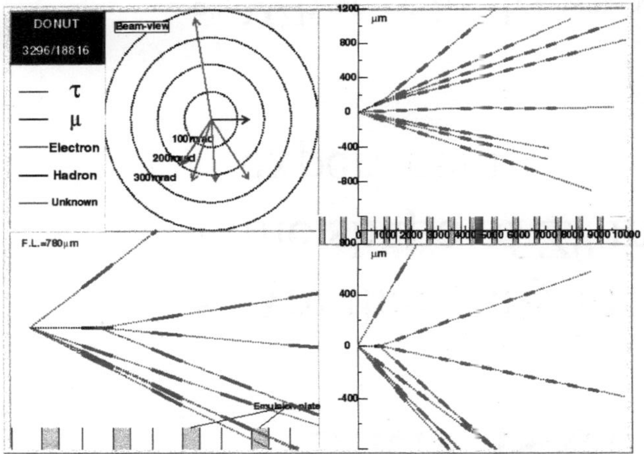

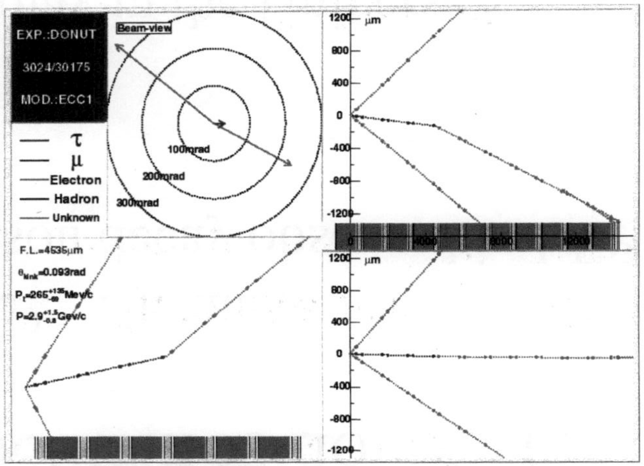

2 examples of ν_τ candidates

3 kinds of neutrinos do exist.

- Are there any more? The answer is **NO** from the shape of Z^0 meson decay observed in LEP experiments.

- What role do they play in the Universe?
- What can they tell us about the Universe?
- What physical properties do they have?
- And so on.

The sun is emitting ν_e.

- In the inner part of the sun the nuclear fusion process, 4p to ^{4}He, is going on. Namely 2 ν_e s per 1 ^{4}He are emitted; p changes to n + e^-+ ν_e. The flux is reliably estimated from the observed solar energy flux.
- The next slide shows the estimated solar neutrino energy spectrum together with the summary of the neutrino flux measurements.

Summary of the Solar neutrino data

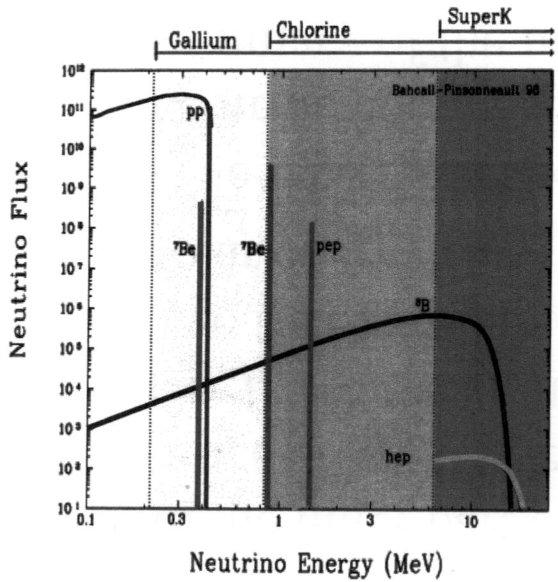

http://www.sns.ias.edu/~jnb/

Rate measurements

@ Neutrino 2002

	Target	Data / SSM (BP2000.2)
Homestake	^{37}Cl	0.34 ± 0.03
SAGE ^{71}Ga		0.55 ± 0.05
GALLEX+GNO	^{71}Ga	0.55 ± 0.05
Super-K	e^- (water)	0.465 ± 0.016
SNO (CC)	d (D_2O)	0.348 ± 0.020
SNO (NC)	d (D_2O)	1.01 ± 0.13

B.T.Cleveland et al., Astrophys. J 496 (1998) 505

J.N.Abdurashitov et al., Nucl. Phys. Proc. Suppl. (Neutrino 2002) 118 (2003) 39

T.A.Kirsten for the GNO collab. Nucl. Phys. Proc. Suppl. (Neutrino 2002) 118 (2003) 33.

S.Fukuda et al., Phys. Lett. B 539 (2002) 179.

Q.R. Ahmad et al., Phys. Rev. Lett. 89 (2002) 011301

R.Davis Jr. at Homestake

Radiochemical solar neutrino experiment

1st prenatal move

Solar neutrino deficit;

Only 1/3 of the theoretically expected flux.

R.Davis Jr., D.S.Harmer and K.C.Hoffman,

Phys. Rev. Lett. 20 (1968) 1205.

B.T.Cleveland et al.,

Astrophys. J. 496 (1998) 505.

Homestake solar neutrino experiment

$\nu_e + Cl^{37}$ to

$e^- + Ar^{37}$

And then observe the decay of Ar^{37}.

...en National Laboratory

Arrival **D**irection: **No.**

Arrival **T**ime: 1 month.

Energy spectrum: **No.**

L.Wolfenstein, A.Yu Smirnov and S.P.Mikheyev

Enhancement of ν_e oscillation in e-rich media.

(This mechanism effective for zero mass neutrinos, however, turned out not to be necessary. The neutrinos do have non-zero masses and they oscillate among themselves.)

S,P.Mikheyev and A.Yu.Smirnov, Sov. J. Nucl. Phys.42 (1985) 1441.
L.Wolfenstein, Phys. Rev.D 17 (1978) 2369.

The astrophysical observation.

- The astrophysical observation, i.e., with time T, arrival direction D and energy spectrum E, of the solar neutrinos was made by observing

  the Cherenkov light emitted in the water by the electron struck by these solar neutrino.
- KamiokaNDE of 3000tons of water was for the feasibility experiment of the astrophysical observation of the solar neutrinos.
- Super-KamiokaNDE was built as the genuine solar neutrino observatory.

KamiokaNDE with 20”ϕ photomultipliers

The KamiokaNDE performance at the beginning of 1987.

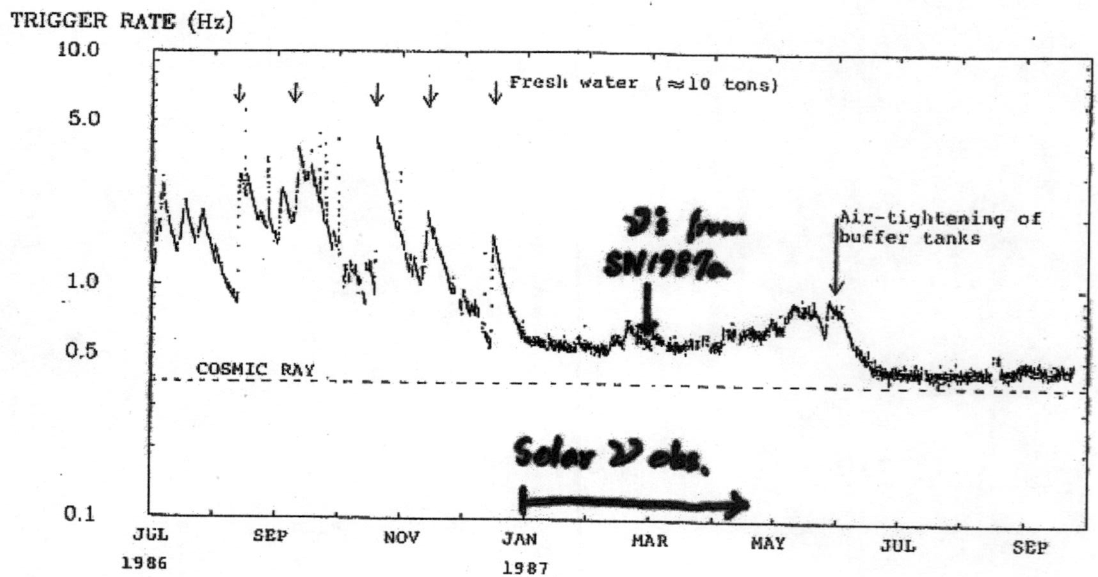

Fig. 3.20. The early performance of the KAM-II detector.

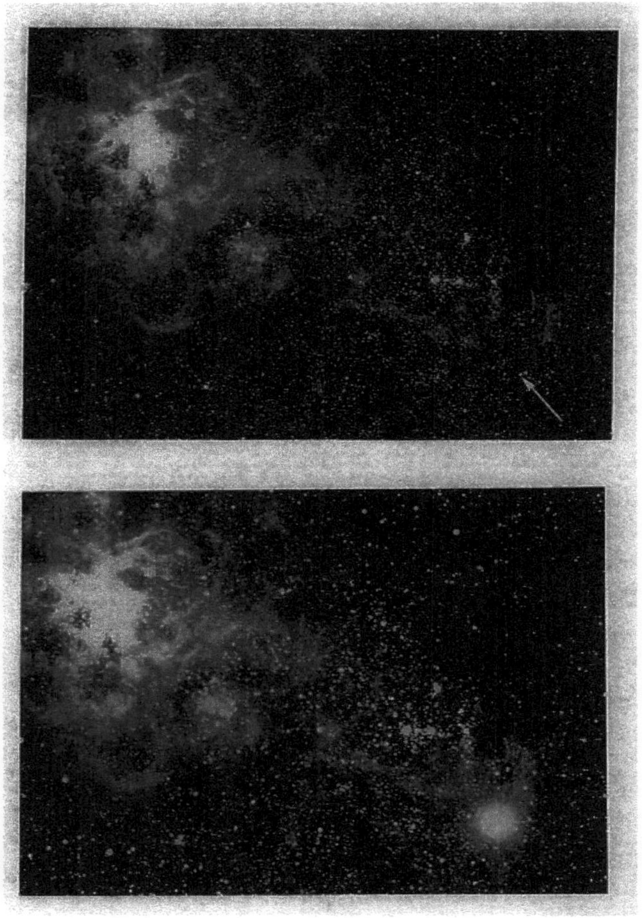

Supernova 1987A in the Large Magellanic Cloud.

Above; before the explosion.

Below; after the explosion.

Anglo-Australian Observatory

2nd prenatal move

The observed signal of the supernova neutrino burst. It was immediately confirmed by IMB experiment in USA. The combined results, T_ν of 4.5MeV and the total ν energy output of 3×10^{53}erg gave strong supports to the theoretical model. The pulse duration of 10msec implies the pre-neutron star density for emmitting these neutrinos. But no directional information.

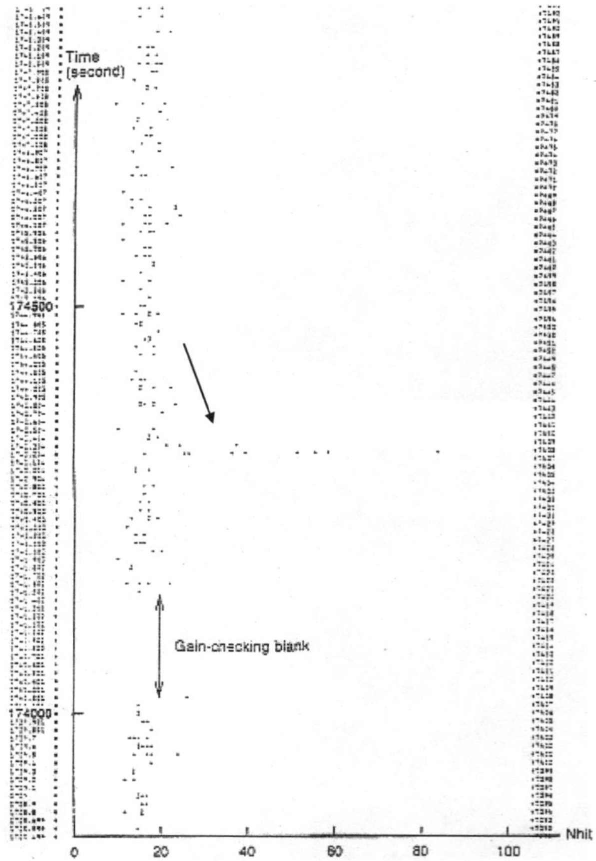

Fig. 5.1. The SN1987A neutrino "signal" in the computer print-out.

Creation of the 92 chemical elements.

- Soon after the Big-Bang, protons and neutrons were created and thus Hydrogen and Helium nuclei with a very minute amount of Lithium.

- Heavier nuclei up to Fe were synthesized inside of large stars.

- In the Type-II supernova explosion, a large number of neutrons are liberated and they stick to the existing nuclei to form up to U nuclei. All these nuclei are spread out by the explosion into the space and they serve as the material for forming next generation stars.

- **The sun and the earth are of this late generation.**

- **Now all the 92 chemical elements are available for creating us human being.**

The neutrinos made this explosion possible by carrying most of the gravitational energy.

We should deeply thank the neutrinos!

KamiokaNDE (1st 450 days)

K.S.Hirata et al., Phys. Rev. Lett. **63** (1989) 16.

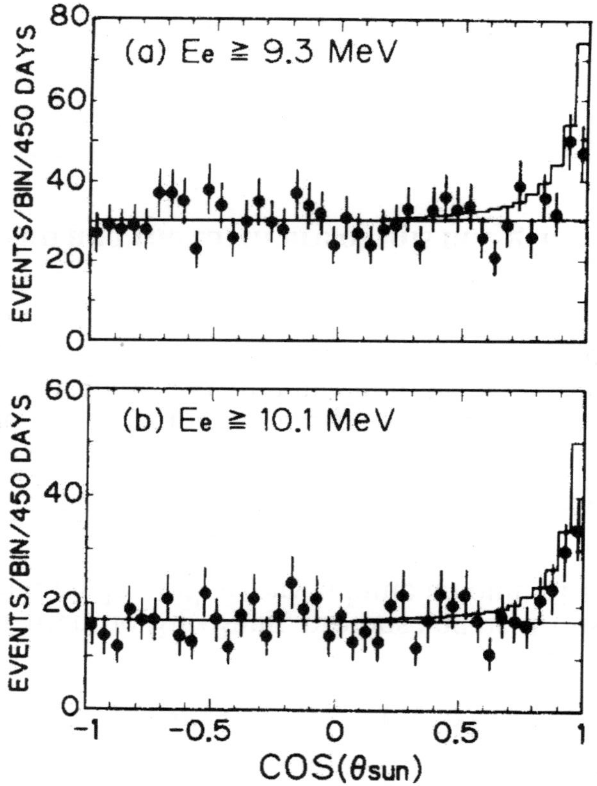

Directional observation, D, of the solar neutrinos.

Timing, T, is better than 10 ns.

Energy spectrum, E, was also observed. See the next slide.

Birth of Neutrino Astrophysics

The observed energy spectra of the recoil electrons. It is consistent with the spectra expected from the B^8 neutrinos. The intensity, however, is only about a half of that expected from the theory. The solar neutrino anomaly discovered by R.Davis's radio-chemical experiment was confirmed.

Energy spectrum of solar neutrino events

Kamiokande II and III (2079 days)

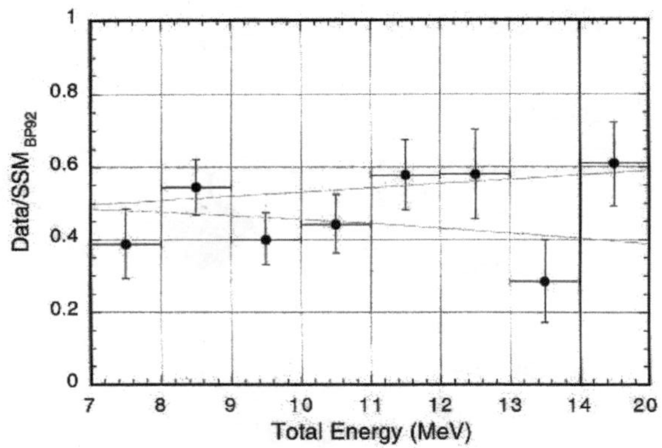

Based on ~600 solar ν events

Y.Fukuda et al., Phys. Rev. Lett. 77 (1996) 1683

ν_μ/ν_e has to be 2 or larger

Look at the ratio of the upward-going events

Atmospheric neutrinos

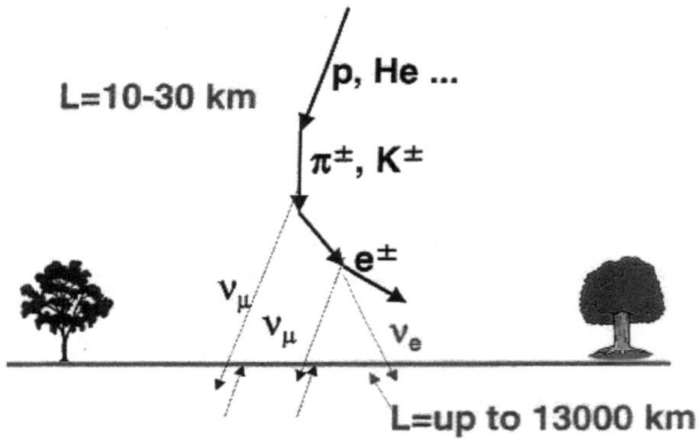

$$\frac{\nu_\mu + \bar{\nu}_\mu}{\nu_e + \bar{\nu}_e} = \sim 2 \quad \text{@ low energy } (E_\nu < 1 \text{ GeV})$$

$$\frac{\nu_\mu + \bar{\nu}_\mu}{\nu_e + \bar{\nu}_e} \nearrow \quad \text{@ high energy}$$

Error in flux~25%, double ratio~5%

Neutrino oscillations :

$$\Rightarrow \left(\frac{\nu_\mu + \bar{\nu}_\mu}{\nu_e + \bar{\nu}_e}\right)_{data} \Big/ \left(\frac{\nu_\mu + \bar{\nu}_\mu}{\nu_e + \bar{\nu}_e}\right)_{MC} \neq 1$$

μ/e ratio measurements

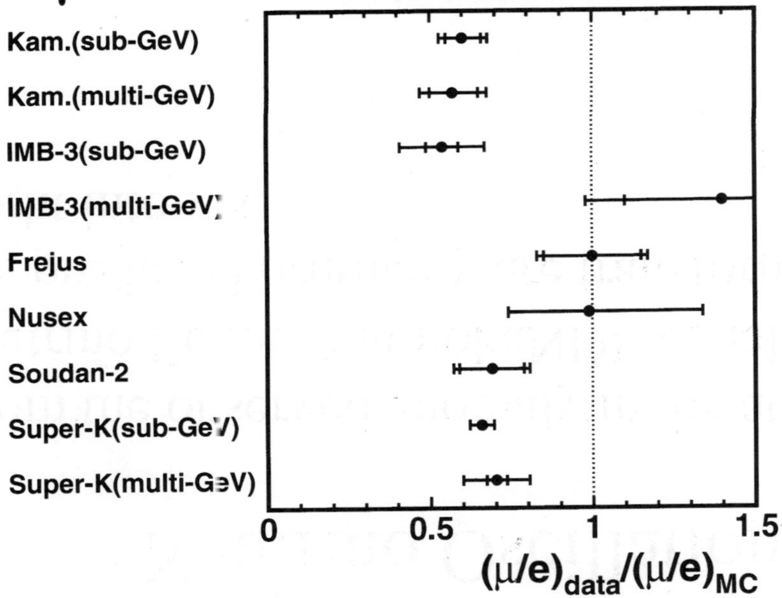

Y.Fukuda et al., Phys. Lett. B 335 (1994) 237

R. Becker-Szendy et al., Phys. Rev. D 46 (1992) 3720.

R.Clark et al., Phys. Rev. Lett. 79 (1997) 345.

M.Aglietta et al., Europhys. Lett. 8 (1989) 611.

K.Daum et al., Z. Phys. C 66 (1995) 417.

M.Sanchez et al., Phys. Rev. D 68 (2003) 113004

Y. Ashie et al. (Super-Kamiokande collaboration), draft in Preparation.

Neutrino Oscillation

- From the observed anomaly in the atmospheric neutrino fluxes, KamiokaNDE concluded that ν_μs produced originally are transforming in flight into ν_τs.

Super-KamiokaNDE, 50ktons H_2O, 11,500PMTs

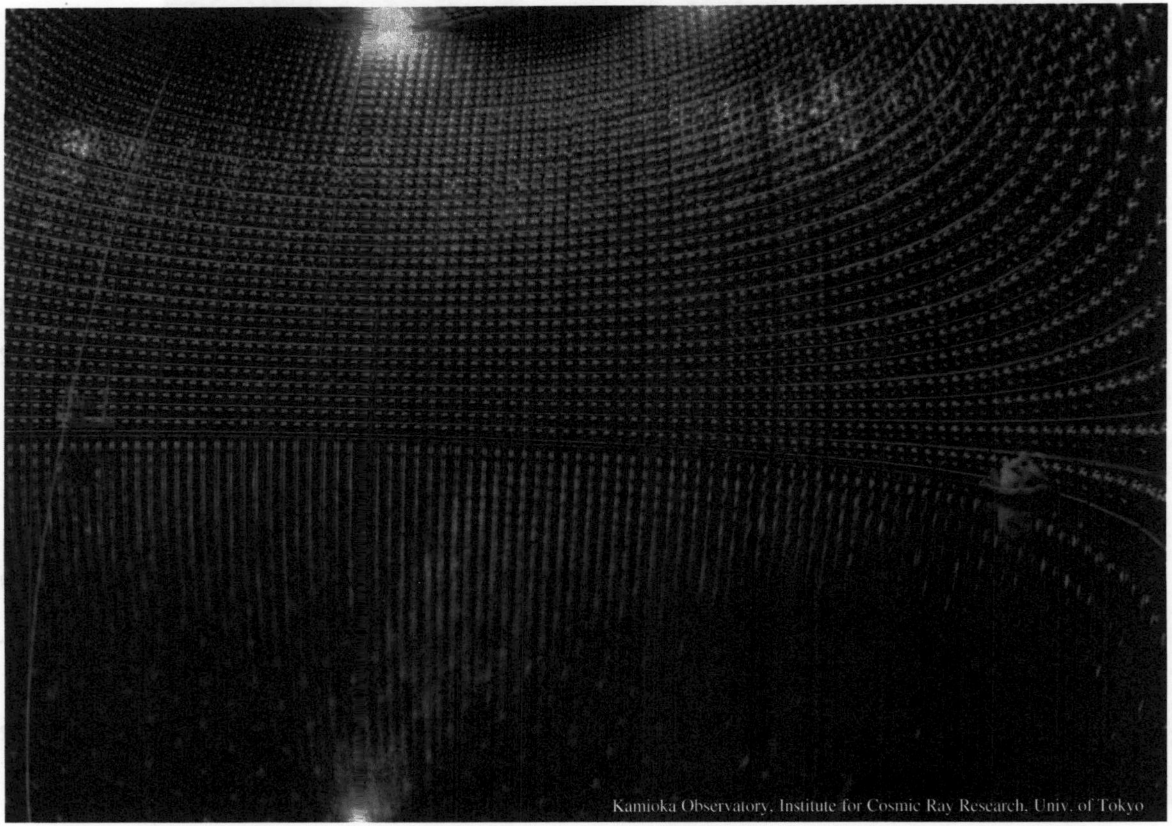

Display of the inner wall of Super-KamiokaNDE

A μ has just entered the detector.

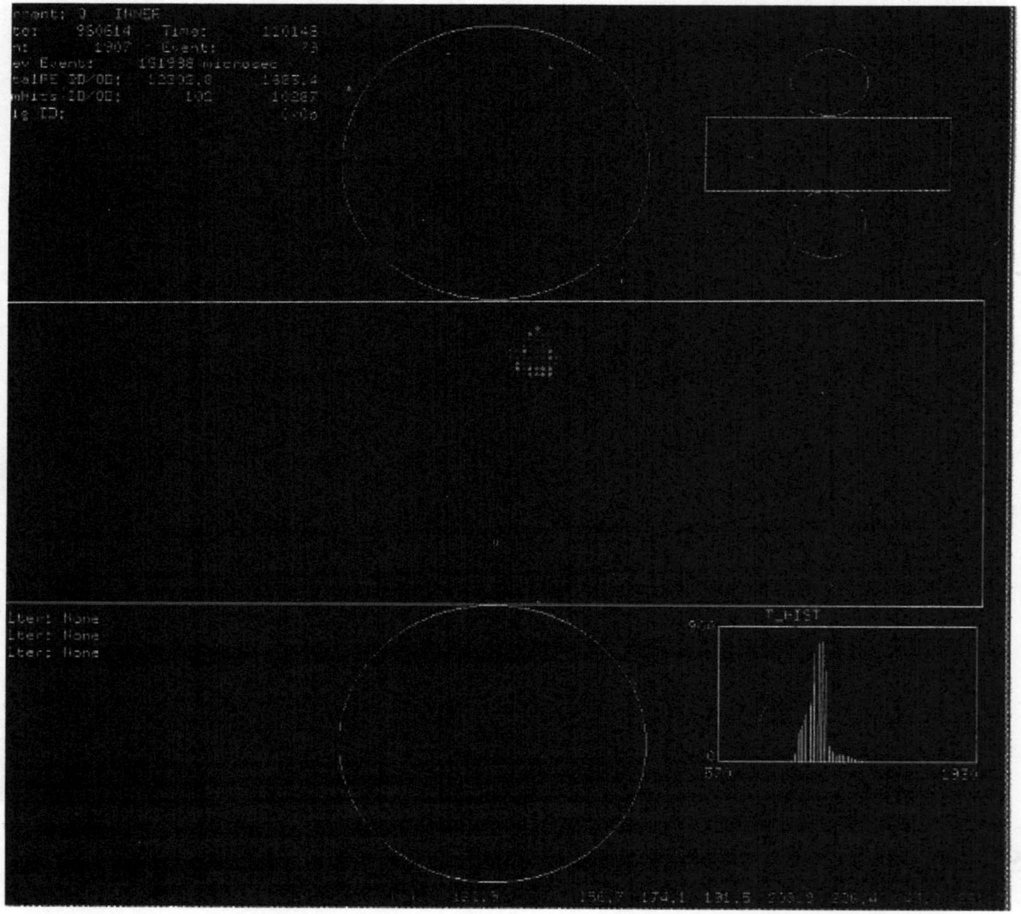

50nsec later:

Cerenkov light proceeds.

Red dots indicate the PMTs of large number of photons received.

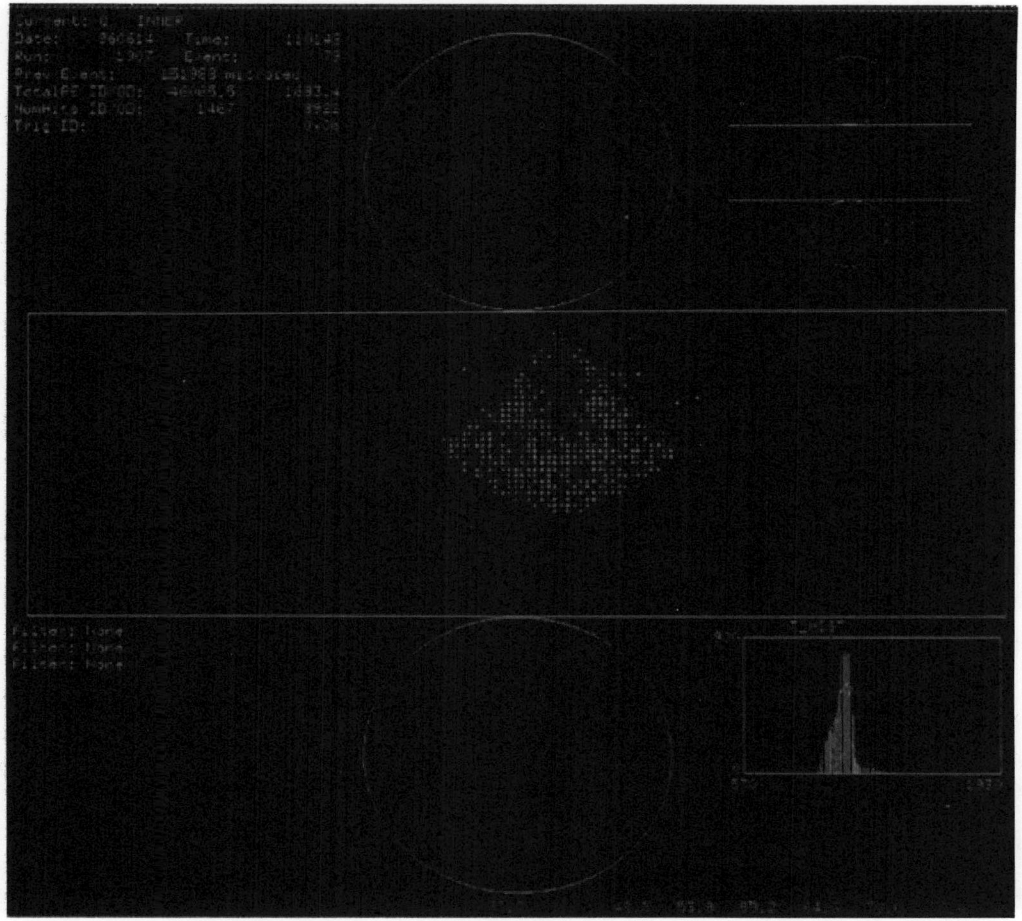

The μ has reached to the bottom of the detector, while the Cerenkov light in water is still on its way.

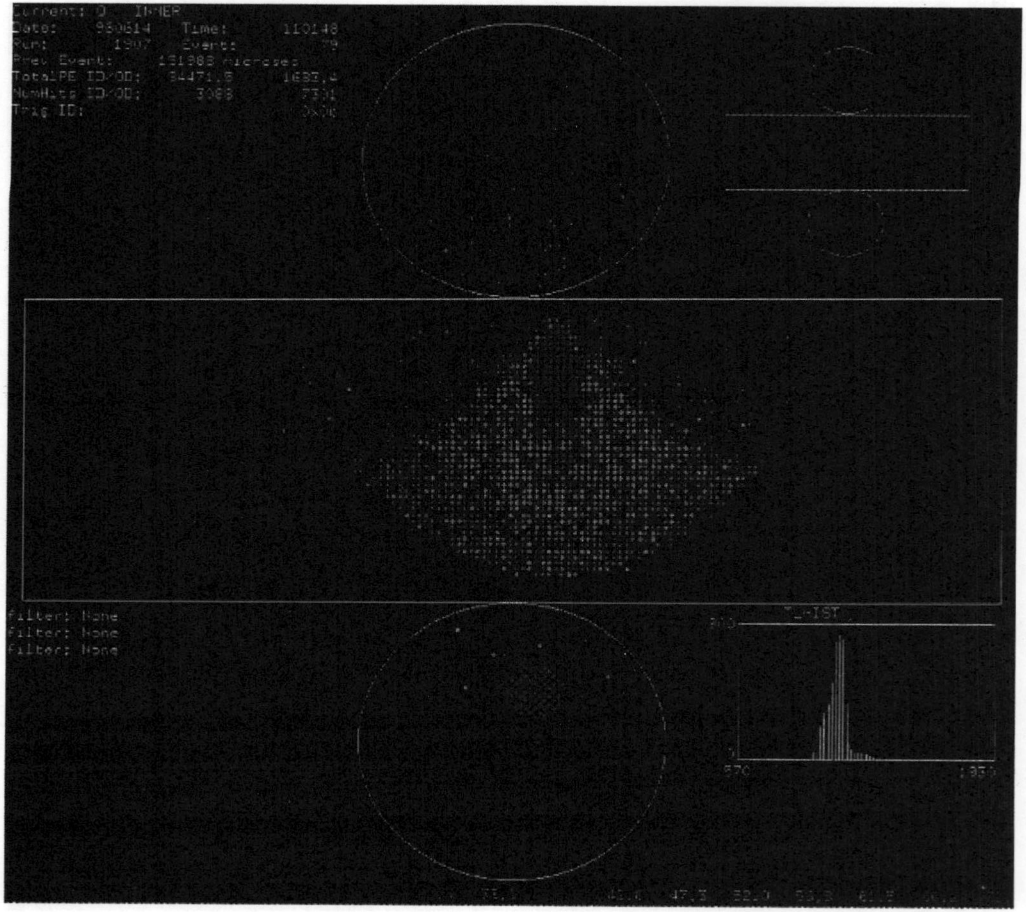

Cerenkov light also reached the bottom.

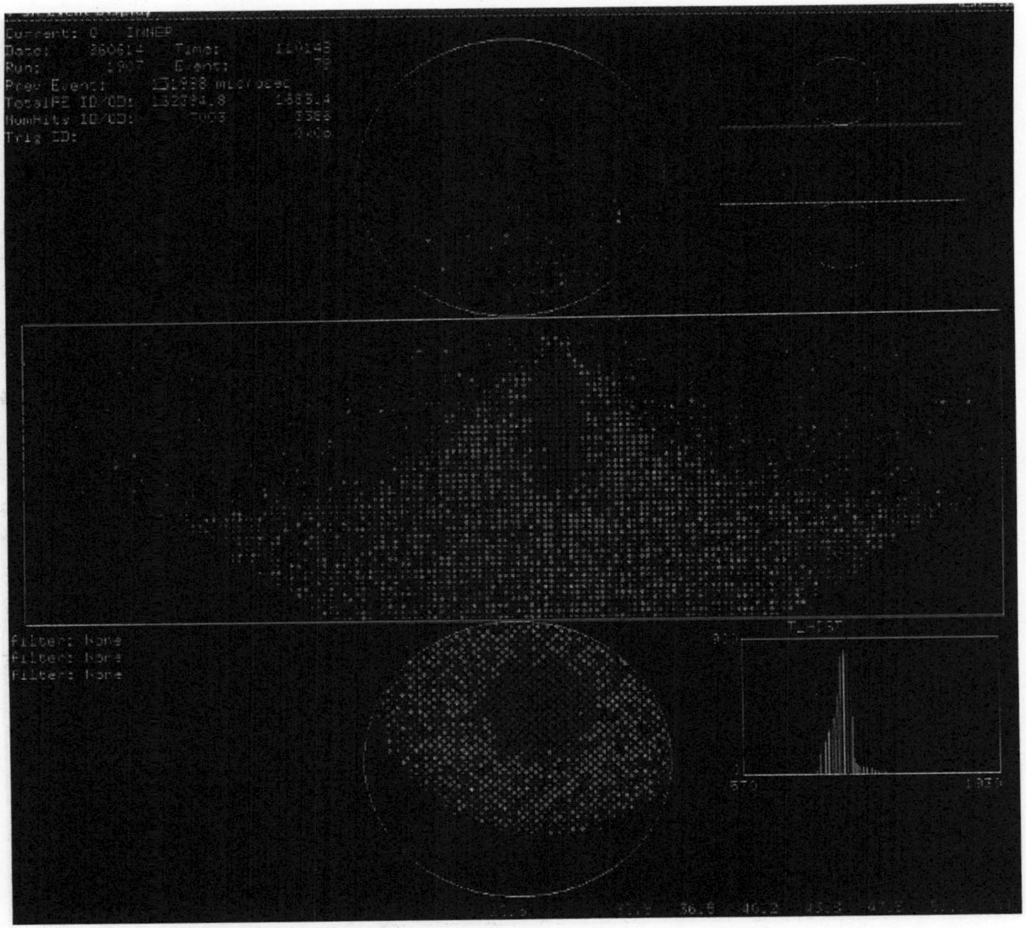

We now see the entire Cerenkov light pattern.

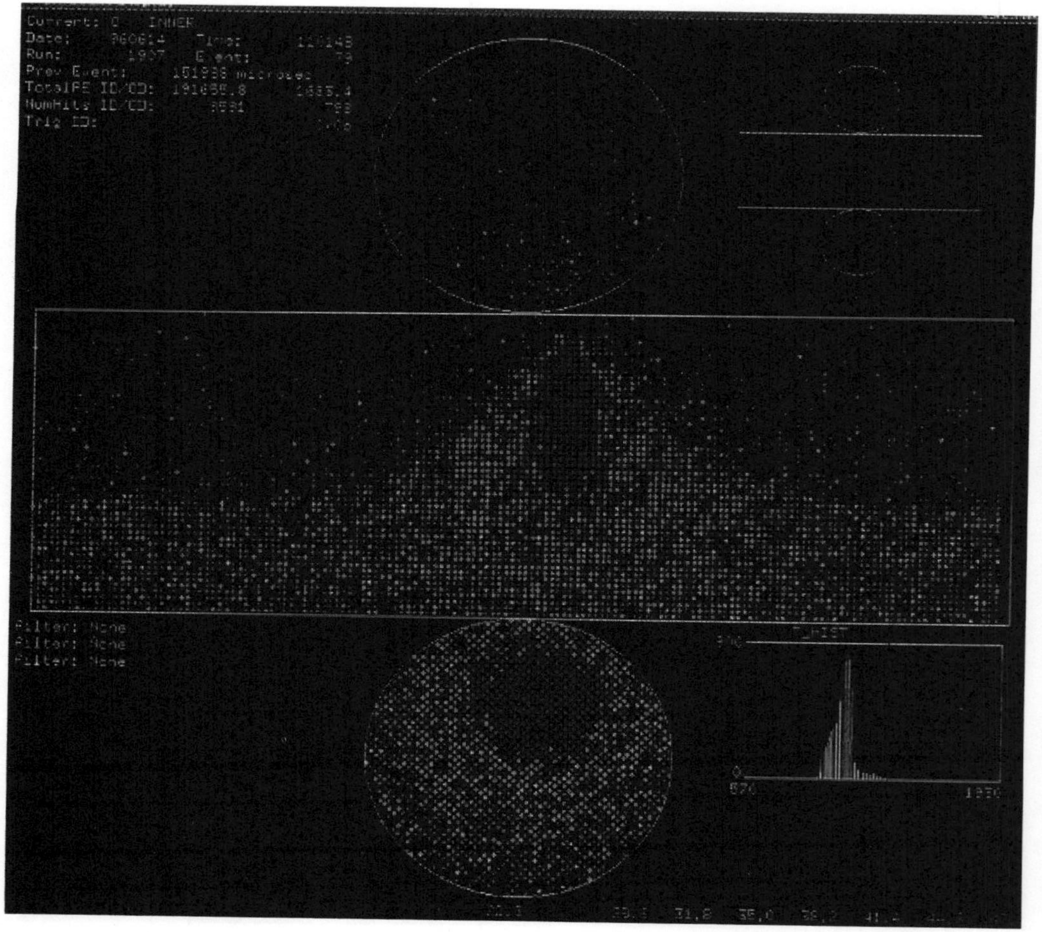

The top e-event has a blurred radial distribution of Cerenkov photons, while the bottom μ-event has a crisp ring image. The discrimination between e and μ is accomplished by the quantitative comparison of the radial distributions; with a misidentification probability of less than 1%.

Note that the μ–event has the decay electron later.

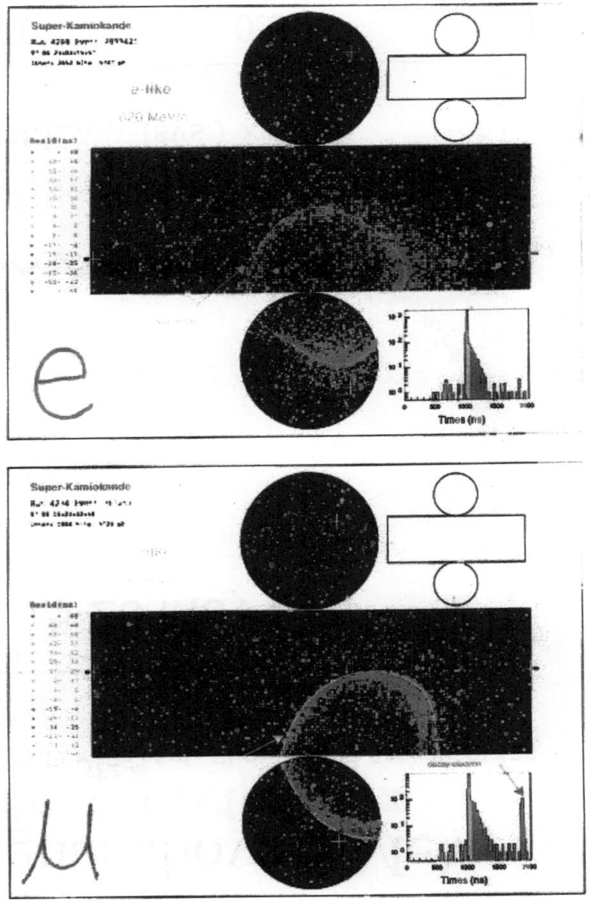

Directional Solar ν signal above 5 MeV

.465 of the theoretically expected flux

Super-KamiokaNDE S.Fukuda et al., Phys. Lett. B 539 (2002) 179.

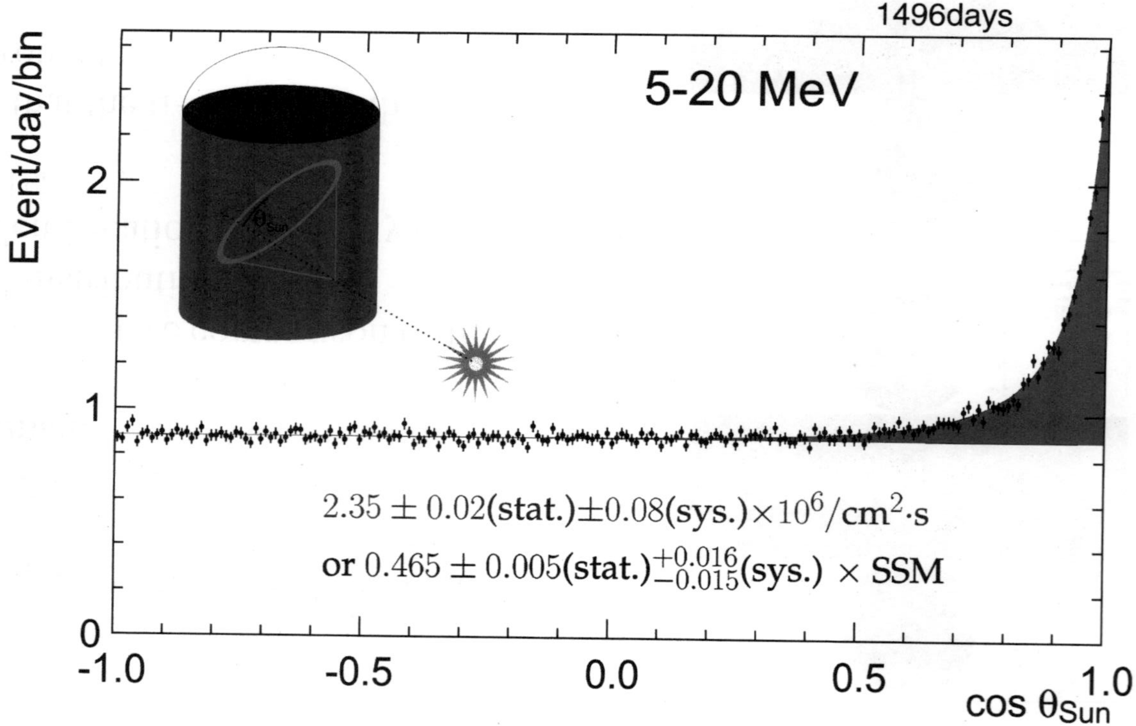

The observed electron energy spectrum compared to that expected for the B^8 decay neutrinos

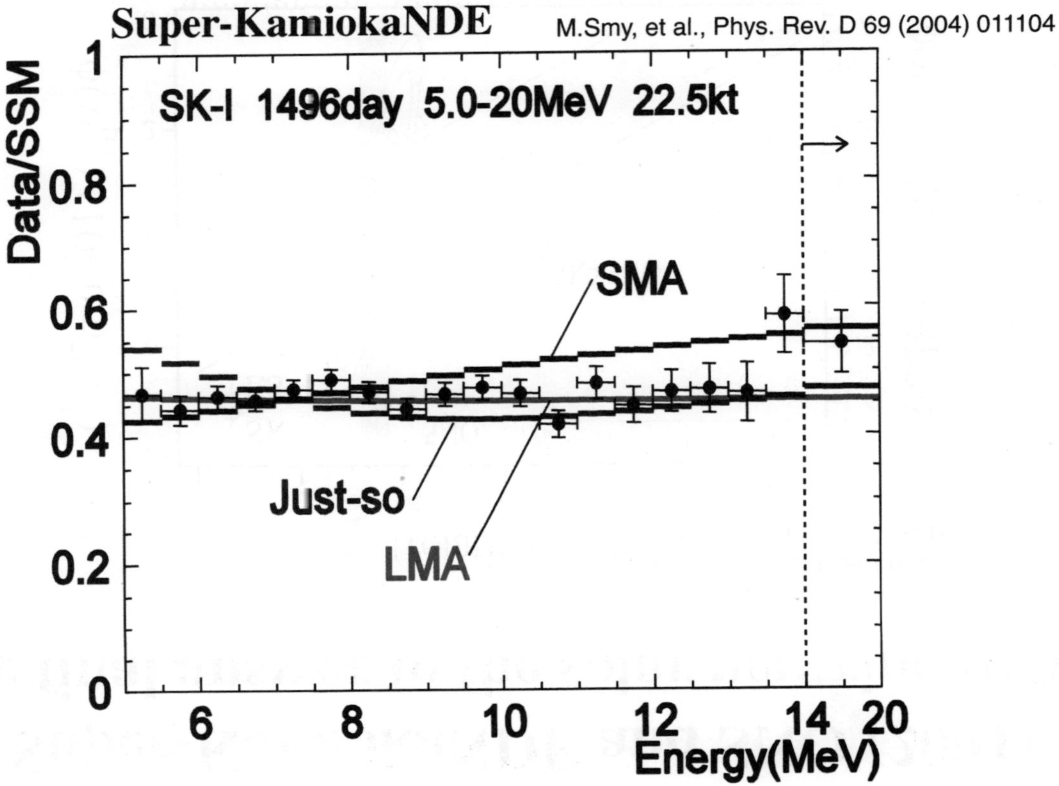

Super-KamiokaNDE and SNO (2001)
The final answer to the solar neutrino deficit

Q.R.Ahmad et al., Phys. Rev. Lett. 87 (2001) 071301.

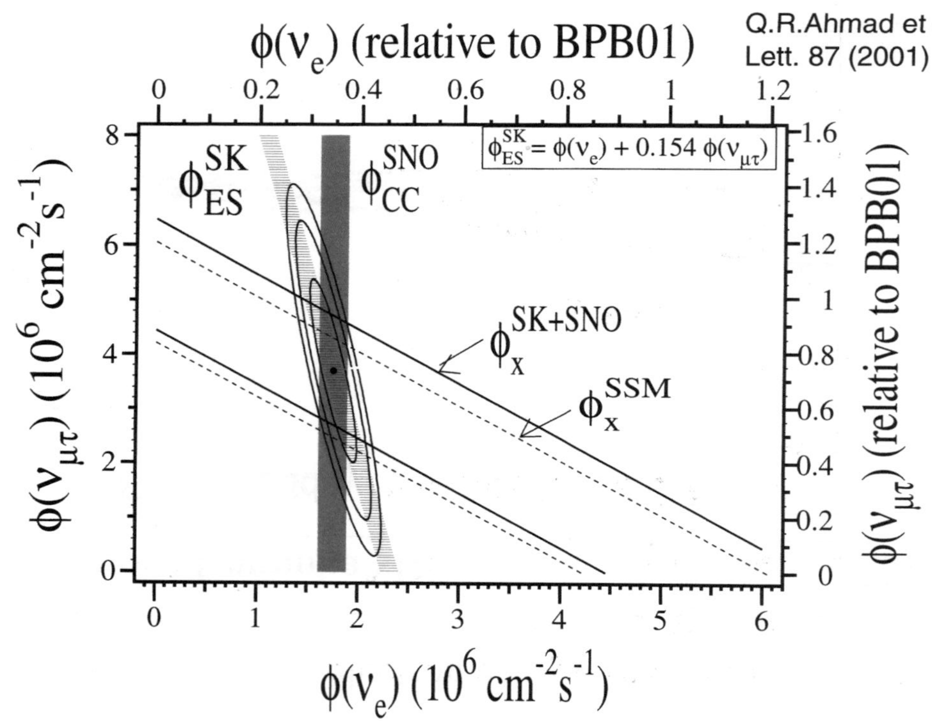

SNO Collaboration

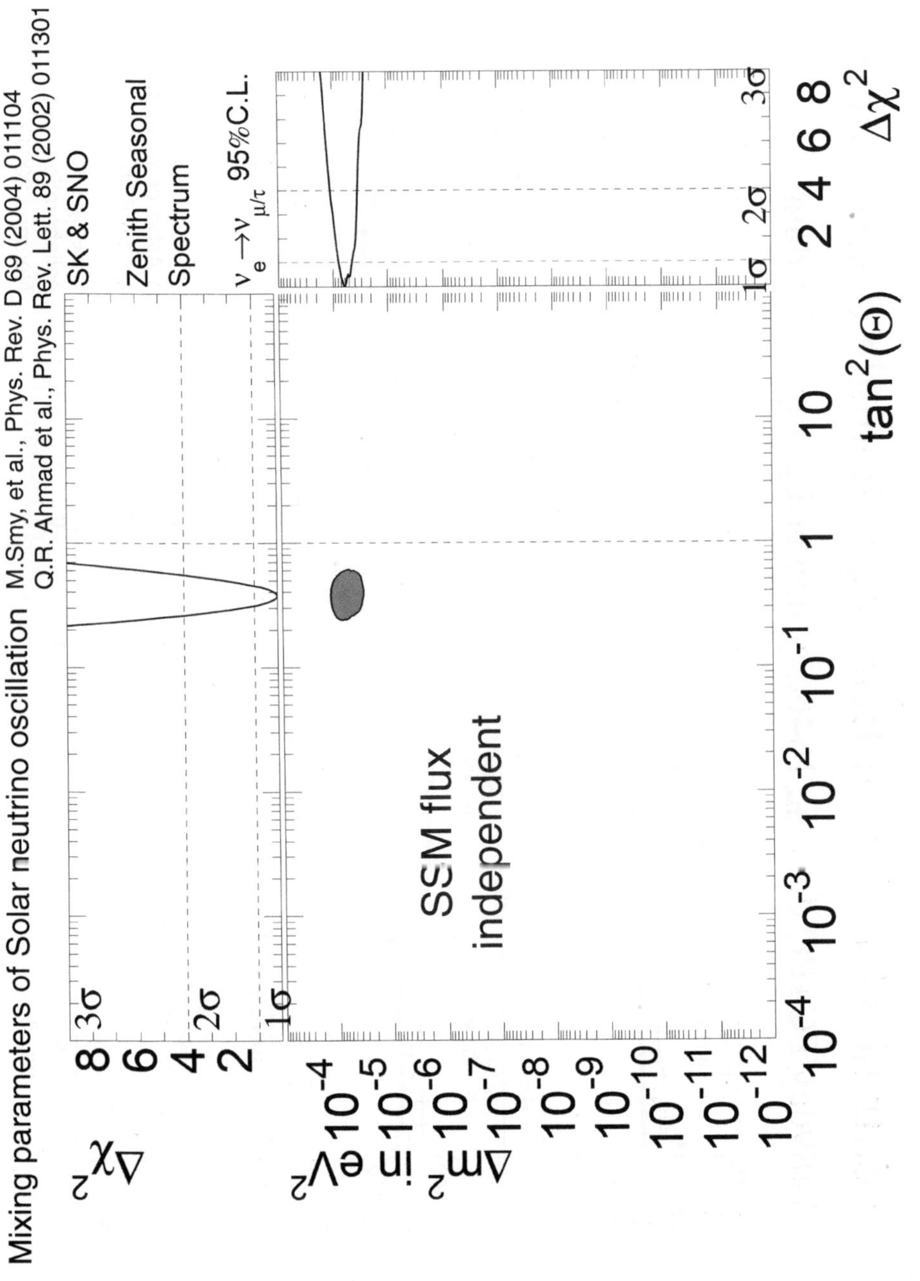
Mixing parameters of Solar neutrino oscillation
M.Smy, et al., Phys. Rev. D 69 (2004) 011104
Q.R. Ahmad et al., Phys. Rev. Lett. 89 (2002) 011301
SK & SNO
Zenith Seasonal
Spectrum
$\nu_e \rightarrow \nu_{\mu/\tau}$ 95%C.L.
SSM flux
independent
$\Delta\chi^2$
Δm^2 in eV^2
$\tan^2(\Theta)$
1σ
2σ
3σ

Neutrino Oscillation from Zenith angle distributions

1489day FC+PC data + 1646day upward going muon data

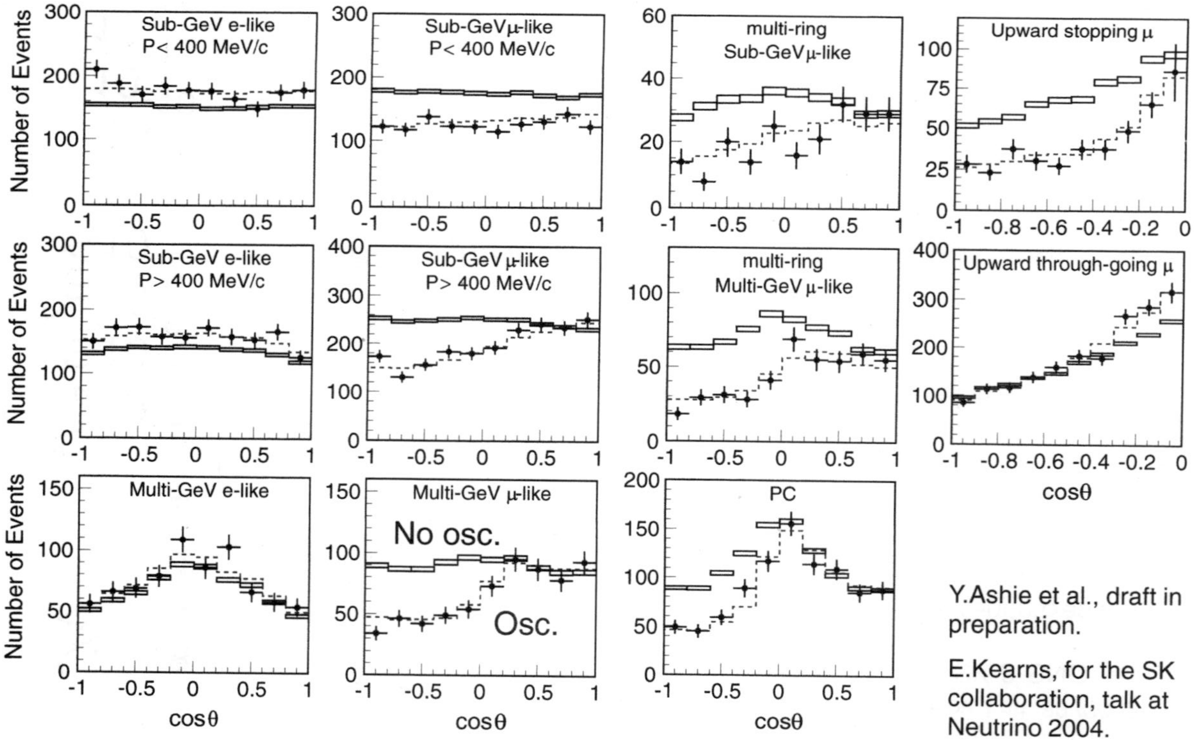

Y.Ashie et al., draft in preparation.

E.Kearns, for the SK collaboration, talk at Neutrino 2004.

Neutrino Oscillation by L/E analysis (1)

1489 days FC+PC (Super-K)

Y.Ashie et al., SK collab.
hep-ex/0404034

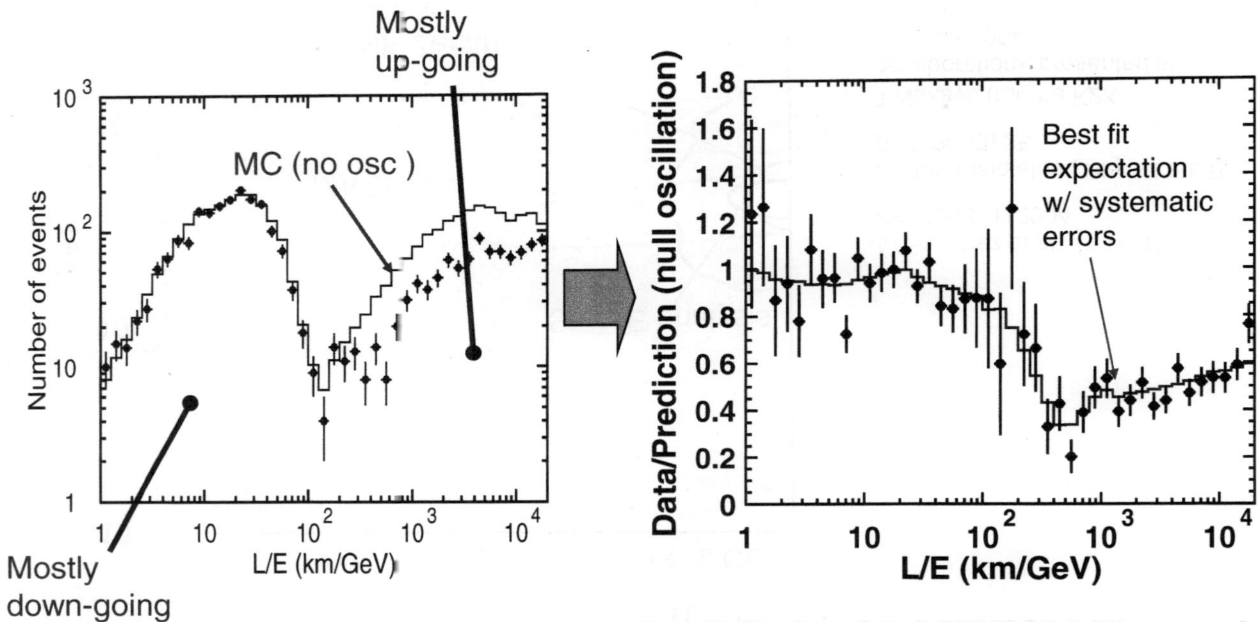

First dip is observed as expected from neutrino oscillation

Allowed regions for ν_μ to ν_τ oscillation

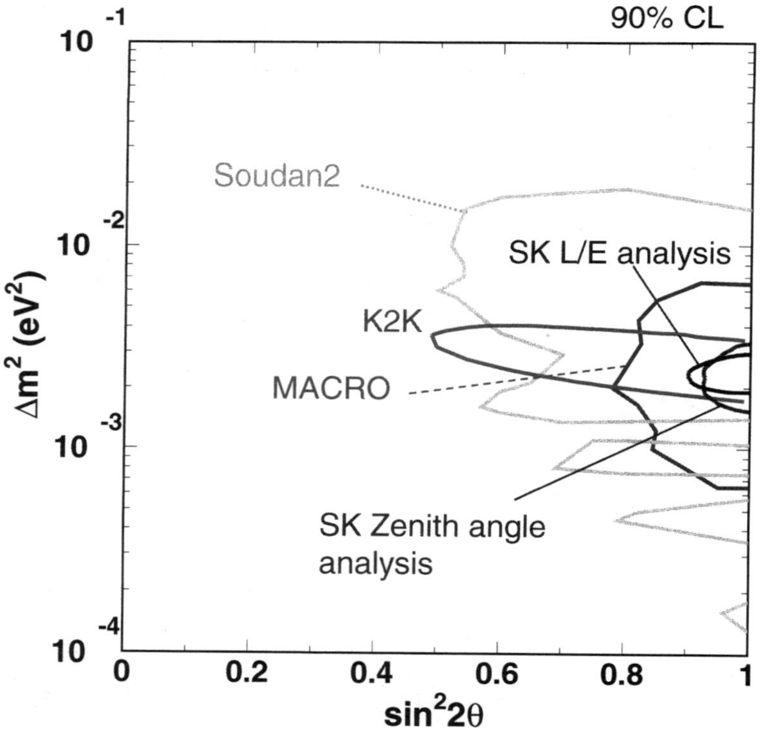

M.Sanchez et al., Phys. Rev. D 68 (2003) 113004

M.Ambrosio et al., Phys. Lett. B 566 (2003) 35.

T.Nakaya (for the K2K collaboration), presented at Neutrino2004.

Y.Ashie et al., hep-ex/0404034

Y.Ashie et al., draft in preparation; E.Keanrs, for the SK collaboration, talk at Neutrino2004.

electronics hut
water purification
liquid scintillator purification
N_2 generator
KamLAND
K a m L A N D
SK
Research Center for Neutrino Science, Tohoku Universi

New KamLAND results.
(anti-ν_e behaves the same as ν_e)

766 ton·year exposure

T.Araki et al., hep-ex/0406035

Energy spectrum

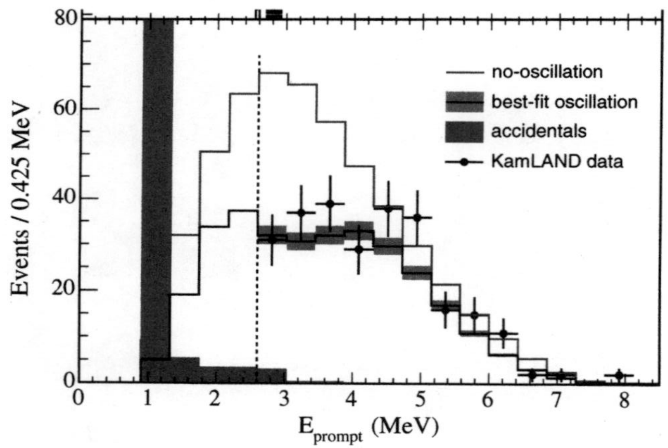

L/E distribution

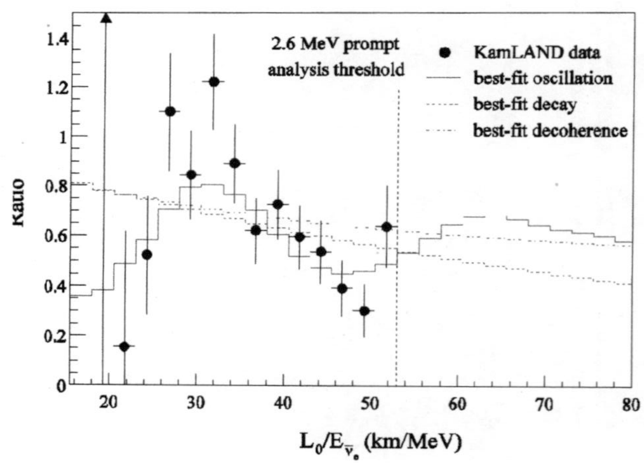

K2K

(Accerator Neutrino Oscillation Experiment)

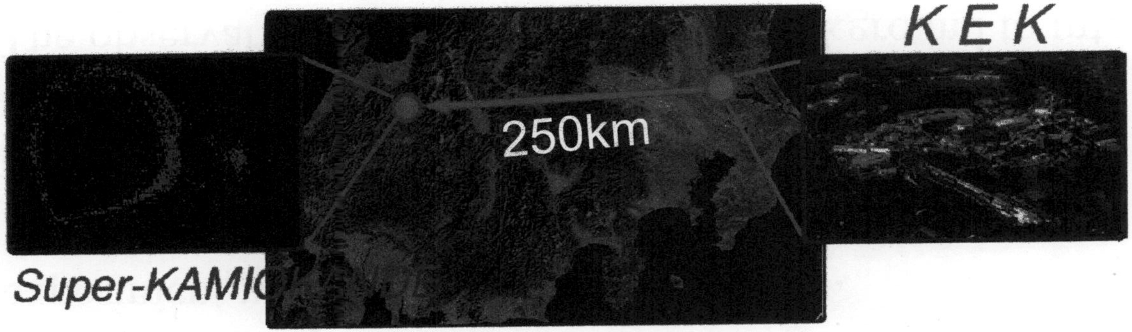

Neutrino oscillation probability for $\Delta m^2 = 0.003 eV^2$ and at 250km.

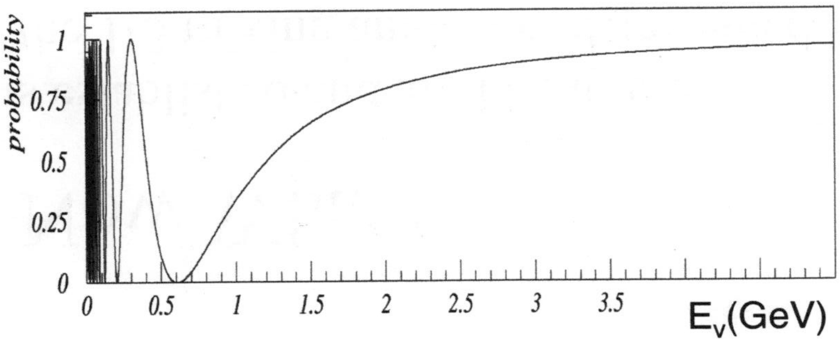

Now What?

- 3 neutrino mixing is established and all the mixing parameters, except the 1-3 mixing angle, are determined. The planned experiment at Gran-Sasso or the future **T2K** experiment will complete the information.
- KamLAND observed the anti-ν_e from inside the earth. Now the **anti-ν_e tomography** is feasible for U,Th and K deposits in the earth core; using several KamLAND type detectors .
- The observation of Cosmic Neutrino Background to infer the state of Universe 3 seconds after Big-Bang is indeed still remainig extremely difficult.

**Thank you for your patience.
For more details see the
Handbook of Particle Physics to
be published this fall by Springer
Verlag where we wrote the latest
summary of neutrino physics.**

M. Koshiba

CHAIRMAN: M. KOSHIBA

Scientific Secretaries: G. Cappello, M. Di Mauro

DISCUSSION

- Marshall:

At the beginning of your talk you mentioned 96 fundamental particles. That includes right-handed neutrinos and separate neutrinos and antineutrinos. Does that indicate a preferred mass model? Is there a reason for?

- Koshiba:

The answer is that because the neutrinos now have non-zero masses.

- Weir:

At the end of the talk you raised the prospect of neutrino tomography, that is looking at neutrino sources from within the Earth; is this realizable in the near future? And could the MSW effect be used to study relative densities of various layers of the Earth in a similar way?

- Koshiba:

I think it is possible if some of the world's experimentalists collaborate. KAMLAND has detected electron anti-neutrino coming from inside the Earth, therefore if similar detectors are distributed over the Earth's surface then it should be possible to find out how many thousand tons of Uranium and Tritium are there and at which depth.

- Weir:

Right, but I was also asking whether we could use the MSW effect due to the matter of the Earth to measure the relative densities.

- Koshiba:

I don't know. Sorry.

- Ni:

If a Supernova explosion should occur within the next years, how many neutrino events would SuperKamiokaNDE detect?

- Koshiba:

Of course it depends on where the supernova event occurs. If it happens near the center of our galaxy, SuperKamiokaNDE will be able to record something like 6000 to 7000 neutrino events. That would be ample enough.

- Tannenbaum:

You skipped over the atmospheric neutrino anomaly in your talk. Could you describe it now? I know that it has been observed in cosmic rays. Has it been observed in an accelerator experiment on Earth?

- Koshiba:

The atmospheric neutrino anomaly has been first observed by old KamiokaNDE, not by SuperKamiokaNDE. The idea was really simple: you make an estimate of the number of muon type neutrino and electron type neutrino. It is a very simple calculation. The pions and the kaons produced somewhere in the atmosphere decay in a short lifetime into muons and muon neutrinos. If this is the only process then the ratio of the number of the muon neutrinos over the number of electron neutrinos would be infinite; but the muon also has a small lifetime, about 1 microsecond, so it decays into muon neutrinos, electron neutrinos and electrons. If this happens, the number of muon neutrinos over the number of electron neutrinos is 2 to 1 even if we cannot calculate the number of those neutrinos produced in the atmosphere. This ratio is never smaller than 2. KamiokaNDE is a water Cerenkov detector capable of making distinction between light electrons and heavy muons produced. Using this capability, we could measure this muon number, and the muon neutrino number against the electron neutrino number, because in water sometimes you observe that electrons and muons appear. Muon must have been produced by muon neutrinos and electrons must have been produced by electron neutrinos, so by taking the ratio of produced muons and produced electrons you can infer the number of incoming muon and electron neutrinos. The result was amazing. Its ratio was almost one. As I said this ratio can not be lower than 2. This was the first observation of this anomaly and this went on to include also the solar neutrino analysis and so forth. However, as mentioned in Prof. Bettini's lecture, there is only one mixing angle still unmeasured because it's so small. There was an accelerator producing neutrinos hitting SuperKamiokaNDE which was about 250 km far, but the neutrino intensity was not big enough to give a finite answer for θ_{13}. Now a much stronger source of neutrinos is being prepared to hit SuperKamiokaNDE and the experiment will start next year. Not only American physicists, but also a number of European physicists are coming to Japan to participate to this accurate θ_{13} measurement.

- *Wen:*

Could you give some comments on the possibility for a future collaboration between China and Japan on a very long baseline experiment or an underground experiment?

- *Koshiba:*

Of course there is this possibility and we welcome collaboration with Chinese physicists, not only in neutrino physics, but also concerning the forthcoming linear collider. My hope is to place next international linear collider center somewhere in Asia. I'm not talking about placing it in Japan. For this purpose Asian scientists should collaborate seriously.

- *Aprile*

Can you comment on the plan of KAMLAND to turn that detector for neutrinoless double beta decay search by adding 100 kg of enriched Xe-136 ?

- *Koshiba:*

The KAMLAND experiment has been busy working on antineutrinos and it was conceived and built by physicist A. Suzuki, so your question should be addressed to him.

P.J. ODDONE

"Highlights from FERMILAB"

The Slides of the Lectures can be found at:

http://www.ccsem.infn.it/issp2008/

CHAIRMAN: P.J. ODDONE

Scientific Secretaries: R. Bernhard, M. Nagy

DISCUSSION I

- Nikiforov:

Ultra-high energy cosmic rays above the BZK cutoff were a big excitement to astrophysicists. Today you have shown Pierre Auger results consistent with BZK cutoff. Is here any possibility that we still do not control the measurement of atmospheric showers at large energies?

- Oddone:

To measure energies you have to make a model of showers, you have a shower detector, you get most of the statistics from the shower detector, and you have to cross-calibrate that with the flourescence telescopes. Experimenters are also measuring the fluorescence at accelerators with particle beams to calibrate fluorescence; in fact, you gradually reduce your systematic errors on that. So there is really one experiment that is not seeing that feature, it's the AGAS experiment, and it's smaller experiment than Pierre Auger. In the end, having the statistics that the Auger has, it will be possible to control the systematics better, and then if you do particle beams to get the real fluorescence, to calibrate the total energy you'll nail the calibration. What you would have to postulate is that the calorimeter doesn't work. I don't think anybody accepts that, it's really a question of calibrating them.

- Marshall:

You showed a plot of WIMP-exclusion cross section vs. WIMP mass with CDMS and XENON-10 on it. In the plot on the left, what is WIMPzilla?

- Oddone:

It's an ultra-massive object.

- Baudis:

10^{15} GeV that's produced at the end of the inflation.

- Oddone:

Well, you have axions at one end and 10^{15} GeV objects at the other end - so there's some real incredible relic out there - I don't think we probe that.

- Baudis:

The cross section is not known. There are Zimpzillas you can probe, but not WIMPzillas. The number of them in the galactic halo would be very small, but in principle they could be inelastically scatter, on the nuclei, like the others.

- Casis:

I'm coming to the proton-antiproton collisions. You have shown this morning among others excellent results of the B-physics programs. My question is if one has/knows any preliminary results from the investigations of heavy quarkonia, like χ_B mesons of about 10 GeV mass?

- Oddone:

The latest particle people observed was the Ω_B made out of one b quark and two s quarks. But at this point, I don't recall any new results of quarkonia around 10 GeV.

- Baudis:

Do you have a good particle physics model in order to explain the excess of events at low energies seen by MiniBoone, especially since you would like to build a new liquid argon detector, in order to test this region?

- Oddone:

That's right. People have a large number of papers of various physics models. Everybody wants to have sterile neutrino, and various other things that you can postulate. None of them are very plausible, quite frankly, the key issue there is to really see, whether it is in fact a single electron production or it is something associated with production of photons. In liquid argon detectors one can discriminate between the two.

- Baudis:

Comment on background-free experiments: the experiment of CDMS is tuned such that we expect < 1 event of background for the given exposure. And CDMS was built using low radioactivity materials. XENON-10, on the other hand, was not a low-background experiment, but a proof of principle. We need to wait for XENON-100 results, in order to make a fair comparison of the backgrounds for the two technologies.

- Oddone:

I think that's just a statement of what has been done. But it's clear that the scaling is very difficult with this solid state detectors, or at least very expensive.

- Baudis:

Well, I mention this, because people use this as an argument, that this is a background-free experiment.

- Oddone:

We will see.

- Patterson:

How is radon (and other non-cosmogenic) background being reduced in the 60 kg model? Does the decision of whether to go to Sudan depend on the performance at the surface?

- Oddone:

Fist we will try it out in the NUMI tunnel at Fermilab at 300 meters equivalent of water.

- Patterson:

Have COUPP given any sensitivity estimates for the 60 kg model?

- Oddone:

I don't have the number in my head, but the 2 kg. chamber already gives you competitive limits just by looking at the shape vs pressure in the chamber. The new chamber is much cleaner since it uses better materials (less radioactivity). With backgrounds the improvement should go like the square root of the ratio of the two masses. With a better fraction in the fidutial volume and less radioactive contamination the improvement should be at least a factor of ten.

- Ni:

About the MINOS results: how much improvement of Δm^2 and $\sin^2(2\theta)$ can MINOS get by continuing running? And how important is it to reduce the uncertainty?

- *Oddone:*

MINOS gets better and better, as of course we collect more data. My recollection is the way results are presented now is at the level of 3.5 of 4. with 5x10^20 protons on target collected. At least another factor of 3 will be achieved. So presumably the errors here will shrink by sqrt(3) in both directions. And if you look at the plots that have been projected up to 16×10^{20} protons on target, it just gets smaller around this region. I think that of course super-K is more mature, in the sense that it has run for many years, so to double the statistics on super-K is going to be difficult. But nonetheless, super-K does have a very good measurement of the mixing, and this should be resolved, whether it's really 1 or away from 1. We probably have to wait for the next generation of experiments like NOVA to do that.

- *Wen:*

You showed the external γ-background as one of the reason for the low energy excess. Why can't this effect be rejected by PID? What is the impact of this excess?

- *Oddone:*

Well, one of course can handle this particular background, by the distribution of the direction where the photons are coming from. So that is done by essentially understanding the distributions. If you see an excess near the tank boundaries, where you presumably have and excess due to the particular effect of the dirt, you can correct for it and eliminate from the data within some small systematic error.

- *Wen:*

Do you need a fiducial volume cut for the analysis?

- *Oddone:*

Either you do the fiducial volume cut or you estimate the number by looking at the distribution; you can do it either way.

- *Wen:*

(for SciBoone): Why choose 1 GeV neutrinos to measure neutrino nucleus cross sections?

- *Oddone:*

If you look at the neutrino spectrum coming out of the booster, it is very similar to the neutrino spectrum that you are going to observe with T2K. The experiment was motivated by really trying to understand the flux of relatively low energy neutrinos, when you make this big detectors. This would give you the topological cross sections in great detail, and then you actually can try to relate that to what a water Cherenkov counter for example would have observed in super-K.

- *Bernhard:*

Given that some of the results from D0 and CDF presented at ICHEP are statistically limited by various reasons. Do you advice the collaborations in focusing the physics output as director?

- *Oddone:*

As the director of the lab, I have very little influence actually, on what physicists should do, nor should I have it. One influence that I have, is trying to flog the two collaborations to combine the results. You're quite right that many of these things are either limited by statistics, because not all the data has been analyzed, or we have not run enough. There is a factor of 2 to gain already by combining the two experiments. The different systematics also helps because it tends to reduce the systematic uncertainties before you can see what happens in the two experiments. In many of the physics goals, where they were doing CP-violation in B-decays, or the Higgs, are really limited by, as you point out, statistics. The biggest issue at this point of the life of the experiment is to combine more and more. Historically, of course, just as it is probably going to happen at CMS and ATLAS, there's a lot of competition between the two experiments, so they don't like to show their cards, until they have their own results, and only then they talk about combination. When I first came to Fermilab, three years ago, I suggested that we merge the collaborations, and have one collaboration with two experiments. That really terrified spokespeople and collaborators, so I did not push it.

CHAIRMAN: P.J. ODDONE

Scientific Secretaries: R. Bernhard, M. Nagy

DISCUSSION II

- *Marshall:*

You showed a plot of DOE budget over about 10 years and the major facilities section started around now and expanded quite dramatically. What projects that you told us about are included in that, and which are not?

- *Oddone:*

Generally, what you see, is the accelerator and detector operations, which is now a fairly big chunk; it really goes way-way down, and it gets substituted by various things. One of them is the LHC and LHC upgrades. This growth you see here, is the LHC increasing in the intermediate time, and that's when we are doing the LHC upgrades. And then, out here, that slice that starts in a few years, is major facility construction; this is presumably project X unless there is a drive to build a linear collider.. The neutrino physics is now relatively small, just operating, and it grows to build NOVA and after NOVA, it grows to build DUSEL, and this grey area as part of it, it is project X.

- *Nikiforov:*

Knowledge of the parton distribution function (PDF) are essential to understand physics at the LHC. What would happen if DGLAP evolution equations are not valid at low and medium x region?

- *Oddone:*

If I understand your question, your question is that in order to extract the physics, you need parton distribution functions in all regions of phase-space, and they may not turn out to be correct in some region, in which case you don't have the right background calculation.

- *Nikiforov:*

Yes, at large q^2 regions.

- *Oddone:*

And then you don't know how to extract physics. I don't really know the answer of what specific physics channels are affected, to go through all the physics that you can do at the LHC, and what subset you are able to do and what subset you are not. You could imagine at some point, if you need more description of this PDF, to do elecron-proton scattering at higher energies, even in an elecron ring, in LHC, but I think we have to wait; it is uncertain what we will find at the LHC, so it's very difficult to plan exactly what you need without having that data.

- *Wen:*

I'm not sure about the three contours of the left pitcure, the different colors, could you explain more about that?

- *Oddone:*

There are essentially three colours, red, green and blue. Those are centered on different presumed values. They show what we would measure if in fact they were the value of those parameters. In another way, if $\sin^2(\theta_{1/3})$ were 0.95, the 1σ confidence level limit would be this circle here, and the dotted one would be the 2σ confidence level limit. Then, as you go from this area, from this plot to this other plot on the right, what is changing is the amount of integrated protons on target, on a 50 kiloton detector.

- *Wen:*

The difference is just because of the difference in the input parameters?

- *Oddone:*

Right.

- *Wen:*

You mentioned that the main goal of NOVA is to measure $\theta_{1/3}$ and the mass hierarchy. If $\theta_{1/3}$ is large, NOVA can give a glimpse of the mass hierarchy. So the question is if $\theta_{1/3}$ is very small, what result could NOVA give?

- *Oddone:*

If $\theta_{1/3}$ is very small, and you're way out here, say at 0.01, then you would not be able to resolve the mass hierarchy. You will get some combination of the parameters of possible mass hierarchies and CP-violation, but it is not very precise, and you will not be able to say definitively what the mass hierarchy is.

- *Wen:*

Could you comment on the mass hierarchy measurement for future very long baseline experiments?

- *Oddone:*

So, here are the plots that one can get in the progressive neutrino experiments with a relatively long baseline. They are not comparing all the same baseline, but you can see, what, for example, NUMI off-axis and NOVA can do in terms of the reach in parameters. The next curves show NUMI off-axis plus a liquid argon of 5 kilotons in Sudan. This gives you better sensitivity than just NOvA but not by much. So if you're coming down in your measurements of $\theta_{1/3}$ at this point, you haven't done anything other than decrease the measured limit for $\theta_{1/3}$. If you then build a liquid argon detector at the second maximum, even further off-axis, you can get the masss hierarchi at a lower value of $\theta_{1/3}$. Then if you build project-X with a wide-band beam to a 100 kiloton detector in the 13 kilometer baseline of DUSEL, you get down to below 10^{-3} in $\theta_{1/3}$. Then you can study the mass hierarchy and CP-violation phase to the limits shown in the picture. We will be able to tell something with NOVA. But if you really want to get the mass ordering, and $\theta_{1/3}$ it is much smaller, then you have to go to the bigger detectors. Similarly, here, in the CP-violation plot, you'll be able to get down to a very low level of $\sin(\theta_{1/3})$ and still tell CP-violation. So when you go to the longer baseline, to much larger detectors and intense beams, you have a much better chance to resolve the mass hierarchy, and get to CP-violation provided that the angle is not absolutely tiny. If it's really very small, then you have to go to a neutrino factory or a β-beam.

- *Wen:*

What do the vertical colour bands mean in the two plots?

- *Oddone:*

The color bands are just guides for the eye at various levels of the angle. And curves represent what they say in terms of what a neutrino factory or a beta beam or a superbeam does.

- *Wen:*

So the sensitivity is calculated by using GLOBES...

- *Oddone:*

These are programs in which people throw in what you know about the type of the detector, and all the oscillation parameters, and the answer is obtained with the program.

- *Wen:*

What kind of detector configuration is assumed here?

- *Oddone:*

For the neutrino factory curves I'm not sure what kind of detectors are assumed here for this calculation, and what baseline, I need to go back and look at the neutrino study. Normally, what happens when you go to a neutrino factory, you want to have detectors that are quite different, because you separate the charge. The muon beams will produce a mixed beam, but if you separate the charge, then you will have a pure beam with one kind of neutrino only. Thus the detectors involved in a neutrino factory are quite different, they have to be magnetized, so that you can distinguish the charge. That's where the power of the neutrino factory comes, in giving you a very pure beam, which you do not get from the superbeams we have today.

- *Caffari:*

Which is the status of the muon collider project? Is there any studies of luminosity, magnetic field etc...?

- *Oddone:*

There is quite a vital program, at least in the US. Over the last few years, it has not been funded very well, roughly the ratio of 10:1 relative to the linear collider, because the linear collider is the accelerator that is perceived we might build relatively quickly on, if the spectroscopy or energy scale or what we find with the LHC is below 0.5 TeV or even between 0.5 and 1 TeV. That's by far the easiest machine to build. What's happening now though, is that there is a greater emphasis on trying to make progress on the muon collider R&D. I showed some of the pieces. There is a MICE experiment, there was the mercury TARGET, and there are theoretical studies of various pieces of the cooling that you need, and the lattices for the machine. One of the things that I'm pushing very hard for our folks to do is to vote themselves the technologies they need to make the machine work, assume some of the technologies that we need for the cooling work, make an end-to-end simulation, because in principle, that should be possible to do, since we just need to track the particles through all these complex electromagnetic fields. That has not been done. And there is an argument saying it's not worth doing unless we know that some of these technologies like big cavities inside magnets work. They don't work now, so why do I bother simulating them. That discussion is going on right now in the community. When I came to Fermilab, I made a task force to work on this which actually doubled the national effort which was about $ 4 million a year of R&D. Now it is $ 8 million, and to really make progress and have a fully designed muon collider by 2012-2013, that number has to double again. We are negotiating with the agencies to ramp up that number now.

- *Baudis:*

You mentioned that the planned satellite experiment SNAP will measure, among other things, thousands of type IA supernovae. Can you explain what we will learn about dark energy from this measurement? For instance, will we be able to distinguish between different models of dark energy, will we be able to measure a time-dependence? How well can we meassure the equation of state?

- *Oddone:*

That's a good question. In principle, you can get to dark energy, or at least the parameters of dark energy, in several ways. What we now know, is just one parameter, which is essentially the present acceleration. All these new probes are aimed to see how this expansion has evolved over time. For example, if there is a model like the cosmological constant, then you would expect this thing just to be absolutely flat as a function of the evolutionary time, whereas other theories predict different curves. Now the precision that you get is not enormous. To first order at least, it's assumed that you get the derivative with these new probes. People say sometimes that these models are too simple, but that's the next thing we can ask of dark energy. We have been surprised before, dark energy was a big surprise. In most of these probes the satellites look at some one patch of sky, and keep coming back to that patch to find supernovae, so that they get the data as a function of the expansion in one direction. But one of the questions you could ask is if the parameters of dark energy in very different regions of the sky are different. In all models they shoud be the same, for instance, if you think it's just the cosmological constant. In principle you can get independent measurements of dark energy with these different methods: evolutions of clusters, baryon acoustic oscillations, through supernovae and through weak lensing. Then one of the questions is, do you get the same answer with all these methods, because the underlying assumptions, such as that the general theory of relativity is correct, imply they should all be the same. If you find a difference you know something is wrong. To tell that something else is wrong, you have to get a high level of precision. But this is an interesting question, whether all these methods measure the same thing and are consistent with each other.

- *Gazis:*

The presentation of the 2nd talk about the future program of Fermilab is impressive. Moreover, the multi-TeV muon-collider is really something which will attract the interest of international community. My question is more technical: the muon beam after the cooling ionization process will be kind of a polarized beam?

- *Oddone:*

I have to think about that.

- *Gazis:*

Because you said that longitudinal part is disappearing...

- *Oddone:*

I haven't asked that question precisely, you have a lot of collisions when you take out energy in the material, and I don't know whether you can preserve the polarization through that.

- *Gazis:*

The same question for ILC: do you plan to produce polarized beams of electrons and positrons? I

- *Oddone:*

Actually, from the beginning. In the ILC you can certainly polarize the beams, they work very hard to preserve the polarization in the whole accdelerator chain. And then if you have a wiggler to produce the positrons, you can also produce the polarization in the positron beam. So the ILC does have this polarization. I'm not sure about the muon collider. It's an odd concept in a way, that you cool this beam by putting it through material. Normally that material heats the beam through multiple scattering, so this combination of acceleration and putting it through material gives, finally, a low emittance.

- *Klopot:*

You mentioned the program for study of K-meson decays. Will you have any kind of study of other light mesons (η, η' and other light mesons)? This is important, because light mesons, due to their simplicity, in comparison for example with barions, allow us to study such fundamental properties of QCD like anomaly.

- *Oddone:*

We have not looked at that. We just looked at the rare K-decays right now.The question we are after is if there are things that are very sensitive to larger mass scales. In the case of K-decays, depending on what theories you have, you can have quite different ratios, very reliably calculated, for example for K^+ and K^0 decays. When you look at the branching ratios of K_{long} to $\pi^0\nu\nu$, these are very clean things you can study in the SM. And then you have the K^+ to $\pi^+\nu\nu$. The SM would predict these numbers to be right out here in the plot that I showed. The numbers that have been excluded experimentally are the blue areas. Beyond the SM, little Higgs, SUSY, etc... lie all in this quadrant here, so you would be sensitive to placing some numbers in that area. Now you could also add additional conditions to SUSY, like minimal flavor violating, which you wouldn't necessarily expect of SUSY. And those areas lie around this area here, at very low decay rates. We've been trying to answer questions that are connected to and complementary to what the LHC might find.

- *Dubinski:*

You mentioned the first observation of New Physics in b to s transitions. Can these data be improved by means of Tevatron before it will b checked by LHCb?

- *Oddone:*

New physics would have to be established with more luminosity and a dirtect analysis by the two collaborations CDF and D0. We cannot claim new physics now.

One of the problems that you have, whenever you look at experiments like μ to e, is that you will get a rate, not absolute values. There are some processes that need new physics, but the new physics model is unknown. The only way you actually can make sense of these things, is that at the end you have to put this picture together, with something else that the LHC might tell you. That would be powerful: you discover something at the LHC, make prediction for this process, you check it if it's right, or you falsify that theory. This also can be useful if you don't find anything at the LHC. (That would be of course a catastrophy, we probably go and open up a restaurant or something). If you don't find anything, then you need a new experiment which can probe scales around ~1000 TeV. There could be intermediate states of ~1000 TeV, that could be never produced at the LHC.

- *Ni:*

About the detector choice of a future DUSEL experiment: The water Cherenkov detector has already achieved very good size and results. But liquid argon detector is still in the few of hundreds of ton level. There seems to be a technical challenge to make a large liquid argon detector. What' the main advantage of liquid argon?

- *Oddone:*

When we look at the liquid argon detector, and how you would actually carry out the oscillation physics, there is roughly a factor of 4 advantage in mass if you just look at the parameters. In the water counter you have to make very tight cuts, in order to be able to actually extract the signals and not be confused by inelastic events, because any time you have a neutral current event, you can be confused. People argue about how well you can do that in a water counter. You put very hard cuts so the efficiency in a water Cherenkov counter is not very high, tipically what we find, is that a 100 kiloton liquid argon counter is equivalent to a 400 kiloton water

Cherenkov, if the phyics is just what we are thinking it is , and we can apply the cuts in the water as ideally as you think it might work. With liquid argon there is no question, you actually see the interaction, you know exactly what's happening, it's almost like a bubble-chamber. Ultimately, it would be great to have liquid argon available as a detector; the caverns that you have to build will be considerably smaller, by a factor of 3 or 4, relative to the water Cherenkov counter. The other aspect for liquid argon is that it's sensitive in principle to different proton decays tha water Cerenkov counters. It's important to do it in water for some of the modes, and in liquid argon for other modes. Of course, super-K has worked for a number of years, so for a water Cherenkov counter to compete it in terms of proton decay, you need to make a much greater mass. By the time one starts this detector, super-K will have integrated 20 years of running. In order to catch up relatively quickly, you need a factor of 5-10 in mass.

- *Bernhard:*

How have the event yields for the rare K-decays (K^+ and K^0) been estimated?

- *Oddone:*

What people have looked at, was an experiment proposed at Brookhaven for measuring those rates. Typically, they would have had a handfull of events. With this number of events those ratios are known to 40% accuracy. We looked at the various factors that would come in. Once you will have a beam of the kinds of intensity produced with project X. One of the advantages, for example in terms of the K^0 to $\pi^0 \nu\nu$mode of the K_{long}, that in the Brookhaven experiment a ribbon beam would have been used, in order to get enough beam to the detector. Therefore, there was not really a good 4π coverage, for example for the calorimeter veto. Whereas, with project X we can produce a real pencil beam going into the detectors. We multiplied all those factors and it looked like what was 10 events in Brookhaven, could easily be 1000 events here, and that would give you a measurement with ~3% accuracy.

Y. WANG

"Highlights from IHEP"

The Slides of the Lecture can be found at:

http://www.ccsem.infn.it/issp2008/

CHAIRMAN: Y. WANG

Scientific Secretaries: A. Bilandzic, Z. Marshall

DISCUSSION

- *Aprile:*

What is the impact of the energy resolution that you showed from the Cesium spectrum? I gather it is about one photoelectron per keV?

- *Wang:*

We have about 200 photoelectrons per MeV, so less than one photoelectron per keV.

- *Aprile:*

If you were to improve this light collection, what would be the impact on the final precision of θ_{13}? Or if you added phototubes to the top and bottom of the detector, rather than using reflectors, how much would that improve things?

- *Wang:*

There are two reasons for using reflectors on the top and bottom. Mainly, it keeps down the cost, since you reduce by a factor of two the number of phototubes. It also makes the engineering simpler, which also saves money. The resolution is determined by the threshold at 1 MeV, and we believe that 10% resolution around 1 MeV is not a problem. We have a resolution of about 5% at the 6 MeV threshold to remove the neutrons, and we believe that is sufficient.

- *Maier:*

Are there any plans to measure the continuum above the charm threshold?

- *Wang:*

Yes, we have plans to. We will run at several peaks, but for each peak we must run in the continuum above and below it to measure the backgrounds.

- *Schwartz:*

You gave a nice overview of the plans for experimental particle physics in China. Could you tell us a bit about expanding the efforts in theoretical particle physics?

- *Wang:*

Theoretical particle physics doesn't need very much money to do what they want to do. We have a group of theoretical physicists in our institute with about 20 people. If we can find good guys, we will hire them; it is not a problem.

- *Zichichi:*

In China, for theoretical physics, there is also CCAST, the Chinese Center for the Advancement of Science and Technology, directed by T. D. Lee.

- *Baldini:*

Do you have plans to increase the luminosity of the storage ring?

- *Wang:*

Yes, we do. We want to use the idea from Frascatti, Crab Waist, to increase the luminosity. We are awaiting the test results from Frescatti. If it is successful and they can increase the luminosity by a factor of four, then we should be able to increase our luminosity by a factor of two. We are also thinking about other upgrades, for example polarizing the beams or reducing the bunch spacing. These are all plans for the next 3-5 years, and for now we will focus on bringing the machine up to spec at 10^{33} cm^{-2} s^{-1}.

- *Patterson:*

What are the calibration capabilities of the final detector design at Daya Bay – can you put a source anywhere, or is it limited?

- *Wang:*

We have a plan to use many sources. In each module there are three holes at three different locations: the middle, the edge of the target, and the normal liquid scintillator. We can calibrate all the different parts of the detector by inserting different sources into these pipes along the z-axis. This is the automatic calibration system which we will use every day. We also have a manual calibration system that is quite sophisticated, so that we can put a source at any location. We can do a manual calibration at the beginning of the detector deployment, and then during daily operation we will do automatic calibration.

- *Dubinskiy:*

You told us about the results of the BES, but you didn't mention the new result near the 3770 MeV resonance, where you found that there could be a composite state.

- *Wang:*

Yes, I gave just a few examples.

- *Dubinskiy:*

Will the analysis of BES III be extended to this region, since there is still a discrepancy between CLEO C and BES in non-DD decays?

- *Wang:*

For the exclusive channels, we have no disagreement. For the inclusive measurement, there seems to be a disagreement in the line shape at 3770. That will be one of our first priorities when we start the machine: we will scan the line shape. It will take us about three months to scan the whole line shape with substantially improved precision.

- *Baudis:*

Could you please compare Daya Bay and Double CHOOZ? What are the relative advantages of each?

- *Wang:*

Daya Bay and Double CHOOZ started construction at about the same time, and now the official schedules say that each will start taking data at the end of 2010. We are on the same schedule. From the point of view of sensitivity, Daya Bay is a factor of three better than Double CHOOZ: Double CHOOZ is 3%, Daya Bay is 1%, and RENO is also 3%.

Fake Supergravity and Black Hole Evolution

Alessandra Gnecchi

Dipartimento di Fisica, Università di Pisa and INFN - Sezione di Pisa, Largo B. Pontecorvo 3, I-56127 Pisa, Italy

Abstract

A first order formalism, originally applied to BPS states, can be extended to non-BPS black holes solution in the case of $\mathcal{N}=2$ minimally coupled, $\mathcal{N}=3$, $\mathcal{N}=4$ pure and $\mathcal{N}=5$ Supergravities. In these cases one can define an effective radius for the black hole, as a combination of two moduli-dependent quantities, but the scalars dependence actually cancel. We describe the symplectic embedding of the scalar manifold to get the explicit dependence on central and (in case) matter charges on the scalar fields, and then present black hole parameters, such as ADM mass, scalar charges and effective radius.

1 Black holes in Supergravity

Black holes are solutions for the spin-2 field in Supergravity spectrum. Like the classical solutions, they can in principle emit thermically through Hawking radiation, this causing instability. Thus, a solitonic description is allowed if and only if the black hole has zero temperature, and they are defined *extremal black holes*.

We consider static and spherically symmetric extremal black holes in asymptotically flat 4-dimensional space-time.

In this case the event horizon manifests an attractor behavior; in general [7] "an extremal black hole attractor is associated to a critical point of a suitably defined black hole effective potential, and it describes a scalar field configuration stabilized purely in terms of conserved electric and magnetic charges, at the event horizon, *regardless* the values of scalar fields at spatial infinity". The main, first papers where this properties were presented and discussed are [1], [2], [3], [4].

The results reported here are based on the work presented [8]

2 Radial Evolution and black hole dynamics

The relevant part of the action that enters in the black hole description is the bosonic part of d=4 N-extended Supergravity, and it is given by

$$S = \int \sqrt{-g} d^4x \left(-\frac{1}{2} R + \mathrm{Im}\mathcal{N}_{\Lambda\Gamma} F^{\Lambda}_{\mu\nu} F^{\Gamma,\mu\nu} + \frac{1}{2\sqrt{-g}} \mathrm{Re}\mathcal{N}_{\Lambda\Gamma} \epsilon^{\mu\nu\rho\sigma} F^{\Lambda}_{\mu\nu} F^{\Gamma}_{\rho\sigma} + \right.$$
$$\left. + \frac{1}{2} G_{rs}(\Phi) \partial_\mu \Phi^r \partial^\mu \Phi^S \right) . \tag{2.1}$$

The first term is the Einstein-Hilbert action, the second one is the kinetic term of n abelian vector fields and, in order to have electric-magnetic duality invariance of the action, according to the Gaillard-Zumino construction [], we must add scalar fields, describing a non linear sigma model, coupled to vectors via the kinetic matrix $\mathcal{N}_{\Lambda\Sigma}(\Phi)$.

Neglecting the contribution of the vector fields, one finds that the equation of motions for the scalars are geodesics

$$\frac{d^2\phi^i(\tau)}{d\tau^2} + \Gamma^i_{jk}(\phi) \frac{d\phi^j}{d\tau} \frac{d\phi^k}{d\tau} = 0 \ , \tag{2.2}$$

where τ is a reparametrization of the radial coordinate, with the constraint

$$G_{ij} \frac{d\phi^i}{d\tau} \frac{d\phi^j}{d\tau} = 4S^2T^2 = 2c^2 \ . \tag{2.3}$$

If we only deal with static and spherically symmetric solutions, we can perform a dimensional reduction on the time coordinate and integrate on angular variables; taking into account the vector fields contribution, one eventually writes the one dimensional Lagrangian [4]

$$\mathcal{L} = \left(\frac{dU}{d\tau} \right)^2 + G_{ab} \frac{d\phi^a}{d\tau} \frac{d\phi^b}{d\tau} + e^{2U} V_{BH} - c^2 \ , \tag{2.4}$$

with $c^2 = 2S^2T^2$; extremal black holes then correspond to the case $c^2 = 0$.

The black hole potential can be written in a manifestly symplectic way as

$$V_B H = \frac{1}{2} Q^{T\,\Lambda} \mathcal{M}_{\Lambda\Sigma} Q^{\Sigma} \ , \tag{2.5}$$

in terms of conserved electric and magnetic charges

$$Q^{\Lambda} = \begin{pmatrix} p^{\Lambda} \\ q_{\Lambda} \end{pmatrix} , \qquad \begin{array}{l} p^{\Lambda} = \frac{1}{4\pi} \int_{S^2} F^{\Lambda} , \\ q_{\Lambda} = \frac{1}{4\pi} \int_{S^2} G_{\Lambda} . \end{array} \tag{2.6}$$

The equations of motion are

$$\begin{aligned} \frac{d^2U}{d\tau^2} &= 2e^{2U} V_{BH}(\phi, p, q) , \\ \frac{D\phi^a}{d\tau^2} &= e^{2U} \frac{\partial V_{BH}}{\partial \phi^a} . \end{aligned} \tag{2.7}$$

In order for the scalar fields dynamic to be regular one requires

$$G_{ij} \partial_m \phi^i \partial_n \phi^j \gamma^{mn} < \infty \tag{2.8}$$

for any value of the radial parameter. The $AdS_2 \times S^2$ horizon geometry of the extremal black hole explicitly emerges once the metric is written in terms of the coordinate

$$\omega = \log \left(-\frac{1}{\tau} \right) , \qquad \text{as } \tau \to -\infty , \tag{2.9}$$

and it results

$$ds^2 \approx -\frac{4\pi}{A} e^{2\omega} dt^2 + \left(\frac{A}{4\pi} \right) d\omega^2 + \left(\frac{A}{4\pi} \right) (d\theta^2 + \sin^2\theta d\phi^2) . \tag{2.10}$$

The regularity condition then requires

$$G_{ij} \frac{d\phi^i}{d\omega} \frac{d\phi^j}{d\omega} \left(\frac{4\pi}{A} \right) < \infty \qquad \to \qquad \frac{d\phi^i}{d\omega} = 0 ; \tag{2.11}$$

this requirement, together with the dynamics description provided by the equation of motions, leading to the near horizon scalar fields expansion

$$\phi^a = \phi^a_{Hor} + \left(\frac{2\pi}{A} \right) \frac{\partial V_{BH}}{\partial \phi^a} \log \tau , \tag{2.12}$$

is equivalent to the condition on the effective black hole potential

$$\left(\frac{\partial V_{BH}}{\partial \phi^a} \right)_{hor} = 0 . \tag{2.13}$$

3 First order formalism for $d = 4$ extremal black holes

The first order (fake supergravity) formalism can be conveniently used to describe non-BPS attractor flows of $d = 4$ extremal black holes [5, 6].

The first law of thermodynamics for a static dyonic black hole in the presence of scalar fields is

$$dM = TdA + \psi^\Lambda dq_\Lambda + \chi_\Lambda dp^\Lambda + \left(\frac{\partial M}{\partial \phi^a}\right) d\phi^a \,, \tag{3.1}$$

where the black hole temperature is $T = \frac{\kappa}{2\pi}$, and ψ^Λ, χ_Λ are electric and magnetic scalar potentials, respectively. The scalar fields contribution to the black hole mass variation is given by the scalar charges Σ^a

$$\left(\frac{\partial M}{\partial \phi^a}\right) = -G_{ab}(\phi_\infty)\Sigma^b \tag{3.2}$$

these are defined from the radial asymptotical expansion of the scalar fields

$$\phi^a = \phi^a_\infty + \frac{\Sigma^a}{r} + O(\frac{1}{r^2}) \,. \tag{3.3}$$

The integral form of the first law gives the constraint

$$M^2 + G_{ab}(\phi_\infty)\Sigma^a\Sigma^b - V(\phi_\infty, p, q) = 4S^2T^2 \,, \tag{3.4}$$

so that in the extremal case the left hand side expression is identically zero.

For $d = 4$ Supergravities one can define an effective black hole potential in terms of central and, in case, matter charges, as

$$V_{BH} = \frac{1}{2} Z_{AB}\bar{Z}^{AB} + Z_I\bar{Z}^I \,. \tag{3.5}$$

Introducing a *fake superpotential* $\mathcal{W}$, one can perform the ansatz

$$\dot{U} = e^U \mathcal{W}(\phi(\tau)) \,; \tag{3.6}$$

taking into account that there is only an implicit dependence of $\mathcal{W}$ on the scalars, one finds that the equations of motion for the scalar fields is the first order BPS-like condition

$$\dot{\phi}^a = 2e^U G^{ab}\partial_b \mathcal{W} \,. \tag{3.7}$$

The black hole potential is, in this formalism,

$$V_{BH} = \mathcal{W}^2 + 2G^a b\partial_a \mathcal{W}\partial_b \mathcal{W} \,; \tag{3.8}$$

the attractor point is thus related to the extrema of $\mathcal{W}$, since

$$\partial_a V_{BH} = 2\partial_b \mathcal{W}(\mathcal{W}\delta^b_a + 2G^{bc}\nabla_a\partial_c\mathcal{W}) = 0 \; . \tag{3.9}$$

Scalar charges and the ADM mass can be written as

$$\Sigma_i = 2 \lim_{\tau\to 0^-} \nabla_i \mathcal{W} = 2 \lim_{\tau\to 0^-} \partial_i \mathcal{W} \; ; \tag{3.10}$$

$$M^2_{ADM} = r^2_H = \lim_{\tau\to 0^-} \left[V_{BH} - 4G^{ij}\left(\partial_i \mathcal{W}\right)\overline{\partial}_j \mathcal{W}\right] = \lim_{\tau\to 0^-} \mathcal{W}^2. \tag{3.11}$$

In the specific $\mathcal{N}=2,3$, pure $\mathcal{N}=4$ and $\mathcal{N}=5$ theories, the *effective radius* for a black hole of entropy S_{BH}, defined as

$$\begin{aligned} R^2_H &= \lim_{\tau\to -\infty} V_{BH} = V_{BH}|_{\partial V_{BH}=0, V_{BH}\neq 0} = \\ &= \lim_{\tau\to -\infty} \mathcal{W}^2 = \mathcal{W}^2|_{\partial\mathcal{W}=0,\mathcal{W}\neq 0} = \\ &= \frac{S_{BH}}{\pi} \, , \end{aligned} \tag{3.12}$$

can be written as a sum of two moduli-dependent quantities, namely the radius of the event horizon and the scalar charges, but the sum cancels the scalar dependence revealing the radius as a function only of conserved electric and magnetic charges, as it comes from the definition

$$\begin{aligned} R^2_H(p,q) &= r^2_H(z_\infty,\bar{z}_\infty,p,q) - G_{i\bar{\jmath}}\Sigma^i\overline{\Sigma}^{\bar{\jmath}} = \\ &= r^2_H(z_\infty,\bar{z}_\infty,p,q) - G^{i\bar{\jmath}}\partial_i\mathcal{W}\overline{\partial}_{\bar{\jmath}}\overline{\mathcal{W}} \; ; \end{aligned} \tag{3.13}$$

thus, depending on the scalar manifold of the supergravity theory we are considering, it is given by

$$R^2_H = |\mathcal{I}_2(p,q)| \qquad \text{or} \qquad R^2_H = \sqrt{|\mathcal{I}_4(p,q)|} \; . \tag{3.14}$$

4 Black holes in $\mathcal{N}=2$ minimally coupled Supergravity

The scalar manifold of $\mathcal{N}=2$ minimally coupled Supergravity is the Kähler symmetric space G/H

$$\mathcal{M}_n = \frac{SU(1,n)}{SU(n)\times U(1)} \quad dim_{\mathbb{R}} = 2n \tag{4.1}$$

defined by the holomorphic prepotential

$$F(X^I) = -\frac{i}{2}\left(1 - \sum_i (z^i)^2\right) \; . \tag{4.2}$$

Because of the vanishing of the C_{ijk}-tensor of special Kähler geometry for the manifold we are considering, there are only two U(1)-Kähler and H invariants

$$\mathcal{I}_2 = |Z|^2 - |D_i Z|^2$$
$$V_{BH} = \mathcal{I}_1 = |Z|^2 + |D_i Z|^2 \,, \tag{4.3}$$

the attractor equations are the criticality condition on the black hole potential that is on the H-, Kähler-gauge- invariant $\mathcal{I}_1$,

$$\partial_i V_{BH} = 2Z D_i Z = 0 \,. \tag{4.4}$$

There are indeed two different solution, the supersymmetric $Z \neq 0$ (BPS-)solution and the $Z = 0$ non-BPS one. In the first case one has the $n+1$ equations

$$D_i Z = 0 \,, \tag{4.5}$$

fixing all the scalars at the horizon in terms of electric and magnetic charges q_i and p^i

$$z^i_{hor} = -\frac{q_i + ip^i}{q_0 - ip^0} \,, \tag{4.6}$$

so the moduli space is empty. In the non supersymmetric solution, the attractor equation is $Z = 0$, thus only one of the n complex scalar fields can be fixed; for exemple, solving for z^1, we find

$$z^1_{non-bps} = -\frac{(q_{\hat{1}} - ip^{\hat{1}})z^{\hat{1}}_{non-BPS}}{q_1 - ip^1} - \frac{q_0 - ip^0}{q_1 - ip^1} \,. \tag{4.7}$$

The case of non-BPS solutions with $n = 1$ vector fields and the BPS one are the only solutions, among all the possible N-extended theories, where the scalar fields of the theory are fixed. In this case, in particular, for $n > 1$ vectors, the unfixed scalars $z^{\hat{1}}$ span the moduli space

$$\mathcal{M}_{\mathcal{N}=2,\,non-BPS} = \frac{SU(1,n-1)}{SU(n-1)\times U(1)} \,, \tag{4.8}$$

whose dimension on the real field is $2(n-1)$, and coincides with the scalar manifold of the $\mathcal{N} = 2$ Supergravity minimally coupled to n vector fields. The black hole charges supporting non-degenerate critical points of V_{BH} split in two branches: the BPS one, defined by $\mathcal{I}_2 > 0$, and the non-BPS (Z=0) one, with $\mathcal{I}_2 < 0$; the entropy of the extremal black hole in $\mathcal{N} = 2$, d=4, minimally coupled Supergravity, is the topological quantity

$$S_{BH} = \frac{\pi}{2}\left[q_0^2 - q_i^2 + (p^o)^2 - (p^i)^2\right] = \frac{\pi}{2}(q^2 + p^2) \,, \tag{4.9}$$

purely defined in terms of the electric and magnetic charges.

5 Black holes in $\mathcal{N}=3$ matter coupled Supergravity

The scalar manifold of the theory is the Kähler coset

$$\mathcal{M}_{\mathcal{N}=3} = \frac{SU(3,n)}{SU(3)\times SU(n)\times U(1)} , \tag{5.1}$$

whose dimension on the real field is $6n$.

Vector fields trasform in the fundamental representation of the U-duality group, as in the previous case of $\mathcal{N}=2$ quadratic Supergravity. The central charge matrix Z_{AB} and the matter charges Z_I are defined as the integral over the 2-sphere at infinity of the dressed graviphoton and matter field strengths

$$\begin{aligned} Z_{AB} &\equiv -\int_{S^2_\infty} T_{AB} = -\int_{S^2_\infty} T^-_{AB} = f^\Lambda_{AB} q_\Lambda - h_{\Lambda\, AB} p^\Lambda , \\ Z_I &\equiv -\int_{S^2_\infty} T_I = \int_{S^2_\infty} T^-_I = f^\Lambda_I q_\Lambda - h_{\Lambda\, I} p^\Lambda . \end{aligned} \tag{5.2}$$

Thus, once the embedding in the symplectic group is determined, one finds that central and matter charges are written as functions of the scalar fields $X(z)$, and electric and magnetic charges as

$$\begin{aligned} Z_{AB} &= \frac{1}{\sqrt{2}}\left[\sqrt{1+XX^\dagger}^{\,C}_{(AB)}(q_C + ip^C) + (X^*)^i_{AB}(q_i - ip^i)\right] , \\ Z_I &= \frac{1}{\sqrt{2}}\left[(X^\dagger)^C_I(q_C - ip^C) + \sqrt{1+X^TX^*}^{\,i}_{\ I}(q_i + ip^i\right] . \end{aligned} \tag{5.3}$$

The attractor equations are determined by the critical condition on the vector potential at the black hole horizon

$$dV_{BH}|_{hor} = P^I_{AB}\overline{Z}^{AB} Z_I + c.c. = 0 , \tag{5.4}$$

which admit, for non-degenerate attractor flows, the $\frac{1}{3}$-BPS and the non-BPS ($Z_{AB}=0$) solution. By introducing the complexified charges as follows

$$\begin{aligned} Q_C &= q_C + ip^C ; \\ Q_i &= q_i + ip^i , \end{aligned} \tag{5.5}$$

the first order fake superpotential [6] takes the form

$$\begin{aligned} \mathcal{W}^2_{\frac{1}{3}BPS} = \frac{1}{2}\Big[&(1+XX^\dagger)^{AB}\overline{Q}_A Q_B + (\sqrt{1+XX^\dagger}\,X)^{\ Ai} Q_i Q_A + \\ &+ (X^\dagger\sqrt{1+XX^\dagger})^{j\,B}\overline{Q}_B\overline{Q}_j + (X^\dagger X)^{kl}\overline{Q}_k Q_l\Big] ; \end{aligned} \tag{5.6}$$

(5.7)

$$\mathcal{W}^2_{non-BPS} = \frac{1}{2}\Big[(XX^\dagger)^{AB}Q_A\overline{Q}_B + (\sqrt{1+X^\dagger X}\,X^\dagger)^{iA}\overline{Q}_i\overline{Q}_A + \\ +(X\sqrt{1+X^\dagger X})^{Bj}Q_BQ_j + (1+X^\dagger X)^{kl}\overline{Q}_kQ_l\Big]\,. \tag{5.8}$$

We eventually write the entropy, which is the invariant of the scalar manifold, explicited in terms of the electric magnetic charges in the relevant representation,

$$\frac{S_{BH}}{\pi} = \mathcal{I}_2 = \frac{1}{2}(Z_{AB}\overline{Z}^{AB}) - (Z_I\overline{Z}^I) = \frac{1}{2}(q^2+p^2)\;, \tag{5.9}$$

with $q^2 = q_\Lambda \eta^{\Lambda\Sigma} q_\Sigma$, $p^2 = p^\Lambda \eta_{\Lambda\Sigma} p^\Sigma$ and $\eta_{\Lambda\Sigma}$ the Lorentzian metric with signature $(3,n)$. We point out that it can be written analogously to that of the previous $\mathcal{N}=2$ theory; in fact, both the $\mathcal{N}=2$ and $\mathcal{N}=3$ scalar manifolds we have considered, are specific cases of the symmetric cosets

$$\frac{SU(p,q)}{SU(p)\times SU(q)\times U(1)}\,, \tag{5.10}$$

and in these theories vectors (and thus electric and magnetic charges) are in the fundamental representation of $SU(p,q)$.

6 Black holes in $\mathcal{N}=5$ Supergravity

In order to have fields with spin no bigger than 2, no vector coupling is allowed, and the salar manifold of the pure $\mathcal{N}=5$ Supergravity is

$$\mathcal{M}_{\mathcal{N}=5} = \frac{SU(1,5)}{SU(5)\times U(1)}\,, \qquad dim_{\mathbb{R}} = 10\;. \tag{6.1}$$

The symplectic embedding is defined by symplectic sections that we can compute from the vectors kinetic term in the Lagrangian and the vector field supersymmetry transformation; the explicit dependence of the kinetic matrix in terms of the scalar fields is then

$$\mathcal{N}_{ij,kl} = \frac{-i}{1-(\phi_i)^2}\left(\frac{1}{2}(1+(\phi_i)^2\delta^{ij}_{kl} - \frac{1}{2}\epsilon^{ijkla}\phi_a - 2\delta_{[i[k}\phi_{l]}\phi_{j]}\right)\,. \tag{6.2}$$

This theory has two H-invariants, $I_1 = \frac{1}{2}Z_{AB}\overline{Z}^{AB} \equiv V_{BH}$ and $\mathcal{I}_4 \equiv S_{BH}$; from the first we derive the attractor equations $\epsilon^{ABCDE}Z_{AB}Z_{CD}=0$, whose solution corresponds to the only $(\frac{1}{5})$-BPS non degenerate attractor flow, which has the first order fake superpotential [6] is

$$\mathcal{W}^2_{(\frac{1}{5}-)BPS} = \frac{1}{2}\left[\frac{1}{2}Z_{AB}\overline{Z}^{AB} + \sqrt{Z_{AB}\overline{Z}^{BC}Z_{CD}\overline{Z}^{DA} - \frac{1}{4}\left(Z_{AB}\overline{Z}^{AB}\right)^2}\,\right] = \\ = \frac{1}{2}\left[I_1 + \sqrt{2I_2 - I_1^2}\,\right] = \mathcal{Z}_1^2. \tag{6.3}$$

where $\mathcal{Z}_1$ (and analogously $\mathcal{Z}_2$) is one of the two skew-eigenvalues of the central charge matrix. The criticality conditions are satisfied for a unique class of critical points

$$\mathcal{Z}_2 = 0 \ , \qquad \mathcal{Z}_1 > 0 \ ; \tag{6.4}$$

this implies that the entropy of the ($\frac{1}{5}$)-BPS black hole is given by

$$\frac{S_{BH,BPS}}{\pi} = \frac{A_{H,BPS}}{4} = V_{BH}|_{BPS} = \mathcal{Z}_{1,BPS}^2 = \sqrt{\mathcal{I}_4} \ . \tag{6.5}$$

Since vector fields are in the threefold antisymmetric representation of $SU(1,5)$, the scalar manifold invariant is a quartic expression of electric-magnetic charges, but it turns out to be a perfect square of a quadratic combination of the central charge skew-eigenvalues:

$$\begin{aligned} \mathcal{I}_4 &= Z_{AB}\overline{Z}^{BC} Z_{CD}\overline{Z}^{DA} - \frac{1}{4}(Z_{AB}\overline{Z}^{AB})^2 = (\mathcal{Z}_1^2 - \mathcal{Z}_2^2)^2 \ , \\ &= 4q_{AB}q^{BC}q_{CD}q^{DA} - (q_{AB}q^{AB})^2 \ , \end{aligned} \tag{6.6}$$

and the entropy is then

$$S_{BH} = \pi(\mathcal{Z}_1^2 - \mathcal{Z}_2^2) \ . \tag{6.7}$$

The $\mathcal{N} = 5$ Supergravity we have considered differs from the previous $\mathcal{N} = 2$ and $\mathcal{N} = 3$ because the invariant of the scalar manifold is quartic in the charges. Even in the case of $\mathcal{N} = 4$ Supergravity, the invariant is a quartic expression of the charges; however, if we consider the *pure* theory (with no matter coupling), the black hole solution is still determined by class of critical points

$$\mathcal{Z}_2 = 0 \ , \qquad \mathcal{Z}_1 > 0 \ , \tag{6.8}$$

it is again supersymmetric ($\frac{1}{4}$-BPS) and we can still define an effective radius as in (3.13). This does not hold for the matter coupled $\mathcal{N} = 4$ Supergravity, for which the moduli dependence in the definition of effective radius does not eventually cancel. The property satisfied by the $\mathcal{N} = 5$ and the pure $\mathcal{N} = 4$ theory, which is not true for the $\mathcal{N} = 4$ matter coupled Supergravity, is that the quartic invariant is the perfect square

$$\mathcal{I}_4 = (\mathcal{Z}_1^2 - \mathcal{Z}_2^2)^2 \ , \tag{6.9}$$

and the effective radius, from the definition (3.13), is the exact expression

$$R_H^2 = \sqrt{|\mathcal{I}_4|} = \mathcal{Z}_1^2 - \mathcal{Z}_2^2 \ . \tag{6.10}$$

7 Summary and outcomes

First order formalism, originally applied to BPS solutions, can be extended to a *fake Supergravity* formalism explicitly verified in the case of $\mathcal{N}=2$ minimally coupled, $\mathcal{N}=3$, pure $\mathcal{N}=4$ and $\mathcal{N}=5$ Supergravity in $d=4$, but this does not hold for other theories such as $\mathcal{N}=4$ matter coupled Supergravity.

moreover, a generalized expression for the effective radius in the non extremal case ($c \neq 0$), where one writes $R^2_{BH} \to R^2_+$ cannot be proved at present, since a first order formalism for non-extremal dyonic black holes is yet under construction: the only way one has to compute scalar charges is by direct integration of scalar fields dynamics, currently known for the axion-dilaton black hole and for the STU model only. It could be interesting to build a fake supergravity formalism and determine black hole parameters for other theories in $d=4$ spacetime dimensions, such as $\mathcal{N}=2$ not minimally coupled, matter coupled $\mathcal{N}=4$, $\mathcal{N}=6$ and $\mathcal{N}=8$.

It is interesting to notice here that there is no complete classification of scalar fields fluxes in a higher dimensional string theory, or a stringy realization of the theories discussed, and to point out that in string theory the attractor mechanism is not a perturbative effect (since it fixes the dilaton) and the black hole metric involves non perturbative D-branes bound states.

References

[1] S. Ferrara, R. Kallosh, A. Strominger: *$\mathcal{N}$= 2 extremal black holes*, Phys. Rev. **D52**, 5412 (1995).

[2] S. Ferrara, R. Kallosh: *Supersymmetry and attractors*, Phys. Rev. **D54**, 1514 (1996); S. Ferrara, R. Kallosh: *Universality of supersymmetric attractors*, Phys. Rev. **D54**, 1525 (1996).

[3] A. Strominger: *Macroscopic entropy of $\mathcal{N}$= 2 extremal black holes*, Phys. Lett. **B383**, 39 (1996).

[4] S. Ferrara, G. W. Gibbons and R. Kallosh, *Black Holes and Critical Points in Moduli Space*, Nucl. Phys. **B500**, 75 (1997), hep-th/9702103.

[5] A. Ceresole and G. Dall'Agata, *Flow Equations for Non-BPS Extremal Black Holes*, JHEP **0703**, 110 (2007), hep-th/0702088.

[6] L. Andrianopoli, R. D'Auria, E. Orazi, M. Trigiante, *First Order Description of Black Holes in Moduli Space*, JHEP **0711**, 032 (2007), arXiV:0706.0712.

[7] S. Ferrara and A. Marrani, "Symmetric Spaces in Supergravity," arXiv:0808.3567 [hep-th].

[8] S. Ferrara, A. Gnecchi and A. Marrani, "d=4 Attractors, Effective Horizon Radius and Fake Supergravity," arXiv:0806.3196 [hep-th].

Track-based improvement in the jet transverse momentum resolution for ATLAS

Z. Marshall[1,2] for the ATLAS Collaboration

[1]*Columbia University, Nevis Laboratory, 136 So. Broadway, Irvington, NY 10533, United States of America*
[2]*California Institute of Technology, Physics Department 103-33, Pasadena, CA 91125, USA*

Correspondence may be sent to zmarshal@caltech.edu

Abstract

We present a track-based method for improving the jet momentum resolution in ATLAS. Information is added to the reconstructed jet after the standard jet energy scale corrections have been applied. Track-based corrections are implemented, and a $10-15\%$ improvement in the jet transverse momentum resolution at low p_T is achieved. The method is explained, and some validation and physics results are presented. Additional variables are described and analyzed for their resolution improvement potential.

1 Introduction

ATLAS, one of the general-purpose detectors at the LHC [1], began taking data in 2008. Many of the proposed physics studies include energetic jets, collimated collections of particles often initiated by a single quark or gluon in a hard scatter collision. The calorimeter's response to hadronic jets will be a significant source of systematic uncertainties for both searches for new phenomena, particularly in hadronic decay channels, and Standard Model studies. Understanding the jet energy response is of vital importance. By improving the jet energy resolution, it is possible to improve mass resolution and decrease combinatorial backgrounds as well.

Jet reconstruction in ATLAS has traditionally relied on calorimeter information alone. During event reconstruction, corrections are applied to raw cell energies due to the non-compensating nature of the calorimeter [2]. These jet energy scale (JES) corrections set the energy scale of jets such that -on average- jets have a uniform response in transverse momentum, $R = p_T^{reco}/p_T^{true} = 1$. However, on a jet-by-jet basis, jets may be over-corrected or under-corrected depending on their topology, particle content, etc.

We describe a technique that uses tracks found within jets to extract information about the jet topology and fragmentation in order to improve the jet transverse momentum resolution [2]. The approach is conceptually different from more traditional energy flow methods [3], where precise track momentum measurements replace calorimeter clusters. Rather, tracks are used to correct the response of jets as a function of the jets' particle composition, using the fraction of jet p_T carried by tracks pointing to the jet, $f_{trk} = \frac{p_T^{tracks}}{p_T^{calorimeter}}$. Here $p_T^{calorimeter}$ refers to the jet p_T

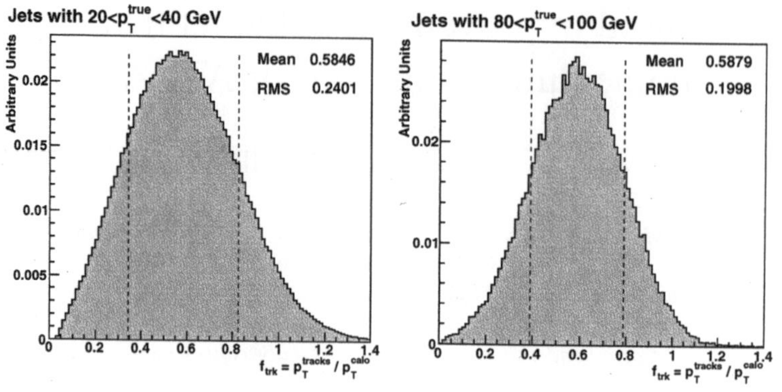

Figure 1: Sample distributions of f_{trk} for jets in two bins of p_T^{true}. Dotted lines are placed at 1 standard deviation above and below the mean of each distribution.

reconstructed using the standard (calorimeter-only) ATLAS reconstruction, and p_T^{tracks} refers to the p_T carried by all the tracks found within the jet.

In general, jets are composed primarily of neutral and charged pions. Charged particles leave tracks in the inner detectors, and so one might naively expect that between half and two-thirds of the jet p_T will be carried by tracks associated with that jet. As shown in Figure 1, Monte Carlo QCD di-jet samples have roughly Gaussian f_{trk} distributions centered near 0.6, with tails extending above one. The tails are more prominent at low energies and include, for example, jets with a true f_{trk} near one and one or more tracks with incorrectly measured momenta. In this all-pion approximation of jet composition, the number of tracks corresponds to the hadronic fraction of the jet, and f_{trk} provides a method for adjusting the jet energy for the non-compensating calorimeter of ATLAS. Additionally, an extreme f_{trk} can point to a very poorly measured jet, since a fluctuation reconstructed jet p_T will cause a fluctuation in f_{trk}.

The transverse jet momentum resolution, $\frac{\sigma(p_T^{reco}-p_T^{true})}{p_T^{true}}$, where σ is the standard deviation of the distribution, is proportional to the width of the jet momentum response in bins of transverse momentum, normalized to the average true jet momentum in a bin. If the response of these jets varies significantly with f_{trk}, the transverse jet momentum resolution will be artificially broadened, as shown in Figure 2. One sees that the total measured transverse jet momentum response distribution is considerably wider than either of the constituent distributions corresponding to jets with different charged particle fractions. By correcting the jet response as a function of jet p_T and f_{trk} we reduce the overall width of the momentum distribution and, hence, improve the jet momentum resolution.

Although the variable f_{trk} demonstrates the method clearly, this principle holds for other track or calorimeter variables as well. Many variables can hold information about jet topology, including the number of tracks associated to a jet, the spread of the tracks within the jet, and the fraction of track p_T carried by the leading track. By using a combination of these variables, one could considerably improve the transverse jet momentum resolution.

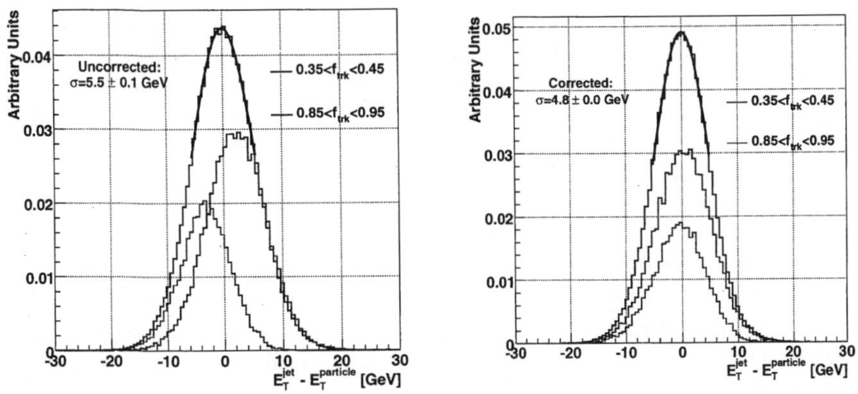

Figure 2: Black: difference between reconstructed and true jet p_T for jets with $|\eta| < 0.7$ and 40 GeV$< p_T <$ 60 GeV before (left) and after (right) a correction based on f_{trk} is applied. The mean (width) of this distribution is proportional to the jet momentum response (resolution). Since jets with different f_{trk} have different responses, the transverse momentum resolution is artificially broadened because of the offset of the distributions for each f_{trk} bin. The f_{trk} bins are centered at ± 1 standard deviation in the f_{trk} distribution. The normalization is arbitrary.

2 Monte Carlo Samples and Event Selection

Track-based jet corrections were determined using QCD di-jet events. A list of the samples used, including cross sections and leading parton p_T ranges, is included in Table 1. Samples were not normalized.

Table 1: Properties of the QCD di-jet samples used in this analysis. The samples are divided into slices according to the transverse momentum of the leading parton from the hard scatter.

Sample	Leading Parton p_T Range	Cross Section	Number of Events
J1	17-35 GeV	1.4 mb	302,000
J2	35-70 GeV	93 μb	1,161,750
J3	70-140 GeV	5.9 μb	254,750
J4	140-280 GeV	0.31 μb	336,000
J5	280-560 GeV	12 nb	321,550
J6	560-1120 GeV	0.4 nb	371,800

Reconstructed tower-seeded, 0.4 cone jets selected from the events ranged in p_T from 7 GeV to 1120 GeV. Only jets with energies between 20 and 600 GeV were considered to derive the corrections. Fits were formed only for central jets with pseudorapidity $|\eta| < 0.7$, however the technique can be extended to other regions of η. Because the ATLAS tracker acceptance ends at $|\eta| = 2.5$, jets beyond $|\eta| = 2.0$ must be examined more carefully and may not be suitable for this technique. We begin from jets with uniform response, $R = p_T^{reco}/p_T^{true} = 1$ on average.

Jets were required to have a "truth" jet (i.e. a jet made from stable particles from the generator, prior to any interactions with the detector) within 0.1 in $\eta - \phi$ space of the jet center. They were also required to not have a hard c- or b-quark in the generator record within 0.1 in $\eta - \phi$ space of the jet center. Such jets, called "heavy-flavor" jets, could have different momentum responses compared to jets that evolved from light quarks. They were, therefore, excluded from the fits and tested later for consistency (see Section 4.2). About 3% of the jets in these di-jet samples were c- or b-jets.

Tracks within a cone of radius 0.4 in $\eta - \phi$ around a jet's axis were included in the calculation of f_{trk}. In order to remove jets with a single poorly measured track, jets were required to be associated with at least two tracks to derive the corrections, and each track was required to have $p_T > 0.5$ GeV and $\chi^2/DoF < 3.0$, both of which are standard ATLAS track quality cuts. No other requirements were placed on the tracks. To determine the track-based momentum corrections, jets were also required to be isolated in a cone of 0.8 in $\eta - \phi$ space, to avoid jet-jet contamination which may skew f_{trk} considerably for low momentum jets. The removal of this isolation criterion will be explored in future studies.

3 Determination of the Track-Based Jet Momentum Response Correction

Jet momentum response is non-Gaussian when plotted in bins of reconstructed transverse momentum but is roughly Gaussian when plotted in bins of true transverse momentum. The momentum response is therefore calculated in bins of true jet momentum. Then, using the definition of jet response, $R \equiv p_T^{reco}/p_T^{true}$, pairs are formed of jet momentum response and reconstructed jet transverse momentum $(R(p_T^{true}), p_T^{reco}(p_T^{true}))$. These pairs are then used to derive corrections for response. A grid was constructed with 20 bins in p_T between 20 and 600 GeV, with finer binning at low p_T, and 12 bins in f_{trk} between 0 and 1.04. Grid points were constructed so that each bin had roughly an equal number of jets. Fits to the jet momentum response are made as functions of f_{trk} and p_T. A two-dimensional fit was then used to fit the data. Slices of the fit surfaces are shown in Figure 3.

4 Algorithm Performance

The track-jet response correction $R(p_T, f_{trk})$ described previously was applied to jets selected for determining the parameterization. The jet momentum response was checked as a function of f_{trk}. Figure 3 shows the dependence of the response on f_{trk} before and after applying the track-based correction. Track-based corrections flatten the response, $R(p_T, f_{trk}) = p_T^{jet}/p_T^{true}$, to a mean value of unity for all $f_{trk} < 1$. Above $f_{trk} = 1$, the fits do not completely correct the jet momentum response. As the f_{trk} distribution in Figure 1 demonstrates, very few jets lie in this region, and so the mean transverse momentum resolution is not significantly affected. When correcting jets, a cut was applied at $f_{trk} = 1.4$, above which jets were not corrected owing to inconsistent variation in response.

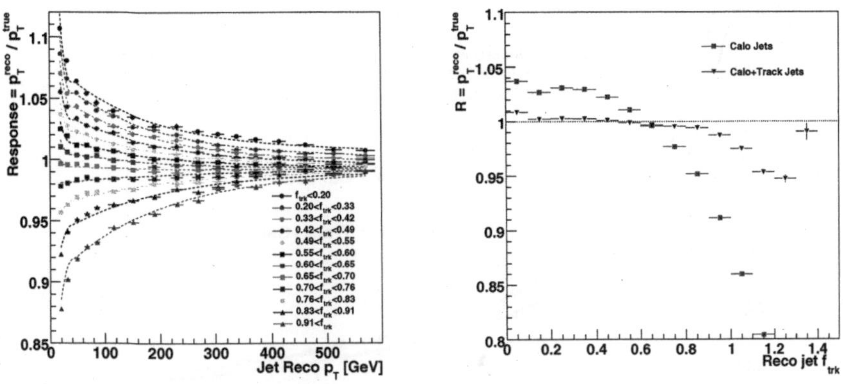

Figure 3: Left, slices of the data and the fit surface. Fits are shown as dotted lines over the data points. Right, jet response as a function of f_{trk}, before (Calo Jets) and after (Calo+Track Jets) applying the track-based response correction. The distribution of f_{trk}, as shown in Figure 1, drops off before one, so few jets are in the poorly corrected high f_{trk} region.

As described earlier and shown in Figure 2, the jet resolution before the $R(p_T, f_{trk})$ correction can be thought of as re-centering several offset Gaussians with different f_{trk}. By re-centering the underlying distributions we improve the jet transverse momentum resolution. Figure 2 shows the overlapping distributions after corrections. The tails of the distribution have been improved without affecting the mean.

4.1 Jet Transverse Momentum Resolution

The jet transverse momentum resolution is considerably improved at low jet p_T, as demonstrated in Figure 4. Fits are shown both before and after the corrections are applied. A $10-15\%$ transverse momentum resolution improvement is achieved at low p_T. The transverse momentum resolution is fit with a function:

$$\frac{\sigma(p_T)}{p_T} = \sqrt{\frac{a^2}{p_T} + \frac{b^2}{p_T^2} + c^2} \tag{1}$$

where a^2/p_T is a stochastic term, b^2/p_T^2 is a noise term, and c^2 is a constant term.

4.2 Heavy Quark Jets

Because of expected differences, due to heavy quark decays, between the fragmentation of a heavy-quark jet and that of a light-quark jet, there was some concern that their properties would be sufficiently different that the same corrections could not be applied. In fact, this is not the case: b-quark tower jet transverse momentum resolution is indeed improved by the same

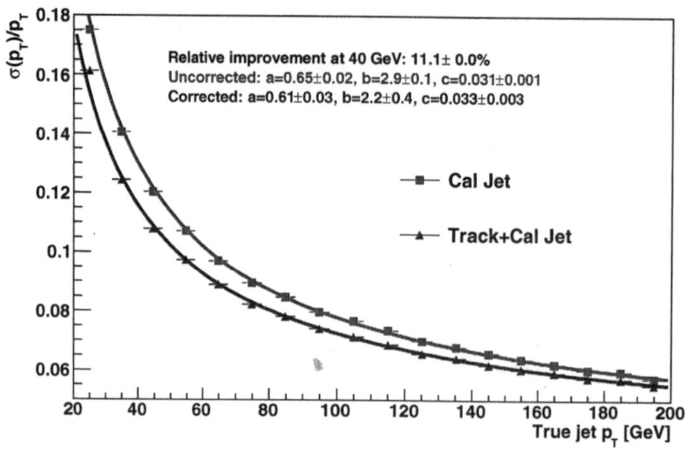

Figure 4: Jet transverse momentum resolution as a function of true jet p_T before (Cal Jet) and after (Track+Cal Jet) correcting for f_{trk}.

corrections derived from and applied to the light-quark tower jets, and the response is corrected properly. The transverse momentum resolution improvement at 40 GeV is approximately 8%.

A great deal of the concern over heavy-quark jets was caused by the assumption that their distributions for f_{trk} would be significantly different from those of light-quark jets. The mean and standard deviation of each the f_{trk} and track multiplicity distributions for light- and heavy-quark jets can be found in Figure 5. The different types of jets show few significant differences at low p_T, where this technique is most useful.

4.3 Missing Transverse Energy Bias

Track-based jet corrections can also be used to improve the scale of the missing transverse energy, $\not{E}_T$. The track-corrected $\not{E}_T$ is computed as

$$\not{E}_{T\,x,y}^{corr} = \not{E}_{T\,x,y} - \sum_{jets}(p_T^{corr} - p_T^{reco}) \quad (2)$$

where $p_T^{corr} = p_T^{reco}/R(p_T^{reco}, f_{trk})$ and the sum is over all jets in the event.

Figure 6(a) shows the mean value of the $\not{E}_T{}^{reco} - \not{E}_T{}^{true}$ for the J4 sample alone (see Table 1) as a function of the f_{trk} difference between the two leading jets in a di-jet sample. $\not{E}_T{}^{true}$ is simply the sum of neutrino p_T. Both leading jets were required to have $|\eta| < 2.0$ and $p_T >$ 20 GeV for this study. Events were required to have no more than three reconstructed jets, and the two leading jets were required to have $\Delta\phi > 2.9$.

Figure 6(a) shows a large imbalance of transverse energy when the two jets have large differences in f_{trk}. Here f_{trk}^1 (f_{trk}^2) refers to the f_{trk} of the first (second) leading jet, and the similarly p_T^1 (p_T^2) refers to the p_T^{reco} of the first (second) leading jet. When $\Delta f_{trk} < 0$, $f_{trk}^2 > f_{trk}^1$

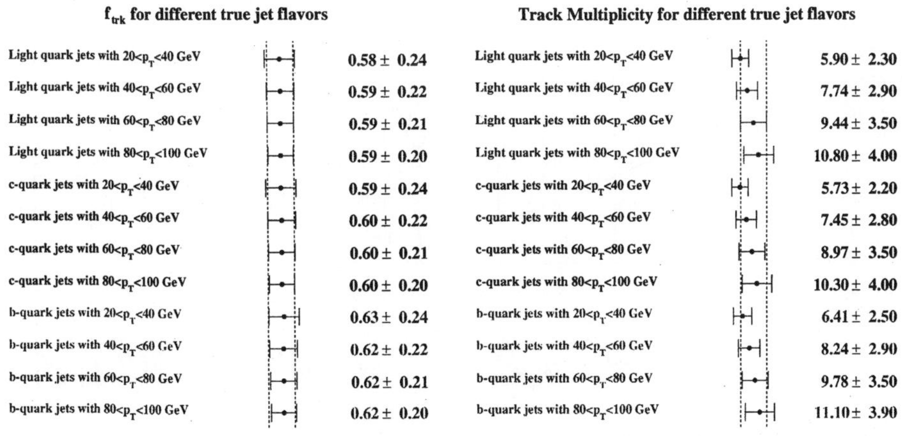

Figure 5: Mean f_{trk} (left) and track multiplicity (right) in light-quark jets, c-quark jets, and b-jets. Widths represent the standard deviation of the distribution. Dotted lines are placed at one standard deviation above and below the mean of the $60 < p_T < 80$ GeV light-quark distribution.

and p_T^2 is underestimated resulting in a positive bias on $\not{E}_T$. A similar argument explains a negative $\not{E}_T$ bias for $\Delta f_{trk} > 0$.

Figure 6(b) shows that the $\not{E}_T$ scale is properly corrected after applying the track-based response correction to the leading two jets, and the $\not{E}_T$ bias has been removed.

5 Evaluation of Additional Variables

Additional variables must be considered to maximally improve jet transverse momentum resolution. The variables should be uncorrelated in order to not introduce correction ordering problems and should be the minimal set that provides maximum information about jet topology.

The improvement provided by a single variable can be evaluated based on the difference in response between those jets at one standard deviation above and below the mean of the variable distribution. The difference should be related to the improvement in jet transverse momentum resolution that will be seen after a correction has been applied. We evaluate the variables both before and after corrections based on f_{trk} in order to understand which variable contains the most useful additional information about jet topology.

If f_{trk} is a measure of the hadronic energy in a jet, the number of tracks pointing to a jet gives some measure of the division of that hadronic energy. A high number of tracks means that hadrons in the jet are each lower energy and may not travel as far into the calorimeter. The softer particles may also result in an increase in non-linearity effects. The variable is highly correlated to jet p_T as well. Distributions of track multiplicity are shown in Figure 7, and the variation in response can be seen in Figure 8. Response shows a 3% variation with track multiplicity. Simply considering the number of tracks pointing to a jet may introduce problems with track

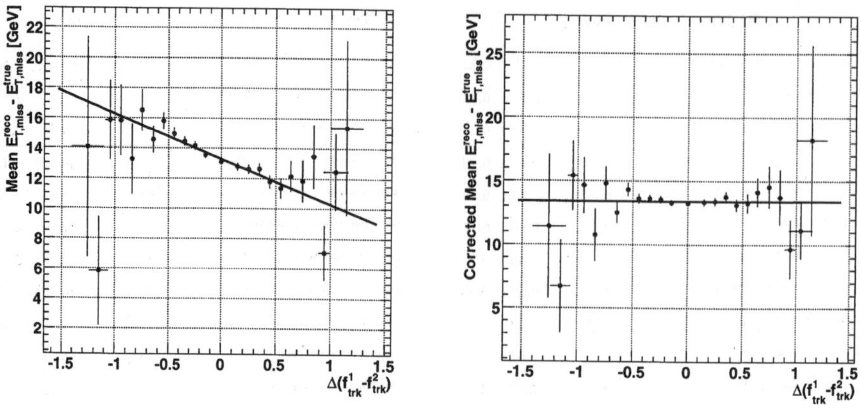

Figure 6: Mean missing transverse energy in a J4 sample as a function of the f_{trk} difference between the two leading jets in a di-jet sample before (left) and after (right) track-based jet momentum corrections.

reconstruction efficiency, ambiguity between low- and high-p_T tracks, and the radius used to associate tracks to jets. Corrections could be explored as a function of a ratio involving track multiplicity in order to resolve some of these issues.

The p_T carried by the leading track may also provide a handle on improving the jet transverse momentum resolution. We have examined the ratio of p_T of the leading track to the p_T carried by all tracks, p_T^1, and the response variation. By using this ratio rather than, for example, the ratio of p_T of the leading track to the p_T of the jet, the variable may be less correlated to f_{trk} and thus independent of corrections based on f_{trk}.

Some measure of track width may also provide useful information about jet topology. A very narrow set of tracks may be included in a single hadronic cluster, whereas several widely spaced tracks may produce several different hadronic clusters. One measure of shower width is the maximum distance in $\eta - \phi$ between any two tracks associated to the jet. One can also construct a measurement of track with using the average distance in $\eta - \phi$ between all tracks, with or without a p_T weighting. This variable is somewhat less sensitive to single low-p_T tracks on the edges of jets when there is a clear group of high-p_T tracks at the core of the jet, a feature that can be particularly important for high-p_T jets which typically contain many tracks. Different weighting schemes may also be explored.

Although these variables are each worth exploring, clearly the optimal set will not include all possibilities. The various ways of describing jet width are likely highly correlated, and correction for one may prove sufficient. There are certainly other variables worth exploring that could provide additional information about jet topology either from track information or from some subset of calorimeter information. Initial corrections can be achieved by applying a second fit on top of the f_{trk} fits already applied. In order to maximize the improvement in response and correctly take into account correlations, full multi-dimensional fits will have to be produced.

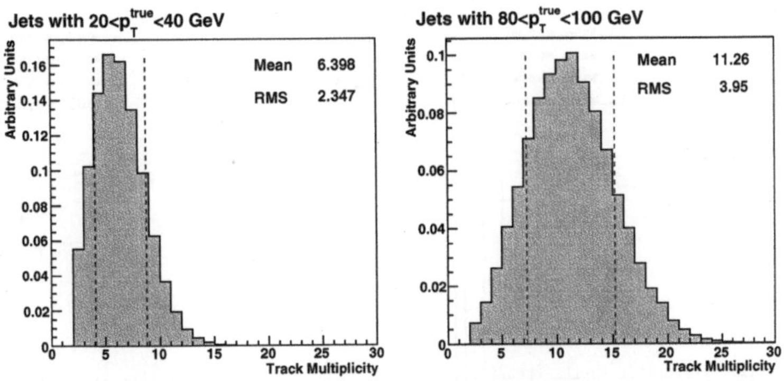

Figure 7: Distributions of track multiplicity for several bins of jet p_T^{true}. Dotted lines are placed at one standard deviation above and below the mean of each distribution.

6 Conclusions and Future Studies

We have introduced a track-based method for correcting the response of jets in ATLAS that provides a 10-15% improvement in jet transverse momentum resolution at low p_T over the standard ATLAS calorimeter jet reconstruction alone. The corrections derived based on light-quark tower-seeded jets are also effective on c- and b-quark jets. The corrections also improve missing energy distributions and biases. These corrections do not require new jet energy scale corrections and can be applied after the standard reconstruction. By systematically adding information from the tracker to jets already reconstructed based on calorimeter information, considerable progress can be made. A tool has been developed in the ATLAS software framework that provides f_{trk} and a correction based on f_{trk} as jet moments for all reconstructed jets.

Additional variables have been examined for their ability to improve jet transverse energy resolution. Several variables are promising and will be explored further. Ultimately, some combination of variables will be used to gain the maximum improvement.

These track-based corrections may be sufficiently sensitive to differences in fragmentation between quark- and gluon-initiated jets, for example, that they may be able to account for differences in jet energy scale between the two. By removing the sample-dependence of the jet energy scale corrections, the jet transverse momentum resolution can be improved.

There are several modifications to the technique being explored to further improve the jet transverse momentum resolution using this and other similar track-based variable methods. The radius used for track-jet association may be adjusted to improve performance, and extrapolated track position may be used rather than track direction at the interaction point. Since pile-up will be a significant issue at the LHC even from early on, these corrections will be tested in combination with a jet-vertex association technique already in progress [2].

References

[1] ATLAS Collaboration, G. Aad et al., JINST3 **S08003** (2008).

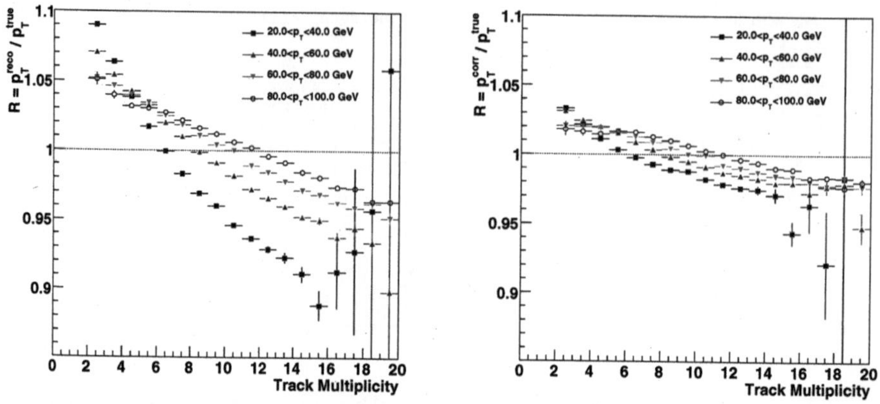

Figure 8: Jet response as a function of track multiplicity, before (left) and after (right) corrections for f_{trk}. A cut required each jet to have at least two tracks associated to it.

[2] ATLAS Collaboration, CERN-OPEN-2008-020 (2008).

[3] O. Lobban, A. Sriharan, R. Wigmans, Nucl. Inst. Meth. **495** (2002) 107–120.

Searches for Supersymmetric Dark Matter with XENON

K. Ni*

(on behalf of the XENON collaboration)

Physics Department, Columbia University,
New York, NY 10027, U.S.A.
E-mail: kn224@columbia.edu

* *Currently at Physics Department, Shanghai Jiao Tong University, China*

XENON is a phased program to search for dark matter directly by using the "two-phase xenon" technique, one of the most promising techniques as demonstrated by the XENON10 prototype detector. Currently, an ultra-low background, large detector with a total of 170 kg of xenon (XENON100) has been installed at the Gran Sasso Underground Laboratory (LNGS) and is in commissioning phase. XENON100 will reach a sensitivity of spin-indepedent WIMP-nucleon cross section at 2×10^{-45} cm^2 after six-month of data-taking. A future ton-scale detector (XENON1T) will improve the sensitivity by another two orders of magnitude, covering the majority parameter space as predicted in supersymmetric models.

Keywords: dark matter, supersymmetry, xenon

1. Introduction

Many astronomical observations (e.g.[1,2]) point to the existence of a non-luminous, non-baryonic component of our universe known as dark matter. This matter could be in the form of a new type of particle predicted in theories beyond the Standard Model, such as supersymmetry (SUSY)[3,4] . Such a particle is very likely to be massive with a weak interaction cross section, and is commonly referred to as a Weakly Interacting Massive Particle, or a WIMP. If WIMPs are the dark matter in the galactic halo of the Milky Way and other galaxies, they can be directly detected in low background underground detectors on Earth. WIMPs will hit the target detector via elastic scattering and create nuclear recoils in the detector materials. The energy deposited by the nuclear recoils will turn into observable signals, such as scintillation photons, ionization electrons and/or heat depending on techniques used in the detector. The signals from nuclear recoils are usually quite different from electron recoils from the background gamma ray interactions. If such a nuclear recoil signal is not observed within a certain exposure (the target mass times the exposure time), an upper limit for the WIMP-nucleon cross section can be calculated. The current upper limit for WIMP-nucleon elastic scattering cross section is at the level of 10^{-43} cm^2 at a WIMP mass of 100 GeV$/c^2$.[5,9]

Direct detection of WIMPs is being pursued by several experiments placed in different underground laboratories around the world. In recent years, experiments based on noble liquids have advanced at a much faster rate than the more established cryogenic experiments. In particular, the XENON10 experiment, with its successful performance in a very fast time, has brought liquid xenon (LXe) detectors to the forefront of the field.[5,6]

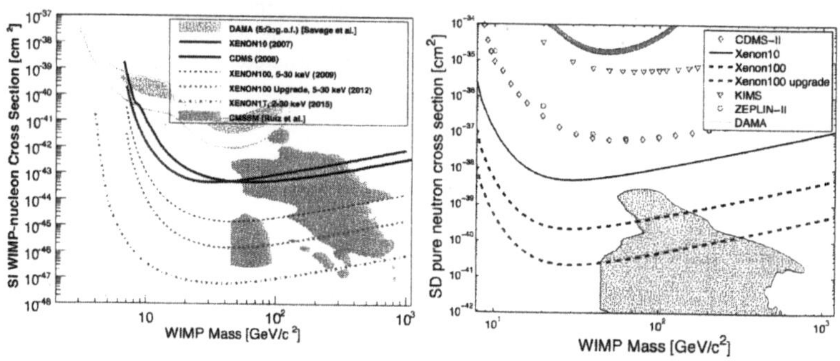

Fig. 1. Spin-independent (left) and spin-dependent pure neutron (right) WIMP-nucleon cross sections as a function of WIMP mass, probed by different stages of the XENON program. On the left, the low mass WIMP region allowed by DAMA,[7,8] and the current most stringent limits from XENON10[5] and CDMS[9] are shown as solid curves. Blue shaded regions are from theoretical models.[10]

XENON10 was the first prototype developed within the XENON dark matter program to prove the concept of a two-phase xenon time projection chamber (XeTPC) for dark matter search. With a phased approach of gradual increase in target mass while reducing the background, the ultimate goal is to detect WIMPs in a detector with 1 ton fiducial mass of LXe, with a sensitivity to spin-independent WIMP-nucleon cross sections better than $\sigma \sim 10^{-47}$ cm^2. WIMPs can also couple to nuclei via spin-dependent (axial-vector) interactions, in which a WIMP couples to the spin of a nucleus with an odd number of nucleons. The spin of a nucleus contains the content from both the proton and the neutron groups. The sensitivity of spin-dependent proton interactions is enhanced if the spin of the nucleus is dominated by the proton component, similar case for the spin-dependent neutron interactions.

The key technologies and performance characteristics relevant for the realization of a larger (ton-scale) XENON experiment are being addressed with the current detector, XENON100, and its upgrade. XENON100 improves on XENON10 with a factor 100 less background in a 10 times larger fiducial target. With drastic reduction in mass and radioactivity of the stainless steel cryostat, low radioactive PMTs from Hamamatsu and a new design of cryogenic system located far away from the proximity of the target, XENON100 is currently commissioning at the Gran Sasso Underground Laboratory and will accumulate data for a dark matter search throughout 2009. With a raw exposure of 6000 kg-days, free of background events, XENON100 will reach $\sigma \sim 2 \times 10^{-45}$ cm^2 for 100 GeV WIMPs, more than a factor of 20 better than the current limits.[5,9] An upgraded XENON100 will further improve the sensitivity by another order of magnitude by 2012 (Figure 1). XENON100 will also test spin-dependent couplings. The spins of ^{129}Xe and ^{131}Xe nuclei are dominated by neutron groups. Thus liquid xenon is mostly sensitivite to spin-dependent neutron couplings,[6] with the sensitivity reach as shown in Fig. 1.

2. The XENON100 Detector

The XENON100 detector uses the same principle of operation and many design features successfully tested in the XENON10 prototype.[11] It is a position-sensitive XeTPC, with the sensitive LXe volume viewed by two arrays of photomultiplier tubes (PMTs), to detect simultaneously the primary scintillation signal (S1) and the ionization signal via the proportional scintillation mechanism (S2). The sensitive target can be "fiducialized" to keep only the inner core free of background. The large S2 is used for x/y position reconstruction by fitting simulated light response matrices to the observed hit pattern on the top PMT array. The z-position is measured from the time difference between S1 and S2 and the constant drift velocity in LXe of 2mm/μs. The TPC 3D-position sensitivity and the self-shielding property of the dense and high Z (54) LXe medium, allows us to select events with a single interaction (as expected from a WIMP interaction) in a central fiducial volume, drastically rejecting the overwhelming background from neutrons and gammas which typically scatter more than once in the volume.

In addition, the high ionization density of nuclear recoils in LXe leads to an enhancement in S1 and reduction in S2, compared to electronic interactions. The S2/S1 ratio therefore discriminates WIMP nuclear recoils from γ and β backgrounds with an efficiency of 99.5–99.9%, as demonstrated with XENON10.[5]

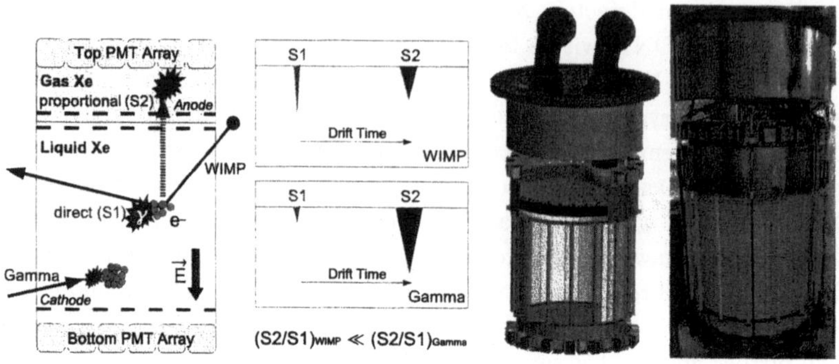

Fig. 2. (Left) Measurement principle of XENON100; (Right) Drawing and a photo of the XENON100 TPC.

A schematic drawing of the XENON100 detector and a photo are shown in Fig.2 (right). The active volume contains 65 kg of LXe and is instrumented with 178 PMTs in two arrays, 98 in the gas and 80 in the liquid. All PMTs are Hamamatsu R8520-06-Al 1" square, optimized for Xe 178 nm light and selected for low radioactivity. The quantum efficiency (QE) of these tubes was also recently improved from $\sim 24\%$ to $\sim 35\%$.

The active target is enclosed in a PTFE cylinder of 15 cm radius and 30 cm height. This PTFE cylinder reflects scintillation light with high efficiency,[12] and optically separates the LXe target from the surrounding LXe which is necessary to separate the TPC with its electric field from the walls of the vessel. 64 PMTs turn this outer LXe volume into an active LXe veto, with a total mass of 105 kg, including LXe layers above the top and below the bottom PMT arrays. The TPC volume is closed at the bottom by a cathode mesh, and at the top by a stack of 3 meshes, a central anode sandwiched by two meshes. Custom-made, low radioactivity, high voltage feedthroughs are used to bias the cathode and the anode, creating 1 kV/cm drift field across the 30 cm LXe gap, and 13 kV/cm field in the 5 mm gas proportional scintillation region. The LXe level is held stable within the stack of meshes, about 3 mm below the anode. The top PMT array is mounted within the bell, a few centimeters above the meshes.

TPC structure, supported by the bell, is enclosed in a double walled ıade of SS 316Ti selected for its low activity, especially in ^{60}Co. All ıal instrumentation was removed from the detector vessels as much ble, and mounted outside the lead/polyethylene shield, in particular rical feed-throughs and the cryocooler. This reduced the amount of ;ivity within the shield significantly. The cooling system is based on / pulse tube refrigerator (PTR), originally developed for the MEG ıent.[13] The PTR is used to liquefy Xe and to maintain the liquid ıture during operation.

ꟿON100 Detector: Initial Calibration Results

time of this writing the detector is filled with 150 kg of Xe which inue to purify with the closed circulation system to improve light ırge yields. The S1 and S2 response of the XENON100 TPC is ·ed with external gamma sources (^{57}Co, ^{137}Cs, ^{60}Co and ^{228}Th). investigating a new calibration source, ^{83m}Kr, which is a decay of ^{83}Rb, has a short half-life of 1.83 h and decays via a cascade of 9.4 keV transitions. ^{83m}Kr will be introduced in the TPC allowing .rry out S1&S2 calibrations uniformly throughout the volume. The ılf-life of ^{83m}Kr allows calibrating the TPC periodically during the ıtter search run. The neutron calibration will be performed in the ɔf the dark matter search run to verify the TPC's response to nuclear Simulation studies show that a one day calibration with a 220 n/s source will produce more than 1000 NR events/kg in the central ırget in the WIMP search energy window.

Figure 3 shows the current S1 light spectrum of ^{137}Cs 662 keV gamma-rays, measured in XENON100, showing a light yield of 2.7 pe/keVee (zero field), limited by the presence of impurities, mostly H_2O. By comparing with the MC studies of the light collection efficiency, the current light yield implies a 1 m absorption length of UV photons in LXe. With continuous circulation and purification of the Xe through the high temperature getter, we expect to achieve an S1 light yield of at least 5 pe/keVee (zero field) at a 5 m absorption length, which was achieved in XENON10.[11] Such a light yield gives an average S1 signal of more than 3 pe for 5 keV nuclear recoils, based on a recent measurement of the relative scintillation efficiency L_{eff},[14] allowing ~100% detection efficiency for NRs as low as 5 keV.

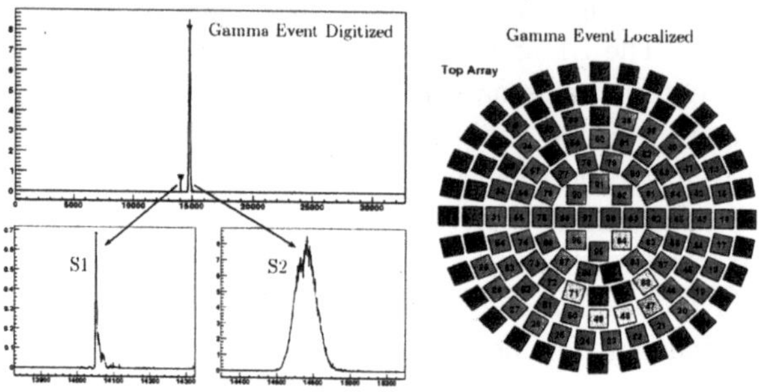

Fig. 4. (Left) A typical event waveform from XENON100, showing both S1 and S2 signals. (Right) S2 hit pattern on the top PMT array from the same event, based on reconstructed XY positions.

Two-phase operation of the TPC has also started. Figure 4 (left) is a waveform from the summed signals, showing clear S1 and S2 signals from a single scatter electron recoil (ER) event. Calibration data from external gamma ray sources (^{137}Cs) are continuously collected to monitor the electron lifetime in the LXe. The top array of the PMTs was designed to optimize the radial position resolution to allow efficient fiducial volume cut for background reduction. Figure 4 (right) shows a typical S2 hit pattern on the top PMT array from the same gamma ray event. From the S2 hit pattern, XY positions can be reconstructed with a radial position resolution of about 2 mm from a few keVee ER events, thanks to the fine granularity of the 1" PMTs.

4. Background and Sensitivity Projections

The estimated background contributions from different sources in the XENON100 detector is based on detailed Monte Carlo simulations and measured radioactivity of detector's construction materials. The materials used in the XENON100 TPC and shield were carefully selected for low intrinsic radioactivity with a dedicated screening facility consisting of a 2.2 kg high purity Ge detector in an ultra-low background Cu cryostat and Cu/Pb shield, operated at LNGS.

The electron recoil (ER) background is dominated by gamma rays from the decay chains of radioactive contaminants, mostly ^{238}U, ^{232}Th, ^{40}K, and ^{60}Co, in the detector materials, mostly from the PMTs. The self-shielding of LXe reduces the ER rate significantly in the central part of the target. The intrinsic ER rate due to β decays of ^{85}Kr will be reduced to 0.1 mDRU (DRU = events/keV/kg/day, mDRU = 10^{-3} DRU), with a Kr/Xe level down to 5 ppt, thanks to the cryogenic distillation column procured for the current experiment. XENON100 will achieve an ER background rate below 10(3) mDRU in the 50(30) kg fiducial mass.

(α, n) and spontaneous fission reactions in the detector, shielding materials, and surrounding rock/concrete are the dominant sources of the neutron background. Based on the measured U/Th activities of the XENON100 materials, we calculated the neutron yield using the modified SOURCES4A code[15] and simulated the nuclear recoil (NR) rates in the LXe target. The dominant neutron flux from outside the shield is from the rock/concrete in the walls of the underground laboratory: the predicted NR rate is more than that from the PMTs but will be reduced by a factor of 10 by adding a layer of 20 cm polyethylene outside of the current Pb shield.

The muon flux at the 3100 mwe Gran Sasso depth is 22 m^{-2}day^{-1}.[16] Neutrons generated by muons in the Pb and polyethylene shields, due to electromagnetic and hadronic showers and direct spallation, yield a total NR background at the level of 2.7±0.7 single NRs/year in the 100 kg target. To reduce this background, a 98% efficient muon veto will be used for the upgraded XENON100 experiment. The high energy muons interacting in the rock can produce highly penetrating neutrons with energies extending to GeV. These neutrons cannot be rejected by a muon veto and can hardly be shielded. The estimated single scatter NR rate from these neutrons is less than 0.3/year/100 kg.

The first background measurement in XENON100 has been performed with the S1 signal. The measured spectrum, shown in Figure 5, is consistent with Monte Carlo predictions.

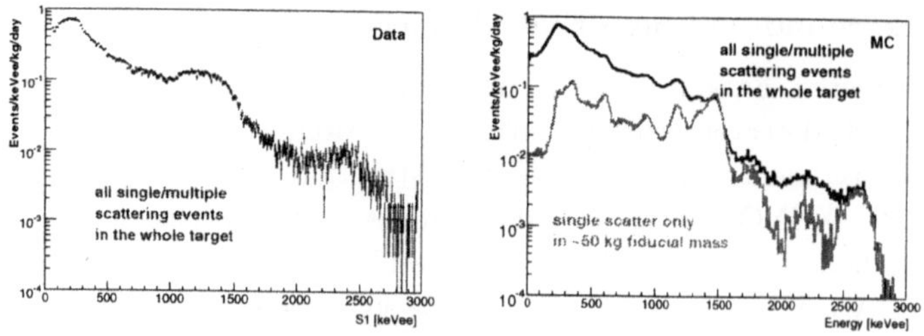

Fig. 5. Measured (left) and MC predicted (right) background spectra in XENON100.

A WIMP search in the current XENON100 will be first performed in 50 kg fiducial target with 40 live-days of exposure. With a nuclear recoil detection threshold at about 5 keV, we can reach a spin-independent WIMP sensitivity of $\sigma \sim 6 \times 10^{-45}$ cm^2 for 100 GeV/c^2 WIMPs. A factor of three improvement will be achieved by taking data for another half a year and searching for WIMPs in 30-kg fiducial target, to be free of background. An upgraded XENON100, with a lower background and larger target mass, will be able to accumulate WIMP search data for two years in 100 kg fiducial target, reaching a sensitivity of $\sigma \sim 2 \times 10^{-46}$ cm^2 (see also Figure 1).

5. The XENON1T Experiment

To improve the sensitivity to explore the major parameter space predicted by SUSY models, a ton-scale detector is needed. Such a detector called XENON1T, as shown in Figure 6, contains total 3 ton liquid xenon with a volume size roughly about 1 m^3. Such a detector requires a long electron drift length (at least 1 m) and ultra-low radioactive detector components. Oxygen-Free High-purity Copper (OFHC) cryostat and QUPID photodetectors will be used. In order to improve the light collection to improve the detector's threshold, a full coverage of QUPIDs is planned. In order to reduce the background from neutrons induced by cosmic muons in the rocks, XENON1T needs a deeper underground laboratory or a very large shield. One possibility is to use the LVD (Large Volumn Detector) liquid scintillator detector as an active shield. The design of such an integration between XENON1T and LVD is currently being studied. Other candidate sites for XENON1T include Modane Underground Laboratory (LSM) in France and Homestake mine in US. XENON1T will search for dark matter in its 1-ton fiducial mass for at least one year without any background.

Fig. 6. A design of the XENON1T detector with full coverage of QUPIDs and OFHC copper cryostat.

This will reach a sensitivity of spin-independent WIMP-nucleon cross section down to 10^{-47} cm^2 (for 100 GeV WIMPs), three orders of magnitude better than the current most stringent limits.

References

1. D. N. Spergel *et al.* [WMAP Collaboration], Astrophys. J. Suppl. **170**, 377 (2007).
2. D. Clowe *et al.*, Astrophys. J. **648**, L109 (2006).
3. G. Jungman, M. Kamionkowski, K. Griest, "Supersymmetric Dark Matter", Phys. Rep. **267**, 195-373 (1996).
4. G. Bertone, D. Hooper, J. Silk, "Particle Dark Matter: Evidence, Candidates and Constraints", Phys. Rept. **405**, 279-390 (2005), arXiv:hep-ph/0404175.
5. J. Angle *et al.* [XENON Collaboration], Phys. Rev. Lett. **100**, 021303 (2008).
6. J. Angle *et al.* [XENON Collaboration], Phys. Rev. Lett. **101**, 091301 (2008).
7. R. Bernabei *et al.* [DAMA Collaboration], Eur. Phys. J. C **56**, 333 (2008) [arXiv:0804.2741 [astro-ph]].
8. C. Savage *et al.*, arXiv:0808.3607 [astro-ph].
9. Z. Ahmed *et al.* [CDMS Collaboration], Phys. Rev. Lett. **102**, 011301 (2009), arXiv:0802.3530 [astro-ph].
10. L. Roszkowski, R. Ruiz de Austri, and R. Trotta, JHEP **07** (2007) 075, [arXiv:0705.2012 [hep-ph]].
11. E. Aprile *et al.* [XENON Collaboration], "The XENON10 Dark Matter Search Experiment", in preparation for Phys. Rev. D.
12. M. Yamashita *et al.*, NIM A **535**, 692 (2004).
13. T. Haruyama *et al.*, in: *Cryocoolers 13*, Springer, New York, 689 (2005).

14. E. Aprile *et al.*, Phys. Rev. C **79**, 045807 (2009), arXiv:0810.0274 [astro-ph].
15. Wilson *et al*, Phys. Lett. B **645**, 153 (2007); M. Carson *et al*, Astrop. Physics **21**, 667 (2004); R. Lemrani *et al*, NIM A**560**, 454 (2006); V. Tomasello *et al*, doi:10.1016/j.physletb.2003.10.071.
16. D. Mei and A. Hime, Phys. Rev. D **73**, 053004 (2006).

Running of Newton's Constant and Quantum Gravitational Effects*

David Reeb†

Institute of Theoretical Science, University of Oregon
Eugene, OR 97403, USA

Abstract

Newton's gravitational constant G_N is shown to be a running coupling constant, much like the familiar running gauge couplings of the Standard Model. This implies that, in models with appropriate particle content, the *true* Planck scale, i.e. the scale at which quantum gravity effects become important, can have a value different from 10^{19} GeV, which would be expected from naive dimensional analysis. Then, two scenarios involving this running effect are presented. The first one is a model which employs huge particle content to realize quantum gravity at the TeV scale in 4 dimensions, thereby solving the hierarchy problem of the Standard Model. Secondly, effects of the running of Newton's constant on grand unified theories are examined and shown to introduce new significant uncertainties in their predictions, but possibly also to provide better gauge coupling unification results in some cases.

1 Running of Newton's constant

We are familiar with the idea that the gauge couplings of the Standard Model of particle physics are running coupling constants, i.e. that the values of these couplings depend on the energy scale at which we perform experiments or, equivalently, on the distance scale at which we probe a certain feature of Nature. The intuitive reason, in the case of the electromagnetic interaction (QED) for example, is that virtual e^+e^- pairs polarize the vacuum and thus screen electric charges at large distances. Formally this comes about because, in an electromagnetic process with momentum transfer q roughly greater than the electron mass m_e, the tree-level diagram for virtual photon exchange receives significant superposition from the one-loop diagram with an electron in the loop, shown in Fig. 1 (left). This modifies the photon propagator and hence effectively modifies the gauge couplings at the vertices, so that the effective coupling becomes dependent on energy scale $\alpha_{EM} = \alpha_{EM}(q)$:

$$\alpha_{EM}(q) = \frac{\alpha_{EM}(m_e)}{1 - \frac{\alpha_{EM}(m_e)}{3\pi}\log\left(q^2/m_e^2\right)} \qquad \text{for } q^2 \gtrsim m_e^2\,. \tag{1}$$

After Fourier transformation into real space this can be equivalently expressed by saying that, at distances r shorter than the Compton wavelength of the electron m_e^{-1}, the force between charges is bigger than expected by extrapolating Coulomb's law from large to small

---

*Based on author's talk given at the 46th Course of the International School of Subnuclear Physics in Erice, Sicily/Italy, on September 1, 2008. For the original research work, see [1, 2].

†Electronic address: dreeb@uoregon.edu

Figure 1: (left) The first-order contribution to virtual photon exchange in QED, having an electron-positron pair in the loop, modifies the effective gauge coupling of the photon to the external particles. (right) In a theory of quantized gravity, the effective graviton propagator is the sum of a tree-level contribution and loop diagrams, in particular with matter particles running in the loop.

distances. The same screening or antiscreening behavior, depending on the particle content of the theory, occurs in non-Abelian gauge theories as well. So, a natural question to ask is whether similar effects might be present in gravity, especially in light of the fact that Newton's force law very much resembles Coulomb's law in form albeit with a dimensionful coupling constant, namely Newton's constant G_N.

When referring to Newton's constant, we usually implicitly think of the value $G_N = (10^{19}\,\text{GeV})^{-2}$ in natural units $\hbar = c = 1$, or similarly, in order to have a mass scale, we talk about the Planck mass $M_{\text{Pl}} = G_N^{-1/2} = 10^{19}\,\text{GeV}$. But these values are only the outcomes of our macroscopic experiments, as shown in Fig. 2: basically, the force F between two masses M_1 and M_2, separated by a distance d, is measured and G_N is deduced via Newton's law $F = -G_N M_1 M_2/d^2$. The smallest distances probed so far are on the order of $d \sim 1\,\text{mm}$ [3], corresponding to an energy scale of $q \sim 10^{-3}\,\text{eV}$ — very much smaller than energy scales at the current frontier in high-energy physics. For this reason it is sensible to indicate that $G_N(\mu \approx 0\,\text{GeV}) = (10^{19}\,\text{GeV})^{-2}$ and $M_{\text{Pl}}(\mu \approx 0\,\text{GeV}) = 10^{19}\,\text{GeV}$ are the values in the far infrared $\mu \approx 0\,\text{GeV}$. But how is this value of G_N at infrared energies related to physics at high energy scales or, equivalently, to physics at short distances where gravity has not been probed yet?

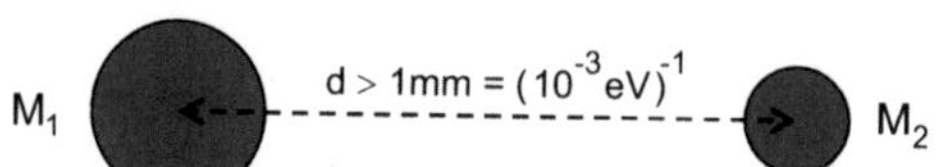

Figure 2: Our experiments measure Newton's constant G_N only at macroscopic distances.

I will first illustrate the scale dependence of Newton's constant in a crude way, using old-fashioned cutoff regularization. The action that describes gravity and matter

$$S_{\text{grav+m}} = \int d^4x \sqrt{-\det g_{\mu\nu}} \left(\frac{1}{16\pi G_b} R(g_{\mu\nu}) + \frac{1}{2} g^{\mu\nu} \partial_\mu \phi \partial_\nu \phi + \ldots + \psi + A_\mu + \ldots \right) \quad (2)$$

constains the Einstein-Hilbert term, which describes the propagation and self-interaction of gravitons $g_{\mu\nu}$, and also couples all matter fields (scalars ϕ, fermions ψ and gauge bosons A_μ) to gravity. As in the case of QED above, these couplings lead to corrections to the graviton propagator, Fig. 1 (right), since matter particles can run in the loop. For momentum transfer q and loop momentum cutoff Λ, the sum of both diagrams is, neglecting the index structure,

$$\frac{iG_b}{q^2} + \frac{iG_b}{q^2} \left(i \frac{c}{16\pi^2} q^2 \Lambda^2 \right) \frac{iG_b}{q^2} = \frac{iG_{\text{ren}}}{q^2}, \quad (3)$$

where the loop contribution follows from dimensional analysis and is proportional to the square of the cutoff. If, as in QED, this one-loop contribution is absorbed into a redefined coupling constant $G_{\rm ren}$ as shown in (3), the relation for the effective coupling is

$$\frac{1}{G_{\rm ren}} = \frac{1}{G_b} + \frac{c}{16\pi^2}\Lambda^2 \,. \tag{4}$$

This cutoff dependence qualitatively proves that Newton's constant is scale dependent and, moreover, that it depends *quadratically* on energy, as opposed to only logarithmic running (1) in the case of the gauge coupling constants.

Regularization with a hard cutoff is very crude in the sense that it violates general coordinate invariance as well as gauge invariance (if gauge bosons in the loop are considered). Nevertheless, the above calculation can be done using heat-kernel regularization [4, 1] which is a method to integrate out (modes of) matter fields directly from the path integral in the presence of a gravitational background $g_{\mu\nu}$, yielding an effective action $S_{\rm eff}(g_{\mu\nu};\,\mu)$ for gravity appropriate to experiments performed at scale μ. This method respects general covariance (and gauge invariance can be ensured with usual field-theoretic methods), so that the coefficient of $\sqrt{g}R$ in the effective gravitational action can be related to a running $G(\mu)$ in the ordinary Wilsonian sense. In the end, if n_0 real scalar fields, $n_{1/2}$ Weyl fermions and n_1 gauge bosons are integrated out, the values of Newton's constant at two different energy scales μ and μ_0 are related via

$$\frac{1}{G(\mu)} = \frac{1}{G(\mu_0)} - \frac{\mu^2-\mu_0^2}{12\pi}\left(n_0 + n_{1/2} - 4n_1\right) \,. \tag{5}$$

This running very much resembles the running of gauge couplings (1) with the significant difference that, as was already obvious from (4), here the coupling constant $G(\mu)$ depends quadratically on energy scale μ, owing to its dimensionful nature.

If several matter fields are integrated out completely, the scale-dependent $G(\mu)$ can be expressed in terms of the measured low-energy $G(\mu_0 = 0) = G_N = (10^{19}\,\text{GeV})^{-2}$ by

$$\frac{1}{G(\mu)} = \frac{1}{G_N} - \mu^2\frac{N}{12\pi} \qquad \text{with } N \equiv n_0 + n_{1/2} - 4n_1 \,, \tag{6}$$

where N counts the field content integrated out. Clearly, having scalars or fermions in a theory increases the strength of gravity $G(\mu)$ at high energies $\mu > 0$, whereas gauge bosons drive $G(\mu)$ in the opposite direction.

For the case of $N > 0$ in particular, this scale-dependence of Newton's constant has one important consequence which will be the basic starting point for both applications in sections 2 and 3: The Planck scale is defined to be the energy scale at which quantum gravitational effects become important. Naively, by dimensional analysis, this is the scale $M_{\rm Pl} = G_N^{-1/2} = 10^{19}\,\text{GeV}$; however, a running Newton's constant can change this picture. The *true* Planck scale μ_* is given by the condition

$$\mu_* = G(\mu_*)^{-1/2} \,, \tag{7}$$

i.e. μ_* is the energy scale so that fluctuations in spacetime geometry at the corresponding length scales $< \mu_*^{-1}$ are unsuppressed. μ_* is the true Planck scale, whereas our value $M_{\rm Pl} = 10^{19}\,\text{GeV}$ should be thought of as derived from this fundamental μ_* via renormalization group running to very low energies. Condition (7) yields, with (6),

$$\mu_* = \frac{M_{\rm Pl}}{\sqrt{1+N/12\pi}} = \frac{M_{\rm Pl}}{\eta} \qquad \text{with } \eta \equiv \sqrt{1+N/12\pi} \,, \tag{8}$$

where $N/12\pi$ in the denominator is the effect of the one-loop diagram in Fig. 1 (right) with particle content N in the loop, correcting the tree result 1. N, or equivalently η, characterizes the matter content of a certain theory. Gravitons (through self-coupling) will also run in the loop and give an additional contribution, but its value is fixed and comparable in size to the running caused by one matter field; and in the applications in the following two sections we are interested in the behavior of μ_* as we vary the matter content N so that an additional small constant contribution would only affect unimportant details.

2 TeV quantum gravity in 4 dimensions

In this section, I will present a model, in four spacetime dimensions, in which the running of Newton's constant pushes the Planck scale all the way down to the weak scale, i.e. to the TeV region, despite the conventional wisdom according to which quantum gravity effects are weak at energies accessible to us.

With formula (8) for the true scale of quantum gravity at hand, it is natural to ask what the particle content of a theory has to be in order to have a fundamental gravity scale of $\mu_* \sim$ TeV; at this scale then new quantum gravitational physics would have to appear, and it would coincide with the scale at which the Standard Model of particle physics becomes finely tuned und likewise with the scale that will be accessible to our particle accelerators in the very near future. From (8), this condition $\mu_* \sim$ TeV implies a particle content of $N \sim 10^{32}$. I.e., if a theory contains $\sim 10^{32}$ scalars or fermions with masses below a TeV (so that they can run in the loop), then our measured low-energy value of Newton's constant would be consistent with a fundamental quantum gravity scale of $\mu_* \sim$ TeV. This behavior, translated into a running gravitational mass scale $M(\mu) \equiv G(\mu)^{-1/2}$, is illustrated in Fig. 3, where the fundamental value $M(\text{TeV}) = \text{TeV}$ is fixed and all lower values are obtained via running (5), in particular also the value $M(\text{eV}) = 10^{19}$ GeV that we observe in the macroscopic world.

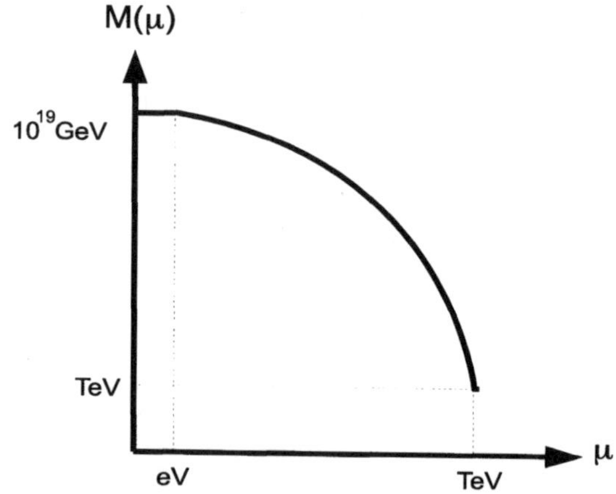

Figure 3: Running of the gravitational mass scale $M(\mu) = G(\mu)^{-1/2}$ in a theory with particle content $N \sim 10^{32}$ from the fundamental scale $\mu_* = \text{TeV}$ to our low-energy value $M_{\text{Pl}} = 10^{19}\,\text{GeV}$ at $\mu < \text{eV}$.

Thus, a model which contains 10^{32} scalars or fermions with masses $m <$ TeV interacting only gravitationally with us could have quantum gravity at a TeV, and yet be consistent with our low-energy measurements. To be phenomenologically viable, this vast amount of new fields has to be contained in a sector hidden from the Standard Model, i.e. must not have any gauge or other interactions with us. Note that such a model could be supersymmetric without contradiction since, according to (6), the bosonic and fermionic components of a chiral supermultiplet give contributions of the same sign to the running of $G(\mu)$. This mechanism solves the hierarchy problem of the Standard Model since new physics describing fluctuations in spacetime geometry (i.e., quantum gravity) then would have to come in at the TeV scale and would cut off the Standard Model there. In this scenario, quantum gravity could be probed by experiments at the CERN LHC. Also, this model is the first one to feature quantum gravity in the TeV region already in four dimensions, which was previously only known to be possible with extra-dimensional models. And although the introduction of $\sim 10^{32}$ hidden particles in our model seems to be very drastic and unattractive compared to the proposition of a few extra dimensions, there is a connection between the two scenarios closely linked to this huge number 10^{32}, as will be shown now.

In a model with $(d-4)$ extra dimensions, the fundamental action

$$S = \int d^4x \, d^{d-4}x' \sqrt{-g_{(d)}} \left(M_{(d)}^{d-2} \mathcal{R} + \dots \right) \equiv \int d^4x \sqrt{-g_{(4)}} \left(M_{(4)}^2 R + \dots \right) \tag{9}$$

is an integral over d-dimensional space with metric $g_{(d)} = g_{(4)} g_{(d-4)}$ involving the fundamental gravitational mass scale $M_{(d)}$ in d dimensions. Since we, confined to a 4-dimensional brane, want to describe physics by a 4-dimensional action integral, we choose the description on the RHS of (9). The mass scale $M_{(4)}$ that seems to be fundamental to 4-dimensional beings is, as per (9), merely induced by more fundamental quantities in the d-dimensional world:

$$M_{(4)}^2 = \int d^{d-4}x' \sqrt{g_{(d-4)}} \, M_{(d)}^{d-2} = V_{d-4} M_{(d)}^{d-2} \,, \tag{10}$$

with the proper volume V_{d-4} of the extra $(d-4)$ dimensions. Now we, in the 4-dimensional world, are wondering why our observed value $M_{(4)} = M_{\rm Pl} = 10^{19}$ GeV is so much larger than the mass scales in all other theories we know (e.g., weak scale). Extra-dimensional models explain this discrepancy by noting that, according to (10), $M_{(4)}$ can be large despite having a comparatively small fundamental scale of $M_{(d)} \sim$ TeV if only the (unaccessible) volume V_{d-4} of the extra dimensions is large. But the number of degrees of freedom hidden in the $(d-4)$-dimensional bulk equals the number of fundamental Planck volumes $V_{\rm Pl}$ in V_{d-4}, i.e. if one wants to achieve $M_{(d)} \sim$ TeV this number invariably is, with (10),

$$V_{d-4}/V_{\rm Pl} = V_{d-4} M_{(d)}^{d-4} = M_{(4)}^2 / M_{(d)}^2 \sim \left(10^{19}\,{\rm GeV/TeV}\right)^2 \sim 10^{32} \,. \tag{11}$$

These 10^{32} hidden degrees of freedom in the bulk are just as unaccessible as the 10^{32} bosonic or fermionic degrees of freedom in our model above, and this huge number is simply dictated, as in (8), by the square of the ratio of scales. Viewed under this perspective, our model does not seem to be much more ugly than extra dimensions.

The most striking phenomenological feature of our model is that it exhibits strong gravity effects at the scale of a TeV. Quantum black holes might be created at the LHC with a (geometrical) cross section as large as $\sigma(pp \to qBH + X) \sim 10^5$ fb [1] and would thus dominate all Standard Model cross sections. There is a sizable cross section $\sigma(pp \to \text{m.e.}) \sim$ 750 fb into missing energy (whereby black holes decay into the 10^{32} hidden fields) which seems to be larger than in extra-dimensional RS- or ADD-models in which almost all energy remains on the brane [5]. The quantum black holes created are likely charged under $\mathrm{SU}(3)_c \times \mathrm{U}(1)_Q$ since they are created from colored and possibly electromagnetically charged partons (color is not confining at such small distances). Thus, although the number of hidden fields by far exceeds the number of Standard Model species, decay of these quantum black holes involving one or two Standard Model particles is likely due to charge conservation, and one particular characteristic signature is a back-to-back hard lepton+jet [5].

In conclusion, the enormous number of particles required below a TeV, which must interact only gravitationally with us, seems unattractive, and other issues, like early cosmology above a TeV, pose difficulties and have not been addressed. Regardless which model might achieve strong gravity at the TeV scale, if any, it at least seems to require drastic means to realize this possibility, be it either the assumption of extra dimensions or the introduction of a large number of extra particles.

A slightly different possibility to overcome the large hierarchy between the weak scale and $M_{\mathrm{Pl}} = 10^{19}\,\mathrm{GeV}$ might be modifications to gravity itself: the coefficients of the higher-derivative terms in the gravitational action

$$S = \int d^4x\sqrt{-g}\left(c_1 R + c_2 R^2 + c_3 R_{\mu\nu}R^{\mu\nu} + \ldots\right) \tag{12}$$

are very weakly constrained, c_2, $c_3 < 10^{61}$ [1]. It is thus conceivable that boundary values $c_i(\mu)$ at the scale $\mu_* \sim$ TeV exist which through large gravitational self-couplings lead to the observed huge value of $c_1(\mu = 0) = M_{\mathrm{Pl}}^2/16\pi$. Incidentally, coefficients c_2, $c_3 \sim (10^{32})^2$ would be expected after integrating out 10^{32} matter fields.

3 Quantum gravitational effects on grand unification

As a second scenario, I want to elucidate what effects the running of Newton's constant can have in more mundane models without unusually large particle content. In particular, this section examines how quantum gravity and running G_N affect grand unified theories, especially supersymmetric ones. Results could influence whether one considers evidence for, or the possibility of, grand unification as plausible, and might also impact expectations of what we will find at the LHC.

The gauge coupling constants of the Standard Model are measured very accurately at the (low) energy scale M_Z (mass of the Z boson), for example by the LEP experiment: $\alpha_1(M_Z) = 0.016887$, $\alpha_2(M_Z) = 0.03322$, $\alpha_3(M_Z) = 0.118(5)$ [6]. If these couplings are evolved to higher energies using the renormalization group evolution equations of the Standard Model, the couplings run towards a common region, but do not quite meet (Fig. 4, left panel). However, if they are evolved with the RG equations of the Minimal Supersymmetric Standard Model using a phenomenologically plausible SUSY breaking scale of $M_{\mathrm{SUSY}} \sim 1\,\mathrm{TeV}$, then the measured low-energy couplings seem to be just right for the evolved couplings to meet at some common value near a unification scale of $M_X = 10^{16}\,\mathrm{GeV}$. This meeting is pretty much exact if two-loop beta functions are used (right panel of Fig. 4), which now has become part of the standard analysis [7]. The fact that the numerical values of the three gauge couplings meet under SUSY RG evolution at a high energy scale is often taken is strong evidence for supersymmetric grand unification. I.e. on the one hand it is taken as evidence for grand unification, the idea that the three gauge couplings arise from one coupling of a unified simple gauge group which is spontaneously broken at M_X via Higgs mechanism; and on the other hand it is also often quoted as evidence for SUSY itself, which would then be expected to be found soon at the LHC.

How do quantum gravity effects influence this picture? First, the alleged unification scale of $M_X = 10^{16}\,\mathrm{GeV}$ is uncomfortably close to the naive Planck scale $M_{\mathrm{Pl}} = 10^{19}\,\mathrm{GeV}$, and the *true* Planck scale μ_* might be even lower (8). Implicit in the above description of grand unification was the unification condition $\alpha_1(M_X) = \alpha_2(M_X) = \alpha_3(M_X)$ of numerical equality in the values of the evolved couplings, and it is this condition that is changed by quantum gravity.

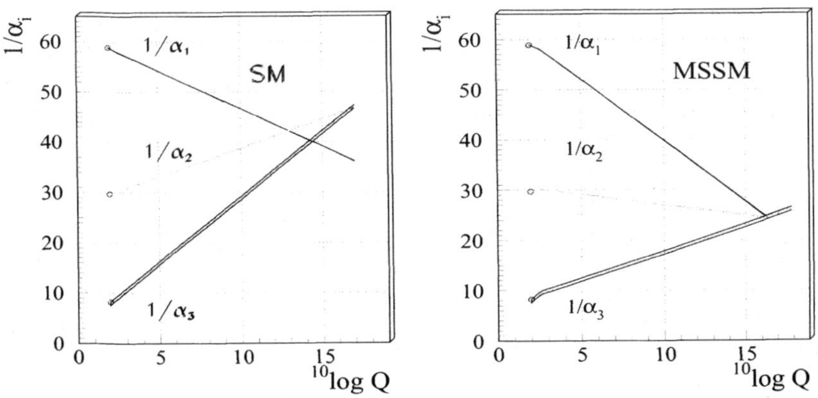

Figure 4: The inverses of the three gauge coupling constants evolved from their precisely measured values at M_Z to higher energies using the renormalization group equations of the Standard Model (left panel) and of the Minimal Supersymmetric Standard Model with SUSY breaking scale $M_{\rm SUSY} \sim 1\,{\rm TeV}$ (right).

Quantum gravity induces (non-perturbatively) a dimension-5 operator in the (effective) grand unified theory [8, 9]

$$\mathscr{L} \sim \frac{c}{\hat{\mu}_*}{\rm Tr}\,(G_{\mu\nu}G^{\mu\nu}H) \qquad \text{with } c \sim \mathcal{O}(1)\,, \tag{13}$$

which is suppressed by the reduced true Planck scale $\hat{\mu}_* = \mu_*/\sqrt{8\pi}$, the quantity that enters quantum gravity calculations, and so naturally has dimensionless coefficient c of order 1. $G_{\mu\nu}$ is the gauge field strength of the grand unified theory and H is one of the Higgs fields needed to break the unified gauge group down to the Standard Model group SU(3) $\times$ SU(2) $\times$ U(1). Below the scale M_X of grand unified symmetry breaking the Higgs field acquires a vacuum expectation value $\langle H\rangle$, and with the replacement $H \to \langle H\rangle$ the operator (13) modifies the gauge kinetic term of the Standard Model vector bosons which are massless below M_X:

$$\mathscr{L} \sim -\frac{1}{2}{\rm Tr}\,(G_{\mu\nu}G^{\mu\nu}) + \frac{c}{\hat{\mu}_*}{\rm Tr}\,(G_{\mu\nu}G^{\mu\nu}\langle H\rangle) \sim \sum_{i=1}^{3} -\frac{1}{2}(1+\epsilon_i)\,{\rm Tr}\left(F^{(i)}_{\mu\nu}F^{(i)\mu\nu}\right)\,, \tag{14}$$

where i runs over the factors of the Standard Model gauge group and the ϵ_i are calculable in each specific model of GUT breaking. In the case of minimal SU(5) unification, which we assume for now for simplicity, the multiplet H in the adjoint representation acquires a vacuum expectation value $\langle H\rangle = M_X(2,2,2,-3-3)/\sqrt{50\pi\alpha_G}$, where α_G is the value of the SU(5) gauge coupling at M_X, so that

$$\epsilon_1 = \frac{\epsilon_2}{3} = -\frac{\epsilon_3}{2} = \frac{\sqrt{2}}{5\sqrt{\pi}}\frac{c\eta}{\sqrt{\alpha_G}}\frac{M_X}{\hat{M}_{\rm Pl}}\,. \tag{15}$$

Thus, the three gauge kinetic terms of the Standard Model are modified in *different* ways. After a finite field redefinition $A^i_\mu \to (1+\epsilon_i)^{1/2}A^i_\mu$ the kinetic terms are canonically normalized below M_X, and it is then the corresponding redefined coupling constants $g_i \to (1+\epsilon_i)^{-1/2}g_i$ that are observed at low energies and that obey the usual RG equations below M_X, whereas it is the *original* coupling constants that need to meet at M_X in order for unification to happen. In terms of the observable rescaled couplings, the unification condition therefore reads

$$(1+\epsilon_1)\alpha_1(M_X) = (1+\epsilon_2)\alpha_2(M_X) = (1+\epsilon_3)\alpha_3(M_X) = \alpha_G\,. \tag{16}$$

It is worth emphasizing that there are two separate effects from quantum gravity playing in here: first, the presence of the gravitationally induced operator (13) influences grand unification simply because M_X is close to M_{Pl}; and second, the *true* scale of quantum gravity might be even lower due to running G_N so that the dimension-5 operator is effectively enhanced by the factor η from (8), which is what our work focuses on [2].

Supersymmetric grand unified theories, which can seemingly better accommodate coupling constant unification (Fig. 4), have typically quite large particle content N (or equivalently η) because most of their fields are Higgses in chiral supermultiplets, which contain scalars and fermions giving contributions of the same sign to η. Also, grand unified groups larger than SU(5) are commonly considered since they can satisfy phenomenological constraints such as proton decay, R parity and fits to the mass spectrum more easily. In the minimal SUSY-SU(5) model with **24**, **5** and $\overline{\mathbf{5}}$ Higgs multiplets the enhancement factor is $\eta = 2.3$, but the "minimal SUSY-SO(10) model" with **126**, $\overline{\mathbf{126}}$, **210** and **10** Higgses already has $\eta = 6.2$. Other models like E8 $\times$ E8 with **248** and **3875** Higgs multiplets clearly have even bigger η, such that $\hat{\mu}_*$ will be significantly lower than $\hat{M}_{\text{Pl}}$. Since minimal SUSY-SU(5) is phenomenologically ruled out, $\eta \gtrsim 5$ is a reasonable assumption for realistic SUSY GUTs.

I will now first illustrate the size of the corrections coming from the modified unification condition (16) by comparing it to the size of two-loop corrections in the running of gauge couplings, which have become part of the standard analysis of grand unification. In a toy model of SUSY-SO(10) one has $\eta = 6.2$ and, with (16), SU(5)-like gauge symmetry breaking then requires a splitting between the numerical values $\alpha_i(M_X)$ of, for example,

$$\frac{\alpha_3(M_X) - \alpha_2(M_X)}{\alpha_3(M_X)} \approx \epsilon_2(c\eta) - \epsilon_3(c\eta) \approx \begin{cases} +9\% & \text{for } c = +1 \\ -9\% & \text{for } c = -1. \end{cases} \tag{17}$$

This is a big uncertainty in our knowledge of the correct unification condition since within the grand unified theory nothing can be inferred about the exact size c of the induced operator, except that it is reasonable to assume $c \sim \mathcal{O}(1)$. In particular, even making this reasonable assumption, the uncertainty of $\pm 9\%$ is much bigger than the corrections to the $\alpha_i(M_X)$ coming from the inclusion of two-loop running. These are smaller than 3.5% for all $i = 1, 2, 3$. Thus, two-loop computations do not improve evidence for grand unification, as is often claimed, since the size of, and the uncertainty in, the effects from quantum gravity is far greater than two-loop contributions.

But not only is two-loop running of the gauge couplings too accurate in view of such large uncertainties from quantum gravity, also do claims of grand unification based on the pretty precise measurements of the low-energy couplings $\alpha_i(M_Z)$ not hold up in this light. For illustration, Fig. 5 shows different low-energy inputs of the SUSY breaking scale M_{SUSY} and the strong coupling $\alpha_3(M_Z)$ at the Z mass, while the relatively precisely measured $\alpha_1(M_Z) = 0.016887$ and $\alpha_2(M_Z) = 0.03322$ are held fixed; M_{SUSY} is expected to be in the region $10^{3\pm1}$ GeV for phenomenological reasons, and the current uncertainty in $\alpha_3(M_Z)$ is indicated by a vertical bar. For each point in the diagram, these low-energy inputs have been evolved to higher energies with the supersymmetric two-loop renormalization group equations, and for each input it has been found that the modified unification condition (16) is satisfied for exactly one pair of values of the coefficient c and M_X. Contour lines for these variables are shown in the figure, and $\eta \gtrsim 5$ is imagined to be fixed by the particle content of the model. The coefficient c, which is only determined by quantum gravity, naturally ranges between $-1 < c < 1$, which covers a pretty large area in Fig. 5. In particular, for many of those values c, grand unification is incompatible with our measurements of low-energy parameters like $\alpha_3(M_Z)$: slight variations in c would require unacceptably large adjustments in low-energy inputs $\alpha_i(M_Z)$ and M_{SUSY} for unification to still happen.

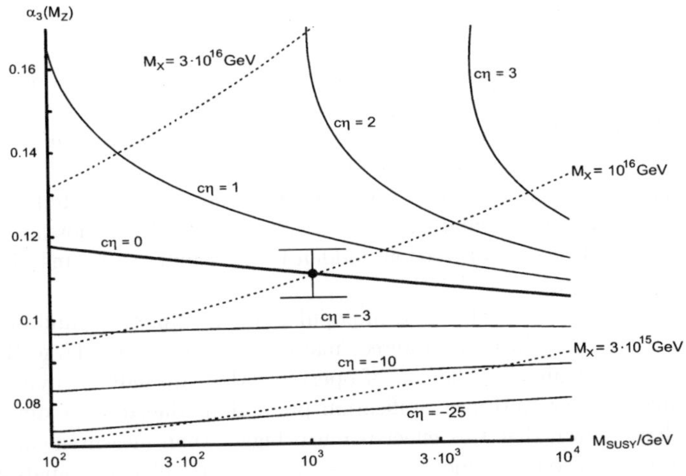

Figure 5: For η fixed by the particle content of the theory, solid lines are contours of constant c such that, under the modified unification condition (16), SUSY-SU(5) perfectly unifies at two loops for given values of the initial strong coupling $\alpha_3(M_Z)$ and SUSY breaking scale $M_{\rm SUSY}$.

On the other hand, if $\eta \sim 5$ then for a large range of low-energy inputs one can easily find a natural $c \sim \mathcal{O}(1)$ such that unification happens; under this perspective, unification does not seem very special. Our ignorance of $c\eta$ is far greater than the uncertainty in LEP measurements. And disturbingly, as can be seen from the figure, if the underlying theory of quantum gravity determines the coefficient of the operator (13) to be $c > 4/\eta \sim 1$, then unification is impossible even far beyond the allowed SM/MSSM parameter range. The commonly quoted evidence for unification (right panel of Fig. 4) arises because the central value of our measurements of low-energy parameters (black dot in Fig. 5) happens to lie almost perfectly on the contour line $c = 0$. But considering the uncertainties, it seems that we cannot either suggest or rule out grand unification with any high degree of confidence based on low-energy observations alone, and thus the idea of grand unification loses much of its beauty. In particular, previous attempts to pin down $\alpha_3(M_Z)$ or $\sin^2\theta_W$ based on the assumption of unification seem invalid. Also, expectations about whether we will find supersymmetry at the LHC might need to be reconsidered if one previously regarded Fig. 4 as strong evidence for SUSY.

But one can also turn this reasoning around and argue that the gravitational operator (13) with $c \sim \mathcal{O}(1)$ might facilitate gauge coupling unification in models where this did not seem to happen previously but which otherwise are phenomenologically viable. The picture in this case would be that the underlying fundamental theory leads to an effective grand unified theory with (strong) remnants from quantum gravity.

4 Conclusion

I have shown that Newton's constant is a running coupling constant, as are the familiar gauge couplings of the Standard Model, and that therefore the strength of gravity depends on energy scale. The value $G_N = (10^{19}\,\text{GeV})^{-2}$ that we measure in the infrared is not directly related to the fundamental scale of quantum gravity, but is derived via renormalization group running from the fundamental value μ_*. The exact relationship depends on the particle content of the theory under consideration, eqn. (8).

A four-dimensional model was presented that contains $\sim 10^{32}$ scalars or fermions with masses below a TeV in a hidden sector, interacting with the Standard Model only gravitationally. This model solves, already in four dimensions, the hierarchy problem of the Standard Model since the large particle content implies that the fundamental scale of quantum gravity is around a TeV; at this scale then, new (gravitational) physics is expected to come in and replace the Standard Model. Popular extra-dimensional models were shown to postulate an equally large amount of degrees of freedom hidden in their bulk, and some phenomenology of our model was outlined.

In a second scenario, it was noted that in popular supersymmetric grand unified theories the true Planck scale is significantly lowered, mainly because of their large Higgs content. This amplifies the uncertainty in the size of operators that are induced by quantum gravity and so enhances uncertainties in unification predictions due to a modified unification condition. We have shown that this effect is much bigger than measurement uncertainties and the two-loop corrections commonly considered. In particular, our analysis might impact whether one considers *apparent* unification of gauge couplings to be strong evidence for grand unification or SUSY. On the other hand, phenomenologically viable GUT models that are only lacking sufficient gauge coupling unification might be revived by the presence of such induced operators along with the running of Newton's constant.

Acknowledgments — I would like to thank Antonino Zichichi and Gerardus 't Hooft for organizing a wonderful 46th Course of the International School of Subnuclear Physics 2008, bringing together students and well-known researchers in a hospitable atmosphere in Erice, and for the opportunity to present this work. Also with great pleasure I acknowledge collaboration with Xavier Calmet and Stephen Hsu. The original research work and the author's travel to Erice were supported by the Department of Energy under DE-FG02-96ER40969.

References

[1] X. Calmet, S. D. H. Hsu and D. Reeb, Phys. Rev. D **77**, 125015 (2008) [arXiv:0803.1836 [hep-th]].

[2] X. Calmet, S. D. H. Hsu and D. Reeb, Phys. Rev. Lett. **101**, 171802 (2008) [arXiv:0805.0145 [hep-ph]].

[3] C. D. Hoyle *et al.*, Phys. Rev. D **70**, 042004 (2004) [arXiv:hep-ph/0405262].

[4] F. Larsen and F. Wilczek, Nucl. Phys. B **458**, 249 (1996) [arXiv:hep-th/9506066].

[5] X. Calmet, W. Gong and S. D. H. Hsu, Phys. Lett. B **668**, 20 (2008) [arXiv:0806.4605 [hep-ph]].

[6] W.–M. Yao *et al.* (Particle Data Group), J. Phys. G **33**, 1 (2006).

[7] U. Amaldi, W. de Boer and H. Furstenau, Phys. Lett. B **260**, 447 (1991).

[8] C. T. Hill, Phys. Lett. B **135**, 47 (1984).

[9] Q. Shafi and C. Wetterich, Phys. Rev. Lett. **52**, 875 (1984).

[10] L. J. Hall and U. Sarid, Phys. Rev. Lett. **70**, 2673 (1993) [arXiv:hep-ph/9210240].

DIPLOMAS

Twenty one Diplomas were open for competition among the participants. They have been awarded as follows:

- John S. BELL Diploma to:

  John Joseph CARRASCO
  University of California at Los Angeles, UCLA, USA

- Patrick M.S. BLACKETT Diploma to:

  Zachary MARSHALL
  Columbia University, New York, USA

- James CHADWICK Diploma to:

  Liangjian WEN
  IHEP, Institute of High Energy Physics, Beijing, China

- Sidney COLEMAN Diploma to:

  Gerben STAVENGA
  Utrecht University, Utrecht, The Netherlands

- Richard H. DALITZ Diploma to:

  Sergiy DUBYNSKIY
  University of Minnesota, Minneapolis, USA

- Paul A.M. DIRAC Diploma to:

  David Franz REEB
  University of Oregon, Eugene, USA

- Vladimir N. GRIBOV Diploma to:

  Stefano NICOTRI
  INFN, Sezione di Bari, Bari, Italy

- Robert HOFSTADTER Diploma to:

  Andrei NIKIFOROV
  DESY, Hamburg, Germany

- Gunnar KÄLLEN Diploma to:

  Henrik JOHANSSON
  University of California at Los Angeles, UCLA, USA

- Yuval NE'EMAN Diploma to:

  Yaroslav KLOPOT
  JINR, Dubna, Russia

- Giuseppe P.S. OCCHIALINI Diploma to:

  Ralf BERNHARD
  University of Freiburg, Freiburg, Germany

- Oreste PICCIONI Diploma to:

  Eike MÜLLER
  University of Edinburgh, Edinburgh, UK

- Bruno PONTECORVO Diploma to:

  Kaixuan NI
  Yale University, New Haven, USA

- Giampietro PUPPI Diploma to:

  Davide CAFFARRI
  University of Bologna, Bologna, Italy

- Isidor I. RABI Diploma to:

  Armen YERANYAN
  INFN, LNF, Frascati, Italy

- Giulio RACAH Diploma to:

  Barbara GUERZONI
  University of Bologna, Bologna, Italy

- Bruno ROSSI Diploma to:

  James B. DENT
  Vanderbilt University, Nashville, USA

- Victor F. WEISSKOPF Diploma to:

  Shamil SHAKIROV
  ITEP, Moscow, Russia

- Eugene P. WIGNER Diploma to:

  Alessandra GNECCHI
  University of Pisa, Pisa, Italy

- Bjorn H. WIIK Diploma to:

  Martin GOEBEL
  University of Hamburg, Hamburg, Germany

- Chien Shiung WU Diploma to:

  Anna KAMINSKA
  University of Warsaw, Warsaw, Poland

AWARDS

- The Awards for the best 'New Talent – Speakers' have been attributed to:

  i) **Alessandra GNECCHI**
  University of Pisa, Pisa, Italy
  for her presentation
  Fake Supergravity and Black Hole Evolution

  ii) **Zachary MARSHALL**
  Columbia University, New York, USA
  for his presentation on
  Jets and the Status of ATLAS

  iii) **Kaixuan NI**
  University of Yale, New havn, USA
  for his presentation on
  Searches for Supersymmetric Dark Matter with Xenon

  iv) **David Franz REEB**
  University of Oregon, Eugene, USA
  for his presentation on
  Quantum Gravity Effects through Running of Newton's Constant

- Award for BEST QUESTION given to:
  John Joseph CARRASCO
  University of California at Los Angeles, UCLA, USA

- Award for BEST STUDENT given to:
  Gerben STAVENGA
  Utrecht University, Utrecht, The Netherlands

- Awards for BEST SCIENTIFIC SECRETARIES given to:
  Martin GOEBEL
  University of Hamburg, Hamburg, Germany
  Anna KAMINSKA
  University of Warsaw, Warsaw, Poland

PARTICIPANTS

Yasuyuki AKIBA Japan	BNL, UPTON, USA
Ignatios ANTONIADIS France	CERN, GENEVA, Switzerland
Elena APRILE Italy	Columbia University NEW YORK, NY, USA
Richard ARNOWITT USA	Texas A&M University COLLEGE STATION, TX, USA
Roberto BALBINOT Italy	University of Bologna BOLOGNA, Italy
Rinaldo BALDINI Italy	Enrico Fermi Centre ROME, Italy
Laura BAUDIS Switzerland	University of Zürich ZÜRICH, Switzerland
Francesca BELLINI Italy	University of Bologna BOLOGNA, Italy
Zvi BERN USA	UCLA LOS ANGELES, CA, USA
Ralf BERNHARD Germany	University of Freiburg FREIBURG, Germany
Alessandro BETTINI Italy	INFN - University of Padova PADOVA, Italy
Ante BILANDZIC Croatia	High Energy Physics Section - NIKHEF-H AMSTERDAM, The Netherlands
Stefano BOLOGNESI Italy	University of Minnesota MINNEAPOLIS, MN, USA
Claudio BONATI Italy	University of Pisa PISA, Italy
Davide BOSCHERINI Italy	University of Bologna BOLOGNA, Italy
Maria BRUNO Italy	University of Catania CATANIA, Italy

Timothy BUDD
UK

Utrecht University
UTRECHT, The Netherlands

Davide CAFFARRI
Italy

University of Bologna
BOLOGNA, Italy

Gigi CAPPELLO
Italy

University of Catania
CATANIA, Italy

John Joseph CARRASCO
USA

University of California at Los Angeles - UCLA
LOS ANGELES, CA, USA

Sebastian CASSEL
UK

University of Oxford
OXFORD, UK

Luisa CIFARELLI
Italy

University of Bologna
BOLOGNA, Italy

Emilio CIUFFOLI
Italy

University of Pisa
PISA, Italy

Fabrizio COCCETTI
Italy

Enrico Fermi Centre
ROME, Italy

Eugenio COCCIA
Italy

LNGS - INFN
L'AQUILA, Italy

Cezar CONDEESCU
Romania

Ecole Polytechnique
PALAISEAU Cedex, France

James B. DENT
USA

Vanderbilt University
NASHVILLE, NT, USA

Abhay DESHPANDE
India

Stony Brook University
STONY BROOK, NY, USA

Marco DI MAURO
Italy

University of Salerno
BARONISSI (Salerno), Italy

Sergiy DUBYNSKIY
Ukraine

University of Minnesota
MINNEAPOLIS, MN, USA

Petro FEDOSENKO Ukraine	National Taras Shevchenko University of Kyiv KYIV, Ukraine
Sergio FERRARA Italy	CERN GENEVA, Switzerland
Elisa FIORAVANTI Italy	INFN - Sezione di Ferrara FERRARA, Italy
Evangelos GAZIS Greece	National Technical University ATHENS, Greece
Alessandra GNECCHI Italy	University of Pisa PISA, Italy
Ivan GNESI Italy	University of Torino TORINO, Italy
Martin GOEBEL Germany	Universität Hamburg HAMBURG, Germany
Barbara GUERZONI Italy	University of Bologna BOLOGNA, Italy
Elisa GUIDO Italy	INFN - Sezione di Genova GENOVA, Italy
Huma HAIDER India	Aligarh Muslim University ALIGARH, India
Peter Ware HIGGS UK	University of Edinburgh EDINBURGH, UK
Philipp HOEHN Germany	Utrecht University UTRECHT, The *@gmail.com*
Henrik JOHANSSON Sweden	UCLA, LOS ANGELES, CA, USA
Anna KAMINSKA Poland	Warsaw University WARSZAWA, Poland
Shinsuke KAWASAKI Japan	The University of Tokyo TOKYO, Japan
Richard D. KENWAY UK	University of Edinburgh EDINBURGH, UK
Kamal Hussain KHAN Pakistan	COMSATS ISLAMABAD, Pakistan

Yaroslav KLOPOT
Ukraine
Joint Institute for Nuclear Research
DUBNA, Russia

Petr KOPNIN
Russia
ITEP - Institute for Theoretical and Experimental Physics
MOSCOW, Russia

Christian P. KORTHALS-ALTES
The Netherlands
CNRS-Luminy
MARSEILLE, France

Masatoshi KOSHIBA
Japan
University of Tokyo
TOKYO, Japan

Beijiang LIU
China
Institute of High Energy Physics - IHEP
BEIJING, China

Nigel LOCKYER
USA
TRIUMF
VANCOUVER, BC, Canada

Renate LOLL
Germany
Utrecht University
UTRECHT, The Netherlands

Cecilia MAIANO
Italy
INFN - Sezione di Milano - Bicocca
MILANO, Italy

Andreas MAIER
Germany
Universität Karlsruhe
KARLSRUHE, Germany

Alessio MARRANI
Italy
Enrico Fermi Centre
ROME, Italy

Zachary MARSHALL
USA
Columbia University
NEW YORK, NY, USA

Adam MARTIN
USA
Yale University
NEW HAVEN, CT, USA

Larry MCLERRAN
USA
Brookhaven National Laboratory - BNL
UPTON, NY, USA

Silvia MIGLIORANZI
Italy
DESY - Zeus

Gregory MORAITIS
Greece / France
Swansea University
SWANSEA, UK

Eike MÜLLER
Germany
University of Edinburgh
EDINBURGH, Scotland, UK

Mauro MUNERATO Italy	INFN - Sezione di Ferrara FERRARA, Italy
Edward MUSAEV Russia	ITEP - Institute for Theoretical and Experimental Physics MOSCOW, Russia
Marton NAGY Hungary	H.A.S. C.R.I.P. Research Institute for Particle and Nuclear Physics BUDAPEST, Hungary
Dmitri V. NANOPOULOS USA	Texas A&M University COLLEGE STATION, TX, USA
Pran NATH USA	Northeastern University BOSTON, MA, USA
Kaixuan NI China	Yale University NEW HAVEN, CT, USA
Stefano NICOTRI Italy	INFN - Sezione di Bari BARI, Italy
Andrei NIKIFOROV Ukraine	DESY - EPJC HAMBURG, Germany
Francesco NOFERINI Italy	INFN - CNAF
Piermaria J. ODDONE USA	FNAL BATAVIA, IL, USA
Pawel PACHOLEK Poland	Warsaw University WARSZAWA, Poland
Thomas PAPAEVANGELOU Greece	CERN GENEVA, Switzerland
Chang Soon PARK South Korea	CALTECH PASADENA, CA, USA
Davide PATITUCCI Italy	Enrico Fermi Centre ROME, Italy
Ryan PATTERSON USA	CALTECH PASADENA, CA, USA
Anastasiia REBESH Ukraine	National Taras Shevchenko University of Kyiv KYIV, Ukraine

David Franz REEB Germany	University of Oregon EUGENE, OR, USA
Igor SAMSONOV Russia	Tomsk Polytechnic University TOMSK, Russia
John H. SCHWARZ USA	CALTECH PASADENA, CA, USA
Shamil SHAKIROV Russia	ITEP - Institute for Theoretical and Experimental Physics MOSCOW, Russia
Andriy SHCHERBAKOV Ukraine	INFN - LNF FRASCATI, Italy
Alessandro SILENZI Italy	University of Bologna BOLOGNA, Italy
Fabio SIRINGO Italy	University of Catania CATANIA, Italy
Artur SLOBODENIUK Ukraine	National Taras Shevchenko University of Kyiv KYIV, Ukraine
Gerben STAVENGA The Netherlands	Utrecht University UTRECHT, The Netherlands
Alicja SUCHECKA Poland	Wroclaw University of Technology WROCLAW, Poland
Leonard SUSSKIND USA	Stanford University STANFORD, CA, USA
Marek SZYDŁOWSKI Poland	Obserwatorium Astronomiczne KRAKÓW, Poland
Pawel TAMBOR Poland	Jagellonian University KRAKÓW, Poland
Michael J. TANNENBAUM USA	Brookhaven National Laboratory UPTON, NY, USA
Gerardus 't HOOFT The Netherlands	Utrecht University UTRECHT, The Netherlands
Panteleimon TZIVELOGLOU Greece	CERN GENEVA, Switzerland

Ita and Aart van der SLOOT The Netherlands	UTRECHT, The Netherlands
Yifang WANG China	Institute of High Energy Physics - IHEP BEIJING, China
David James WEIR UK	Imperial College London LONDON, UK
Liangjian WEN China	IHEP - Institute of High Energy Physics BEIJING, China
Horst WENNINGER Germany	CERN GENEVA, Switzerland
Armen YERANYAN Armenia	INFN - LNF FRASCATI, Italy
Chiara ZAMPOLLI Italy	INFN - Sezione di Bologna BOLOGNA, Italy
Antonino ZICHICHI Italy	CERN GENEVA, Switzerland *and* INFN and University of Bologna BOLOGNA, Italy *and* Enrico Fermi Centre ROME, Italy
Robert ZIEGLER Germany	SISSA - ISAS TRIESTE, Italy